PUBLICATIONS

OF THE

NAVY RECORDS SOCIETY

VOL. 136

THE COLLECTIVE NAVAL DEFENCE OF THE

EMPIRE, 1900–1940

The NAVY RECORDS SOCIETY was established in 1893 for the purpose of printing unpublished manuscripts and rare works of naval interest. The Society is open to all who are interested in naval history, and any person wishing to become a member should apply to the Hon. Secretary, Department of War Studies, King's College London, Strand, London WC2R 2LS. The annual subscription is £30, which entitles the member to receive one free copy of each work issued by the Society in that year, and to buy earlier issues at much reduced prices.

SUBSCRIPTIONS and orders for back volumes should be sent to the Membership Secretary, 5 Goodwood Close, Midhurst, West Sussex GU29 9JG.

THE COUNCIL OF THE NAVY RECORDS SOCIETY wish it to be clearly understood that they are not answerable for any opinions and observations which may appear in the Society's publications. For these the editors of the several works are entirely responsible.

THE COLLECTIVE NAVAL DEFENCE OF THE EMPIRE, 1900–1940

edited by

NICHOLAS TRACY
Adjunct Professor, Department of History,
University of New Brunswick

PUBLISHED BY ASHGATE
FOR THE NAVY RECORDS SOCIETY
1997

Published by
Ashgate Publishing Limited
Gower House
Croft Road
Aldershot
Hants GU11 3HR
England

Ashgate Publishing Company
Old Post Road
Brookfield
Vermont 05036–9704
USA

British Library Cataloguing-in-Publication Data
The collective naval defence of the Empire, 1900 to 1940. (Navy records series; v. 136)
1. Great Britain. Royal Navy – History – 20th century
2. Great Britain – Colonies – Defences
I. Tracy, Nicholas II. Navy Records Society
359′. 009′171241

Library of Congress Cataloging-in-Publication Data
The collective naval defence of the Empire, 1900 to 1940 / edited by Nicholas Tracy.
(Publications of the Navy Records Society; vol. 136)
Includes bibliographical references and index.
ISBN 1–85928–402–7 (hbk.: acid-free paper)
1. Great Britain—Military policy. 2. Great Britain. Royal Navy—History—20th century. 3. Navies—Commonwealth countries—History—20th century. I. Tracy, Nicholas. II. Series. VA454.C6485 1997
359′.03′09410904—dc21   96–40238
CIP

Printed on acid-free paper

ISBN 1 85928 402 7

Typeset in Times by Intype London Ltd and printed in Great Britain at the University Press, Cambridge.

CONTENTS

INTRODUCTION

The development in the twentieth century of a collective system of naval defence for the British Empire was determined by an imperfect compromise between strategic ideals and political realities. The prevailing strategic conception, that unity of control and freedom of movement was essential to ensure that overwhelming force could be brought to bear at the decisive battle which would determine the outcome of a war, was tremendously influential. The necessity for the utmost efficiency of strategic direction was reinforced by the belief that forward deployment for offensive action was the imperative means to achieve naval success. Centralisation was also valued because it was feared that small fleets could not be efficient. In conflict with the strategic and administrative ideals of centralised command and a single, great naval service was the political reality of the need to recognise the constitutional autonomy of the constituent members of the Empire.

The absolute right of the Dominions and Colonies to control their own budgets was undisputed, and this budgetary independence had to be accommodated with strategic requirements, which in practice were perceived differently from the vantage points of London and the subordinate capitals. Until 1926 the Empire maintained at least the fiction of a common foreign policy, but in reality the differing strategic situations of the constituent parts of the Empire affected their views on foreign relations, and influenced their willingness to devote financial resources to a common purpose they imperfectly controlled. Efforts to develop a federal structure which would have made the common foreign policy politically realistic were frustrated, in part by the sheer size of the Empire, and in consequence the Dominions developed their own naval forces paid for out of their national budgets and controlled locally. This necessitated the development of a compromise defence structure which would enable the several navies to train and fight together. It was eventually accepted at the Admiralty that it was more important to foster political support for naval effort than it was to have unfettered centralised command, and that only national navies were likely to receive that support.

The compromise system of co-operative naval defence was imperfect, certainly as an instrument of deterrence. On the other hand, the structure of Imperial defence co-operation retained strong enough political support to keep the Empire intact in the crises of 1940 and 1941, and right through the Second World War. The inadequacy of the system to sustain an effective defence against Japan while fighting Germany and Italy, however, was revealed when Singapore fell in early 1942. That failure set the Empire on its course for eventual dissolution.

The papers collected in this volume are drawn entirely from the archives of the British government in London. Accordingly, they support study of the problems of collective defence more effectively than they do that of the development of Dominion naval forces. Operational matters are rigorously excluded, which inevitably means that the collection is thinnest during the First World War. Towards the end of the First World War, thought began to be given to the structure of post-war defence. Constitutional and legal considerations, administrative control, training, strategic consultation, and to a limited extent naval procurement, form the foci of attention. Selection has been determined by the need to present papers representative of the principal themes which dominated Imperial defence, and by the need to indicate the nature of the debate by which decisions were reached. No collection of papers can hope to be more than illustrative, but by presenting a relatively complete paper trail for the most important discussions it is hoped to avoid giving any impression of governance by a succession of monolithic fiats.

In this introductory finding aid, some effort is made to put management of collective defence into the context of the early histories of the Royal Australian, Canadian, New Zealand, South African and Indian navies. The archival references to this material, and to additional papers for which there was not room in this volume, are noted in the references to documentary sources. Among these are references to papers already published in volume 111 of the Navy Records Society, A. Temple Patterson's second volume of the Jellicoe papers.

BEFORE THE GREAT WAR

At the 1897 Colonial Conference, the First Lord of the Admiralty, Mr Goschen, had said that the Admiralty welcomed the limited financial support received from the colonies because it produced 'certain ties which we value'. However, he continued: 'From the strategical point of view, we should be glad that the Admiralty should have a free hand.' Five years later, at the 1902 Imperial Conference which convened the year following the confederation of Australia, the First Lord of the Admiralty, Lord Selborne, made a presentation which contrasted the per-capita expenditure on naval defence made by the people of Great Britain with the smaller effort made in the rest of the Empire, especially with the poor showing made by Canada. The assembled prime ministers were also given a memorandum on 'Sea-Power and the Principles Involved' in which they were advised that centralised strategic control of the navy was essential. Reconciliation of the objectives of sharing the expense of naval defence, while retaining centralisation of control, was to dominate Imperial defence consultation during the years before the Great War. [6]

Defence of Imperial Naval Stations

In the first decade of the twentieth century the problem of the defence of overseas stations attracted considerable attention. The question of the size of Malta's garrison, which depended on the question of whether the Royal Navy could itself defend its principal Mediterranean base, was just one of many similar which created controversy between the Admiralty and the Army Council. At the other end of the scale was St Helena, where a small battery was required to deny the anchorage to commerce raiders, but which was so isolated that a garrison would be inclined to deteriorate during years of peace. In August 1905 the Colonial Defence Committee completed for the Committee of Imperial Defence a general policy paper on 'The Defence of Colonial Ports' in which it stipulated that 'The strength of the garrison of naval bases is governed by the consideration that in no case could a greater attacking force than a few thousand men be collected and conveyed over sea without such arrangements and preparation as would bring the operations under the category of an "organised invasion" which the navy has undertaken to prevent.' Halifax and Esquimalt, however, were vulnerable in the event of war to American action on a scale which, in the latter instance at any rate, was entirely beyond the ability of the Empire to contain. With the concentration of Royal Navy forces to face the German naval threat in home waters, the Canadian acceptance of responsibilty for the bases on its soil was welcome, except by those such as the former Governor-General, Lord Minto, whose experience of Canadian politics led them to believe that political patronage appointments would undermine dockyard efficiency. The British garrisons at Halifax and Esquimalt were withdrawn. [2, 16–18, 20–3, 26–7, 30–1, 54, 57, 62] In a major Committee of Imperial Defence paper of 7 July 1910 entitled 'The Principles of Imperial Defence', the limited requirements for local defence of naval and commercial harbours was laid down. [75]

The Australasian Squadron

As a result of discussion at the Colonial Conference of 1887, an agreement had been reached by which the Admiralty was to provide a squadron of 'five fast cruisers and two torpedo-gunboats' for 'the protection of floating trade in Australian waters' in return for an annual contribution of £126,000 of which New Zealand contributed £20,000. At the 1902 Imperial Conference discussions were conducted concerning the renewal of the naval agreement. Selborne was concerned to protect the freedom of the Admiralty to use those ships in wartime without reference to the Australian and New Zealand governments, but recognised the need to develop public interest in naval defence in those colonies. He was also prepared to suggest similar agreements with Canada and Newfoundland, but recognised the inability of Newfoundland to find the money. Sir Wilfred Laurier had already made it clear that Canada had other priorities. Cape Colony and Natal, recognising the importance of naval defence to the Empire, agreed to contribute to the cost of the Royal Navy without insisting on participation in its control. This was satisfactory, but the sums involved were small. [7, 10]

At the 1902 conference the Commonwealth of Australia, Minister of State for Defence, John Forrest, made it clear that he would not advise the development of an Australian navy. Instead, a new agreement was signed in 1903 by the British government with the Australian administration of Alfred Deakin, and the New Zealand government of R. Seddon, increasing the size of the Australasian squadron, and determining that Australians and New Zealanders should be trained to man four of its ships. Officers would be provided by the Royal Navy and Royal Naval Reserve, but Australians and New Zealanders would be trained as cadets in the Royal Navy. Australia was to pay five-twelfths of the cost of the squadron, and New Zealand one-twelfth, provided the sums did not exceed £200,000 and £40,000, respectively. The arrangement, however, did not satisfy Australians, and failed to satisfy the Admiralty's views on effective strategic deployment. It was difficult to obtain political support for expenditure on naval forces if that money was not spent locally, as it predominantly was by the land forces of the Crown, and if the ships and men were not distinctively identified with the dominion paying the bills. In August 1905 Deakin wrote the Governor-General, Lord Northcote, that the criticism recently made of the agreement by Admiral Sir Arthur Fanshawe, Commander-in-Chief Australia Station, that it did not adequately address Imperial needs, suggested that it was time to review its provisions. In November 1905 Captain Creswell, Commonwealth Naval Director, presented to Deakin a proposal for a small Australian navy. [11–13, 19, 29, 32]

The conflict was intensified by the concentration of naval forces in home waters by Admiral Fisher in response to the growth of the German Navy, and to technical changes in naval warfare. In April 1906 Deakin complained bitterly to the Commonwealth Governor-General that Britain had not lived up to the spirit of the agreement, sending ships of weaker force than had been agreed to. Worse, ships on the Australasian station had been deployed as far away as South America. [33–5] In August a 'Report of the Committee of Imperial Defence upon the General Scheme of Defence for Australia' was submitted to the Commonwealth senate, which strongly deprecated the idea of the formation of a local navy. [36–7, 41] The Commonwealth parliament, however, preferred Captain Creswell's 'Report of the Director of the [Australian] Naval Forces'. [38–40, 42]

The strategic value of unity in wartime continued to dominate Admiralty policy. In the winter of 1906–7, Admiral Ottley, Director of Naval Intelligence, wrote several assessments of the problem of developing an effective system of collective naval defence of the Empire in which he was critical of the influence on Australian policy enjoyed by local naval officers who stood most to gain by the creation of an Australian navy. On the other hand, he admitted that the Admiralty did not like the agreement, and he sympathetically noted that the Empire was a voluntary association of peoples which derived its strength from the freedom of their choice. The creation of local navies was a necessary attribute to the ability of colonies to leave the association, should they so wish. Ottley even suggested that it would be a reasonable distribution of the defence burden if the United Kingdom carried the whole cost of naval defence and the colonies concentrated on training men for army service. [43–4, 47]

The issue was recognised to be politically explosive, and when in March 1907 the Colonial Office requested that the Admiralty prepare a memorial for a forthcoming Imperial Conference, the Admiralty Permanent Secretary, Sir Evan MacGregor, demurred. The Admiralty had concluded that the better part of valour would be to conform to Australian and New Zealand views. [48] Following the conference, Deakin wrote to the First Sea Lord, Admiral Sir Jacky Fisher, to confirm the convergence of their attitudes, that the Australasian Naval agreement was inefficient in its use of naval resources, and that its termination would not lead to the abandonment of Australia by the Royal Navy. Fisher's marginal notes and underscores indicate that he was in general agreement. The subsidy was inadequate compensation for the misuse of the ships of the Australian squadron: 'No, nothing like! Never was there such an extravagant waste of money, ships and men as this agreement entails on the Admiralty.' And later: 'I simply made no secret of my detestation of the Agreement.' [49–51]

Formation of the Royal Australian Navy

The Australian government took the decision to form their own navy, and formally requested the British government to consider means of linking the Australian naval force for training purposes to the Royal Navy. W. Graham Greene, Assistant Secretary of the Admiralty, indicated 10 February 1908 that 'my Lords do not see any insuperable difficulty in coming to an agreement' on the arrangements to train the Australian Navy, although the Judge Advocate, R.B.D. Acland, pointed out in November the need for Australia's code of naval discipline to be brought into line with that of the Royal Navy. On the other hand, Greene could not agree to the Australian Navy being controlled in wartime by the Australian government: 'Under international law there is only one executive authority in the British Empire capable of being recognised by foreign States, Colonial ships of war cannot operate independently of the Royal Navy.'[53, 56]

Adherence of the Empire to a common foreign policy meant that the Dominions had to accept Admiralty control of all warlike naval operations outside their own territorial waters. The Dominions need not contribute to the Royal Navy, but at the same time they might not establish a naval force under their own control for war purposes. The Australian government quickly conceded this point, but, as Greene wrote on 20 August, reserved the right of 'the responsible Government at the time' to make the decision to put the Australian fleet under the control of the Senior British Naval Officer, 'it being understood that the vessels should not be moved out of Australian waters without the approval of the Commonwealth Government'.[55] When Lord Dudley, the Governor-General of Australia, reported Australia's understanding of the decisions which had been made about the development of Australia's naval role, the Under-Secretary of State for the Colonies, H.W. Just, commented that 'their Lordships will observe that in effect the proposals now made by the Commonwealth Government as to the control of the naval force of the Commonwealth concede everything desired by the Admiralty'.[61]

Contributions to the Royal Navy

The decisions leading to the formation of an Australian navy had been influenced by the relative tranquility of international affairs during the first eight years of the century, but the latitude this gave for constitutional development came rapidly to an end with the 1908 German Naval Law which was perceived in Britain as a challenge to her isolation from European crises, and to her Empire. On 22 March 1909 the Government of New Zealand telegraphed an offer to bear the cost of the immediate construction of a battlecruiser of the latest type, HMS *New Zealand*, and a second of the same type if necessary. On 26 March 1909 G. Simpson, Administrator of New South Wales, wrote to the Secretary of State for the Colonies that a citizens' meeting unanimously voted that 'the time has arrived for the Commonwealth to take an active share in the Naval Defence of the Empire' and that 'in view of the expressed determination of Britain's rivals to challenge her Naval supremacy Australia should present a Dreadnought to the British Navy as the immediate expression of her invincible resolve to stand by the Mother Country and take her place in the Empire's fighting line'. The New South Wales and Victoria governments passed the idea of purchasing a battleship for the Royal Navy on to the Commonwealth government, promising to pay their share. On 15 April Mr Fisher, the Commonwealth Prime Minister, telegraphed that Australia really ought to make its contribution to Imperial defence by developing its own naval forces, but his administration fell soon after and on 5 June Deakin telegraphed formally offering the gift of a battleship. Nevertheless, on 22 July Lord Dudley telegraphed that Australians were having cold feet about spending money on a battleship which would 'never be seen in Australian waters'. [58, 60, 63, 67] Eventually HMAS *Australia* was built, but for service in the Australian Navy. The Federated Malay States participated in the patriotic movement, offering to pay the cost of the construction of a battleship, HMS *Malaya*. [105]

Formation of the Royal Canadian Navy

Three days after the public meeting in New South Wales, the Canadian House of Commons passed a resolution that 'this House fully recognises the duty of the people of Canada as they increase in numbers and wealth to assume in larger measure the responsibilities of national defence'. [59] The Laurier administration, however, continued to resist the idea of contribution to the Royal Navy. Instead it brought forward legislation to create a Canadian navy.

The Canadian government was no less concerned than was Australia about the constitutional considerations affecting collective Imperial defence, and in addition Ottawa had to deal with the intractable problem that the United States was the only country in a position to challenge Canadian security. Before the end of the nineteenth century it had become apparent that the Empire lacked the resources for an effective close defence of Canada. There was some desultory discussion at the beginning of the decade of the possible development of the Canadian Fisheries Protection Flotilla into a force capable of active participation in Canada's defence, largely motivated by a Canadian wish to back its fishery law with more effective force, but the past and present Directors of Naval Intelligence warned that the United States would be likely to react negatively. The difficulty of the situation was underlined when in February 1905 the Admiralty advised the Committee of Imperial Defence that the Royal Navy could no longer guarantee sea control in the western Atlantic. The danger that the United States might abrogate the Rush-Bagot Agreement of 1817, which virtually disarmed the Great Lakes, added to the difficulties. [14–15, 24–5, 52]

As a result of Laurier's announcement, it was proposed that the Canadian Defence Minister visit London for technical discussions, and on 30 April 1909 the Secretary of State for the Colonies circulated among the dominion governments a proposal for an Imperial defence conference in July. On 11 May Lord Grey, Governor-General, reported that Laurier still maintained that Canadian money should be put into transcontinental railways and canals so as to build up Canada's economic and demographic strength. However, he added, 'although Sir Wilfred is opposed to the idea of holding a subsidiary Conference in July, his reply to your cable of April 30th is couched in terms which will enable the Canadian Ministers of Defence to take part in it. Pressure of public opinion here, as well as in England, Australia and New Zealand, will probably make the co-operation of the Canadian Ministers of Defence with the Ministers of Defence of the other self-governing Dominions, in the July Subsidiary Conference, necessary.' [64–6]

Fleet Units

When the conference convened 20 July 1909, the First Lord, Sir Reginald McKenna, presented a plan for dominion naval development which could satisfy both the constitutional and political needs of the Dominions, and the requirements of the naval service: 'In the opinion of the Admiralty, a Dominion Government desirous of creating a navy should aim at forming a distinct fleet unit; and the smallest unit is one which while manageable in time of peace, is capable of being used in its component parts in time of war.' The fleet unit would be based on an 'armoured cruiser' which would be of the *Dreadnought* type, supported by unarmoured cruisers, destroyers and submarines. The latter two types would be suitable only for local defence, but the cruisers could act in defence of trade as part of the imperial fleet. Each fleet unit would require about 2,300 men and cost £3,700,000 plus an annual charge of £600,000. [66]

When detailed discussions were held a month later at the Admiralty, the Canadian representative rejected the proposed 'fleet unit' as unsuitable for a country with two widely separated coasts, and requested the Admiralty to provide them with a plan for a navy of small ships divided between the two coasts, and costing £400,000, or £600,000 including the cost of maintaining the Halifax and Esquimalt dockyards. Australia decided to pursue the fleet-unit plan, constructing a squadron which would constitute part of the Eastern Fleet of the Royal Navy, and New Zealand proposed to contribute to it through continued subsidy. A total of three fleet units, one provided by Australia and one partly paid for by New Zealand, would make up the Pacific Fleet. As early as 25 January 1910, however, the Director of Naval Intelligence was asking whether in fact it would be possible to spare so many ships from the European theatre. [68, 71]

Operational Control

The issue of Imperial control had not, in fact, been settled. The optimism felt in April 1909 was premature. At the end of the year Greene set up an interdepartmental conference on the status of Colonial Ships of War. On 1 June 1910 the Canadian Minister of the Marine and Fisheries, G.J. Desbarats, enquired of the Admiralty whether the new Canadian navy would be permitted to fly the White Ensign, and use Royal Naval signals. When the Interdepartmental Committee returned its report on 4 July, however, it recommended the formulation of a legally binding structure for a united Imperial fleet with all officers subject to, and entitled to participate in, British courts martial. The Australian High Commission was advised on 16 July that the matter of the flag was integral to that of control, and on 15 August the Secretary of State for the Colonies forwarded a suggestion to the Canadian Governor-General that advantage should be taken of the presence of the Canadian Minister and Deputy Minister of Justice in Europe to confer on the legal and constitutional issue. A memorandum prepared by the Admiralty, August 1910, for the Committee of Imperial Defence noted that the problem of arranging for peace-time control was the more difficult, but that it had to be borne in mind that 'wars arise out of acts done in times of peace'.

On 8 February 1911 an internal Admiralty memorandum on the 'Status of Commonwealth Naval Forces' was drawn up following a discussion with the Australian High Commissioner in which it was proposed that the requirement that Australian forces should be under the control of the Australian government for ordinary pur-poses be facilitated by giving the commander of the Commonwealth fleet unit an independent command under the Admiralty. To make possible central control in wartime, the Aus-tralian government should enact its own Naval Discipline Act framed in identical terms with that of the United Kingdom. It should remain to the Australian government the discretion to put Australian forces under a senior British commander, but in wartime it would be desirable to do so by order-in-council to reinforce the seriousness of the development. This memorandum formed the basis of the report of a Committee of Imperial Defence sub-committee assembled 'to formulate questions connected with naval and military defence of the Empire to be discussed at the Imperial Conference, 1911'. In order to build up a feeling of unity, the Admiralty eventually recommended the use of the White Ensign throughout the Imperial fleet. On 3 March 1911 the Canadian government forwarded its request to the King that the new Canadian fleet, authorised by Laurier's 1910 Naval Service Act, should be called 'The Royal Canadian Navy'. [70, 72, 74, 76–81]

The Canadian government agreed that R.C. Smith, K.C., of Montreal could discuss with the British government the legal relationship of the new Canadian navy with the Imperial government. Ottawa was unhappy with a requirement that they communicate through the Colonial Office deployment of ships abroad, and were reluctant to conform to the Admiralty's position that once engaged in hostilities the Dominion was no longer free to withdraw its forces. However, Lord Haldane expressed the view in the Committee of Imperial Defence that 'there was very little doubt' that the Dominions would participate in a serious war, and in any event, they 'were quite alive to the fact that they would really have very little choice, none unless the enemy so willed ...' On 30 May the Committee of Imperial Defence finalised its position on 'Co-operation between the Naval Forces of the United Kingdom and Dominions'. On 20 and 29 June the final 'Memorandum' prepared for the Imperial Conference was discussed at the Admiralty. The text of the memorandum consolidated the agreements reached on command, training, discipline and co-operation among the fleets of the Crown, and established the limits of the stations under Australian and Canadian command.

Borden's Naval Aid Bill

Laurier's Liberals were defeated in the Canadian 1911 general election, but Sir Robert Borden, who formed a Conservative administration, was interested in Canadian naval development. On 25 July 1912 Winston Churchill, First Lord of the Admiralty, agreed to visit Canada, if properly invited, to discuss naval requirements. At the same time, Borden, then visiting England, requested from the Admiralty 'an unanswerable case for an immediate emergency contribution' to the Imperial navy; a confidential one he could present to the cabinet and one for public consumption. This latter, published as a command paper, concluded with the ringing words that Canada's contribution to the Imperial fleet 'should include the provision of a certain number of the largest and strongest ships of war which science can build or money supply'. In acknowledging the Canadian wish to contribute to the central strategic forces of the Empire, Churchill observed that neither dreadnoughts nor submarines could be built in Canadian yards, but Borden suggested that if Canada paid for the construction of dreadnoughts in British yards, these shipbuilding firms could off-set the problem of fiscal transfer by building smaller ships in Canadian yards. Somewhat cautiously, Churchill agreed to the idea in a letter of 4 November. [90–5, 101, 107]

Borden was in political difficulties, however, and in the end was not able to fulfil his offer of ships for the Royal Navy. The Canadian senate threw out the bill to fund ships for the Royal Navy. Borden suggested to Churchill that the United Kingdom government start work on the ships on an undertaking that a new bill would immediately be put before the Canadian parliament. Churchill, however, thought that constitutionally unacceptable, and cautioned that it would only be possible for Canada to provide the funding for the new battleships if she acted quickly, because the Admiralty had to make its contracts with builders within the fiscal year. He also warned that Britain's 1914 budget might replace dreadnoughts with torpedo boats because of their increasing value, the soaring cost of battleships, and because their multiplier effect on naval strength in home waters could release dreadnoughts for Imperial service. Finally, he warned that Canada's eventual decision could have an impact on negotiations with Germany for a 'naval building holiday'. The Canadian government abandoned its efforts to subsidise the Royal Navy. [121–4, 129]

Imperial Squadron

Churchill wrote to the First Sea Lord, 17 January 1913, suggesting that an idea he was promoting for the establishment of an Imperial squadron might help to sell the proposed naval contribution to the Canadian parliament. His proposal was, in effect, that the Dominions should pool their naval resources and form a single Imperial squadron which would cruise 'about the Empire giving each part a turn and going to any threatened point; its influence would be felt everywhere . . .' The idea had no appeal, and perhaps it really had little merit. A staff note of 22 January admitted that the Canadian ships, if stationed in Gibraltar, would serve only to reinforce the already adequate provision which had been made to contain the Australian fleet. The staff work which went into the problem of an Imperial squadron, however, did have significance for the longer term. A list of steaming times to the various strategic ports of the Empire which was drawn up may be looked on as the basis of post-war staff plans for deployment of the main fleet to Singapore. [110, 112]

Control of Foreign Policy

The basis for agreement over command and control of the Australian and Canadian navies, and for the contributions made by the colonies and smaller dominions to the Royal Navy, was the willingness to accept the idea of a common foreign policy for the Empire. That, in turn, depended upon the development of some means of influencing common policy. At the 1907 Imperial Conference, Sir Henry Campbell-Bannerman, the Liberal Prime Minister, had said that 'the cost of naval defence and the responsibility for the conduct of foreign policy hang together', but in fact there was no means by which a common policy could be developed which served the diverse needs of the constituent parts of the Empire. Although Canada did benefit from the strength of the British Empire, Canada was also peculiarly vulnerable to the consequences of Imperial policies which might lead to conflict with the United States. In defending the 1912 naval appropriations bill Prime Minister Borden made the statement in parliament that 'When Great Britain no longer assumes sole responsibility for defence upon the high seas, she can no longer undertake to assume sole responsibility for and control of foreign policy, which is closely, vitally and constantly associated with that defence in which the Dominions participate.' Opposition, especially that of Henri Bourassa in Montreal, objected that Canada would in fact have little control over Imperial policy. Even a purely national navy, in his view, would inevitably commit Canada to the military consequences of policies made in London. A memorandum sent to the Admiralty by Lional Curtis of the Round Table Movement, dated 9 January 1914, succinctly stated the Canadian constitutional objective of obtaining control of foreign policy, either by separation from the Empire, or by establishing a more 'Imperial' foreign policy separate from British control, in effect reducing the United Kingdom to the status of a Dominion. [130] In fact, however, the forces of nationalism in Canada were stronger than were those seeking a system of Imperial federalism.

The constitutional relationship between the Royal Navy and the Dominions' navies which had been established by the agreements of 1911 was too much of a lawyers' deal which did not quite satisfy anyone. Churchill had made waves in Australia when in a speech of 9 November 1912 he said that the Royal Australian Navy would come under Admiralty control in wartime, without mentioning that Australia retained the right to decide at the time to authorise or not the transfer of control. Churchill also made waves by approaching the Australian Prime Minister through the High Commissioner requesting that Australia co-operate in the Admiralty decision to expedite completion of all ships under construction. The Governor-General suggested to the Secretary of State for the Colonies that 'the usual channel of communications should be employed by the Admiralty'. [103–4]

New Zealand

Seddon had assured the 1902 Colonial Conference that New Zealand believed in the necessity of a single Imperial navy. In the subsequent years, however, concentration of resources in the North Sea led to demands in New Zealand that some local naval capability should be developed. Vice-Admiral Fawkes, Commander-in-Chief Australasian Squadron, had in 1907 discouraged the acquisition of destroyers and torpedo boats, as 'not strategically necessary', although he conceded they could be useful to give 'confidence to the mercantile community, and to foster a growth of a maritime spirit in the people'. He reported to the Admiralty that the new administration of Sir George Ward would 'probably be no party to a local Navy in any shape or form but will be ready to adhere to the present Naval Agreement with any necessary modifications in the Regulations'. [45–6]

In October 1912 the Commander-in-Chief Australian Station, Admiral King Hall, got the bit between his teeth and wrote the Admiralty that there was a good prospect New Zealand would join their naval forces to those of Australia. When his letter reached Whitehall it met with swift condemnation from the Lords Commissioners. Rear-Admiral Beatty, the Naval Secretary to the First Lord, felt that 'No one can yet say whether the Commonwealth Navy will be a success or not[.] There are many people who predict not[.] It w[oul]d there[fore] be a misfortune for New Zealand's Navy to be absorbed in that of the Commonwealth.' King Hall was telegraphed to refer all discussion of the subject to London. When a meeting was arranged at the Admiralty on 10 January 1913 to discuss the question, however, the decision was made that the Admiralty should express its willingness to implement whatever policy was decided upon by the New Zealand government. [99, 102, 109]

Colonel Allen, New Zealand Minister of Finance, visited the United Kingdom in early 1913 and had consultations with Churchill which led to an extensive correspondence. In his letter of 18 March Allen asserted that there was no conflict in fact between the idea of local navies and that of a central Imperial fleet, because the former stimulated the political support in the Dominions which enabled them to assist the Imperial fleet. He suggested practical measures to meet New Zealand's naval requirements, especially the development of the New Zealand naval reserve, and urged the development of a means of consultation at cabinet level by which the Dominions could learn how to handle defence matters. [114, 116–17]

The idea of Dominion representation on the Committee of Imperial Defence had been canvassed at the 123rd meeting of Committee of Imperial Defence, 11 April. At the meeting Churchill read a memorandum on 'Imperial Naval Policy' in which he took the line that the strategic needs for unity of command were absolute whereas the political need to satisfy public opinion in the Dominions was a temporary problem. In the discussion which followed, however, Colonel Allen made it clear that New Zealand needed to acquire greater control over its naval defences in anticipation of future trouble with Japan, and hoped it could do so in a way which reinforced the Imperial fleet. [115, 118]

In a letter of 24 April, written after his departure from the United Kingdom, Allen explained the means by which New Zealanders could make a career of the navy through training and service in the Royal Navy. Churchill made careful marginal notes on it and replied on 31 July with a letter and draft memorandum, setting out the proposed organisation and training of New Zealand Naval Forces. This crossed one from Colonel Allen which stated that New Zealand political reasons required that the 1909 agreement to station *Bristol*-class cruisers in New Zealand waters be honoured. Churchill patiently answered on 25 August that the German situation no longer permitted deploying bristols to the Pacific where there were no ships of similar force belonging to other powers. This had been explained in person to Colonel Allen, but it was a point the New Zealand government was reluctant to concede. On 24 September the Governor-General of New Zealand forwarded another request by telegraph asking for the bristols to be sent to New Zealand, and Oswyn Murray, Assistant Secretary of the Admiralty, minuted on it that it was evident Colonel Allen had not passed on the detailed explanation which had been sent him. [120, 125–6, 128]

Churchill's proposal for an Imperial squadron died of many cuts. The Admiralty was the first assassin when it requested the New Zealand government to allow HMS *New Zealand* to form part of the battlecruiser force in home waters. New Zealand expressed concern that the other provisions of the agreement to deploy British forces to the Pacific theatre were not being carried out. In an Admiralty memorandum sheet the Chief of Staff indicated in October 1912 that the Admiralty objective must be to 'appear to be adhering to the arrangement' even if thin excuses had to be made about delayed deployments. The 1909 intention to station submarines in Auckland was reconsidered on the grounds that New Zealand waters were unsuitable, although the Chief Hydrographer believed Auckland to be quite capable of taking submarines. A suggestion substituting an additional destroyer was embodied in the Admiralty's official reply to the Colonial Office in January 1913. The Secretary of State for the Colonies, Viscount Harcourt, asked why the undertaking to send HMS *Defence* to the China Station, made when New Zealand offered HMS *New Zealand* for home service, had not been honoured, a weaker ship being substituted. The Admiralty replied that they had not thought it 'necessary' to advise the New Zealand government of the change. [88–9, 96–7, 100, 108, 111]

India

India contributed £100,000 a year to the maintenance of the Royal Navy in eastern waters, but also had a navy of sorts, the Royal Indian Marine, which was part of the military department of the Indian government. Its principal tasks were marine survey work, the transport of soldiers, harbour defence and the maintenance of aids to navigation. Royal Indian Marine patrol vessels operated in the Persian Gulf. In March 1912 a committee which had been struck under the presidency of Rear Admiral Edward Slade to consider the future of the Royal Indian Marine brought down its report, in which it was recommended that the Royal Indian Marine should cease to do transport work, and that India's dockyards should be leased out to civilian contractors. No change was implemented, however, until after the First World War, during which the service was engaged in active operations. Difficulties were experienced when British soldiers and sailors refused to obey the orders of the Royal Indian Marine officers in charge of transports, and it was agreed to accord the unusual privilege of placing the King's sign-manual on Royal Indian Marine commissions. [106, 131, 135, 137–9, 143–4, 146, 158]

Newfoundland

Newfoundland, the first colony and now a Dominion, was in no position to contribute money to the Royal Navy, let alone develop naval forces of its own. In the age of sail, Newfoundland fishermen provided a valuable source of naval manpower, but by the twentieth century the contribution of men for the Royal Naval Reserve was more valuable to the Newfoundland economy than it was to the navy. Its value to the Empire as a whole lay chiefly in its fostering of 'Imperial Sentiment'. In 1900 the first fifty men of the newly established Royal Naval Reserve in Newfoundland were embarked for training. The Treasury objected, however, that the Admiralty did not have authorisation to pay Royal Naval Reserve men outside of Great Britain unless they were seamen in registered British merchantmen. Approval was only obtained for the payments when the Admiralty undertook not to incur any further liabilities until the Colonial Conference could discuss the whole question of defence contributions. [1, 3–4] Sentiment in Newfoundland was directed exclusively across the Atlantic towards Britain. No representative of the government of Newfoundland had been present during the Admiralty discussions of June 1911, and the Newfoundland Prime Minister, E.P. Morris, wrote to the Administrator of Newfoundland to make it clear that Canada's naval station must not include Newfoundland territorial waters. [82–7]

The outbreak of war in August 1914 put an end to the wrangling about constitutional matters. Thousands of men and women from the colonies and Dominions served in the Royal Navy, and the navies of the Dominions, and still more thousands fought as soldiers. HMS *New Zealand* had a distinguished career in the Grand Fleet, serving as flagship of the Second Battle Cruiser Squadron. HMS *Malaya* was one of the four fast battleships in the Fifth Battle Squadron. HMAS *Australia* was initially retained by Australia in the Pacific but served later with the Grand Fleet. Hostilities, however, ultimately tended to reinforce conviction in the Dominions that they should control their own foreign policy and their own fleets.

CONFIRMING THE POST-WAR STRUCTURE

At the Imperial War Conference called together in early 1917 consideration began to be given to the form which should be given to post-war defence co-operation. New Zealand remained strongly committed to defence centralisation, and Sir Joseph Ward framed a resolution requesting the Admiralty to provide a formal assessment of requirements. When it did so in May 1917, its advice was an uncompromising advocacy of a single Imperial navy, controlled by an Imperial naval authority, and supported by local Naval Boards in each Dominion which would take care of local dockyards and fuelling depots. When discussed at the War Cabinet meeting of 24 May, however, the Dominions, except for Newfoundland, rejected the concept of a single navy as impracticable, but approved of the idea of common standards of training. At a meeting of the Imperial War Cabinet, 27 June 1918, the First Sea Lord, Admiral Wemyss, drew attention to the memorandum, which he acknowledged had been approached at the Admiralty 'with some trepidation in so far as it touched upon political ground'. On 24 July the War Cabinet decided not to vote on the Admiralty memorandum but to reserve the subject for individual discussions by Dominion governments with the Admiralty. [136, 140, 142]

The Admiralty recommendation that oil-fuel depots be established world-wide to enable the main fleet to deploy anywhere in the world, was more acceptable to the Dominions. A memorandum of 31 October 1918 began the process of setting standards for facilities in the Dominions, in Crown Colonies, and in foreign countries. Two months later the Admiralty drew up for the War Cabinet a memorandum proposing a 'Telegraphic Communications Board' to co-ordinate the war communication of the Empire, especially by ensuring effective wireless telegraphy. [145, 151]

The Admiralty did not despair of securing its ends. Politics are fickle and there was some hope that the life of the administrations of Australia and Canada might be short, and then followed by more accommodating regimes. In the meantime, it responded to the idea that an Admiralty representative should be sent to the Dominions to advise on naval policy. Shortly before the outbreak of war in 1914 the Canadian government had suggested that the Admiralty send Admiral Jellicoe to advise on Canadian naval policy. There had been some concern in the British cabinet that his personal opinions would be interpreted as the policy of London, but Churchill reassured it that Jellicoe would be fully briefed. The outbreak of war put a stop to the idea, for the moment, but six months after the battle of Jutland, in November 1916 when Jellicoe was about to retire as Commander-in-Chief Grand Fleet, Joseph Cook, the Commonwealth of Australia's Navy Minister, expressed a desire for Jellicoe to visit Australia to advise on naval policy in the light of technical developments during the war. The appointment of Jellicoe as First Sea Lord again prevented the trip, but when Lloyd George ousted Jellicoe, the idea of an Empire tour was revived, possibly as a way of getting Jellicoe out of England. Sir Eric Geddes, First Lord of the Admiralty, suggested that New Zealand should be included in the tour. [132, 141, 147]

Sir Oswyn Murray, Permanent Admiralty Secretary, drew up a remit for Jellicoe's tour on 30 November 1918 which made it clear his task was to 'take any proposals made by the Dominions seriously, and not merely to use them as a means of carrying on a crusade on behalf of the Admiralty scheme'. A revised draft was completed 17 December along the same lines, but the final one of 23 December added the proviso that 'the Admiralty do not depart from their declared views'. However, they accepted the importance of the objective of reaching 'uniformity in naval organization and training and types of naval material throughout the Empire'. India and Canada were added to the itinerary, instructions being sent to Jellicoe by telegram on 3 and 5 March 1919. [148–50]

In the light of the declared intention to accord the government of India the status of a Dominion, Jellicoe believed that India would want to accept greater responsibility for naval defence. Accordingly, he recommended that the Royal Indian Marine should be transformed into the Royal Indian Navy. It should be under the operational control of the Admiralty: 'Unity of command is of the first importance.' Being a sympathetic imperialist, Jellicoe appreciated that the strategic objective of unity would necessitate changes in the 'future composition of the Admiralty and the representation therein of opinions from the different parts of the Empire'. That he was out of touch with at least a part of Admiralty opinion, however, can be judged by the reaction of the Fourth Sea Lord, Rear-Admiral Sir Hugh H.D. Tothill, who expressed great skepticism about the ability to man a fleet with Indian nationals. Such attitudes to race at the Admiralty were to be a major obstacle to the development of a true partnership with the Royal Indian Navy. [156]

Jellicoe found upon reaching Australia that unity of command based on the Admiralty simply was not practical politics. In a letter to the First Lord of 20 August, Jellicoe explained that Australian public opinion precluded his carrying out the Admiralty's exhortation to canvas for a single Imperial navy. However, he did recommend that there should be direct communication between an Australian Naval Board and the Admiralty on matters of intelligence and operations, and that all officers should be on a general list. By the time he reached Canada, Jellicoe was so convinced that centralised control at the Admiralty was an impossibility that he gave the impression to William Lyon Mackenzie King, who was soon to become Canada's prime minister, that 'even in the offices of the Admiralty they had rejected the idea of a centralized navy'. Mackenzie King was favourably impressed by Jellicoe, whom he met at a Government House reception. [162, 164]

Even before Jellicoe's departure from England, the post-war shape of Dominion navies had begun to develop. The Admiralty had agreed in January 1919 to Sir Joseph Cook's request that destroyers and submarines be transferred as a gift to the Royal Australian Navy. On 10 February W.F. Massey, Prime Minister of New Zealand, had written to the new First Lord, Walter H. Long, that in the interest of developing the naval instincts of New Zealanders, new cruisers must be stationed there. Furthermore, he urged that the strategic situation in the Pacific gave reason for concern. An internal Admiralty memorandum of 25 February suggested that cabinet approval be sought to give New Zealand a modern, light cruiser and two sloops, which were held to be more suitable for policing the islands, and a month later Long wrote to Massey extending the offer. Massey asked for a training cruiser in addition to the service unit, which latter would initially be manned by the Royal Navy. The Admiralty approved. Its offer of HMS *Canterbury* for the service unit had to be rejected by New Zealand, on Admiral Jellicoe's advice, because it was a modern oil-burning unit and New Zealand had no oil-storage facilities. [152–5]

In preparing a memorandum on 'Imperial Naval Defence' for the 1919 Imperial Conference the Admiralty had gone so far as to admit that 'The policy of forming Dominion Navies has the great advantage of stimulating national pride and effort in naval affairs, and it is therefore recommended that Canada, New

Zealand and South Africa should gradually build up navies of their own on the Australian model.' On the other hand, the Admiralty believed that effective defence by a coalition of British navies depended upon 'the creation of an Imperial Council or Cabinet'. The development of war plans should be undertaken by the Naval Staff to which Dominions naval officers would be attached. (The idea of uniting Navy, Army and Air Force Staffs into an Imperial War Staff was rejected.) Executive command of British and Dominion forces should be local, possibly in the hands of a Dominion Navy Commander-in-Chief. Administrative command of Dominion navies should be in the hands of Dominion naval authorities, and the Dominions should provide their own naval bases for Imperial naval use. On 20 August 1920 an Admiralty memorandum was completed on 'Co-operation of the Dominions and Colonies in a System of Imperial Naval Defence' which made only slight alterations to these ideas. Such changes, as they were recognised, that there was no prospect of a return to the single-navy concept. [156, 160]

During the 1921 Imperial Conference, called to prepare the British position before the Washington naval conference, the representatives of the United Kingdom, Dominions and Colonies decided that their naval experts should meet at the Admiralty under the chairmanship of Lord Lee, the new First Lord. The Admiralty once again returned to the idea that a unified Imperial navy was 'the ideal policy', but the paper which it prepared after the meeting providing 'A Brief Summary of Recommendations . . . 11 July' conceded that 'Each Dominion should maintain such naval forces as their resources admit.' The Admiralty's true opinion was recorded in the summary of proceedings: the 'coordination of independent naval forces . . . is not considered the ideal policy, which would be a unified navy under a single command'. [178, 180]

The degree of unity which in fact existed in higher direction of Imperial defence is evident from the appointment of the Canadian, Sir Robert Borden, to be one of the delegates to the Washington conference. As late as 1929 he was still actively involved in the problems of collective defence, but as a distinguished citizen of the Empire rather than as a Canadian statesman. [238] The cultural ties of Empire were so strong that the constitutional problems were manageable, although they did create uncertainties which adversely affected the ability of Britain to confront the dictators in the 1930s.

The problems of developing separate Dominion fleets which had common standards with the Royal Navy were not easily solved. C.C. Ballantyne, Canada's Naval Service Minister, proposed that there should be a common officer list, but in an Admiralty Minute Sheet of 21 April 1922 it was pointed out that the numbers of junior officers being trained for the Australian and New Zealand navies would lead to pressure to promote them unduly rapidly. There was an additional problem caused by the desire of the Dominions to have their officers serving near to home. It would be difficult to resist the pressure, but compliance would affect training and service. Finally, the Admiralty considered that Canadian midshipmen were 'a rougher type than ours or the Australians'. The arrangements which were made to call up naval reservists throughout the Empire had to be adapted to suit the differing legal relationships of the colonies and Dominions with the UK government. [168, 180]

A paper prepared by the Admiralty Statistical Department (12 March 1923) accounting the money spent per capita on naval defence by the UK and by the several Dominions since 1901 showed how great was the discrepancy. In 1922 the United Kingdom government spent £1 10s 1½d per capita on the navy, Australia £0 9s 10¼d, New Zealand £0 4s 1d, Canada £0 1s 9½d and South Africa £0 0s 8½d. Nevertheless, when in May 1923 Captain Dudley Pound, Director of Plans, completed a paper outlining the progression of Admiralty policy *vis-à-vis* the Dominions, and the means which had been established for coordinating the naval policies of the Empire, he struck a positive note. Although not the unified force which strategic planning suggested was ideal, in Pound's view the arrangements as they had developed over the years had the great strength that they recognised that 'wars are no longer waged by Navies and Armies alone, but by nations in arms. The success or failure of any measure of defence rests ultimately on the sanction of the people; . . . It is for this reason that Public interest is placed first in the list of requirements to be fulfilled.' At the thirty-third meeting of the Committee of Imperial Defence, 19 June 1923, it was agreed to ask the Colonial Secretary to forward to the Dominions the Admiralty's paper CID 196-C on 'Empire Naval Policy and Co-operation' which the First Lord, Leo Amery, said was intended to make clear 'that the Admiralty were now not merely accepting as a second best, but whole-heartedly endorsing the idea of separate

Dominion naval units'. Attention at the Admiralty was not focused on developing the means of cooperation. [189, 191, 193]

On 31 July Captain L.H. Haggan, Director of Training and Staff Duties, drew up a minute on 'Proposed Methods for Obtaining Naval Staff co-ordination between the British and Dominion Navies' in which he urged the importance of Dominions' representatives being called to meetings of the Committee of Imperial Defence when matters of general Imperial defence were being discussed, and the importance of establishing a right of direct communication between the Chief of Staff Royal Navy and those of the Dominion navies. He also recommended the appointment of Dominion naval officers to Admiralty staff positions, not as liaison officers but as regular members of the staff, who could also speak for their own services. A minute added by Captain Pound went further, and suggested that there ought to be an Imperial Naval Staff parallel to the Imperial General Staff. This was supported by Captain Hugh Tweedie when he was appointed Director of Training and Staff Duties, 28 September 1923. Additional parts of the Committee of Imperial Defence memorandum on 'Empire Naval Policy and Co-operation', were prepared in 1923 and 1925–6. However, it recommended only that staff officers be given uniform training, and that Dominion naval officers be appointed to the Admiralty Navy Staff. [194, 196, 212].

The response in the Dominions to these proposals was on the whole favourable, although the Canadian government wanted to make clear that there would not be a fusion of the naval staffs. For fear that back-bench pressure might force its hand, the government blocked the idea that the memorandum might be published openly. The concluding resolutions of the Imperial Conference in November 1926 contained a reference to the standing invitation to the Dominions to be represented on the Committee of Imperial Defence when appropriate. Mackenzie King's objection to Canadian representation on the Committee of Imperial Defence, however, was that 'it would be regarded as a surrender to the supposed centralising proclivities of Whitehall'. 'CID,' he said to Sir William Clark, the British High Commissioner, in July 1929, 'despite its imperial name, is really a British rather than an Empire institution.' Instead, arrangements were made to facilitate the circulation of copies of Committee of Imperial Defence files in the Dominions. [195, 206, 213–14, 217, 219, 237]

Control of Dominion naval forces in wartime continued to be a somewhat vexed question. The concluding resolutions of the

Imperial Conference in November 1926 suggested that each part of the Empire should take primary responsibility for defence of its own immediate area, and provide a contribution to the defence of maritime communications. It was generally recognised that only the Admiralty had the capacity to control merchant shipping world-wide. On 11 July 1927 Lord Stonehaven, Governor-General of Australia, forwarded to the Dominions Office a memorandum by the Australian Naval Board, 'Protection and Control of Mercantile Marine in War', indicating the requirements for implementation of this arrangement. However, when a memorandum, prepared in November 1927 by the Oversea Sub-Committee of the Committee of Imperial Defence on 'General Principles of Imperial Defence', spoke wistfully of the efficiency of maintaining a 'single Navy', it went too far. It went on to talk about the importance of uniformity of training and procedure so that the Dominion fleets could operate easily with the Royal Navy, but Captain H.J. Feakes, RAN, Australian naval representative in London, felt that it had been very unwise to refer at all to the politically unacceptable concept of a single navy for the Empire. 'The phrase "single Navy" is a dangerous one to revive, also it is subversive of both the spirit and the letter of principles as expressed to the Dominions by the Admiralty of their wishes relative to Dominion Naval contribution ... It is a theory that leaves the Dominion taxpayer cold.' The offending phrase was deleted, and was replaced by an essay on naval strategy which insisted on the primary importance of the main fleet disposed to contain enemy forces, but which also allowed the need for local forces to be provided against minor enemy raids. [217, 230–1, 233]

DOMINION NAVAL DEVELOPMENTS

Australia

At the Imperial Conference in November 1926, Mr Bruce, Prime Minister of Australia, was able to point to a fleet of three cruisers, three destroyers, three sloops and one repair ship, with an additional cruiser, a flotilla leader, eight destroyers and a sloop in reserve. 'Our guiding principle,' he said, 'on which all our defence preparations are based ... is uniformity in every respect ... with the fighting services of Great Britain.' A 10,000-ton, floating dock was being built and oil-tanks were being erected at Darwin as an auxiliary base. He also took the opportunity to

iterate the importance Australia attached to the effective development of Singapore. [217] Financial constraints made it difficult to
keep to the intended establishment. In December 1929 Captain
R.M. Bellairs, the Director of Plans, suggested that there was
'much to be said for reconsideration of our recommendation to
Australia to maintain submarines as well as cruisers', and suggested that this change of position could be justified by the
probable outcome of the Five-Power London Naval Conference
of 1930, and be discussed at the 1930 Imperial Conference. [240]
The inconsistencies in British policy were the subject of discussion
which took place at that conference.

The naval limitations agreements, necessary as they were to
Anglo-American relations, created problems for those parts of the
Empire most exposed to Japanese naval power. A consequence
of the 1930 London Naval Disarmament agreement was that the
number of cruisers and destroyers in British Empire navies had
to be reduced, at a time when the Australian Navy needed to
replace its ageing destroyers but was prevented by the international financial crisis. The Admiralty was reluctant to supply
Australia's requirements, but the alternative of scrapping servic-
able 'V'- and 'W'-class destroyers made little sense. The Admiralty
was also concerned that Australia replace the *Brisbane* with a
Leander-class cruiser, but recognised that financial constraint
would prevent any precipitate decision. In June 1932 the Attorney-
General of the Commonwealth of Australia, Sir John Latham,
attended a meeting of the Chiefs of Staff Sub-Committee, Committee of Imperial Defence, and introduced the idea that there
might be a better way for Australia to participate in Imperial
defence, perhaps by providing crews for ships supplied by Britain.
In January 1933 the First Sea Lord, Admiral Sir Frederick Field,
and his successor, Admiral Sir Ernle Chatfield, discussed Australia's building plans with Mr Bruce, Australian High
Commissioner in London, and a proposal was again suggested
that Britain might build the cruiser at her own expense and in
return Australia would increase its annual expenditure by a similar
sum which would be used to maintain Royal Naval ships under
Australian control in the Australian station. Captain H.K. Moore,
the Director of Plans, however, pointed out that the effect of the
deal would be to enable 'Australia to control and maintain 6
cruisers at the cost of maintenance and period replacement of her
existing force of 4 cruisers; whereas the United Kingdom would
lose control of two cruisers without any financial relief.' [248,

255–6] The Australian offer was rejected, but on 20 July 1933 the Admiralty agreed to lend the Australian fleet five destroyers. At the end of the summer, on 19 September, the First Lord, the Secretary of the Committee of Imperial Defence, Sir Maurice Hankey, and the Dominions Secretary, Sir H. Batterbee, had a meeting with Mr Bruce and Mr Shedden from the High Commission in which Australia's naval plans were discussed. This was followed three days later with another in the office of the Director of Plans. It was not until March 1934, however, that Australia finally decided to replace *Brisbane* by purchasing one of the *Leanders* being built in Britain. [261–3] It was agreed that payment should be spaced over five years. In a memorandum prepared by the Admiralty and the Foreign Office prior to the 1935 naval conference, the Admiralty indicated that it would be more convenient to obtain a separate quota for each of the Dominion fleets. [264]

New Zealand

On 10 August 1922 the Admiralty repeated its advice to the New Zealand government that 'however perfect local defence measures may be these cannot in war protect any portion of the Empire adequately'. New Zealand was advised to maintain the nucleus of a seagoing squadron which could be expanded when financial circumstances improved, to establish a reserve of oil fuel in New Zealand, and to contribute money or materiel to the development of appropriate naval bases. [186] For all its small size, New Zealand vigorously supported the needs of collective Imperial defence during the inter-war years. On 1 July 1920 the Admiralty had advised the Colonial Office that the posts of Naval Adviser to the New Zealand government, the Commodore commanding the New Zealand Division, and the Captain of HMNZS *Chatham*, should be one and the same man. The experience of the New Zealand government, however, suggested to them that they needed someone permanently on shore. [166–7] Admiral Hotham recommended (24 November 1923) that New Zealand send cadets to Royal Naval officer training schools, and work towards making the personnel of the navies interchangeable. [195, 206]

At the time of his retirement in December 1925 Acting Rear Admiral Beal, the New Zealand Naval Secretary, sent to the Director of Naval Intelligence, Rear Admiral A.G. Hotham, a paper outlining Naval Board policy for the New Zealand naval division, 'deprecating strongly any idea of a separate Navy on the Australian or Canadian lines'. He did not perceive a need for any major change. Officers, he felt, should continue to be seconded from the Royal Navy, New Zealand cadets should be trained in the United Kingdom, and New Zealand ratings should experience service in the Royal Navy. The division should continue to accept the loan of a cruiser from the Royal Navy but should also seek to acquire sloops, which would relieve the Royal Navy of the task of patrolling the Pacific Islands, and should acquire minesweeping and training trawlers. However, he considered that the acquisition and maintenance of submarines was an expense which could not be justified. Arrangements should be made for the conversion of armed merchant cruisers and for defensively arming merchant ships. [209] In October 1926 the Admiralty Director of Plans, Captain W.A. Egerton, submitted a memorandum on 'New Zealand Naval Policy' which recommended that, if New Zealand was prepared to spend £250,000 per annum on the naval division, priority should be given to contributing to the Singapore base, to taking over the two Royal Navy sloops attached to the division, and to developing docking accommodation and other preparations for modern cruisers in New Zealand. At the Imperial Conference in November 1926 Mr Coates, Prime Minister of New Zealand, undertook to increase his government's contribution to naval defence. [217]

Canada

The small scale of Canada's naval effort reflected the philosophy expressed by the Nationalist party in 1910 that the existence of a Canadian military establishment which could be valuable to London in a crisis, such as the Turkish one of 1920–2, could make it difficult for Ottawa to retain control of its national policy. In 1920 Admiral Jellicoe's most modest proposal, costing $4 million per year, had been implemented with modifications by the Borden administration, but, only after pressure in the Union caucus, had nearly led to the Canadian navy being abolished altogether. When on 28 July 1922 the Admiralty drew up for the Committee of Imperial Defence a paper on the effects of the Washington Conference upon Imperial Naval Policy, the Canadian explanation that 'overhead charges would have been out of all proportion to the defence value obtained' produced the tart comment that 'The Admiralty cannot help feeling that this bears out the view . . . that the ideal form of Dominion co-operation lies in a unified navy with [a] quota of men and ships supplied by the Dominions and India.' The British government nevertheless provided two old destroyers, *Patriot* and *Patrician* and the cruiser *Aurora* which joined a small force of trawlers and two submarines. Following the election of Mackenzie King as Prime Minister of Canada in December 1921, the budget for the Royal Canadian Navy was reduced to $1.5 million, and the fleet was reduced to reserve training. Commodore Walter Hose, Director of the Canadian Naval Service, took the radical step of reducing the seagoing fleet to the two destroyers and four minesweepers. [184]

At the Imperial Conference in November 1926 the Canadian Prime Minister, Mackenzie King, apologised for the small size of the Canadian navy, but was able to point out that in the last five years it had been 'transformed from 450 officers and men borrowed or specially engaged from Great Britain, and 50 Canadians, to 40 borrowed ranks and ratings and 460 Canadians. The Royal Canadian Naval Volunteer Reserve has been organized in the last three and a half years, and is up to its full authorized strength of 1,000 officers and men.' [217] The Mackenzie King administration decided in early 1927 that *Patriot* and *Patrician* had to be replaced, and telegraphed the Admiralty for advice. 'M' branch recommended the construction of two new ships, or, failing that, the purchase of two good ones from Britain's reserve list, and suggested that they might be sold at scrap value. Captain W.A. Egerton, Director of Plans, disagreed with saving Canada money at the expense of the British taxpayer, as it only appeared to be encouraging Canadian parsimony, and as the insistence of the Canadian government that it retain the right to decide whether the Royal Canadian Navy would be placed under Admiralty control in wartime greatly reduced the strategic value of any ship transfer to Canada. The advice eventually sent on 9 June was that the Canadian government should consider building two minesweeping sloops, which would be 'particularly suitable for the general training of personnel and would it is thought form an essential part of any naval force which the Canadian Government might eventually decide to maintain'. If the Canadian government decided to do so, the British would be prepared to lend it two 'S'-class destroyers at the cost of bringing them out of reserve. However, Commodore Hose, Chief of the Canadian Naval Staff, let it be known at the Geneva naval disarmament conference that the Canadian government was not interested in building sloops. Considering its own advice had been that destroyers were unsuitable to Canadian needs, the Admiralty were no longer willing to sell the 'S'-class destroyers at less than commercial terms, but eventually offered a price of about one third of the new construction cost. [220, 223-4, 226-7] The Admiralty also recommended that the Canadian navy make preparations to convert four *Empress*-type liners to armed merchant cruisers on the Pacific coast on the outbreak of war. [221] The Canadian government agreed to the storage of equipment at Esquimalt for this purpose. In the end, furthermore, it decided not to purchase the 'S'-class destroyers, but to take them on loan for three years while two new destroyers were built for Canada in Britain. [229, 232] In August 1930 Commodore Hose repeated his advice to the Canadian government against implementing the Admiralty's recommendation that Canada contribute to Imperial defence by acquiring cruisers.

In 1933, for the second time, the Royal Canadian Navy was all but abolished by the Conservative government of R.B. Bennett. This was staved off, however, and in November 1934 Commodore P.W. Nelles, Acting Chief of the Canadian Naval Staff, prepared a paper on 'The Naval Defence Policy of Canada' which provided a convincing rationale for expenditure. He wrote that 'Canada's two responsibilities of major importance are: (a) *War*. The protection of trade to and from Canadian ports, and sea communications in our coastal areas; (b) *Peace*. The preservation of our neutrality by preventing the perpetration of un-neutral acts by belligerents in our territorial waters.' The wartime role of defending Canada's coasts had to be given top priority.

The Canadian navy now consisted of four destroyers and a trawler, but according to the provisions of the London Naval Conference, and the terms of loan of *Vancouver* and *Champlain* to the Royal Canadian Navy, these ships should be scrapped before the end of 1936. In any case, their lives could only be extended by a major refit. In the circumstances, the Canadian navy welcomed the informal suggestion made by Admiral C.J.C. Little, Deputy Chief of the Naval Staff, that two Royal Navy reserve 'C'-class destroyers might be transferred to Canada. As Prime Minister R.B. Bennett was setting out to visit England it was hoped that the Admiralty would help persuade him to accept the gift which would increase Canada's annual personnel costs by $100,000. This task was undertaken, and Bennett was informed that 'we had formed a high opinion of the efficiency of the Canadian Destroyer Division as a result of reports we had received from the Commander-in-Chief, Home Fleet, when they had joined up with the Home Fleet in the West Indies in the Spring of 1934 and of 1935'. There was some discussion at the Admiralty as to whether 'two "C"'s in full commission in the Canadian Navy are of less value than 2 "C"'s in Reserve as Flotilla reliefs at home. This may be true but the Canadians might not appreciate it.' The Treasury, on the other hand, were happy that the ships would be well used, but wanted to get cash value of £80,000 per ship plus the scrap value of the old ships. 'The Public Accounts Committee have been very critical of the recent transfers to Australia of the destroyers and the Cruiser SYDNEY. They suspect that we were not getting an adequate quid pro quo, and that the changes were being made at the expense of the British taxpayer.' 'This,' Captain Nelles sadly replied, 'had to put an end to the idea.' The High Commissioner to Canada, F.L.C. Floud, had a meeting with Grote Stirling, Minister of National Defence, and Nelles, and reported to E.J. Harding, the new Secretary of State for the Dominions, that 'there is [not] the least prospect of this or any other government being prepared to agree to the proposals of the Admiralty'. [279–81, 283–5]

The Admiralty continued to consider the proposal of loaning *Crescent* and *Cygnet* to the Canadian navy and in Ottawa a paper was produced for the Department of National Defence on 'The Naval Service of Canada' which noted that 'in the event of a threat of war . . . we in Canada can only expect a reduction, not an increase, of British naval forces in the vicinity of our coasts.' The minimum requirement was thought to be 1 leader, 8 destroyers, 4 submarines and 3 minesweepers. It was recognised that submarines would be too expensive, but it was argued that the complement of the navy should be increased immediately because it took six years to train men, but only two to build ships. [294] Evidently the arguments of the Department of National Defence were persuasive, and the increasing tensions internationally no doubt also played a part in developing the attitudes of the Canadian government, which had returned to the control of the Liberal party under Mackenzie King. In August 1936 the Minister of National Defence, Ian Mackenzie, called on the First Sea Lord, Admiral of the Fleet Sir Ernle Chatfield, at the Admiralty and indicated that the Canadian government wanted a 6-destroyer navy, and were interested in minesweepers. 'My impression,' Chatfield noted, 'was that he was anxious to be up and doing and is an active minded man who will not let the grass grow under his feet.' Indeed, within a month the Canadian government had undertaken to purchase the *Crescent* and *Cygnet* at the full Treasury price, and had asked if it would be possible to borrow two others. [308–9] In fact, the Canadian government mistakenly agreed to pay too high a price, but the Admiralty thought it best to stick to the lower one agreed to by the Treasury. By saving the Canadian government £98,000 now, there was greater chance that they would find the cash to pay for two more 'C's a few years later. Should Canada not have the money at the time, however, the Admiralty strongly, but without success, urged the Treasury to agree to lend her the destroyers none the less. 'The Board was very gratified at Canada's acceptance of the *Crescent* and *Cygnet* especially as it is not long since the Canadian Government was inclined to do away with their Navy altogether. We are therefore very anxious to encourage this fresh sign of Canada's desire to strengthen and develop her naval forces.' [311–13, 339] At the 1937 Imperial Conference the Admiralty recommended that Canada develop fuel-storage facilities for war use, especially on the west coast, and that Halifax, Vancouver and Esquimalt should be defensively armed and supplied with Anti-Submarine Warfare equipment.

In January 1938 Rear Admiral Percy Nelles was able to advise Admiral Chatfield that the Canadian government was almost certainly going to vote the money to purchase the second two 'C'-class destroyers. And in June he asked whether the Admiralty would be able to support his request that *Kempenfelt* be purchased as a squadron leader. The new First Sea Lord, Admiral Sir Roger Backhouse, offered *Keith* instead, but Nelles was unhappy that she might not be able to receive up-to-date instruments until 1941. [335, 342, 365] In December 1938 the Canadian High Commission formally requested permission to purchase *Kempenfelt*, and on 27 February 1939 the Treasury approved the transfer at the price of £150,000. The Canadian government indicated on 5 June that they would like the transfer to take place on 1 August 1940, but the transfer was brought forward to the autumn of 1939, and then delayed by the outbreak of war and the immediate need for anti-submarine forces in the Channel.

Newfoundland

At the 1926 Imperial conference Mr Monroe, Prime Minister of Newfoundland, made a plea for the return of a Royal Navy training ship to St John's so that Newfoundland could continue to provide seamen to the Royal Naval Reserve. The idea of facilitating recruiting in the Domininions was eventually taken up by Churchill in 1936. At first the Admiralty was not interested, but by May 1937 it had been persuaded. The scheme to allow each station commander a quota, however, was considered 'a Dominions Office baby', and in consequence Newfoundland would have to be included in the quota for the North American station. 'There is at present little Naval need for Dominion recruits; Newfoundlanders particularly are reported to be dull, whereas our only present requirement is for men intelligent enough to be specialists in one line or another.' [296, 299, 301, 323, 327] Orders were only sent to the station commanders in July 1939. Canada agreed to permit recruiting, but did not want its own agencies involved. Australia and New Zealand, however, refused to permit recruiting for the Royal Navy as it could interfere with the build-up of their own navies. [378]

Union of South Africa

In 1921 Cape Colony and Natal ceased to pay their subsidies to the Royal Navy, and the Union government took over two Royal Navy trawlers for training purposes, and undertook to maintain Simonstown naval base for Imperial use. [172] On 28 March 1923 Prime Minister Jan Smuts prepared for the Governor-General a statement of the Union of South Africa Navy's activity, in survey work, training and port development. Although the naval effort was very small in scale, progress in constructing an efficient force was felt to be evident. [192]

Financial difficulties led to the Union of South Africa returning its ships to the Royal Navy in 1935, and the Royal Navy accepting responsibility for naval training of reservists. South Africa undertook to establish a reserve of minesweeping and anti-submarine gear which could be fitted into requisitioned vessels in an emergency. The Admiralty recommended that 72,000 tons of oil be maintained at Cape Town, but Prime Minister Pirow preferred that the operation be kept secret and be carried out through a holding company. He also preferred to establish 15-inch guns for the defence of Cape Town and Simonstown, but agreed to the mooring of the monitor *Admiral Soult* in the harbour. [300, 354–7, 363]

India

The questions raised by Jellicoe's recommendation that the Royal Indian Marine be transformed into a Royal Indian Navy were not soon reconciled. On 25 May 1921 the Deputy Admiralty Secretary, Sir Charles Walker, addressed the Secretary of the Committee of Imperial Defence on the subject of the relative merits of an extension of the annual subsidy the Government of India had paid for the services of the East Indies Squadron since 1869, the formation of a Royal Indian Navy, or of India contributing to the task of local naval defence. He felt that this was a matter for cabinet decision, and must await a definite proposal from the Government of India. [172, 245] In 1922 the government of India established a committee under the chairmanship of Lord Inchcape which, as part of a general effort to reduce governmental expenditures, was to consider the future of the Royal Indian Marine. Its report recommended that the Royal Indian Marine be 'drastically curtailed and reorganized', in particular that its troopships be sold and its harbour works be privatised. [188] In 1925, however, a committee presided over by Lord Rawlinson proposed the reorganisation of the Royal Indian Marine into the Royal Indian Navy along the lines proposed by Admiral Jellicoe. [208] The Naval Discipline bill intended to implement these proposals was thrown out by the Indian legislature by one vote, but most of the changes were undertaken in any case, although the name remained unchanged. On 11 November 1928 the White Ensign was raised on all Royal Indian Marine ships which were fully equipped with guns, and the Royal Indian Marine was given combat status. In 1931 the force consisted of 11 minesweeping and 5 anti-submarine vessels, and it was anticipated that in the event of war a further 16 vessels in each category would have to be taken up. [264]

The Secretary of State for India recommended to the Committee of Imperial Defence that the Imperial government agree to the request of the Indian government that control of the Royal Indian Marine not be transferred to the Commander-in-Chief East Indies Squadron without its prior consent. This was understood to be politically important. [247] So that the Royal Indian Marine could take control of merchant shipping, steps were taken to establish a Royal Indian Marine Voluntary Reserve. [252] Financial problems were so acute, however, that the formation of its Voluntary Reserve had to be deferred, and the fleet had to be reduced to 8 vessels. [254]

The process of Indianisation was necessarily a slow one, because of the time required to find and train suitable officers. In 1933 there was only one Indian serving officer in the executive branch and two in the engineering with four and seven under training, but the rate of Indianisation was double that anticipated in 1928. [267] In 1934 the Indian Navy (Discipline) Act was finally passed in the Indian Legislature and Council of State, the Royal Indian Marine was abolished on 2 October 1934, and the King Emperor presented his King's colours to the Royal Indian Navy which was officially brought into existence. [268–74, 277–8]

In contrast to the policies of the Australian and Canadian governments, the principle of unity of command was pressed for by the India Office, but the Admiralty resisted. Attitudes to race made it difficult for the Royal Indian Navy to take its place alongside the Royal Australian Navy and the Royal Canadian Navy. India Office felt it was important that discrimination should be eliminated at least in form, even if practical arrangements had to be accepted which provided for effective white superiority. In August 1935 J.A. Simpson, Secretary of the Military Department of the India Office, requested that the Admiralty express its views on whether officers of the Royal Indian Navy should be granted commissions by the Governor-General 'in his Majesty's Fleet' in the form employed by the Australian and Canadian navies. 'Agreement,' he said, 'had already been reached that RIN officers would rank and command with, but immediately below, Royal Naval officers of the same rank and seniority.' On 3 October 1935 Maurice Hankey, Secretary to the Cabinet, submitted to the King a request for an order-in-council granting commissions to officers of the Royal Indian Navy from the Secretary of State for India. It remained to be decided, however, whether the wording of the commission should resemble that used for Dominions officers. [275–75, 286–7]

The Royal Navy was determined that Indian officers should not acquire a right to command Royal Navy personnel. On 24 August 1936 Rear Admiral G.C.C. Royle, Naval Secretary, noted to the First Lord that 'the Board do not wish that Indian Naval Officers should have command over, or sit on Courts-Martial on, white Naval personnel'. Simpson continued to try and find a formula which ensured for Indian naval officers an equality of rank and command in the Royal Navy, but excluded them from courts martial of white personnel. This the Admiralty Board was prepared to accept. Nothing definite had been ordered, however, as late as 23 August 1939 when the India Office declared that it was no longer prepared to wait and was acting on its own to grant commissions 'in His Majesty's Fleet'. The Admiralty did agree to grant equality of status to those officers manning ships transferred to the control of the Commander-in-Chief East Indies. [304, 310, 318, 332, 384–6]

Racial attitudes impeded training programmes. In November 1938 India Office complained that Royal Indian Navy cadets were being kept in isolation when sent to England for training, not competing in sports with Royal Navy cadets, or meeting anyone who might be a good influence on their careers. With a renewal of war in Europe an ever growing menace, the Admiralty in February 1939 agreed in principle that Royal Indian Navy junior officers should receive the same training as Royal Navy junior officers. [364, 372]

In June 1934 the government of India proposed to the Secretary of State for India that the £100,000 subsidy paid to off-set the cost of the Royal Navy's East Indies Squadron should be reduced temporarily so that money could be made available for the development of local naval defences, requiring guns and minesweeping equipment and the establishment of the Royal Indian Navy Voluntary Reserve, and for the acquisition of seagoing ships for the Royal Indian Navy. Admiralty support was sought, and obtained. Arthur Bedford, in the Navy Office, Bombay, wrote that 'the attitude of the Government of India is that they already pay enough for Imperial Defence by maintaining a large part of the British Army as much on Imperial account as on their own and they do not see why they should spend more on Imperial defence by a subsidy towards the R.N.' The attitude of the Treasury, as expressed by the Chancellor, Neville Chamberlain, was entirely unsympathetic, and again reference was made to the unsympathetic attitude of the Public Accounts Committee. The Admiralty pleaded the importance 'to the United Kingdom that provision should be made for Indian local naval defence. If it is not made by the Government of India the liability will in the long run have to fall on the United Kingdom.' Despite Admiralty advice not to make the stipulation, the only terms on which the British government would agree to letting part of the subsidy be used for construction and maintenance of sloops was that they should be made available unreservedly in wartime for use by the Royal Navy anywhere in the world. India Office continued to press the proposal, and in January 1936, the Chancellor agreed that half of the cost could be taken from the subsidy. [265–6, 282, 287–292, 295]

The growing tension with Italy added a sense of urgency. Confusion arose when the Commander-in-Chief East Indies ordered the implementation of local defence arrangements, which could only be carried out by employing the crews of seagoing units of the Royal Indian Navy, and then requested the deployment of two sloops to Aden. [297–8, 306] The government of India could not agree to the Chancellor's compromise. [293, 315–16] The Admiralty Director of Plans, Captain T.S.V. Phillips, initiated a minute in February 1937 which urged again that 'the position with regard to the Indian Navy is . . . even in the worst way of looking at it, better than that of the Dominion Navies', since the Dominion governments might choose to stay out of a war. [317, 326, 328]

In April 1937 the Government of India sent to London its draft 'Nine Year Plan' for the development of the Royal Indian Navy, and urged that it be paid for by the complete abolition of the Indian subsidy, in return for which the Royal Indian Navy would relieve the Royal Navy of the obligation to maintain three sloops in the Persian Gulf. [321] In July 1937 Simpson discussed the matter 'demi-officially' at the Admiralty, and on the twelfth Duff Cooper, Secretary of State for War, wrote to Sir John Simon, the Chancellor, urging that it was in Britain's interest to give up the subsidy if in return India made a commitment to develop the Royal Indian Navy. [329] On 26 July 1937, Sir John Simon wrote to the Secretary of State for India, the Marquis of Zetland, agreeing to the policy, but he sought guarantees that political conditions in India would not prevent transfer to the Admiralty of tactical control of the Royal Indian Navy in the event of war. [330–1, 333] Once the financial problem had been dealt with, it was possible for the government of India to establish naval reserves, but legislation was necessary for the purpose. In protest at the reservation in the Indian constitution of defence from control by the Indian Legislature, the bill was rejected there and had to be 'certified' by the Viceroy. [369, 377]

In September 1938 Simpson wrote to the Admiralty outlining the circumstances in which the Royal Indian Navy would be placed under the command of the Commander-in-Chief East Indies, noting that 'As result of this agreement . . . the Government of India are bound in the event of an emergency to place all their sloops "at the disposal of" the Admiralty, whether or not the purposes for which they are being used in connection with that emergency are the purposes of Indian Defence. It is,' he added, 'implicit in the agreement that action of this kind would not be taken unless the emergency were geographically related to the defence of India.' [345] Four days later, 28 September 1938, the Munich crisis led to the Royal Indian Navy being placed under Admiralty control, thus putting the lie to the concern that Indian politics might impede effective employment of the Royal Indian Navy. The only complaint was that it took three days to carry out the routines of government authorising the transfer of operational control, and it was agreed that in future the Indian Government would authorise the transfer on the receipt of the warning telegram. [347, 352, 376, 384–6, 388–90] In January 1939 Vice-Admiral Sir James F. Somerville, Commander-in-Chief East Indies Station, reported to the Admiralty that the Royal Indian Navy appeared to be well established, although the ships were not adequately armed. [367] The War Organisation of the Royal Indian Navy, which included the revised statement of the responsibility of the Flag Officer Commanding, Royal Indian Navy, to co-operate with the Commander-in-Chief, East Indies, was finalised in August 1939. [382]

Crown Colonies

In contrast to the Dominions, none of the Crown Colonies maintained substantial naval forces. In June 1938 the Oversea Defence Committee of the Committee of Imperial Defence prepared a paper which included a report on 'The Co-operation of the Colonial Empire in Imperial Defence'. It stated that the role of colonial forces in securing the bases from which the navy might operate was central to Imperial defence. Furthermore, 'At every defended port local naval forces have to be maintained for minesweeping and other local defence duties. For these purposes Naval Volunteer Reserve units are particularly valuable, since they enable a corresponding economy to be made in regular naval personnel, who can be more profitably employed in other duties requiring a higher standard of technical training. The uses of local Naval Volunteer Reserve units are by no means confined to defended ports.' Besides minesweeping, there was a requirement for 'Detaining', and 'Naval Reporting' officers, for contraband control and control of shipping, for defensive and offensive purposes. [340]

Imperial Naval Strategy

The decision that the bureaucratic and operational organisation of the fleets of the Empire and Commonwealth after the Great War should continue on a national basis, but optimised for effective co-operation, made it essential to address collectively the strategic problems of the post-war world. This had begun with Jellicoe's far from encouraging threat assessments during the Empire tour. On his way to India he had drawn up a general paper on 'Post-War Naval Requirements' which looked at the none-too-attractive prospect of having to match post-war United States naval expansion. In Australia he had felt obliged to raise the spectre of a Japanese war because, as he explained in a letter of 20 August to the First Lord, Lord Long, otherwise he could not have advised any standard for Australian fleet strength. In doing so, of course, he was echoing opinion already held in the antipodes. [153, 162–5]

Jellicoe's indication in his reports to the Australian and New Zealand governments in 1919, that a British Pacific fleet would have to be formed, was highly embarrassing to a government committed to economy. He was warned to clear with the Admiralty any strategic advice he intended to offer the Canadian and South African governments. However, in a paper prepared for the War Cabinet 21 October 1919 on 'The Naval Situation in the Far East' the Admiralty had anticipated the non-renewal of the Anglo-Japanese alliance. In the light of Japan's naval building programme, the Admiralty laid the basis of the inter-war crisis planning by suggesting that, in the event of anticipated hostilities, naval forces in the Eastern theatre should be concentrated at Singapore, there to await the arrival of the main fleet dispatched from Europe. [161, 163] In March 1921 the Commanders-in-Chief of the China, East Indies and Australasian stations met at Penang to discuss the need for a higher organisation to co-ordinate their efforts in time of war. [169] The problem, as Prime Minister Hughes of Australia explained to the Commonwealth House of Representatives, was compounded by the United States' hostility to Japan which made it necessary to let the Anglo-Japanese alliance lapse. Churchill, as Colonial Secretary, forwarded Hughes's statement to the Committee of Imperial Defence on 28 April 1921. [170]

The oil-fuel requirements to ensure fleet mobility, especially so that the main fleet could be deployed to Singapore, were outlined in an Admiralty memorandum of 24 May 1921 for the Committee of Imperial Defence. On 7 June 1921 the Oversea Sub-Committee of the Committee of Imperial Defence reported its recommendation that Singapore be developed as a base, the Treasury submitting a dissenting opinion. On 13 June the 141st meeting of the Committee of Imperial Defence agreed to present to the forthcoming Imperial Conference the plan to develop Singapore as British governmental policy, and on 21 June the First Lord presented a memorandum to the Committee of Imperial Defence on 'Reserves of Oil Fuel' which considerably elaborated on that of 24 May. [171, 173–5]

The Imperial Conference considered the problem of naval defence, especially in the Pacific, on 4 July 1921, and Earl Beatty, now the First Sea Lord, iterated the importance of developing a defended dockyard at Singapore. The next day the leaders met at the Admiralty and the fundamentals of inter-war strategy were discussed and agreed to. The main fleet would have to be stationed at home because of its need for developed bases, but fuel supplies would be built up to facilitate deployment to the Pacific, and a base would be built at Singapore.

In 'A Brief Summary of the Recommendations by the Admiralty' (11 July) it was suggested that 'Australia, New Zealand and India should be asked to contribute a considerable portion of the expenditure necessary to develop Singapore as a fleet base, since they are intimately concerned therein.' At the twenty-sixth meeting of the Imperial Conference, however, the Secretary of State for India made it clear that the money India spent on its army, which was important throughout the region, precluded any subsidy of the Singapore project. If any money came available, India would prefer to use it to develop a Royal Indian Navy. The Prime Minister of Canada, Arthur Meighen, rejected Mr Massey's appeal for a common perspective on Imperial defence in the Pacific, on the grounds that the Canadian public could never be brought to agree. [176–8]

The Director of Plans, Captain Barry Domvile, completed a memorandum on 'Plans for War in the Far East', 3 November 1921, which outlined the problems of conducting war in the Pacific, and suggested that 'the best way to provide a skilled and an adequate war staff at Singapore would be to transfer a large part of the Admiralty to Singapore'. He even suggested that the Chief of Naval Staff should himself be prepared to relocate. The Admiralty paper for the Committee of Imperial Defence, 28 July 1922, on the effects of the Washington Conference and the treaties it led to, noted that the undertaking by the United States not to build any fortifications west of Hawaii effectively precluded the United States' involvement in a naval war in the western Pacific, 'and leaves the British Empire the sole Power to counter, with

Naval Forces, any aggressive tendencies on the part of Japan'. In March 1923 the Secretary of State for the Colonies, the Duke of Devonshire, drafted a letter to the Dominion governments advising them of London's decision to proceed with development of the base, 'on the understanding that no considerable expenditure will be incurred during the financial year 1923–24 and only a moderate expenditure during 1924–25'. [181, 184, 187, 190]

In June 1922 the Admiralty provided the Committee of Imperial Defence with a memorandum indicating the naval threat which, in the event of war with naval powers, would have to be provided against at British ports around the world. [183]

The New Zealand government voted £100,000 to support the construction of a base at Singapore but British policy was inconsistent, tormented by considerations of economy and the hope that the League of Nations would obviate the need for substantial defence initiatives. The Singapore plan was shelved, producing a very strong objection, 11 March 1921, from Admiral Jellicoe who had been appointed Governor General of New Zealand and was unofficially acting as that government's defence adviser. The Australian government was also strongly against the decision. The Canadian government, however, refused to make any comment, and the government of South Africa approved of 'the great cause of appeasement and conciliation', which might be undermined by the pursuit of security by military means. Prime Minister Ramsay MacDonald made good use of these latter in presenting his decision to the House. [198]

The three services continued to put as much pressure as they could on the government to proceed with the Singapore base. Their position was strengthened when in December 1924 the government of Hong Kong offered a gift of £250,000 to support the work at Singapore. [197, 205] The provision of adequate defence for the dockyard was delayed by a dispute with the Royal Air Force which insisted that money intended for 15-inch guns should instead be used to provide torpedo bomber squadrons and a chain of air stations. [207] But further Imperial support in February 1926, this time from the Federated Malay States, helped to keep the project on course, if not strongly. [210–11, 235, 241]

The development of the Singapore base, and the arrangements for the main fleet to operate in the Pacific, were central to the defensive requirements of Australia and New Zealand, and hence a litmus test of the value of collective Imperial defence to them. Defence strategies, however, must always acknowledge fiscal realities. In April 1929 Churchill, now Chancellor of the Exchequer, evoked the principle of the ten-year rule, based on Foreign Office assessment that war was not to be foreseen within the next ten years, to curb expenditure on Singapore and on the fleet. [235, 241, 243, 250, 253]

Presuming that the Imperial fleet would be able to survive the assaults of hostile navies, it was necessary to consider how it could be used to protect the British Empire's interests. In February 1921 the Admiralty prepared for the Committee of Imperial Defence a paper, for consideration by 'the Imperial Cabinet', on 'Empire Naval Policy and Co-operation' which indicated that the Admiralty regarded the enforcement of world-wide trade control as a primary function of the Imperial navies in a future war. Another paper for the Committee of Imperial Defence, 30 May 1923, on 'Trading, Blockade and Enemy Shipping', emphasised to what 'enormous importance to the economic side of any future war is the co-operation of the Self-governing Dominions (including the Irish Free State) and India'. A draft paper on 'The Law of Prize in Relation to the Dominions', c. 21 June 1926, emphasised the importance of the Empire adhering to a common code of prize law. In the struggle to resist American demands that the right of belligerents to intercept trade be curtailed, Canada, the Dominion most affected by Anglo-American quarrels, was briefed, if not exactly consulted. [168, 215]

The decision to reduce the scale of effort in developing Singapore into a naval base was endorsed at the 1930 Imperial Conference, but only after a strong protest from the New Zealand government. [243] The Royal Navy certainly did not agree with British government policy, which was based on economic rather than strategic considerations. In the 'War Memorandum (Eastern)' which was completed in July 1931 it was stated that 'British strategy, in the event of war with Japan, must commence with the successful transfer of such forces of the Empire as are available, particularly the British Main Fleet, from the various parts of the Empire in which they are stationed to the scene of hostilities. Since our fleet cannot operate in distant waters without adequate bases, in this instance Singapore and Hong Kong, our strategy in the first instance must be directed towards the preservation of Singapore and Hong Kong.' [244] It was not until 1933, after the Japanese occupation of Shanghai, that serious work was recommenced on the Singapore base. [258–60]

In a memorandum of March 1934, jointly prepared by the

Admiralty and the Foreign Office for the 1935 Naval Conference, it was pointed out that the Empire was losing its capacity to provide adequate naval defence. As a minimum, 'We should be able to send to the Far East a fleet sufficient to provide "cover" against the Japanese fleet; we should have sufficient additional forces behind this shield for the protection of our territories and mercantile marine against Japanese attack; at the same time we should be able to retain in European waters a force sufficient to act as a deterrent and to prevent the strongest European naval Power from obtaining control of our vital home terminal areas while we can make the necessary redispositions.' [264] In August 1936, in response to a request for advice from Britain, the Secretary of State for the Dominions passed on to the Governor-General of New Zealand a general appreciation of the importance to New Zealand of the maintenance of British naval supremacy. In that light, the continuation of New Zealand's efforts to support the Royal Navy were deemed to outweigh the utility of increased investment in air forces. [302, 320]

When the 1937 Imperial Conference convened, New Zealand continued to press its questions. Uncertainties about the capability of the Royal Navy to deploy a superior naval force to the Pacific in the event of war with Japan, uncertainties fully shared by the Admiralty Board, cast doubt on the value to New Zealand of her cruiser force which could not on its own do much to block a Japanese attack. [322] The opinion of the Committee of Imperial Defence, however, was that 'the primary role of the New Zealand cruisers in the initial stage of a war against Japan will be to assist in providing local protection in New Zealand waters. Once the Fleet is established at Singapore the situation can be reviewed and any necessary change in dispositions can be effected. This does not necessarily mean that New Zealand cruisers would even then be required to leave home waters, as their reinforcement in that area for purposes of trade protection might conceivably prove to be necessary.' [324]

In response to questions raised by the Australian delegation, the Committee of Imperial Defence reviewed its strategy for a Pacific Ocean war, which remained 'the establishment of our Fleet at Singapore at the earliest possible moment after the outbreak of hostilities'. In the event of war in Europe and the Pacific at the same time, a defensive strategy would have to be adopted, but nevertheless Australia was not judged to be at risk of invasion. 'A Japanese overseas expedition aimed at Australia may consequently be said to be a highly improbable undertaking so long as our position at Singapore is secure, and the fleets of the British Empire are maintained at such a strength as to enable a force capable of containing the Japanese fleet to be despatched to the Far East, should the occasion arise. Although ... our battleship strength will be at a low ebb in 1939 compared to Germany and Japan, at no time up to 1940 will we be unable to send a fleet to the Far east; and with the completion in 1940 of five new battleships (already laid down) for this country the period of weakness will be past.'

THE MUNICH CRISIS AND THE OUTBREAK OF WAR

As the prospect of war increased in the late 1930s the British government became increasingly alarmed by the informality of the defence commitment by the Dominions. At the 1923 Imperial Conference Mackenzie King had commented on the influence the United States had on Canadian opinions, but added: 'If a great and clear call of duty comes, Canada will respond, whether or no the United States responds, as she did in 1914.' This prediction proved to be entirely correct, for all the Dominions, and the wiser heads in London were prepared to trust to the wisdom and loyalty of Dominion governments. Sir Maurice Hankey, the Secretary to the Cabinet and to the Committee of Imperial Defence, wrote to Colonel Ismay, 7 January 1938, that his 'own very strong view is that we should continue to assume that the Dominions will come into a war in which we are engaged. If once we begin to assume the contrary, and especially if we start talking to them about it, we shall give the whole show away point by point.' His visit to Canada in 1936 had convinced him that Canada could be relied upon. In April 1938 he discouraged any alteration to the War Book to make provision for a possible refusal of any of the Dominions to participate in a major war. [335, 337] Sir Edward Harding, the Permanent Undersecretary of State for Dominion Affairs, warned Francis Floud, the British High Commissioner in Ottawa, that 'it is regarded as most important that no indication should be given to the Canadian Government that the possibility of their standing aside in an emergency is being seriously considered here'. [342] Floud replied that Mackenzie King was determined 'to keep his eye on what he considers to be the main objective, viz. the preservation of Canada's unity', which he believed required his refusing any commitment to Imperial defence until, in the event of war, the matter could be decided in 'Parliament as the interpreter of the people's wishes'. This was 'unsatisfactory', but there was nothing to be done about it. He added, however, that Mackenzie King was sympathetic to the policy of the British government. 'All I myself really fear is a period of hesitancy, and I am afraid that we cannot necessarily count on Canada being in with us from the very beginning.' [343–44] The Admiralty advised the Dominions Office 'that My Lords have reluctantly reached the conclusion that it is impracticable to devise further alternative plans'. [337–8]

The Munich crisis brought these worries into close focus. War is an alarming business and not everyone had the courage to be calm in the eye of the storm. The South African government was reported to object to posting call-up notices for reservists on the eve of bye-elections. The Acting High Commissioner in Canada reported that Naval Service Headquarters was unable to implement an Admiralty request to establish the Naval Control Service in the absence of an official request from the United Kingdom government. The Viceroy of India, however, informed the Admiralty that the Royal Indian Navy would be placed under the control of the Commander-in-Chief East Indies when requested. [347–9, 351]

Concern about Dominion reactions proved to be unfounded. Shortly after the settlement at Munich had given British forces a reprieve the Commander-in-Chief North America and West Indies, Vice-Admiral Sir Sidney Meyrick, reported to the First Sea Lord that he and the governor of Bermuda had spent the evening with Mackenzie King, who made a very good impression and informed him that during the crisis Admiral Nelles had consulted with him as soon as it was known that the Royal Navy had mobilised during the crisis. He 'fully agreed' that the Royal Canadian Navy should have pre-arranged war stations, and that they could not protect Canadian interests by staying in harbour. [353] The Head of 'M', C.B. Coxwell, concluded that it had been the British Government which was really at fault because the Dominions Office had not wanted to send a warning telegram, while the Admiralty arrangements to establish a naval control service had no authority outside London's jurisdiction. [358] The Dominions Office informed the Admiralty of the great activity the Canadian government had shown in meeting the crisis, and reported that 'both Chiefs-of-Staff and the Senior Air Officer have assured me privately that the emergency has shaken up the Department of National Defence itself and still more the Department of External Affairs, and the Chief of the General Staff told me that he had been able, though very much at the last moment, to assure his Minister that every nut and bolt was in place, so far as they existed at all'. Vice-Admiral G.H. D'O. Lyon, Commander-in-Chief Africa Station, reported to the Admiralty that in fact the South African government also had been supportive of the preparations for war, which had gone smoothly. [359–61]

The Admiralty prepared a memorandum on 'Co-operation of Dominions in Defence Action Required during Precautionary Period', in November 1938. [362] The Dominions Office wanted to tighten up the legalities of defence relations with the Dominions, and proposed a pre-warning telegram, but the Admiralty preferred the existing machinery. The Committee of Imperial Defence was adamant that tinkering could only be counterproductive. [366, 368, 371] In July 1939 a letter was drafted at the Admiralty, addressed to Rear-Admiral P.W. Nelles RCN, making it clear that Admiralty warning telegrams should only be considered as advisory, or to be acted on only if authorised by the Canadian government. [378]

On 25 August 1939, with war only weeks away, the Commander-in-Chief North America and West Indies was given instructions about call-up of reservists in Canada in the differing contingencies of Canadian participation in a war, or of Canada remaining neutral. [383] In March 1939 the Government of India had made it clear that, while local naval defence would have to remain under the Government of India, naval movements would be controlled by the East Indies Station. On 4 September 1939 the Commander-in-Chief East Indies was advised that the government of India had agreed to place Royal Indian Navy forces unreservedly under his command. [384] A case was made six months later, however, for retaining under the control of the government of India the ships taken up from merchant service for local naval defence. [389]

The Munich crisis made it all too apparent that the Singapore strategy was built on sand. The Australians and New Zealanders became increasingly skeptical of Britain's ability, or will, to deploy naval forces to the Pacific with an armed and hostile Germany and Italy making threatening noises, and doubted that Britain would really be willing to leave her interests in the Eastern Mediterranean undefended so as to respond to dangers in the Pacific.

[373] A year later, the bankruptcy of naval strategy for the Pacific was transparent, and the best that could be done following the outbreak of war in Europe was to increase the provision allowance for the Singapore garrison in the hope that it could hold out for six months, should Japan go to war. [375–6, 387]

Although far from a perfect solution to the problem of Imperial defence, the system of defence relations which was developed within the British Commonwealth and Empire during the first four decades of the century was to prove itself adequate to the crisis of the Second World War. The navies of the Dominions and of India played a full part in the conflict. Although it is possible that their teething troubles reduced their technical effectiveness, it is also certain that the independent services profited from the willingness of Dominion peoples to commit great resources to them, which helped to provide the naval services of the Crown with the forces needed to fight the war. One of the great virtues of the relatively loose structure of Imperial defence was that precipitate belligerent action, or Imperial adventures, were unlikely. At the same time, historical and cultural constraints would insure unity in the event of attack. It is probably true to say that the lessons which were learnt in the development of an effective system of Imperial defence co-operation formed the basis for the North Atlantic Treaty Organisation which in the post-Second-World War period replaced the Empire as Britain's main system of collective defence.

[1] *Sir Henry E. McCallum, Government of Newfoundland, to*
Arthur Mews, Deputy Colonial Secretary
16 October 1900

Send a circular to all newspapers for the information of the Public that a branch of the Royal Naval Reserve has been approved of by Her Majesty's Government for Newfoundland and that a first detachment of 50 men will embark on board H.M.S. *Charybdis* on November 10th for a six months' cruise.

Add that full information as to the advantages offered can be obtained on application to a Stipendiary Magistrate who will receive the names of any suitable men who have not yet applied to join the force.

Minuted by Arthur Mews: 'Your Excellency, Circular has been sent as directed.'

[2] *Permanent Admiralty Secretary [Sir Evan MacGregor] to the*
Joint Naval and Military Committee of Defence
4 October 1901

I have received and laid before my Lords Commissioners of the Admiralty your letter, 266/Malta/51 (D.G.M.L.), dated the 10th August, in which it is stated that his Excellency the Governor of Malta has asked for the addition of a brigade of infantry to the garrison; on the plea that there is some doubt whether the supremacy of His Majesty's fleet in the Mediterranean is so fully assured as to afford a reliable guarantee that Malta is not liable to the attack of a hostile fleet accompanied by a large expeditionary force.

2   In reply to the Secretary of State's request for assurances on this point, I am directed to inform you that their Lordships are fully impressed with the importance of making such a distribution of the British fleet that no attempt against the Island of Malta of

1

the kind contemplated shall be reasonably possible. They anticipate no difficulty in making the arrangements necessary to secure
this end, unless there is a general combination of the Maritime
Powers against this country. Should a crushing defeat of the British
fleet, arising from this or from some other unforeseen cause,
render possible an attack on the scale contemplated by the Governor, their Lordships are of opinion that the island must in time
be captured, and no addition to the garrison could possibly do
more than delay its inevitable fall.

[3] *Francis Mowatt, Permanent Treasury Secretary, to the
Secretary of the Admiralty*

11 December 1901

'Confidential'
I have laid before the Lords Commissioners of His Majesty's
Treasury your letter (N) of the 25th ultimo, on the subject of
raising in British Colonies men for the Royal Naval Reserve, –
from which it appears that, in the case of Newfoundland, a limited
number of fishermen were enrolled last year in the Reserve and
trained on board the Commodore's ship in Newfoundland waters,
and that arrangements have been made to embark a similar
number this winter.

As it has transpired that neither the Colonial Government nor
the Admiralty have power to enrol men of the Reserve entered
in the Colonies, and as the Law Officers of the Crown have advised
that further Imperial legislation is required, I am to request you
to move the Lords Commissioners of the Admiralty to cause My
Lords to be informed how and under what authority the expenditure involved has been charged to Navy Votes 1900/1 and will be
charged to Navy Votes 1901/2.

With reference to the proposal that My Lords should instruct
the Parliamentary Draughtsmen to prepare a Bill to carry out the
policy of entering Colonials in the Royal Naval Reserve, I am to
state that, before this can be done, a very important question of
Policy must be considered. No Colony now contributes any
adequate sum to the heavy cost (nearly 32 millions gross for 1901/
2) of the Royal Navy; and, however expedient it may seem to
recruit from Colonial sources the Reserve for which the British
taxpayer pays, it is obvious that every step by the Imperial Government in this direction must tend to weaken local effort to provide

at local expense for its own defence on the lines contemplated by the Colonial Naval Defence Act, 1865.

Adverting to the views expressed on behalf of the Board of Admiralty in the House of Commons on 22nd March last, I am to state that, in the opinion of This Board, the whole subject of Colonial contributions to the Naval Defence of the Empire deserves the very early attention of His Majesty's Government, and that, though My Lords are unable to judge from the letter before Them what would be the relative cost of this particular Newfoundland scheme to the mother-country and to the Colony, yet They think it most inadvisable that a precedent should be made which would appear to justify reluctance on the part of the more important Colonies than Newfoundland to accept anything like their fair share of the total cost, when the full discussion takes place which the subject ought to receive. On this ground the Admiralty scheme appears to Their Lordships of such importance in principle that it must be brought before His Majesty's Advisers before any final arrangement is made with the Newfoundland Government or before any attempt is made to alter the law.

[4] *Lord Selborne, First Lord, to Graham Greene, First Lord's Private Secretary*

22 January 1902

The Chancellor of the Exchequer [Sir Michael Hicks Beach] has agreed that the Bill to validate the enrolment of the Newfoundland fishermen may also validate for the future enrolments of all subjects of the Crown, no matter in what part of the Empire they may be domiciled. He has asked me to communicate with Sir Courtenay Ilbert [Parliamentary Council] on the subject and I am doing that privately.

Sir Courtenay Ilbert will arrange with me the terms of the new Bill and when it is officially settled we will send it to the Treasury with a letter drafted on lines suggested by the Chancellor himself. It is this letter which I want you to draft. Point out that it seems a pity not to utilise this opportunity of settling the question once for all, and that not to do so and to introduce the Bill with the very limited scope of a validating Act would probably provoke comment and debate; and therefore the Admiralty propose that the question should be settled once for all by the passing of such an Act as is contained in the accompanying proposed draft. Point out that the principle of the enrolment of subjects of the Crown other than those domiciled in the United Kingdom, the Isle of Man and the Channel Islands, which was the original limitation of the Act of 1859, has been already conceded by the Act of 1896 under which the limitation of domicile of the Act of 1859 was repealed and under which we are enabled to enrol and retain in the Royal Naval Reserve subjects of the Crown serving on the steamers of the Canadian Pacific Railway Company, which ply between Vancouver and Hong Kong; and that the only reason why we are able to enrol those men and not the Newfoundland fishermen lies in the fact that the steamers which the former men serve upon happen to be registered in the United Kingdom, whereas the fishing boats of Newfoundland are not.

You may add a paragraph to the effect that supposing, which is not likely, that the Act were passed before the Conference with the Colonial Delegates in the summer, no use will be made of the powers under it in respect of further enrolments until after that conference.

[5] *Selborne to MacGregor and to Rear-Admiral Reginald Custance, Director of Naval Intelligence*

8 July 1902

I wish an official letter drafted to the Colonial Office suggesting that from the commencement of the Civil Administration of the two new Colonies of the Orange River and the Transvaal the principle of a Naval contribution should be asserted.

I wish also to know what is the exact position of our Agreement with the Government of India in respect of a Naval contribution. When does the present Agreement terminate? And have we commenced negotiations with reference to a new Agreement?

[6] *Papers relating to a Conference between the Secretary of State for the Colonies and the Prime Ministers of Self-Governing Colonies*

June 30–11 August 1902

Joseph Chamberlain [Colonial Secretary]   In this connection I would venture to refer to an expression in an eloquent speech of my right honourable friend, the Premier of the Dominion of Canada [Sir Wilfred Laurier] ... 'If you want our aid, call us to your councils.' Gentlemen, we do want your aid. We do require your assistance in the administration of the vast Empire which is yours as well as ours. The weary Tital staggers under the too vast orb of its fate. We have borne the burden for many years. We think that it is time that our children should assist us to support it, and whenever you make the request to us, be very sure that we shall hasten gladly to call you to our councils. If you are prepared at any time to take any share, any proportionate share, in the burdens of the Empire, we are prepared to meet you with any proposal for giving to you a corresponding voice in the policy of the Empire . . .'

Lord Selborne [First Lord of the Admiralty] The danger to the Empire which I fear is that Canada, South Africa, and Australia, being in fact continents, should become too much continental and too little maritime in their aspirations and ideas. The British Empire owes its existence to the sea, and it can only continue to exist if all parts of it regard the sea as their material source of existence and strength. It is therefore desirable that our fellow subjects in the Dominions beyond the seas should appreciate the importance of naval questions . . . If they will undertake a larger share of the naval burden, well and good. But I regard it as of even more importance that they should cultivate the maritime spirit, that their populations should become maritime as ours are, and that they should become convinced of the truth of the proposition that there is no possibility of the localisation of naval force and that the problem of the British Empire is in no sense one of local defence.

Minute to the Right Honourable
the Prime Minister [of the
Commonwealth of Australia] as
to Naval Defence, Mr John
Forrest, Minister of State for
Defence, Melbourne
15 March 1902

1 It will, I think, be generally conceded that it is the duty of the Commonwealth to adequately contribute to the defence of Australia and its floating trade. We admit this obligation in regard to our local military defence, but we must remember that naval forces require to be even more efficient than military forces, which have the great advantage of local knowledge to assist them in any active operations in their own country . . .

14 I am not prepared to recommend under existing conditions the establishment of an Australian Navy. Even if it were estab-

lished, I am afraid it would not be very efficient, for besides the enormous cost of replacing the fleet from time to time with more modern ships, there would be no change for the officers and crews, who would go on year after year in the same ships, subject to the same influences, and, I fear, with deteriorating effects.

18   Our aim and object should be to make the Royal Navy the Empire's Navy, supported by the whole of the self-governing portions of the Empire, and not solely supported by the people of the British Isles, as is practically the case at the present time. It is, I think, our plain duty to take a part in the additional obligations cast upon the Mother Country by the expansion of the Empire, and the extra burdens cast upon her in maintaining our naval supremacy . . .

Rt. Hon. R.J. Seddon [Prime Minister and Minister of Defense of New Zealand]   Coming to the naval side of the question, I would say at once, speaking for New Zealand, that I repudiate any idea of providing or setting up a navy or the purchase of ships for ourselves. The conclusion that we have arrived at, and I do not think there is the slightest chance of it being modified, is that we must depend alone upon the Imperial Navy . . .

Sir Wilfred Laurier [Prime Minister of Canada] If we were to put the figures – I have not had time to do it – on one side what is expended by the Government of Great Britain on military and naval expenditure, and, on the other hand, what Canada is spending every year for public works, I think the difference would not be very, very great, so that at the very outset we are met with a difficulty which to my mind seems to be very nearly insuperable, if not absolutely insuperable, to the contemplation of the Colonies – I speak for my own, at all events – coming into making a direct contribution to the Imperial Exchequer either for naval or military expenditure . . .

The Rt. Hon. Joseph Chamberlain [Secretary of State for the Colonies]   What would you do; where would you be, if you were absolutely independent nations? In spite of what you are spending for old age pensions, and railways, and all those other things, you would have to spend a vast deal more than you are spending to make an adequate, or any kind of proportionate, preparation for your own defence. To a large extent your liability is relieved by the action of the Imperial Government; but I am quite sure that recognising this fact you would be prepared to meet us in as generous a spirit as is possible.

APPENDIX IV: *Memorandum on*
Sea-Power and the Principles
Involved.

To any naval Power the destruction of the fleet of the enemy must always be the great object aimed at. It is immaterial where the great battle is fought, but wherever it may take place the result will be felt throughout the world . . .

Our possible enemies are fully aware of the necessity of concentrating on the decisive points. They will endeavour to prevent this by threatening our detached squadrons and trade in different quarters, and thus oblige us to make further detachments from the main fleets. All these operations will be of secondary importance, but it will be necessary that we should have sufficient power available to carry on a vigorous offensive against the hostile outlying squadrons without unduly weakening the force concentrated for the decisive battle, whether in Europe or elsewhere.

The immense importance of the principle of concentration and the facility with which ships and squadrons can be moved from one part of the world to another – it is more easy to move a fleet from Spithead to the Cape or Halifax than it is to move a large army, with its equipment, from Cape Town to Pretoria – points to the necessity of a single navy, under one control, by which alone concerted action between the several parts can be assured.

In the foregoing remarks the word *defence* does not appear. It is omitted advisedly, because the primary object of the British Navy is not to defend anything, but to attack the fleets of the enemy, and, by defeating them, to afford protection to British Dominions, supplies and commerce. This is the ultimate aim.

To use the word *defence* would be misleading, because the word carries with it the idea of a thing to be defended, which would divert attention to local defence instead of fixing it on the force from which attack is to be expected.

The traditional role of the Royal Navy is not to act on the defensive, but to prepare to attack the force which threatens – in other word to assume the offensive.

[7] *Memorandum [by Selborne] 'The Colonies and the Navy,'*

20 May 1902

Australia and New Zealand

The current agreement in round figures costs the Commonwealth of Australia £105,000 per annum, New Zealand £21,000 per annum, and the United Kingdom £193,000 per annum. In addition the sole cost of the maintenance of the Naval Brigades of Australia is borne by the various States of which the Commonwealth is composed.

I have no doubt that this agreement was the best that could have been made at the time it was made, and has been of great value as commencing an application of the principle that the Colonies should contribute something towards the maintenance of the Navy. It has, however, two cardinal defects which I desire to avoid in any fresh agreement. In the first place, the hands of the Admiralty are tied in war, as the ships which are the subject of the Australian agreement cannot be moved from the Australian Station in time of war except with the consent of the Commonwealth and New Zealand Governments. The safety of the Empire can only be secured by seeking out and destroying the ships of the enemy wherever they are to be found, and it is practically certain that on the outbreak of war, say, with France and Russia, the first thing that the Australian Squadron ought to do would be to join the East Indian or China Squadron and hunt down and crush the French and Russian Squadrons in Far Eastern Seas. The cruisers which are the subject of the present agreement are small and are rapidly becoming obsolete. The Australians press, and naturally, that the ships to replace them should be first and second class cruisers of the newest type. Under no circumstances will I consent to an arrangement by which the hands of the Admiralty would be tied in respect of the orders they might wish to give to such ships on the outbreak of war. I would rather forego any direct contribution to the Navy from Australia and New Zealand than consent to this. The other flaw in the current agreement consist in the fact that it has had no effect in drawing the Australians and New Zealanders more towards the sea, which I take it, ought to be an object of great importance to us. They are not at present in any sense a maritime people, and the Naval agreement has had no effect in making them more so. They have felt no personal interest in the Australian Squadron, but have regarded it as an article which they have bought and of the use of which by anybody else they are jealous.

The Naval Brigades of Australia are formed of excellent material, but they are Naval only in name. They have no sea training, and it has not been possible to give it them, because to embark men of these brigades and mix them with the crews of the ships on the Station would not answer. The discipline to be administered to these Australians must be on different lines to that of the ordinary blue-jacket, and it would be most unfair to the ordinary blue-jacket to put beside him in the ship a man who as a sailor would be inferior but would be paid a much higher rate of wage. I am, nevertheless, very anxious to develop these Naval Brigades into a real and serviceable Naval Reserve.

The principle on which I propose to proceed is to prepare a scheme in which the Admiralty will ask from Australia and New Zealand a substantial portion of what Australia and New Zealand ought to give, but I shall be prepared to take eventually whatever Australia and New Zealand are prepared to give, subject, of course, to corresponding modifications in the position of the Admiralty, and to the reservation mentioned above.

The Intelligence Department of the Admiralty is at this moment working out what should be the composition of an Australian Squadron in first and second class cruisers in view of the composition of our East Indian and China Squadrons and of the French and Russian Squadrons in the far East. Till that has been accurately worked out and accepted by the Board, I shall not have complete or exact figures to give, but the following are the lines which my proposal will take.

The new Australian Squadron to consist of, say, 1 first class armoured and 6 second class cruisers, and four sloops. The annual cost of this squadron in round figures will be £618,000, to which the Commonwealth of Australia and New Zealand will be asked to contribute 1/9 [1 shilling and 9 pence £] per head of population, or the total sum of £326,000 and £67,000 respectively. This Squadron to be free to move at the orders of the Admiralty within the limits of the present Australian, East Indian, China, and Pacific Stations, but not to be taken beyond those limits without the concurrence of the Commonwealth and New Zealand Governments. Under no circumstances can the Admiralty agree to an arrangement which would bind them not to move this Squadron or any part of it from the limit of the present Australian Station without the concurrence of these two Governments.

In time of peace, of the ships named above, one first class armoured and four second class cruisers and four sloops to be always in commission, and two second class cruisers in reserve.

Of the ships in commission, one shall, as soon as the Commonwealth and New Zealand Governments are in a position to supply the men, be manned exclusively by Australians and New Zealanders at local rates of pay, these men to join the Navy for non-continuous service for such a period as shall be agreed upon between the Admiralty and the Commonwealth and New Zealand Governments, and officers to be supplied from the Royal Navy. The ships in reserve to be manned in war with Active Service ratings drawn partly from the cruisers in commission and partly from the sloops, which will be kept in commission in peace for F[oreign] O[ffice] and C[olonial] O[ffice] work in the islands in the Pacific, but not used in war, and partly from the Royal Naval Reserve men of Australia and New Zealand to be mentioned presently.

The cruisers built in accordance with the present Australian agreement, viz: – the five ships of the 'Katoomba' class – to be used as training ships for an Australasian branch of the Royal Naval Reserve to be formed from the Colonial seamen trained and passed through the ships of the Squadron to be manned exclusively by Colonials and if possible from the present Naval brigades. If the numbers are sufficient and the occasion demand these ships to be manned in war by the Royal Naval Reserve and maintained in peace and mobilised in war at the expense of the Commonwealth and New Zealand Governments; the whole pay and cost of training of the Australian and New Zealand Governments; the Admiralty to have the right to draw on these reserves to complete the complements of the Australian squadron in time of war; the men so embarked in time of war to receive ordinary rates of pay, any difference between that and local rates of pay being made up to them by the Commonwealth and New Zealand Governments on demobilisation: Whenever embarked these Australasian Royal Naval Reserve men to be under the Naval Discipline Act: during peace the Australasian Royal Naval Reserve and their training ships to be inspected by the Commander-in-Chief on the Station: if mobilised in war these ships to be under the orders of the Commander-in-Chief on the Station but not to be removed from the Australian Station without the consent of the Commonwealth and New Zealand Governments. The annual cost of the maintenance of the five 'Katoombas' and of a reserve of 1,500 men at Australian rates would be approximately £100,000.

The aggregate and ultimate financial effect of these proposals which of course could only be carried out gradually, would be in the case of the Commonwealth a new contribution of (say) £410,000, as opposed to the present contribution of £105,000 and the cost of pay and training of the present Naval brigades and the cost of maintenance of their present training ships. In the case of New Zealand the cost would be £83,000 as opposed to the present contribution of £21,000. The total annual contribution per head of population would be in each case about 2/2 [two shillings and tuppence] as opposed to the present 15/1 [fifteen shillings and a penny] per head of the inhabitants of the United Kingdom.

The annual cost to the Imperial Government would be £225,000 as opposed to the present cost to them of the Australian Squadron, and they would have to find 3,200 officers and men (including those raised in Australia and New Zealand) Active Service Ratings as opposed to 2,050 officers and men Active Service Ratings at present.

The Dominion of Canada

I consider that it will be right for the Admiralty to make a proposal to the Dominion Government exactly on the same lines as that to be made to the Commonwealth Government in respect of a new contribution, and the manning of one ship of the Squadron, the calculations applied to the Australian Squadron and the Australasian population to be applied mutatis mutandis to the North American Squadron and the population of Canada. This can be worked out by the Director of Naval Intelligence and Accountant General as soon as the Australian proposals are definitely settled.

I am further prepared, if the Dominion Government decline to entertain any such proposal, to ask them to pay and bear the whole cost of training, on lines to be laid down by the Admiralty, as many Royal Naval Reserve men as can be raised either on the Atlantic or Pacific Coasts.

In the case of Newfoundland, the Admiralty are quite prepared to ask the Newfoundland Government for a contribution per head of population on the same basis as that asked for the Dominion of Canada, but it is practically certain that the Colony is not in a position to afford this contribution, and in the event therefore of their declining this arrangement, the Admiralty will propose that they should bear as large a proportion as they can afford of the cost of pay and training of Newfoundland Royal Naval Reserve men. In this connection, a memorandum will be prepared showing the aggregate cost of the batch of Newfoundland Royal Naval Reserve men proposed to be raised under Mr Goschen's scheme and how that cost was to be divided between the Imperial and Colonial Governments according to the provisional arrangement arrived at.

South Africa

Cape Colony now contributes £30,000 and Natal £12,000 a year to the cost of the Navy. Proposals can be made to both these Colonies on the same basis of a pro rata contribution as that suggested in the case of Australia and Canada.

It is doubtful if the time has yet come to ask for any contribution from the Transvaal or Orange River Colonies or from Rhodesia.

I am informed that there are no materials as yet in South Africa for the formation of a Royal Naval Reserve. If there are, they can be utilised if properly trained.

[8] *J. Gordon Sprigg [Premier of Cape Colony] and A.H. Hime, [Premier of Natal Colony] to the First Lord of the Admiralty*

London
8 July 1902

Handed to the First Lord personally by the signatories

The enormous preponderance of the Native population in the Colonies of the Cape and Natal renders it impossible for us to agree to any proposal involving the obligation to furnish a Military Contingent in the event of a war in which the Imperial Government may be involved in any other part of the Empire.

Recognizing, however, as we do, the absolute necessity for the maintenance of an efficient Navy if the Empire is to be held together, and regarding the Navy as the first security for the preservation of the people of South Africa, and especially of those residing in the two Coast Colonies, we are prepared to submit to the Legislatures of our respective Colonies Bills providing for Contributions towards the cost of maintaining the Imperial Navy.

At present the Cape Colony Contributes under an Act of Parliament passed in 1898 the sum of £30,000 a year for that purpose.

So soon as Parliament assembles a Bill will be submitted raising that sum to £50,000.

To the Legislature of Natal a Bill will be submitted providing for a yearly Contribution of £35,000.

At a later period, when a Confederation has been established in South Africa, the question of training men residing in the two Coast colonies for service in the Imperial Navy will be brought under the consideration of their respective Legislatures who will, we believe, be prepared to entertain such a proposal most favourably.

[9] *Frederick Graham, Under-Secretary of State, Colonial Office, to the Secretary to the Admiralty*

13 August 1902

I am directed by Mr Secretary Chamberlain to acknowledge the receipt of your confidential letter M.0556 of the 13th ultimo, regarding the question of Naval contributions from the Transvaal and Orange River Colonies.

2   In reply, I am to say that, in Mr Chamberlain's opinion, it would be inadvisable in present circumstances to require Naval contributions from these Colonies. If His Majesty's Government should decide to obtain contributions from all Crown Colonies, the case would be different: but as that question has not yet been considered, Mr Chamberlain would deprecate any proposal to treat the Transvaal and Orange River colonies in an exceptional way.

[10] *Memorandum, prepared for the 1902 Colonial Conference by Selborne*

7 August 1902

Accompanied by the Senior Naval Lord [Admiral Lord Walter Kerr] and the Financial Secretary [Hugh Arnold-Forster], my colleagues of the Board of Admiralty, and assisted by the Director of Naval Intelligence [Captain, Prince Louis of Battenberg], I have now had interviews with the Premiers of His Majesty's Government in Cape Colony [Sir J. Gordon Sprigg], Natal [Sir A.H. Hime], Newfoundland [Sir Robert Bond], and New Zealand [Richard J. Seddon], and with the Premiers and Ministers of Defence of His Majesty's Government of the Commonwealth of Australia [Sir Edmund Barton and Sir John Forrest] and the

Dominion of Canada [Sir Wilfred Laurier and Sir Frederick W. Borden]; and, as the result, the Board of Admiralty have received the following offers of assistance towards the Naval Expenses of the Empire:

Cape Colony: 50,000£ per annum to the general maintenance of the Navy. No conditions.

Commonwealth of Australia: 200,000 per annum to an improved Australasian Squadron, and the establishment of a branch of the Royal Naval Reserve.

Natal: 35,000£ per annum to the general maintenance of the Navy. No conditions.

Newfoundland: 3,000£ per annum (and 1,800£ as a special contribution to the fitting and preparation of a drill ship) towards the maintenance of a branch of the Royal Naval Reserve of not less than 600 men.

New Zealand: 40,000£ per annum to an improved Australasian Squadron, and the establishment of a branch of the Royal Naval Reserve.

I wish to tender to the Colonial Ministers the hearty acknowledgment of the Board of Admiralty for the manner in which they have assisted them to arrive at the satisfactory result given above. As might be supposed, we have not always been able to see eye to eye on the questions discussed; but the interchange of views has been naturally frank, and governed on all sides by a desire to treat the defence of the Empire on a business footing, and to abandon the discussion of theories for the attainment of results immediately practicable. It is true that the Board of Admiralty have not in these conferences achieved all the results for which they might previously have hoped; but, on the other hand, it has been a great pleasure to them to hear from the Prime Ministers of all the self-governing Colonies a generous appreciation of the work of the Navy.

If the Parliaments of the above-named self-governing Colonies accept and endorse these arrangements, a real step in advance will have been made in the organisation of the Naval Forces of the Empire.

In the first place, an increase in the number of modern men-of-war maintained in commission will have been promoted by the aid of British subjects in the Dominions beyond the seas.

In the second place, the conditions attached to the various agreements will show how keen is becoming the appreciation throughout the Empire of the peculiar characteristics of Naval warfare, and of the fact that those local considerations which find their natural place in military organisations are inapplicable to Naval organisation. I would draw special attention to the improved composition and organisation of the Australasian Squadron when the new agreement will have come into force, and to the fact that Cape Colony and Natal have made their offer unfettered by any conditions, a mark of confidence and of appreciation of the Naval problem of which the Board of Admiralty are deeply sensible.

Thirdly, I hail with much satisfaction the establishment of a branch of the Royal Naval Reserve in the Colonies. The circumstances of the population of Newfoundland are peculiar, and nowhere else probably within the Empire can so large a proportion of fishermen be found. A branch of the Royal Naval Reserve has already been founded there, and I look forward confidently to its steady growth. In Australia and New Zealand, although Naval Brigades have for some time been in existence there, and did excellent service in the recent China War, the formation of a branch of the Royal Naval Reserve is a new departure. The formation of a branch of the Royal Naval Reserve is not the only, or indeed the chief, step in advance, in connection with the personnel of the fleet, for it has been agreed that if possible one of the ships of the Australian Squadron in permanent commission shall be manned by Australians and New Zealanders under officers of the Royal Navy, and that 10 cadetships in the Royal Navy shall be given annually in Australia and New Zealand.

I have accordingly great pleasure in recommending to His Majesty's Government that the offers of assistance which have been received should be accepted.

Sir Wilfred Laurier informed me that His Majesty's Government of the Dominion of Canada are contemplating the establishment of a local Naval force in the waters of Canada, but that they were not able to make any offer of assistance analogous to those enumerated above.

I have alluded to the fact that our interchange of views at the Conferences has been characterised by mutual frankness, and I desire to put on record the facts and arguments which I thought it my duty to set forth.

In the first place, I pointed out that even after the Colonial Parliaments had ungrudgingly increased the proportion of their assistance towards the Naval Defence of the Empire, as recommended by their Ministers, the taxpayers of the British Empire would, in respect of Naval Expenditure, still be in the following relative positions:

	Population (White)	Naval Contribution per Caput per annum
United Kingdom	41,454,621	15/2
Cape Colony	538,000	1/10 1/2
Commonwealth of Australia	3,765,805	1/0 3/4
Dominion of Canada	5,338,883	Nil
Natal	64,954	10/9 1/4
Newfoundland	210,000	0/3 1/2
New Zealand	772/719	1/0 1/4

For the year 1902–3 the Navy Estimates amounted to 31,423,000£ after deducting appropriations-in-aid other than contributions from the self-governing Colonies. Of this sum the self-governing Colonies would, on the basis of the new agreements, have paid 328,000£.

This table shows how large a proportion of the burden falls upon the taxpayer of the United Kingdom, and although it is true that by far the larger portion of the money which he provides is spent within the United Kingdom, yet it so happens that more of the money provided by the taxpayer of the United Kingdom is spent in the British Dominions beyond the seas than the British Dominions beyond the seas contribute to the maintenance of the British Navy. Additional interest is lent to the fact by an analysis of the trade which the British Navy has to protect. In the year 1900 the seaborne trade of the Empire may be roughly stated to have been worth between eleven and twelve hundred millions sterling; but of this vast sum a proportion of certainly not less than one fourth was trade in which the taxpayer of the United Kingdom had no interest either as buyer or seller of the particular goods represented by these values. It was either inter-Colonial trade or trade between the British Dominions beyond the seas and foreign countries. The taxpayer of the United Kingdom has therefore the privilege, not only of taking upon himself the lion's share of the burden the interest in which is shared between himself and his fellow subjects in the Dominions beyond the seas, but also a not less share of the burden in respect of interest which are not his own, but exclusively those of his fellow subjects beyond the seas.

I expressed my hope that this simple statement of fact would help the Colonial taxpayer to realise the extent of the advantage he reaps from the existence of one Navy for the whole Empire. He is not only relieved thereby of a heavy burden of taxation in time of peace, but in time of war he knows that to him, if faced by any possible coalition, is furnished the concentrated strength of one navy under one command.

In reply to an inquiry, I undertook to endeavour to form an estimate of the Naval obligations which would be forced upon the British Dominions beyond the seas if they were nations with a separate existence, and not, as now, partner nations of the British Empire, and it was suggested that the proper comparison would be between the Commonwealth of Australia and New Zealand or the Dominion of Canada and some State with a population of about the same size.

I pointed out that if such a basis of comparison were chosen the annual Naval expenditure of Holland is 1,400,000£, and that of Argentina 920,000£, not to mention a past capital expenditure of several millions which must have been incurred in the creation of fleets and for the proper equipment of dockyards and Naval bases. These countries were only taken because their populations roughly correspond in size with those of Australasia and Canada, and not because they are in other respects in any way comparable. Indeed, I submitted that this was not an adequate or satisfactory comparison. Each great group of Dominions beyond the seas would, so it seemed to me, have to face the Naval position in which it found itself, and the governing conditions of that position would be the political and geographical environment of the group. As a matter of fact, each of these groups would find itself within the orbit of a great Naval Power. The Dominion of Canada would have to frame its Naval policy with a view to the Navy of the United States. The Commonwealth of Australia and New Zealand would be forced to remember that France in New Caledonia, and Germany in New Guinea, are near neighbours. Cape Colony and Natal would constantly find themselves reminded of the fact that France is their neighbour in the Indian Ocean, in Madagascar, and that the greater part of Western South Africa is German. It is not easy in either of these cases to see how it could be possible to avoid the influence of those forces which have forced Japan to become a Naval Power. The Naval Budget of Japan for the year 1901–2 was 3,700,000£, and this sum included nothing for interest, sinking fund, or depreciation account in connection with former capital expenditure on stores, ships, dockyards or Naval bases.

These are the facts and arguments which I felt it my duty to lay before the Premiers, and to which, whether agreeing or disagreeing, they have listened with that friendly courtesy which has made all our interviews so agreeable a recollection. I recapitulate them now because the danger to the Empire which I fear is that Canada, South Africa, and Australia, being in fact continents, should become too much continental and too little maritime in their aspirations and ideas. The British Empire owes its existence to the sea, and it can only continue to exist if all parts of it regard the sea as their material source of existence and strength. It is therefore desirable that our fellow subjects in the Dominions beyond the seas should appreciate the importance of Naval questions. If they will undertake a larger share of the Naval burden, well and good. But I regard it as of even more importance that they should cultivate the maritime spirit; that their populations should become maritime as ours are, and that they should become convinced of the truth of the proposition that there is not possibility of the localisation of Naval force, and that the problem of the British Empire is in no sense one of local defence.

The sea is all one, and the British Navy therefore must be all one; and its solitary task in war must be to seek out the ships of the enemy, wherever they are to be found and destroy them. At whatever spot, in whatever sea, these ships are found and destroyed, there the whole Empire will be simultaneously defended in its territory, its trade, and its interests. If, on the contrary, the idea should unfortunately prevail that the problem is one of local defence, and that each part of the Empire can be content to have its allotment of ships for the purpose of the separate protection of an individual spot, the only possible result would be that an enemy who had discarded this heresy, and combined his fleets, will attack in detail and destroy those separated British squadrons which, united, could have defied defeat.

[11] *Decipher, Telegram, Lord Tennyson, Governor-General of the Commonwealth of Australia, to Joseph Chamberlain, Secretary of State for the Colonies, received Colonial Office*
24 April 1903

Prime Minister and Minister of Defence wish me to say that probable passing of Naval Agreement difficult and doubtful unless something is done to satisfy the strong feeling for local defence, and to allay the fear that the Australian Squadron is not sufficiently strong, and that when the Australian Squadron leaves these waters there will be no local defence left to deal with stray vessels of enemy. They therefore suggest that six locally officered and manned small torpedo boat destroyers (one for each capital port, crews and officers to be paid out of £200,000 per annum by the Imperial Government) should be substituted for, or be lent in addition to third class cruiser manned from the United Kingdom. If they had destroyers they would be maintaining a properly equipped force of vessels which could be used for Defence of Australian Coast. Great Barrier Reef is singularly adapted for destroyers and for these alone. Prime Minister cannot ask Parliament for more opulent subsidy on account of great drought, and he hesitates to apply for destroyers unless he has assurance that it will be favourably received. Two destroyers even, manned as described, would go far to calm popular feeling. Governor of New South Wales and other experts are strongly in favour of destroyers, to be under the control of the Admiral.

Prime Minister wishes to be advised as soon as possible. Reid, Leader of Opposition, has promised me support of Naval Agreement, but asked me not to inform Cabinet of this, before which he assured me that Agreement would have much better chance of ratification if local defence of kind suggested could be introduced. Prime Minister fears that proposed manning active and reserve service by Australians will be difficult unless the condition (is) relaxed as to previous avocation of applicants. Many men anxious and able to make good sailors outside sea-going and quasi-sea-going population.

[12] *Telegram, Vice-Admiral Arthur Fanshawe, Commander-in-Chief Australia, to Admiralty*
24 April 1903

New Zealand certain to accept Navy Agreement as drawn up, but Governor General informs me passing by Federal Parliament very doubtful unless strong feeling for local defence is satisfied. Governor General is asking Home Government for Destroyers either in addition to, or instead of, third class cruizer(s). Submit I strongly deprecate reducing number of ships, and suggest to balance percentage on cost of destroyers, if granted article 5 of the draft agreement be cancelled. Destroyers to be maintained and manned by colonies. Admiralty would save about £20,000 per annum wages if article 5 cancelled. Consider that payment to colonials necessarily more than double imperial rates will cause great discontent in other ships, and also consider press and members of parliament will cause much trouble and make mischief, with regard to discipline these objections not applicable to reserve which might be morcased [?]. Believe that federal government propose to abandon present ship *Cerberus* &c and naval forces thereby saving most of the increase in contribution under new agreement. Therefore strongly urge destroyers if sent be manned and paid for by colony. Consider rejection of agreement most deplorable. Therefore submit local sentiment be satisfied. Consider moreover destroyers most suitable for forming local defence. New Zealand not yet consulted on proposed change in agreement.

[13] *Memorandum by Captain [Prince] Louis Battenberg,*
Director of Naval Intelligence

2 May 1903

Much could be said on such points as stationing torpedo boat destroyers around the coast to destroy torpedo boats operating from bases 4000 miles away.

On the obvious impossibility of finding Colonial Officers and men to efficiently fill the various special ranks & ratings, which make up a destroyer crew.

On the number of R.N. Officers & men to be locked up in the three Drill Ships.

On the large initial outlay in building the Destroyers – Even if the value of one *Mildura* be taken as a set off . . . &c &c.

All this is mere detail compared to the principles involved.

The existing Australasian agreement has for years been condemned as tying down a portion of the Royal Navy for the so called 'defence' of certain localities, in direct opposition to accepted Admiralty policy.

The underlying principle of the agreement about to be laid

before the local legislatures is that the whole of these naval forces, towards which Australasia is to contribute in money, are free to be employed where the Admiralty think fit, in war.

If these Destroyers are now added as a 'Defense Mobile' in the French sense, it can only be taken as an admission that the seagoing fleet is insufficient and that the great principle, on which this new Agreement is based, is unsound.

In connexion with this, it may be mentioned that a few days ago I read a leading article in an influential Australian paper, urging the necessity of a 'Defense Mobile' on the ground that the Home Government evidently considered such Coast defence necessary in England, seeing that a large force of Port Guard Ships & Coast Guard ships were stationed around the coast of the United Kingdom.

These erroneous ideas are not likely to be dispelled if the Governor of a State, who is a flag officer, expresses his approval of such inept proposals as a perpetual Destroyer Patrol in War in the many hundred miles of water inside the Great Barrier Reef.

As regards the 2 telegrams under discussion it will be seen that Lord Tennyson & Admiral Fanshawe are by no means in agreement as to how this new proposal is to be effected.

[14] *Remarks on the Inland Waters Naval Defence of Canada*
c. March 1903

'Secret'

The Naval Position on the Lakes

In the Defence Scheme of 1898. Kingston is recommended as the naval base of operations, and a scheme of fortification for that purpose is given.

I have carefully inspected the position of Kingston and have come to the following conclusions:

Though Kingston is in some ways suitable for the purpose of a naval base, owing to the possession of a good dry dock, a convenient harbour, and a certain amount of fortification, the position of Wolfe (or Long) Island makes it unsuitable.

This island is quite near the American shore, and any troops occupying it could make it impossible to use the harbour and dock of Kingston. To prevent them from occupying it, it would be necessary to lock up on it a considerable land force, which could not be spared for the purpose.

There is, however, a good place for a dry dock a few miles along the coast to the west of Kingston.

Proposals for a Canadian Naval force

I had a conversation with the Minister of Militia on this question, and as a result of it I made an application to him for the services of a British Naval Officer to assist me in dealing with the question. Later we had a second conversation and came to the following conclusions:

The Militia Department cannot get even the money needed for bare military requirements, and there is no prospect of its being able to get any more for naval purposes. There is no means in the Department for dealing adequately with the question.

We both felt strongly that the authority which controlled the estuary of the St Lawrence (i.e. the Department of Marine and Fisheries) should also control the whole floating defence. The Militia Department has, therefore, given up the idea of attempting to organise a Naval force.

The new Minister of Marine and Fisheries (who has charge of the Fisheries Protection Fleet) – M. Préfontaine is a powerful man who is naturally bent on increasing the activity and influence of his Department. He announced, immediately after his appointment, when he sketched his intended policy, that he proposed to take up the question of naval development. A small appropriation has already been made to this Department for the purpose – about £25,000.

Public opinion leans towards doing something in this direction as an atonement for the position taken up by the Canadian representatives at the Colonial Conference – a position which was very unpopular with the pronounced Imperialistic minority.

I understood from Commander Spain [see below] when he talked over the question of naval defence with me that the Marine Department expects that the British Admiralty will be willing, when asked, to present them with a suitable training ship as a free gift. But in my opinion it would be far better to give young Canadians nominations to the Britannia than to establish a Naval School here. But failing this, I am strongly of opinion that any Naval College should be controlled by Imperial Officers, and that the Cadets should if possible serve in the Imperial Navy, as it is very important that the officers who may control a future Canadian Navy should be imbued with the right spirit when young.

Should a Naval School be started here, it is very important that the Admiralty should, if possible, have some voice in its control. Conditions can be laid down now while the Naval force is in embryo and while money is an object which later on might never be agreed to. It must be remembered in this connection that if any portion of the Militia is taken from the control of the Minister of Militia, it is taken from the control of the Imperial officer who commands the Militia, and this Imperial control should be re-asserted in the person of a Naval Imperial Officer.

Those in whose opinion I have confidence say that the only possible Commander of the embryo Navy is Commander Spain (late R.N.), the Officer Commanding the Fisheries Protection Fleet. He is a man with a certain following and stands well with his Minister, though his drink habits are not such as to commend him to many others. At the same time he is a capable officer, thoroughly British in sentiment, and keenly loyal to the British connection. His position is such that it would be difficult to pass him over, at the same time in the interests of the Empire I strongly recommend that it be laid down that the Officer Commanding the Canadian Naval Force shall be an Imperial Officer appointed by the Governor General with the advice of the Admiralty.

Coal supply in war

Attention should be drawn to the important fact that not only is a great deal of Canada's ordinary coal supply derived from the United States, but the Canadian shipping on the Lakes is supplied entirely from the other side.

[15] *Earl of Minto, Governor-General of Canada, to Chamberlain*
12 June 1903

'Secret'

With regard to the reference in the Minute enclosed in my despatch No. 195 of even date, as to the 'proposed establishment by the Dominion of Naval training ships in the Maritime Provinces,' I have the honour to say that I to-day asked my Minister of Marine and Fisheries [J.R.F. Préfontaine] as to the present position of such a proposal, and he informed me that he had to some extent considered its possibility and that the matter would be further dealt with. In reply to my enquiry as to the system upon which he contemplated the formation of a Naval Reserve, by means of such training ships he told me that it would undoubtedly be on the lines of putting a certain number of ships in commission the property of the Canadian Government, and on my enquiring where he expected to obtain competent naval officers, he replied that he believed he would have to apply to His Majesty's Government for assistance in that direction, though Canada itself possessed a few such officers who would be available.

I cannot but think that the possibility of any definite action towards the formation of a naval reserve by my Government is still somewhat remote; if any such idea genuinely exists, I am afraid it would seem to be in the direction of an enlargement or reorganisation of the present small force of vessels used for coast-guard and other purposes, the officers and crews of which are appointed almost entirely for their political qualifications, the patronage for each vessel belonging to the Member of Parliament representing the district from which it is manned.

I am much afraid that the commissioning by my Government of any vessels for the training of naval reserves would in all probability be on the above lines.

[16] *Memorandum on the Standard of Defence at Esquimalt (Prepared for the Consideration of the Committee on Imperial Defence)*

May 1903

Esquimalt is the naval base of the Pacific Squadron, and has a Government dock, towards the construction of which the Admiralty contributed 50,000£. The harbour has been fortified on the scale estimated to be necessary to resist the attack of one or two cruisers. In accordance with an agreement made in 1898 between the Canadian Government and Her Majesty's Government, a garrison of 358 regular troops (Artillery and Engineers) is maintained in the fortress permanently by the War Office, and this garrison would be supplemented, on the outbreak of war, by about 700 local Militia. The latter are very imperfectly trained, and as the greater portion of this Militia are resident on the mainland, there is some doubt whether the local authorities might not detain them for the defence of Vancouver City, if that place were threatened.

The British Pacific Squadron at present consists of one 1st-class and two 2nd-class cruisers. The United States' Pacific Squadron includes one battle-ship, one 1st-class cruiser, two 2nd-class, and four 3rd-class. Russia has no ships allotted to this station. France maintains one 2nd-class cruiser and a gun-boat. The only other naval force to be reckoned with in these waters is that of Chili.

2   Esquimalt is in close proximity to the United States' naval base of Seattle. It is estimated that, given sea command, the United States could, ten days after the declaration of war, despatch against Esquimalt a force of 3,000 men and 24 guns; a month after declaration of war, a force of double that strength could be despatched. In a war with the United States, Esquimalt could not be reinforced by troops from the east, as the Canadian Pacific Railway would certainly be cut by the Americans at the commencement of hostilities.

3   The O[fficer] C[ommanding] Troops, Esquimalt, having recently represented the inadequacy of the garrison to meet an attack from the United States, the Admiralty were requested by the War Office (*vide* Appendix No. 1) to state their views as to the strategical position of Esquimalt in the event of war with that country.

In a letter dated the 12th April, 1903 (*vide* Appendix No. 2), the Admiralty replied:

(a.)   That their Lordships are aware of no reason for departing from the opinion which was expressed in 1882 by the Royal Commission on the Defence of British Possessions Abroad, that, in the event of war with the United States, Esquimalt could not be defended.

(b.)   That its capture by the United States would not be a matter of serious strategic importance.

(c.)   That the Navy could render no assistance in the defence of Esquimalt.

(d.)   That in such an eventuality our Navy would be concentrated on decisive points, among which no position in British Columbian waters could be included.

4   As neither France nor Russia have any coaling station in the Eastern Pacific, it appears improbable that any portion of the fleets of those Powers would frequent Columbian waters in time of war. It is concluded, therefore, that the dictum that the Columbian waters are not one of the decisive positions at which our fleet would be massed, applies also to the case of war with France and Russia.

5   If this be so, it would appear that, from a naval point of view, the garrison and fixed defences of Esquimalt will be useless in war, and could be dispensed with.

Furthermore, the present scale of defence is absolutely inadequate to hold the fortress against the only Power by whom an attack on it is at all probable.

The garrison is, therefore, from a military point of view, an isolated outpost, certain to be captured in the case of war with the United States, and, unless required for other than purely naval purposes, useless in the case of war with any other Power.

6 The logical conclusion to be drawn from these considerations would appear to be that the garrison should be removed and the works dismantled; but before any decision to that effect is arrived at it is necessary to review the history of the establishment of Esquimalt as a naval base, and our obligations to Canada in connection with its defence. . . .

[17] *Memorandum for the Committee of Imperial Defence on the Standard of Defence for the Naval Bases of Halifax, Bermuda, Jamaica and St Lucia*

September 1903

'Secret'

15 The points at issue in this case may be summarised as follows:

(a.) The War Office has recently incurred an expenditure of nearly 3 1/2 millions, exclusive of the cost of barracks, and has allotted and maintains a force of some 8,000 regulars, besides local troops, for the defence of Bermuda, Halifax, Jamaica, and St Lucia. This has been done to meet the requirements of the Navy, the Admiralty having declared that certain ports, among which the above ports are included, are 'essential to the Navy as bases and coaling stations.'

The scale of defence adopted has been based on the Admiralty assurance that they accept the responsibility of protecting all British territory at home and abroad against organised attack by sea.

(b.) In view of the growth of the American navy, the Admiralty now say that, under certain contingencies, they could not protect Bermuda, Halifax, Jamaica, and St Lucia against organised attack by sea; they state that these bases are of comparatively minor Imperial value; they remark that the strategic value of naval bases is dependent on our having command of the sea; and they conclude that it is unnecessary to reconsider the standard of defensive power at the points in question.

The line of reasoning by which this conclusion is arrived at is not very clear, for our loss of sea command must be either permanent or local and temporary. If the former, it is obvious that we should have to make peace on the best terms we could obtain. If the latter, it is equally obvious that a strongly defended fortress or naval base could hold out longer than a weakly defended one. Moreover, it may be doubted whether the strategic value of our naval bases abroad is not rather increased than diminished by the growth of the navies of other Powers, provided always that the British navy is maintained at a strength capable of upholding or eventually acquiring that sea supremacy which in their 'Instructions on Defence Matters' the Admiralty rightly declare to be the basis of the defence of the Empire.

16   It will be observed that the Admiralty have not answered the question raised in paragraph 7 of War Office letter No. 266/West Indies/2, dated the 4th May, 1903, as to the effect of the capture by the United States of our naval bases in the Western Atlantic and Caribbean Seas upon our sea-borne food supplies from Canada and South America.

17   Whether, if the Admiralty did not assume exclusive responsibility for protecting the United Kingdom against organised attack by sea, it might not be possible to place a stronger naval force in the Western Atlantic and Caribbean, and thus protect our trade routes and naval bases in that quarter, is a point which seems to deserve consideration.

[18] *'Strategic Position of British Naval Bases in Western Atlantic and West Indies', memorandum for the Committee of Imperial Defence by Battenberg*

November 1903

The Military Intelligence Division's Memorandum under consideration deals with the correspondence which has passed between the Admiralty and the War Office on the subject of the standards of defence for the naval bases of Halifax, Bermuda, Jamaica, and St Lucia.

In the course of the correspondence the Admiralty, in their letter M. 0567 of the 29th June, 1903, after reviewing the strategical situation in the Western Atlantic and the Caribbean Sea in the event of uncertain or hostile relations with a European Power at the time of an outbreak of war with the United States, stated that in the conditions named it would not be possible for Great Britain to deplete her squadrons in European waters to an extent sufficient to place her on anything like an equality with the American fleet, and came to the conclusion, for reasons fully stated in the above-quoted letter, that it is unnecessary to reconsider the standard of defensive power at the points in question.

The Military Intelligence Division's Memorandum states that it is gathered from this letter 'that if war should break out with the United States at a time when relations between this country and an European Power were uncertain, the Admiralty propose to abandon the sea command in the Western Atlantic to the American fleet, and that, under these conditions, they consider that the strategic value of the naval bases in those waters would disappear, and that their loss would not materially affect the main issue of the war.'

This deduction is practically correct. The Admiralty, in their letter M.0109 of the 24th February, 1903, pointed out that by the end of 1905 the United States will have a fleet of twenty battleships and eight armoured cruisers ready for sea, smaller vessels being left out of consideration, and in view of the increasing naval strength of the European continental Powers, it does not seem probable that Great Britain would be able to spare a force sufficiently powerful to blockade the Atlantic portion of the United States' fleet in its own ports.

They pointed out also that the ultimate fate of Halifax, Bermuda, Jamaica, and St Lucia would depend on the results of the main issues of the war.

In the Military Intelligence Division's Memorandum stress is laid upon an expression made use of in Admiralty letter M. 0109, in which Halifax, Bermuda, Jamaica, and St Lucia are referred to as positions of 'comparatively minor Imperial value,' an expression which appears to have been somewhat misunderstood.

These positions are only of 'minor Imperial value' in comparison with something greater. It is only in the event of a war with the United States, coupled with complications in Europe, which prevented us from maintaining the command of the sea in American waters, that our bases become comparatively of minor Imperial value.

The Admiralty did not intend, in employing this expression, to deprecate their value, but the fact has to be faced that under certain eventualities these stations in Western seas are at the mercy, sooner or later, for a time at least, of the forces of the United States, if hostile.

The Admiralty's opinion is that we cannot hold the bases without the fleet, and as the fleet may, and probably will, be engaged elsewhere under the conditions contemplated, their strategic importance, which is contingent on our holding the command of the sea in those waters, temporarily declines.

So long as we can hold the command of the sea in the Western Atlantic and the Caribbean Sea the bases are safe; should we be unable to do so, a contingency which is recognised as possible, the bases are no longer safe, but at the mercy of the Power which holds the command of the sea.

The Admiralty are, moreover, of opinion that, under the circumstances above mentioned, no increase of fixed defences or garrison will make these bases permanently safe. In their isolated position they must sooner or later capitulate or starve.

The precarious position of these bases, if we should temporarily have lost the command of the sea in surrounding waters, cannot be cured by large and costly additions to the armaments and garrisons.

Their safety, like their strategic value, is inseparably connected with the command of the Western Atlantic and Caribbean seas by the British fleet.

The Admiralty is of opinion that it is unlikely at the outset of a war the United States would trouble themselves about the bases, which to them would be strategically of minor value. They would turn their attention to what would really touch us, viz., the grain trade with Canada. The latter must derive its protection, not from the bases, but from the fleet.

The Admiralty's assumption that the ultimate fate of the places in question would depend on the results of the main issue of the war is one which the history of naval warfare in general would appear to justify.

As they have already pointed out, the maintenance of an outlying position is in the end dependent upon the security of its communications with the mother country, and if British sea supremacy were in the end to be re-established, American troops in a British island would be in a similar position to that of the Spanish troops in Cuba, with their communications severed.

[19] *Agreement between His Majesty's Governments of the United Kingdom, the Commonwealth of Australia, and the Colony of New Zealand*

1903

The Commissioners for executing the Office of Lord High Admiral of the United Kingdom of Great Britain and Ireland, & c, and the Governments of the Commonwealth of Australia and of New Zealand, having recognised the importance of sea power in the control which it gives over sea communications, the necessity of a single navy under one authority, by which alone concerted action can be assured, and the advantages which will be derived from developing the sea power of Australia and New Zealand, have resolved to conclude for this purpose an Agreement as follows:

Article I

The Naval Force on the Australian Station shall consist of not less than the under-mentioned sea-going ships of war, all of which shall be from time to time throughout the terms of this Agreement of modern type, except those used as drill ships:

One armoured cruiser, 1st class;
Two second class cruisers;
Four third class cruisers;
Four sloops;
And of a Royal Naval Reserve consisting of 25 officers and 700 seamen and stokers.

Article II

The base of this force shall be the ports of Australia and New Zealand, and their sphere of operations shall be the waters of the Australia, China, and East Indies Stations, as defined in the attached schedules, where the Admiralty believe they can most effectively act against hostile vessels which threaten the trade or interests of Australia and New Zealand. No change in this arrangement shall be made without the consent of the governments of the Commonwealth and of New Zealand; and nothing in the Agreement shall be taken to mean that the Naval Force herein named shall be the only force used in Australasian waters should the necessity arise for a large force.

Article III

This force shall be under the control and orders of the Naval Commander-in-Chief for the time being appointed to command His Majesty's ships and vessels on the Australian Station.

Article IV

Of the ships referred to in Article I, one shall be kept in reserve and three shall be only partly manned and shall be used as drill ships for training the Royal Naval Reserve, the remainder shall be kept in commission fully manned.

Article V

The three vessels used as drill ships and one other vessel shall be manned by Australians and New Zealanders as far as procurable, paid at special rates, and enrolled in proportion to the relative population of the Commonwealth and New Zealand. If a sufficient proportion of men from either Colony should not on the aforesaid basis be forthcoming, a sufficient number of men to complete the complements of the ships may be enrolled from the other Colony.

They shall be officered by officers of the Royal Navy, supplemented by officers of the Royal Naval Reserve.

Article VI

In order to ensure that the naval service shall include officers born in Australia and New Zealand, who will be able to rise to the highest posts in the Royal Navy, the undermentioned nominations for Naval Cadetships will be given annually:

Commonwealth of Australia   8
New Zealand   2

Article VII

The branches of the Royal Naval Reserve established in Australia and New Zealand shall be called into actual service by His Majesty in Council, acting on the advice of his Governments of the Commonwealth of Australia and New Zealand respectively.

The men forming the Royal Naval Reserve shall be divided into two classes:

(a) Those who have served for three years on board one of His Majesty's ships.

(b) Those who have not so served.

These men shall be trained on ships specially provided for the purpose.

The officers of this reserve force shall be included on the list of officers of the Royal Naval Reserve.

Article VIII

In consideration of the service aforementioned the Commonwealth of Australia and New Zealand shall pay the Imperial Government five-twelfths and one-twelfth respectively of the total annual cost of maintaining the Naval Force on the Australian Station, provided that the total amount so paid shall in no case exceed 200,000£ and 40,000£ respectively in any one year. In reckoning the total annual cost a sum equal to 5 per cent on the prime cost of the ships of which the Naval Force of the Station is composed shall be included.

Article IX

The Imperial Government recognise the advantages to be derived from making Australasia a base for coal and supplies for the squadrons in Eastern waters.

Article X

1 This Agreement shall be considered to become actually binding between the Imperial Government and the Commonwealth of Australia and New Zealand so soon as the Colonial Legislatures shall have passed special appropriations for the terms hereinafter mentioned, to which Acts this Agreement shall be attached as a First Schedule.

2 The Agreement shall be for a period of ten years, and only terminate if and provided notice has been given two years previously, viz., at the end of the eighth year, or at the end of any subsequent year, and then two years after such date.

[20] *Colonel Sir E.W.D. Ward, Permanent Under-Secretary of State, War Office, to the Under-Secretary of State, Colonial Office*
29 November 1904

With reference to your letter, No: 23133/1904, dated 27th July last, enclosing a despatch from the Governor General of Canada on the subject of the garrisons of Halifax and Esquimalt and a draft of the proposed reply thereto, I am commanded by the Army Council to state that in view of the recent decision of the Committee of Imperial Defence, it is their intention to immediately withdraw the battalion of regular infantry now stationed at Halifax. The defences of that fortress are now under revision and a reduction in the strength of the Artillery personnel will be effected shortly.

As regards Esquimalt, the Committee of Imperial Defence have decided that in view of the existing arrangement with the Dominion Government no withdrawal of the garrison of that station can take place previous to 1909, unless the Government should agree to assume charge of the defences, but they recommend that the Secretary of State for the Colonies should open negotiations with this object in view if and when he considers the time opportune.

The Council hope that Mr Secretary Lyttelton [Secretary of State for the Colonies] may see his way to press for the withdrawal of the regular Garrison from this station at an early date.

[21] *Paraphrase of Telegram, Earl Grey, Governor-General of Canada, to Alfred Lyttelton, Secretary of State for the Colonies, received at the Colonial Office*
15 December 1904, 8.20 a.m.

'Confidential'

With further reference to your telegram of the 9th of December, Sir R. Cartwright assures me that matter referred to will be considered by my responsible advisers on return of Sir W. Laurier in about a fortnight. He has confidentially expressed his individual view that no objection will be urged against taking over Esquimalt and Halifax if no conditions as to expenditure are attached but he warns me that the opinion of my responsible advisers will probably be in favour of a reduced garrison and stoppage of further works.

[22] *Charles Inigo Thomas, Assistant Admiralty Secretary, to the
Under-Secretary of State, Colonial Office*
16 December 1904

I am commanded by my Lords Commissioners of the Admiralty
to state, for the information of the Secretary of State for the
Colonies [Alfred Lyttelton], that as a logical consequence of
the scheme of re-distribution of His Majesty's Fleet recently sanctioned by the Cabinet and published as a Parliamentary paper in
the Memorandum of the First Lord, it has been decided to
reduce the Naval Establishments at: Halifax, Jamaica, Esquimalt
and Trincomalee to 'cadres', on which no money will be spent in
Peace, but which can be developed in war according to necessity.

[23] *'Transfer of the Defences of Halifax to the Canadian
Government', memorandum by Minto*
29 December 1904

As regards the transfer of Halifax to the Canadian Government,
I do not know the proposals made by His Majesty's Government, but assume that it is intended to hand over the command
of the harbour and fortifications, and that though conditions may
probably be made as to the strength of the garrison and upkeep
of works, that in future it is intended that Halifax should form
part of the military system of the Dominion independent of
Imperial control.

It must rest with naval and military experts to decide as to the
Imperial value of Halifax in case of war with the United States.
It is difficult, however, to imagine that under such circumstances
it would not be of immense value to us, not only as a naval base,
but in a commercial sense also, as it is our only fortified harbour
on the American coast, our only other harbours of any importance
being Sydney – closed by ice in the winter – and St John [New
Brunswick], in close proximity to the American frontier, whilst in
case of war Halifax would of necessity become the eastern terminus of the whole trans-Continental railway. Personally, I cannot
imagine anything much more disastrous to the defence of Canada
than leaving the eastern termini of her trans-Continental railways
at the mercy of a foreign Power.

The authority into whose charge it is proposed to transfer this
important position is the Canadian Government.

I may say that, when I took up my appointment in Canada, I

was myself inclined towards such a course, believing that it would encourage a sense of Imperial responsibility, and would appeal to the sentiment of the population. Such a belief is natural for those not closely acquainted with Canadian political life, and the capabilities of its military resources, but six years' residence in Canada has entirely changed my opinion.

I am now aware that the authority to whom it is proposed to commit this valuable trust is perfectly unreliable. I have no hesitation in asserting that if the offer in question is accepted it will be accepted with no realisation of its Imperial importance; that the fulfillment of conditions will be evaded, and that Halifax will be treated merely as a valuable political asset – its military appointments to be politically filled, its building and supply contracts to be given to political supporters.

In addition, however, to these grave objections to the proposed transference, there is the question of the actual force upon which Canadian Ministers can rely to justify themselves in accepting such a charge. If they are anxious to accept it, they will no doubt minimise the difficulty of obtaining a garrison, but it is as well to consider the actual military strength of the Dominion, and the possibility of increasing it.

[24] *'Naval Notes upon the Defence of Canada', signed by Rear-Admiral Sir Charles Langdale Ottley and Rear-Admiral Battenberg, present and retiring Directors of Naval Intelligence*
6 January 1905

It is stated in the Press that the Canadian Minister of Fisheries [J.R.F. Prefontaine] contemplates a visit to England to consult the Admiralty on the question of the renewal and reorganisation of the Canadian Naval Defences.

In this connection a proposal has been put forward to replace the existing vessels of the Canadian Fishery Protection Flotilla by ships discarded from the Royal Navy.

The advocates of this change frankly recognise that the ships of the Canadian Fishery Protection Flotilla, as thus remodeled, could not, under existing treaty arrangements, enter Lake Ontario in peace time. But it is urged that if they were retained at Montreal as a centre, they would be so near the entrance to Lake Ontario that the moment the relations became strained they could be pushed up on to the lake at very short notice.

Their presence on Lake Ontario would undoubtedly be of the

utmost value, and would go far to secure for the Dominion flag that command of the lake which (as is clearly pointed out in para. 5 of the General Staff Memorandum) is so essential for the defence of Canada proper.

If, in addition, a torpedo-craft and submarine boat flotilla could be maintained at the mouth of the St Lawrence, nominally as a defence mobile, but in reality for service on the Lakes, the force would be enormously strengthened.

Attractive as this scheme is as thus stated it is open to several objections.

In the first place we have to consider what would be [the] attitude of the United States in [the] face of any such fundamental alteration of the character of the Canadian Fishery Protection Flotilla, carried out as it would be with the connivance of England.

As at present constituted[,] that Flotilla consists of 16 small craft, whereof two are sailing ships. These little vessels vary in tonnage as follows: – six of them are under 200 tons, four of them are from 200 to 400 tons, and of the remaining six, not one exceeds 1,100 tons. Eight of them carry no guns at all, and the remainder carry one or two boat guns only, usually a Gatling or 7-pr M.L.R. The Canadian Government, in fact, maintains these vessels merely for police duties and for such hydrographic purposes as the maintenance of harbour bouys and lighthouse services, with, of course, the added duty of protecting the sea coast and inland fisheries.

The proposal is that these old and valueless craft are to be replaced by war vessels of suitable type from the list of ships now being discarded from H.M. Navy, regard being had to the size of the locks through which they must pass in order to reach the theatre of operations upon the Lakes. The latter consideration limits their dimensions to a maximum length of 270 feet with a draft of 14 feet.

The following 16 vessels would, however, all fall with[in] this category and would undoubtedly form a most valuable force for employment on Lake Ontario: – *Rosario, Fantome, Merlin, Clio, Catmus, Rinaldo, Espiegle, Mutine, Vestal, Bramble, Britomart, Thistle, Alert, Torch, Phoenix, and Algerine.*

None of these vessels are more than five years old, except the last four, which were built ten years ago. None of them have a speed of less than 13 knots and their armament (for vessels of their class) is sufficiently formidable. Most of them carry six 4-inch Q.F. guns and four 3-pounders, with two or three Maxims. In a word[,] the above proposal would give Canada a force of 16 effective warships in place of her existing 16 practically valueless vessels.

The effect of this exchange upon the minds of statesmen in the United States would be particularly unfortunate, however, at the present time, in view of the fact that (by the recently concluded Convention with France regarding the Fisheries) one of the principal Canadian excuses for the maintenance of a strong fishery protection flotilla has been removed.

What, it may be asked, would be the conclusions likely to be drawn in the United States from such an action on the part of Great Britain and Canada at the present moment?

Is it not clear that any measure of the kind, however we might attempt to explain it away, would be rightly construed as a threat to which the United States, in self-defence, would be bound to reply by a corresponding augmentation of their own preparations on the Canadian frontier.

It may be rejoined that Canada's commanding geographical position at the mouth of the St Lawrence is such that no adequate reply is possible on America's part.

This, however, is far from being the case . . .

[25] *'Defence of Canada', memorandum by the Admiralty for the Committee of Imperial Defence*
24 February 1905

To summarise the foregoing conclusions. The view of the Admiralty is that Canada must primarily rely upon her own resources for defence against invasion by the United States: Firstly, because any action on the part of the Mother country is conditional upon British sea supremacy in the Western Atlantic, which cannot, even now, under all circumstances, be guaranteed in such a war, and, in the near future, in view of the steady increase of the United States' navy, could probably not be counted on at all. Secondly, because the greatest interest of this country being to maintain friendly relations with the United States, any military preparations for the defence of Canada on the part of Great Britain obviously aimed at that country are greatly to be deprecated; and Thirdly, because Canada's knowledge that, in a war with the United States, she must mainly rely upon her own resources, must tend to inculcate in her statesmen a wholesome caution.

As regards the First reason, it is clearly a matter for high policy to determine whether it is the duty or the interest of the Mother Country to add in peace time to the heavy burdens already imposed upon her in order to secure this condition by maintaining a larger navy. But it appears to the Admiralty that it can hardly be seriously maintained that Canada, contributing as she does nothing towards the maintenance of the British navy, can have any moral or material claim to such additional naval effort on our part for her sole benefit. Naval support would, of course, be freely given as far as available in war, but it could only be placed at Canada's disposal after the command of the sea had been fought for and won. The basis upon which the strength of the British navy is assessed has not hitherto contemplated additional burdens upon the British taxpayer for the purpose of providing a flotilla of His Majesty's ships or of officers or men to man them for the defence of the Great Lakes.

On the assumption that the duty of defending her own frontier against invasion devolves primarily upon Canada, it is useful to consider what forces Canada has available and how far they can be supplemented. At present Canada's only naval force is the Fisheries Protection fleet, which is commissioned by 'scratch' crews in summer and laid up for the most part in winter. With the exception of a small cruiser built last year and still smaller vessels on Lake Erie, none of the vessels composing it have any fighting value . . .

As regards the command of the Atlantic, the Admiralty stated last year, in a paper submitted to the Committee of Imperial Defence, that, in the event of war with the United States at a time when the relations between this country and European Powers were uncertain, it did not seem probable that Great Britain would be able to spare a force sufficiently powerful to blockade the Atlantic portion of the United States' fleet in its own ports. In short, they contemplated that under such circumstances the command of the Western Atlantic could not be retained by Great Britain . . . Generally, the more carefully this problem is considered, the more tremendous do the difficulties which confront Great Britain in a war with the United States appear. It may be hoped that the policy of the British Government will ever be to use all possible means to avoid such a war. – 6 January 1905

Enclosure 'Defence of Canada
– Remarks by the Admiralty on
Paper 24C [General Staff memo]
March 1905

The growth of the American Navy has fundamentally altered the strategic situation in the Western Atlantic.

The conditions which have prevailed in the past exist no longer. Changes which materially affected the balance of naval power in the Western Atlantic are in rapid progress, and are all such as tend to make the direct intervention of the British Navy in the local defence of Canada impracticable. The Admiralty cannot, as they have already stated, guarantee that in the event of war with the United States the command of the Western Atlantic will in future be always maintained.

[26] *'Strategic Conditions of Halifax and Esquimalt',*
memorandum by the Colonial Defence Committee for the
consideration of the Committee of Imperial Defence, signed by
J.E. Clauson, Secretary

4 May 1905

'Secret'

The Committee of Imperial Defence have referred to the Colonial Defence Committee the question of the extent to which the strategic position of Halifax and Esquimalt, as defined in the existing local Defence Schemes, has been modified by the recent changes in the distribution of our fleet, and in the naval strength of other Powers in neighbouring waters.

Halifax

2   In the case of war with Powers other than the United States, Halifax has little strategical importance. Owing to its remoteness from the probable scene of major operations, its defences would exercise no influence on their conduct. It is, of course, an important commercial port; but it is little liable to any form of raiding attack, on account of its great distance from the naval bases of European Powers.

It has been laid down by the Committee of Imperial Defence that, as regards Powers other than the United States, it is not necessary to contemplate anything more serious than a raid by one or two cruisers.

3   In the event of war with the United States the strategical position of Halifax would be very different. Should the naval base at Bermuda fall, or its resources be destroyed before our fleet had time to establish itself in the North Atlantic, Halifax would probably become, for a time, at any rate, our principal naval base in these waters. It would also possess great importance as the terminus of the Intercolonial Railway, and the port of disembarkation for military reinforcements for Canada after the St Lawrence had become ice-bound. Its defence then, if practicable, is a matter of great importance both on naval and military grounds.

4   If the British navy were not able to be employed in full strength in the Western Atlantic, there is little doubt that the United States would hold local command of the sea for a period of indefinite duration. For if the British naval forces were in such local inferiority of strength as to be wholly outmatched, it would be very false strategy to run the risk of having our fleets defeated in detachments, and naval action would have to be postponed until superior forces could be brought to bear. For purposes of calculation, then, it should be assumed that the United States might be in local command of the sea for an uncertain and possibly considerable period, and consequently in a position to land an organised expedition in Nova Scotia for the attack of Halifax.

Even under these conditions Halifax could be defended on land by a moderately small garrison, such as that hitherto approved, or its future equivalent, which may be taken as one Canadian permanent battalion and two militia battalions on the establishment of 1,060, all ranks, laid down in 1904. It is a position of extraordinary natural strength. The approaches to it are narrow defiles. The soil is most unfavourable for siege operations. As it would, if effectively garrisoned, be impregnable by the United States except at the cost of a lengthy land siege, it will be justifiable to secure for it a similar immunity against attack from seaward.

5   A deliberate naval attack on the coast works is not considered at all a probable operation. Until our fleet had been met the United States would be anxious to keep their ships intact, and would not be likely to risk them within effective range of the forts. The position of their fleet would be somewhat analogous to that of the Japanese fleet outside Port Arthur, before the fall of that fortress, and, as in that case, the naval operations would probably be confined to harassing the defence by long-range bombardments from positions where this entailed small risk, and to feints and attacks by light craft, with a view to keeping as large a

proportion of the garrison as possible at the sea defences, and correspondingly diminishing the force on the land front.

But if the coast defences were too weak, a large fleet might possibly succeed in overwhelming them by a concentrated fire without itself suffering serious loss; and since battleships would be available if a naval attack were made on Halifax, the coast defences should be strong enough – not to fight a duel to a finish – but to deter from attack a fleet, including battleships, which could not afford to incur serious losses.

6   Having regard to the naval and military considerations stated above, and on the assumption that war with the United States must be contemplated as a possible contingency, the Colonial Defence Committee submit that the Defence Scheme should be based on the assumption that Halifax, as the probable naval base in the North Atlantic in the event of certain contingencies, and as the port of disembarkation for military reinforcements and stores when the St Lawrence is ice-bound, should be defended to deter a fleet, including battleships, which could not afford to incur serious losses, from attacking the sea defences, and that the garrison must be able to resist for a considerable period the attack of an organised expeditionary army landed in the vicinity.

7   The strategic conditions, as defined in the existing Defence Scheme of Halifax, do not differ very materially from the above statement, and any necessary modifications could be pointed out in the Colonial Defence Committee's Remarks on the annual revision of the Defence Scheme, which is now under their consideration.

Should the above standard of defence be approved, the recent decision to reduce the approved armament of Halifax will require reconsideration.

Esquimalt

8   The Colonial Defence Committee recommend that Esquimalt be no longer retained as a fortified port. The reasons for this recommendation are given in the accompanying draft Memorandum No. 347 M, which has been prepared for communication, if approved, to the Dominion Government.

[27] *Minutes, 73rd Meeting, Committee of Imperial Defence*
June 28, 1905

Conclusions   The standard of defence of Halifax should be sufficient to deter a fleet, including battleships, which could not afford to incur serious losses, from attacking the sea defences, and to enable the attack of an organized expeditionary army, landed in the vicinity, to be resisted for a considerable period . . .

[28] *'Defence of Canada', 'Naval Aspects', General Staff Memorandum*

3 July 1905

2   Both papers speak of the 'ultimate command' of the Lakes. I would point out that ultimate command is of far less moment from the military point of view than early command . . . What we want is to delay invasion as long as practicable in the hopes that the Navy will gain definite command of the sea and thereby bring sea pressure to bear against the United States and at the same time enable reinforcements to reach Canada before the best part of it has been overrun . . .

Finally I cannot help pointing out that the inability of the Royal Navy to assist Canada at the most critical time is a strong argument in favour of the Dominion starting a flotilla of its own in the Gulf and Nova Scotia waters. The arguments always used against local fleets in Australasia cannot possibly be applied to Canada. The Admiralty guarantee Australasia against invasion. In the case of Canada they will guarantee nothing that is of any practical use to the Dominion in its hour of need.

[29] *Alfred Deakin, Prime Minister of the Commonwealth of Australia, to Lord Northcote, Governor-General (forwarded to the Secretary of State for the Colonies, 29 August)*

28 August 1905

As Your Excellency is aware, under an agreement with His Majesty's Government, the Commonwealth contributes five-twelfths of the annual cost of the Naval Forces on this Station, whose base is in Australasian Ports, but whose sphere of operations includes the China and East Indies Stations. The protection of Australasia and its commerce, and of great Imperial interests in China and India upon the high seas, are its principal duties in this very extensive area.

The Naval Commander-in-Chief, Admiral Sir A. Fanshawe, has recently criticised this agreement on several public occasions, in order to support his contention that our contribution, as there fixed, is altogether insufficient. Since as yet only three payments have been made according to an agreement arrived at in 1902, which is to have effect for ten years from 1903, this appeal for an alteration of its terms might be deemed premature. But as it may indicate a dissatisfaction with the agreement shared by the Lords of the Admiralty, as well as by their Official Representative; and as a similar dissatisfaction, though upon other grounds, exists here, it may be advantageous to commence its reconsideration without delay.

The paramount importance of the Navy to the British Empire and to Australia may be taken to be freely admitted. Nothing in this despatch is intended to question it. Indeed our obligations to share in the general defence of the Empire have been already recognised in practice and in principle. Beyond this the defence of Australia and of its coasts is accepted as a duty and as a necessity of our national self-respect. Yet even under these circumstances the present Naval Agreement is not and never has been popular in the Commonwealth. It has been approved only in default of a better means of indicating our acceptance of Imperial responsibilities. Whatever may be the assumed basis upon which our contribution is there determined it is regarded as merely an arbitrary proportion of an existing expenditure. Whatever the intention may have been, this attempt at joint naval action has failed to enlist a fraction of the support that was spontaneously accorded in all the States to the despatch of military contingents to South Africa.

On this account the question why the Naval Agreement is coldly regarded here appears serious enough to merit careful scrutiny. There is much truth in the customary interpretation that its want of popularity is due to the fact that, except to the small extent permitted by Articles V, VI, and VII, none of our grant is applied to any distinctively Australian purpose. When the squadron is pointed to as a justification for our subsidy, it must be remembered that a similar squadron, more localised than the present, was maintained prior to our first agreement with the Admiralty in 1887, and would be maintained now if there were no subsidy. What has been obtained by us in return for an annual appropriation has been simply an increase of its strength, coupled with an extension of its sphere of operations.

The British Man-of-war and the British seaman awaken enthusiasm whenever they visit our ports because, being British, they are inseparably associated with our race and history; but the particular squadron supposed to be paid for in part by us is not specially Australian any more than it is Anglo-Indian or representative of the Straits Settlements, to which it may be called at any time. What is really required is that any defences, if they are to be appreciated as Australian, must be distinctively of that character. At present we are without any visible evidence of our participation in the naval force towards which we contribute. Our £200,000 a year would seem a part repaid if we were enabled to take a direct and active part in the protection of our shores and shipping. But as we have no identification with the squadron, except in the articles already alluded to, there is so far nothing naval that can be termed Australian, or even Australasian. No Commonwealth patriotism is aroused while we merely supply funds that disappear in the general expenditure of the Admiralty. The Imperial sentiment languishes too since the squadron is rarely seen in most of our ports, and then only by a small proportion of the population.

Having regard to the obvious lack of public interest upon the part of the people of this country in our present naval defence, several means have been suggested that would assist to awaken and render it deeper and more permanent. An acceptable expedient ought not to be hard to find. The Admiralty probably desires naval and coaling stations in Australia other than those already or likely to be hereafter established at such or our seaports as may be defended by local works, or it may need other accommodation established here that would earmark the moneys expended. There would then be something to show for our contribution which at the same time would be a real assistance to His Majesty's Navy. Although object lessons of this kind would have a healthy influence, they are not the only, nor, when our remote situation is remembered, are they the wisest means of popularising our grants. They might possibly be criticised as devices for the spending of money upon our own shores or for local benefit only.

Imperial purposes can be served to which no such objection could be taken but which would be at the same time of conspicuous value both to the Admiralty and the Commonwealth . . . Both the naval power and mercantile shipping of the Empire would be materially re-inforced if the sum at present paid by us towards the local squadron were applied in securing up-to-date ships usefully engaged in commerce during times of peace but capable of being employed economically and at the shortest notice in the event of war. The French and German Governments are understood to have important agreements of this nature now in force.

[30] *'St Helena, Proposal by the Army Council that the Garrison should be provided by the Admiralty', memorandum for the Committee of Imperial Defence*

October 1905

The Army Council desire to bring the question of the garrison of St Helena again before the C[ommittee of] I[mperial] D[efence].

2   At their 63rd Meeting the Committee recommended that provision should be made for manning the armament contemplated by the Admiralty, and that the regular infantry should be withdrawn. These recommendations have been accepted by the Army Council, and the necessary action has been taken to put them in force. But a further consideration of the subject, and correspondence which has taken place with the Admiralty, has satisfied the Army Council that the arrangements are far from satisfactory, and that some misapprehension exists as to what will be their effect.

3   The war garrison is to be limited to approximately 100 officers and men of the R[oyal] G[arrison] A[rtillery] and R[oyal] E[ngineers], who represent the proper force to man the two 6-inch guns and two 6-prs which the Admiralty declared to be sufficeint for purposes of defence when the question came before the C[ommittee of] I[mperial] D[efence]. A military detachment so small as this, in such an isolated position, is most objectionable from every point of view. It cannot be regularly inspected, its efficiency and discipline are likely to suffer, and it will indeed be impossible to carry out ordinary peace routine with such a limited force. The Army Council therefore fear that, unless some other arrangements can be arrived at, it will be necessary to maintain a garrison in peace time in excess of war requirements. This is contrary to the fundamental principles governing the allotment of troops to foreign naval stations and defended ports, and cannot be justified on public grounds if any other solution of the difficulty can be devised.

4   In the opinion of the Army Council the most satisfactory plan, alike from the point of view of efficiency and of economy, is that responsibility for the defence of St Helena be taken over by the Royal Navy. This transfer of responsibility was proposed to the Admiralty in a letter of the 5th June, 1905, of which a copy is

attached (B). It will be seen that in the reply to this communication, dated the 28th June, their Lordships did not see their way to falling in with the views of the War Office. The Army Council cannot, however, but feel that, in rejecting the suggestion, the Admiralty were largely governed by their adhesion to the accepted principle (to which attention is called in their fifth paragraph), that naval personnel should be exclusively employed in manning the fleet; in view, however, of the proposal recently made to the C[ommittee of] I[mperial] D[efence] on behalf of the Royal Navy that Gibraltar should in future be garrisoned by naval personnel, it appears to the Council that this question of the manning of coast fortresses can no longer be governed by abstract principles, but that each particular case should be considered on its merits . . .

[31] *Colonial Defence, 'Defence of Colonial Ports', memorandum prepared by the Colonial Defence Committee and approved by the Committee of Imperial Defence, signed by Clauson*
2 August 1905

'Confidential'

From time to time the Colonial Defence Committee have drawn up, for the information of colonial Governments, estimates explaining some of the strategic principles which have been adopted by His Majesty's Government in determining the application of the naval and military resources at their disposal to the defence of the Empire, and which have therefore guided the Committee in their recommendations on questions of local defence referred to them for consideration.

In the present Memorandum, after quoting in paragraph 2 the unchanged general principles formulated in their Memorandum No. 57 M, dated the 19th May, 1896, the Committee propose to explain certain changes in the classification of defended ports abroad which recent measures of naval redistribution have entailed . . .

2   The maintenance of sea supremacy has been assumed as the basis of the system of Imperial defence against attack from over the sea. This is the determining factor in shaping the whole defensive policy of the Empire, and is fully recognised by the Admiralty, who have accepted the responsibility of protecting all British territory abroad against organised invasion from the sea. To fulfill this great charge, they claim the absolute power of disposing of their forces in the manner they consider most certain to secure success,

and object to limit the action of any part of them to the immediate neighbourhood of places which they consider may be more effectively protected by operations at a distance.

The resources of places which, in the opinion of an enemy, would justify the very considerable risks which a raid on them would involve, are generally sufficient to admit of the provision of local defence by local means, and where the liability to attack and the resources to resist attack coexist, it has been held to be the duty of the Colony to make provision for adequate defence.

3 The peace distribution of the fleet is determined primarily by the strategic requirements of war. The recent comprehensive review and readjustment of this distribution has been carried out, partly in view of alterations in the strength and disposition of foreign fleets, but mainly in order to satisfy modern naval conditions resulting from the evolution of war vessels and the improvement of telegraphic communications throughout the world. The new conditions involve a large measure of concentration of naval force in peace, and a higher degree of preparedness for war. Each fleet or squadron will, therefore, as far as possible, be kept together, ready at any time for instant action in any required direction.

One consequence of this redistribution has been a reduction in the number of naval bases which it is necessary to maintain permanently in time of peace. Gibraltar is now the permanent base of the Atlantic fleet, with its affiliated cruiser squadron. Malta remains the base of the Mediterranean fleet and the attached cruiser squadron. Hong Kong will in war be the primary base of the Eastern fleet, which in peace is distributed among the China, Australia and East Indies stations. Simon's Bay affords the facilities for refitting and repairs required by the Cape of Good Hope squadron. At Bermuda the cruisers employed in the Atlantic find certain facilities for repair when needed, although the ordinary refitting of these vessels takes place elsewhere.

Coast defences and garrisons are required to protect these naval dockyards and establishments, on which our warships depend for munitions of war and repairs, against sudden capture or destruction by such naval and expeditionary land forces as may temporarily evade our naval forces. The strength and character of the attacks to be provided against vary at the different naval bases, according to the proximity and resources of the nearest bases of possibly hostile Powers, and the troops there available. From time to time alterations in the strength of foreign fleets and changing political combinations may affect the possible scales of attack. The object of the coast defences is to deter attack by a hostile fleet not supreme at sea, and therefore not in a position to risk serious loss of fighting efficiency. Such defences must therefore be strong enough to be able to inflict substantial damage upon a squadron suddenly attacking them; but they are not required to sustain a deliberate duel between forts and ships for a prolonged period.

Naval bases within striking distance (300 miles) from a possibly hostile torpedo craft station should be provided with boom defences, guarded by quickfiring guns and electric lights, to close the entrances to the special portions of the harbours required for the use of His Majesty's ships.

The strength of the garrison of naval bases is governed by the consideration that in no case could a greater attacking force than a few thousand men be collected and conveyed over sea without such arrangements and preparation as would bring the operations under the category of an 'organised invasion' which the navy has undertaken to prevent.

4   In addition to the permanent naval bases which are required as main sources of supply, a fleet operating on the offensive – as the British fleet must always do – may require advanced bases of a temporary and auxiliary nature to enable it to keep close touch with the enemy. This necessity arises from the fact that the systematic transhipment of coal, ammunition, and other supplies constantly needed by a fleet, is an operation requiring that ships shall anchor in smooth water. The distance of the permanent bases may be inconveniently great for the purpose. The essential qualifications for these temporary bases will, therefore, be suitable position, sheltered anchorage, and security from attack by the enemy.

A suitable position may, in some cases, be found in the nearest British fortified commercial port (see paragraph 5 below), which will also afford a sheltered anchorage and a certain measure of protection from attack. Such positions, however, may not be the most advantageous, and the occupation of an unfrequented natural harbour – such as that utilized at the Elliott Islands by the Japanese fleet off Port Arthur – has the advantage that secrecy can more easily be maintained. The continued proximity of the fleet itself will constitute the chief safeguard of such a base, and any local defences – especially against attack by torpedo craft – will, when required, be organised by the navy. In the case of a fortified commercial port the scale of armament to be provided is not affected by the possibility of the utilization of the port as a temporary naval base.

5   Certain commercial ports require fixed defences, but, before discussing the objects and standard of these defences, it is necessary to consider briefly the extent of protection, direct and indirect, which will be afforded to British commerce and commercial ports by the navy. Such protection must be regarded as the main security of these ports from attack in war, and all defensive measures on land must be regarded as merely ancillary.

The first duty of British fleets and squadrons on outbreak of hostilities will be to seek to engage the enemy's naval forces with a view to securing the command of the sea. The vital importance of sea command is now so well understood that an enemy possessing a powerful battle fleet is unlikely to undertake organised attacks on commerce or commercial ports until an attempt at least has been made to cripple our naval power, for which latter purpose his cruisers are likely to be required in the first instance to act in conjunction with his battleships. Isolated attacks on merchant vessels met during the progress of some strategic movement may indeed occur, but regular attacks on commerce, if they take place at all, are more likely to develop during a later phase of the war, when the enemy's battleships have been reduced to inactivity and his cruisers liberated. The British naval reply to such attacks on commerce would probably involve extended operations with cruiser squadrons and single ships taking full advantage of the facilities afforded by our numerous commercial ports as coaling places and as centres for the collection and distribution of intelligence relating to the movements of the enemy. The hostile raiding vessels would in all probability resort to methods of evasion in order to inflict the maximum of damage on our trade with the minimum of risk to themselves, and, in order to avoid observation and opposition alike, they would probably frequent open water on the great trade routes rather than the vicinity of our coasts and commercial ports where their presence would be quickly reported. It is obvious, however, that the British naval forces could not, by their action alone, guarantee the absolute immunity of our merchant shipping from interference, nor prevent the possibility of hasty predatory raids on undefended ports if the injury capable of being inflicted would justify the risks involved.

Certain fortified commercial ports on frequented trade routes are, therefore, advantageous as harbours of refuge, where merchant vessels can, in case of need, seek protection from capture or molestation, and await a favourable opportunity of continuing their voyages. Fixed defences may also be required at great mercantile ports if valuable and vulnerable resources of national importance, such as docks and extensive plants of machinery, are collected within such a limited and exposed space that they might be seriously damaged by the gun fire of an evading [sic] cruiser if no defences existed. The object of fixed defences, in such cases, is to keep the enemy's cruisers at a distance.

The measure of defence required should be such that the risk of injury to the attacking cruiser would not, in the opinion of a naval commander, be justified by the possible advantages to be obtained. Thus, while the destruction of the naval resources at a primary base might exercise a considerable influence on the course of a war, the damage effected by a predatory raid on shipping and other private property at a commercial port could rarely be commensurate with the risks involved in coming under the fire of even light defences on shore. Foreign cruisers carry only about 100 rounds of common and semi-armourpiercing projectiles for their heavier guns, and it is reasonable to assume that no commander would care to expend a large proportion of this small stock in engaging forts when he might afterwards have to meet a British warship with his magazines depleted. To a naval commander, the strong probability of sustaining injury or even disablement to his ship, at a distance from a port where repairs could be effected, would constitute another powerful deterrent. It follows, therefore, that the strength of the armaments at fortified commercial ports should be governed, not so much by the remote contingency of having to carry on an engagement with raiding vessels, as by the necessity of deterring the latter from closing to positions from which they could deliver, with immunity to themselves, a deliberate and effective fire against British ships lying in a harbour of refuge, or against docks and other similar property of value at a great commercial port.

No special local defences against torpedo craft are required at commercial ports. Apart from the doubtful legality of sinking merchant vessels without at least warning and examination, and the difficulty of distinguishing neutral ships, it may be assumed that the mere destruction of a few merchantmen would not be considered a sufficient inducement for such an employment of torpedo craft. Dock and lock gates, where they exist, should however be protected by suitable means against injury by Whitehead torpedoes or explosive charges applied in any form.

6   Fixed defences are not in all cases required for the security of commercial ports. The inducements to attack the comparatively small amount of shipping likely to be found in ports remote both from the main trade routes and from hostile naval bases, or to bombard, may easily be outweighed by considerations of ammunition supply and by the knowledge that such attacks would reveal the position of the raiding vessels, which would probably find, in open water, better opportunities and greater security for their predatory operations.

In some cases, hydrographical conditions, such as the existence of a long, narrow, and easily blocked river approach, may prove an effective deterrent to raids which, owing to the indirect influence of naval operations, must, for the ships of a Power inferior at sea, necessarily be of a hasty and fugitive character.

On the other hand, cases may occur in which the configuration of an open harbour, and the depths of the water, render it useless to attempt to protect shipping lying at anchor by means of coast defences. It is obvious that no secure port of refuge exists if an enemy's ship can easily destroy vessels lying in a roadstead while herself beyond the effective range of shore batteries.

7   Even in the case of commercial ports where fixed defences are not required, the presence of an alert and well-organised, though not necessarily large, military force can afford effective protection to local interests. In the presence of such a force, skillfully disposed, the risks involved in landing men from raiding vessels are greatly increased. Without landing men, no substantial injury could be inflicted to property on shore, except by resorting to an expenditure of ammunition which may be regarded as prohibitive when no object of strategic importance is involved. A threat of bombardment, with a view to extorting an indemnity, might seem likely to serve the purpose of an enemy, assuming that British authorities could be found to comply with such a demand; but such a requisition should unhesitatingly be rejected. Surrender would entail more moral, if not material, loss to the place than would result from the few shells which a raiding vessel could afford to expend, and by yielding the supplies that it had been incapable of defending, a Colony would assist the enemy in inflicting damage elsewhere on British vessels or neighbouring Colonies. If on the other hand the requisition were refused, landing parties would have to be sent ashore to obtain payment. If the defenders were properly disposed, the use of the ships' guns against them would be ineffective, and the risks involved in

attempting a landing, opposed by a force intimately acquainted with the ground and armed with long-range rifles using smokeless powder, could scarcely be commensurate with the possible advantages to be secured.

Again, coaling could not be carried out in a port, nor could coal on shore be seized, until the defending force had been driven away, and this would scarcely be possible without landing greatly superior numbers. Even boat attacks with a view to cutting out colliers or lighters lying close inshore would be difficult under the rifle fire of the garrison. To deny coal in this manner to an enemy, while preserving it for the use of British shipping, will, in time of war, be one of the chief duties of the local forces.

[32] *'Australia General Scheme of Defence, Note by the Secretary and Correspondence', Committee of Imperial Defence*
February 1906

Enclosure 1 in No. 6: Alfred Deakin, Prime Minister of the Commonwealth of Australia, to the Governor-General, Melbourne, December 28, 1905

'Secret'
3   With regard to the local Naval Forces, a Memorandum on the subject, that has been received from Captain Creswell, the Director of the Naval Forces, is forwarded for the information of the Committee. It may be pointed out that there is a strong public opinion in the Commonwealth in favour of some action in regard to local naval defence for ports, harbours, and coastal trade. It is desired that consideration be given to these matters, as the sentiment in favour of the development of the maritime resources of the Commonwealth is one which, in the opinion of the Government, deserves and will repay encouragement.

Enclosure 2 in No. 6: Reports by Captain Creswell, Naval Director, (a) in reply to questions asked by Minister of Defence as to the formation of an Australian Navy, Melbourne, 10 November 1905

What the Commonwealth should have in the way of a navy? Three cruiser-destroyers, sixteen torpedo-boat destroyers, and fifteen torpedo-boats, first and second class. Of course it cannot be expected that these vessels will be provided at once or in one year and the provision will be extended over a period of seven years at an average cost of 330,000£ per annum.

2   Estimated cost? Cost of vessels, 1,768,000£; maintenance of vessels, commissioned and in reserve during seven years, 532,000£; total 2,300,000£

3   Cost of up-keep? 120,000£ per annum in peace time, including an addition of 456 to the permanent forces and 466 to the naval militia.

4   What vessels it is proposed to get first? Four torpedo-boat destroyers and four first class torpedo-boats.

5   What vessels at present in commission could be first dispensed with? *Cerberus* to be withdrawn from commission, and to be a depot for torpedo-boat crews within the Heads. Queensland gunboats to be resurveyed, withdrawn from commission, and relegated to such service as may be deemed suitable. *Protector* to be resurveyed and probably used as a tender to gunnery school. This will provide a defence not designed as a force for action against hostile fleets or squadrons, which is the province of the Imperial fleet, but as a line necessary to us within the defence line of the Imperial fleet – a purely defensive line, that will give security to our naval bases, populous centres, principal ports, and commerce.

[33] *Deakin to Northcote, copy*

26 April 1906

The recent correspondence between the Imperial and Commonwealth Governments on the subject of the substitution of the H.M.S. *Powerful* for the H.M.S. *Euryalus* as flagship on the Australian Station has once more directed public attention to the manner in which the Naval Agreement, to which the Commonwealth is a party, is being carried out, and I now have the honour to request that Your Excellency will be so good as to invite the consideration of the Imperial Government to the following circumstances.

2   Our Act was assented to on the 28th August, 1903. A similar measure was passed by the New Zealand Legislature on the 12th November 1903. On the latter date the Agreement was complete,

and it only remained for the parties to it to fulfil its terms. What was expected from the Commonwealth was evidently the most prompt discharge of the whole of its undertaking. That obligation, however, does not seem to have been considered reciprocal.

3   Even before the Act was passed by our Parliament the Admiralty made enquiries in respect of the payments to be made under it. After the measure had become law they took full advantage of article 11 pressing that payment should be made as from the 28th August, 1903, notwithstanding that the Agreement, which was tripartite, remained incomplete until the passing by the New Zealand Legislature of their Act on the 12th November 1903. The Commonwealth have duly paid their share as prescribed by article 8 since the latter date. Nothing on their part has been delayed, altered, or left undone.

4   Attention is invited to the manner in which the Admiralty have performed their part of the contract, which came into force on 12th November 1903. No increase in the strength of the Squadron took place at that time. It was not until 1904 that the *Euryalus* arrived on this Station in pursuance of the Agreement and by way of a commencement to comply with it. In July of the same year the first of the 2nd class Cruisers reached Australia. The despatch of the additional 2nd Class Cruiser required by the Agreement as it originally stood was marked with even less promptitude than was evidenced in the case of the *Challenger*, for it was not until December 1905 – over two years from the date on which Australia had made payments on the increased scale provided by the Agreement that the *Cambrian* joined the Squadron.

5   In the meantime, in pursuance of a suggestion made by the Admiralty in January 1905 – fourteen months after the Agreement had been in force – the character of the Squadron was improved to some extent. The assent of this Government was granted to the proposed alterations within a few days from the receipt of the cable embodying the suggestion of the Lords of the Admiralty, but it was only quite recently that, by the arrival of the *Encounter* the Squadron was at last brought up to its full strength. Even in respect of the 3rd Class Cruisers, the arrival of three out of the five did not occur until November and December of last year.

6   Full particulars are not available in the records of this Government concerning the fighting and navigable qualities of the ships composing the Squadron but considerable public disquietude has been caused by reports of the frequent mishaps which have

befallen them, both on their way to Australia and since they have been in these waters. For instance the 'Encounter' does not appear to be fitted for a voyage of any length.

7   It is not the desire of this Government to continue the correspondence with respect to the fighting qualities of the *Powerful* and the *Euryalus*. Whatever these may be, it is clear that a departure was made from the terms of the Agreement without the consent of Australia and New Zealand having been invited to changes in the most important unit of the Squadron.

8   A further instance in which the terms of the Agreement have been departed from without any reference to this Government has just been brought to my notice. Article 2 provides that the base of the force shall be the ports of Australia and New Zealand, and limits the sphere of operations to the Australia, China and East Indies Stations. That Article expressly provides that no change in that arrangement is to be made without the consent of the Governments of the Commonwealth and New Zealand but the Second Class Cruiser *Cambrian* has recently been detached from the Australian Station and detailed for service on the west coast of America, a locality far removed from the stations named in the Agreement.

9   It is therefore respectfully pointed out that the Agreement sanctioned by an Act passed by our Parliament has been interpreted differently by the Lords of the Admiralty accordingly as it related to themselves or to us. We were considered bound by the very letter of our promise to pay while they were free to supply the Squadron towards which that payment was being made by degrees at much later dates, with vessels of different denominations, some of them either possessing doubtful records, or acquiring them on their way here.

10   The change proposed in January 1905 constituted an improvement, but it is submitted that a review of the whole of the circumstances suggests that the conditions of the agreement have from the first been loosely accepted by the Admiralty, and that the Commonwealth is entitled to ask for their closer observance in future. This is especially desirable in the event of any proposal being made for a continuance or amendment of the present Agreement or for the acceptance of an agreement on similar lines hereafter.

[34] *Minute by Alfred Eyles, Accountant General of the Navy, on Australian Naval Agreement*

6 July 1906

The enquiries made prior to the passage through the Colonial Legislatures of the Acts approving the Naval Agreement were made solely with a view to ascertaining what sum might be anticipated to be received by the Admiralty in the financial year 1903–4, a matter having an important bearing upon the preparation of the Navy Estimates for that year. That reason was clearly stated in the telegram of enquiry dated 6th February 1903, from the Colonial Office to the Governor General of the Commonwealth.

[35] *Minute by Thomas, on Australian Naval Agreement*

9 July [1906]

The delay in completing the Squadron was owing in a great measure to the alteration in its composition, which was considered and decided on in the course of 1904, and which was carried out with the concurrence of the Commonwealth and New Zealand Governments.

[36] *Extract from letter from the Committee of Imperial Defence to the Colonial Office*

30 July 1906

'Confidential'

The Committee of Imperial Defence having been requested by the Australian Government to advise as regards the local defence matters, memoranda were prepared by the Colonial Defence Committee which were approved and forwarded to the Governor-General. The Memorandum dealing with the 'General Scheme of Defence' in Section I – 'Strategical Considerations', and in Section IV – 'Local Naval Defence', lays down the general principles of the naval defence of the Empire in accordance with the views of the Admiralty.

It appears most probable that the Australian Representatives at the Conference will raise the question of a local Navy, which has assumed a political complexion. In this case, I assume that the Admiralty, while adhering to the principles laid down in the Memorandum, would (as intimated by Lord Tweedmouth, Minutes of 88th Meeting) state that, if the Commonwealth

Government decided to maintain a separate naval force, they would be prepared to advise as to its nature, and as to the best means of obtaining and maintaining it.

The question of the working of the existing Naval Agreement will also doubtless be raised by the Australian Representatives. It, therefore, appears desirable that the Admiralty should be asked to prepare a Memorandum for submission to the Conference explaining the working of the Naval Reserve system up to date, and indicating any arrangements which might lead to the improvement of that system.

Such an explanation would have the effect of calling the attention of the Canadian Representatives to what is being done in Australia, and might lead to the establishment of a similar system in Canada, if the Admiralty consider that this would be an advantage.

A statement should also be prepared showing the whole of the Colonial contributions, direct or indirect, to the naval defence of the Empire.

A Memorandum dealing with the employment of wireless telegraphy in peace and in war should also be prepared, embodying the views of His Majesty's Government and indicating the principles which should be observed in Colonies in order that any extension of this system of communication may secure the maximum of Imperial advantage.

[37] 'Report of the Committee of Imperial Defence upon the General Scheme of Defence for Australia', Parliamentary Papers, Commonwealth of Australia, Senate
15 August 1906

IV – Local Naval Defence

19   The subject of the provision of local naval defence for ports, harbors, and coastal trade is discussed in a Memorandum (Commonwealth Parliamentary Paper No. 66 of 1905) by Captain Creswell, Director of Naval Forces, which is forwarded for the information of the Committee of Imperial Defence, who are requested to consider these matters as the sentiment in favour of the development of the maritime resources of Australia is one which, in the opinion of the Commonwealth Government, deserves and will repay encouragement . . .

The protection of Australian floating trade, whether on the high seas or in local waters, demands for its effective accomplishment, as explained in paragraphs 3 to 5, the closely concerted action of powerful sea-going ships. Localised vessels of the destroyer type could play no effective part in securing this object.

There is therefore no strategical justification from either point of view for the creation at great expense of a local force of destroyers – a type of vessel designed for totally different uses . . .

It may be added that the employment of a naval force as 'a purely defensive line' is a misapplication of maritime power opposed to every sound principle of naval strategy. To act deliberately on the defensive, and to organize naval forces with this object in view, is to adopt voluntarily the policy which is of necessity forced upon the weaker naval Power. Australia need not be reduced to assuming such a role so long as she is a member of an Empire which is the strongest naval Power in the world, and which extends naval protection not only to the home land and to the most distant of the King's dominions beyond the seas, but also to all commerce sailing under the British flag.

The policy of devoting the entire naval forces of the Empire to seeking out and destroying the ships of the enemy where ever they may be is that which will best insure not only the safety of floating trade, but also the immunity from attack of coast towns and harbours, and, if this policy is to be properly and efficiently carried out, the Royal Navy must be one and undivided. Unity of training and unity of command can alone insure that thorough co-operation which is essential. A separate Australian navy could not find in any effective organisation of the naval forces of the Empire a role commensurate with the cost of its creation and maintenance or worthy of the aptitude for sea service of the inhabitants of the island-continent . . .

[38] *'Report of the Director of the [Australian] Naval Forces',
Captain W.R. Creswell, C.M.G.*

c. August 1906

Australia's Need of Sea Force

76   In recent years, the Royal Navy has been trebled in strength, has been completely re-organised – one might say, revolutionised – to meet the newer conditions of war.

It is this spirit of advance and widening field of view which will bring about a better understanding of the Naval question in

Australia, and a change from the old attitude of negation to one of positive aid and encouragement. This is, I think, certain.

77   This great advance in Naval armaments and methods and the general standard to-day of Naval science show up in unfortunate contrast the effect on our own forces of the years of stagnation and retrenchment forced upon them since 1903.

78   Prior to that time Australian services were in line with Naval armaments and methods of their day in the Royal Navy. Since that time we have fallen steadily behind the standard of the day, and even our own of thirteen years ago, when the State Naval Forces were efficient and, placed as they were at the entrance doors to the States, were fair specimens of national efficiency. There is, consequently, much lost ground to recover.

79   The opportunities of this two and a half month's tour have, I hope, been availed of to the fullest.

The information gained has confirmed and strengthened the opinions set forth in my previous reports as to the defence value of torpedo craft and their fitness to meet the special conditions of our defence. These it may be well to re-state:

(a)   Against invasion or any expeditionary force we are protected by the Navy of the United Kingdom. We are also assured that no heavy squadron of enemy's cruisers will be permitted to assail or even approach our coasts.

On the other hand we are warned that no fleet, however powerful, can guarantee against some losses in war; that the raiding cruiser, in pairs or light squadrons, will attack commerce whenever and where ever this can be done with fair prospect of immunity, and against such loss there can be no guarantee by the Navy.

(b)   The exigencies of war and major operations require that the Australian Division of the Pacific Fleet shall be away from the Australian Station.

At such time there will be nothing between commerce of an annual value of £170,000,000, on which depends the whole business and industrial life of the Commonwealth; and it is open to the attack of the most insignificant enemy.

The meanest extemporised cruiser with a few guns is completely master of the situation, and can capture, destroy, or dislocate commerce; and it is attack of this description that is regarded as alone probable in war.

(c)   It is noteworthy to point out parenthetically that, with Estimates totaling approximately £1,000,000, 75 percent is expended on defences to meet attack against which we are guaranteed by the British Navy, 20 per cent. to meet attack against which we are partially guaranteed by the British Navy, and 5 per cent. for defence against attack which is regarded as most probable, and involves the whole business life of the State.

80   The defence by torpedo craft offers the following:

(a)   The most valuable auxiliary to the fleet stationed in these seas.

(b)   In the absence of the fleet the most effective substitute.

(c)   It prevents blockade of our ports.

(d)   It compels attack on our ports to be by daylight, when the most effective reply can be made by land defences.

(e)   It compels retreat beyond striking distance from ports to torpedo craft.

(f)   It keeps touch with an enemy, communicates intelligence of his position, thus preventing unnecessary cessation of traffic at parts beyond reach of attack.

(g)   It puts absolutely beyond possibility of attempt any landing from raiding cruisers near our ports.

(h)   The possibility of attack by destroyers from any of our shallow draught inlets or rivers on our coast would keep cruisers off our coastal tracks at night.

(i)   The night watchfulness enforced on a cruiser once within the sphere of destroyer attack is a strain that will not be endured except under necessity.

When the fleet is away from Australia for service elsewhere there is no effective provision for services (c) to (i) inclusive.

81   Besides meeting present requirements, the proposal put forward is the best preparation for future needs.

The Navy of the United Kingdom – our main defence – is, in Australia, 12,000 miles from all those building establishments, arsenals, construction works, as well as training establishments and schools that supply it with a trained personnel and material.

The Navy may be withdrawn to other seas.

To make up for this we should gradually create sea power in Australia. There could not, under our present conditions, be a better means of initiating such a policy than a torpedo defence as proposed.

82   The estimate put forward for a torpedo defence as supplemen-

tary to the Royal Navy was based on the data obtainable at the time.

There are some changes necessary, based on my later information. The destroyers best for Australian service are more powerful vessels and cost more. The same is the case with the first class torpedo boats. The second class boats, of a motor type, are not sufficiently advanced to be included.

There is an increase of £82,000 on the original estimate of the scheme – details attached. The whole scheme will include:

4 Ocean Destroyers,
16 Destroyers, 'River' Class,
4 First class Torpedo boats.

83   I beg to thank you sincerely for the opportunity accorded me of gaining the information and experience in England, which I trust will be of value to this country.

[39] *Admiralty Secretary [MacGregor] to the Commander-in-Chief, Australia, draft*

17 August 1906

If the Board of Admiralty were free to determine the class and numbers of the ships on the Australian Station, as they are in every other quarter of the globe, they would undoubtedly greatly reduce the present squadron, thus setting free ships and men for service in waters where they are more required. Even if this resulted in the reduction or abolition of the Australian Naval contribution M[y] L[ords] believe they would still be pecuniarily the gainers.

[40] *Admiralty Secretary [MacGregor], to Under-Secretary of State Colonial Office, draft*

25 August 1906

In signing the original Agreement, and in proposing a modification of the numbers and types of ships in January 1905 M[y] L[ords] did not contemplate as possible, or suppose that the Commonwealth Gov[ernmen]t contemplated, the immediate replacement of the existing Squadron by a Squadron of the type proposed. Even in home waters, where facilities for equipment and manning are close at hand, changes in the constitution of a squadron have to be gradual, and this necessity is accentuated in the case of a distant Station taking into account these difficulties of administration, M.L. do not think that undue delay can be alleged . . .

M.L. must express their regret that it was decided to employ the *Cambrian* temporarily in the Pacific Ocean outside the limits of the Australian and China Stations as laid down in the schedule of the Naval Agreement, without first communicating with the Commonwealth Gov[ernmen]t. But at the same time they desire in explanation to set out the following facts. At the time when the Agreement was signed the whole sea surface of the globe was parceled out into local Stations of the British Navy. At the present time this system has been definitely abandoned and the station limits (as for instance the line between the Australian and China Stations) are merely retained as a matter of administrative convenience. No part of the American coast is included in a so-called Station. The policy of the Admiralty Board is that only occasional visits of H.M. Ships shall take place on both the Atlantic and Pacific Coasts of the New World, and when it was desired this year that the Pacific Coast should be visited by two ships in company they were naturally detached from the Eastern Fleet. In view of the chronic uncertainty of affairs in China M[y] L[ords] desired to avoid any undue weakening of the Naval Force in Chinese waters, and had recourse to the Australian Squadron of the Eastern Fleet for the second vessel, without foreseeing that thereby they would be held to be transgressing the Naval Agreement.

8   So far M[y] L[ords] have dealt specifically with the various points raised in Mr Deakin's letter, but they are aware that there lies in the background a more serious question still, and that is whether the three parties to the Naval Agreement are or are not satisfied with it. M[y] L[ords] have already, in commenting on another despatch of Mr Deakin, had occasion to dissociate them-selves from any expression of dissatisfaction with its terms. But this further indication of something resembling dissatisfaction on the part of the Commonwealth Gov[ernment]t leads them to state their position in regard to it somewhat more explicitly.

9   The section headed 'Strategical considerations' in the 'General Scheme of Defence' for Australia recently drawn up by the Col-onial Defence Committee, endorsed by the Committee of Imperial Defence, and forwarded to the Commonwealth Gov[ernmen]t, indicates the views held by H.M. Gov[ernmen]t on the principles of Australasian Defence which must lie at the root of any Naval

Agreement. These views would not of themselves justify the retention in Australasian waters of a Squadron of the present size, and it may at once be admitted that if the Board of Admiralty were free to determine on purely strategic grounds the class and numbers of ships on the Australian Station, as they are in every other quarter of the globe, they would not maintain the Squadron at its present strength, but set free some of the ships and men for service in waters where they are more required.

10   M[y] L[ords] recognize however that the Naval Agreement is of great value in so far as it gives the fullest expression practicable at the moment to the principle of the unity of Sea Power, and also on account of other more purely political considerations with which the Admiralty Board is not immediately concerned. They are prepared loyally to carry out any obligations which they or their predecessors have incurred, and they will always try so far as possible to meet the views of the Commonwealth and New Zealand Gov[ernmen]ts, as regards the fulfillment of those obligations.

[41] *Lord Plunket, Governor of New Zealand, to the Earl of*
Elgin and Kincardine, Secretary of State for the Colonies
7 September 1906

3   It is not necessary for me to go into details, but the present Colonial subsidy is not likely to be much increased. It is of little monetary assistance; it leads to criticism of what England supplies Australasia in return for the financial aid given; and it compels the Admiralty to keep a Squadron in Australasian waters, which is hampered by certain stipulations and less useful to the Empire than such provision as the Admiralty would otherwise make for these seas . . .

7   Could not the colonies of Australia, New Zealand and Canada combine to build a class of commercial ships, which would be of real use both as unarmoured fast cruisers in war and speedy mail steamers in peace? Is it not possible to design a passenger vessel, which would have a high rate of speed, carry in the hold or have at certain ports guns which could at short notice be mounted, and be manned by a considerable percentage of men who could shoot and have the training of the Royal Naval Reserve? . . .

[42] *'Australian Maritime Defence', The Brisbane Courier*
21 September 1906

The decision made to form the nucleus of an Australian navy is the most important step that has been taken in the assertion of nationality since the inauguration of the Commonwealth. There is here no disparagement of the magnificent defence which the British navy has been to the Australian colonies, nor any disparagement of the protection which that navy will continue to give, but Australians owe it as a duty to themselves, as well as to Empire, to make some provision for their own naval forces other than a monetary contribution to a general Imperial service. It is nearly twenty years since the first agreement was entered into with the British Admiralty for the formation of an Australian Auxiliary Squadron; but to-day Australia is practically in the same position as then, having no modern war vessels of her own. Yet the agreement was intended by Sir George Tryon to be the first step towards an Australian navy, the vessels of the Auxiliary Squadron being used as a means of drilling and training Australian seamen. The vessels, however, were not commissioned as training ships but as regular ships of war with station duties, being mostly kept at Sydney; and in the new agreement entered into the provision for including Australian seamen in the auxiliary vessels is no more an acceptance of our national responsibilities than would be the enlistment of a few Australian recruits in the Imperial army. The Commonwealth Government have wisely decided to ignore the recommendations of the Colonial Defence Committee against the acquisition of a local navy; though it might have saved a good deal of misunderstanding if these recommendations had never been solicited. At the same time it is interesting to note that the opinion of some of the British defence experts is gradually veering round to the approbation of an Australian navy, the true position of the Commonwealth never having been better stated than by Admiral Fitzgerald. At present, he said, none of the colonies add anything to the strength of the British navy. 'They do – those that subscribe – relieve the British taxpayers by the very small amount of their donations. Yet the fact remains that our navy would be exactly the same strength as it is at present, whether the colonies subscribed or not. Whereas, if they were now to start navies of their own, on ever so small a scale, they would in a few years add materially to the maritime strength of the Empire.' . . . One important step has been taken, which will receive general approval, in relieving Captain Creswell of the local command in Victoria so that he may devote the whole of his time to the development of the Australian naval scheme. This is a recognition of services whose value can only be estimated by those who have had an opportunity of following the work he has done both as State and Federal Commandant in combating the prejudices of the Admiralty and in pleading for the recognition of Australian nationality in providing manhood and naval skill as well as money for local maritime defence. It is largely due to his representations that men like Admiral Fitzgerald now favour an Australian navy, and some of his influence may also be traced in the changing opinion of the English newspapers.

[43] *Minute by Langdale Ottley, 'Naval Questions for Discussion at the Forthcoming Colonial Conference; proposed answer to Lord Elgin'*
18 October 1906

The Board desire to take advantage of this opportunity to inform Lord Elgin in confidence that although in answer to questions raised by the Commonwealth Government in consequence of a misunderstanding of certain speeches of the late Naval Commander-in-Chief of the Australian Squadron, they have recently had occasion to dissociate themselves from any official expression of dissatisfaction with the Australian Naval Agreement, yet, nevertheless, at a time when a considerable section of the Australian public are demanding the denunciation of that Agreement, My Lords feel bound to point out (in confidence) that, so far as the immediate convenience of the Naval administration is concerned, the abolition of the Australasian Naval Contribution would give them great satisfaction. The chief motive that weighs with the Board in rendering them reluctant to give any sort of public expression to this view, is an earnest desire to promote to the utmost a cordial understanding between the Australasian Colonies and the Motherland.

[44] *Memorandum by Langdale Ottley, 'Australian Defence Policy'*
10 December 1906

Enclosed with the Commander-in-Chief's letter of 16th October (M 01430) are a series of Australian Parliamentary papers shewing the latest developments of the question of the naval and military defence of the Commonwealth. The following summary will perhaps be of interest to Their Lordships.

On receipt of the Report by the Committee of Imperial Defence on a 'General Scheme of Defence for Australia' the non-confidential portion of the report was published as a Parliamentary paper. At the same time two committees, composed, one of local naval officers and the other of local military officers, were appointed by the Government to consider the C[ommittee of] I[mperial] D[efence] report.

For a body of officers of the local Australian Navy to have recommended the abolition of the force from which they draw their livelihood, would have argued almost superhuman altruism on their part. So far as I can ascertain, the Australian Government has never held out any hope to these officers that (if the local Australian Navy is abolished) they will receive any sort of retiring allowance, pension or any pecuniary acknowledgment whatever of the fact that their occupation would thereby be gone. Surely, therefore, the inference is clear that, when the Australian Government called upon the Commandant of the local Australian Navy to report upon the C[ommittee of] I[mperial] D[efence]'s recommendation that that Navy should cease to exist, it was putting to Captain Creswell an unfair dilemma.

As might have been expected the committee of local naval officers reported adversely to the suggestion that the local navy should be abolished. Their report contains a lengthy and detailed criticism of the C[ommittee of] I[mperial] D[efence] report. Though specious and well calculated to take the fancy of a public very superficially versed in naval matters, the local naval officers' report is based on a narrow conception of naval strategy, and bristles with misleading statements and false deductions. Thus a speech by Mr Balfour advocating the use of destroyers for the defence of the British Isles is quoted as an argument for the creation of a similar force in Australia. The difference between the correct strategy which localises the employment of destroyers in British Home waters, where they are at all times within easy striking distance of the foci of naval activity of our potential enemies, and the strategical error implied in stationing vessels of this type in Australian Harbours 4,000 miles from the nearest foreign naval base, is to the 'local naval officers in Australia' apparently a matter of insignificance.

The report of the military officers has no special interest for the Admiralty, though it is worthy of notice that they question the power of the Navy to prevent the possibility of a raid on Australia on a large scale.

Other papers enclosed by the Commander-in-Chief contain in tabular form a summary of the various schemes both naval and military which have been proposed for the defence of Australia.

At present the Commonwealth Government has not announced the definite adoption of any of these schemes, though, in a speech delivered on 26th September 1906 the Prime Minister indicates that the local navy scheme will eventually be adopted without ceasing the contribution to the Imperial Navy. He states, however, that 'an election must intervene, and a new Parliament must be returned, before any appropriation can be made for giving effect to the scheme which I have outlined'. These elections are now in progress.

Apparently no good purpose would be served by any detailed criticism of these reports at the present stage. The Commonwealth Government have placed both sides of the question before the public, and it is for them to decide what course is to be adopted. Regarding these reports, therefore, it is submitted to take no action.

A submission dealing with the whole question of policy with regard to the Australian naval agreement will be put forward in due course.

[45] *Vice-Admiral Sir Wilmot Hawksworth Fawkes, Commander-in-Chief Australasian Squadron, to the Secretary of the*

Admiralty   *Powerful* at sea
12 January 1907

'Confidential'

In answer to Admiralty telegram No. 2 of 2nd January 1907, I have the honour to request that you will inform Their Lordships that:

(1) The Prime Minister of New Zealand [Sir Joseph George Ward] will probably be no party to a local Navy in any shape or form but will be ready to adhere to the present Naval Agreement with any necessary modifications in the Regulations. It is said that he will propose an increased contribution for a stronger fleet.

Should the question arise, he would probably not insist so strongly as Mr Seddon [former New Zealand Prime Minister] did on the constant presence of two ships in New Zealand waters, and would perhaps waive that Article of the Agreement altogether if Their Lordships were to order me to make New Zealand my Headquarters for a couple of months or more every year.

I shall have no ship in New Zealand waters during February and March this year, but hope to take some of the Squadron there in April. To this the New Zealand Government have raised no objection.

(2)   It is much more difficult to judge what Australia will do . . .

[46] *Hawksworth Fawkes to Northcote*
Powerful at Fremantle
4 January 1907

They [ie the Colonial and Imperial Defence committees] consider Destroyers and Torpedo Boats were not strategically necessary, and it is a pity to spend money in unnecessary objects, but if they are thought to give security to the residents of the Ports, confidence to the mercantile community, and to foster a growth of a maritime spirit in the people, the Governments concerned may no doubt think it better to provide them.

[47] *Memorandum by Langdale Ottley, 'Admiralty views on the working of the Australian Naval Agreement'*
27 February 1907

Prior to the Colonial Conference of 1902 a 'Memorandum on Sea-Power and the principles involved in it' was prepared by the Admiralty. As a statement of strategic principles this Memorandum cannot well be improved on and were it possible now to revert to the situation of 1902 before the Naval Agreement was concluded it does not appear that the first half of the Memorandum would require any alteration. The latter part of the Memorandum, however, which deals with the incidence of the financial burden of Naval Power on British and Colonial taxpayers and the comparative interest which Great Britain and the several Colonies have in the maintenance of the different squadrons stationed abroad, appears, in the light of fuller information and further consideration, to be a superficial and incomplete examination of a very complicated financial question which mars the effect of the admirable statement of the strategical problem with which the Memorandum opens.

The Agreement which resulted from the 1902 Conference has in spite of considerable administrative difficulties been loyally carried out by all parties and although the Admiralty are fully aware of its many defects they would not of their own initiative have proposed its termination or modification; but Australia having proposed that the terms be reconsidered at the coming Conference, it is now necessary that the Admiralty views should be fully and frankly stated.

That a single Imperial Navy under one control is the most efficient and economical means of maintaining the 'pax Britannica' on the high seas is now generally admitted and at first sight it would seem desirable that all parts of the Empire should contribute a just proportion of the cost of upkeep of the ships, and supply a due quota of the men required to man them.

The question however cannot be reduced to these simple terms; there are other factors which cannot be ignored.

Under present conditions ships can be constructed and manned far more cheaply and effectually in Great Britain than any other part of the Empire and it therefore follows that nearly the whole sum required for the maintenance of the material of the Navy is and must be expended in this country; to ask the Colonies to contribute their due proportion would therefore be unjust, and would practically amount to a tribute paid by them to the Mother Country.

Even supposing that the above objection to an equal incidence of taxation did not exist such a method of raising the necessary funds is quite incompatible with the principle of unity of control, which is considered to be absolutely essential to the fighting efficiency of the Navy.

There is practically no difficulty at present in manning the navy in peace or war from British sources and there is therefore no necessity to ask for Colonial assistance in providing either Active Service or Reserve men especially as the Colonial seamen are much more expensive, very far removed from the storm centre of naval danger and can never be trained so efficiently as the men enlisted at home. The only reason for the enrollment of Colonial R.N.R. men in the past has been the desire to foster Colonial interest in the navy and to develop the maritime instinct of the Colonies.

The economic objections to an Imperial Navy paid for by Imperial contributions and manned partly by Colonial seamen would seem unanswerable and it is thought that the common end of uniting the forces of the Empire against foreign aggression can be better attained by other measures which would not involve invidious financial transactions between the Colonies and the Mother Country.

The still recent experience of the South African War affords convincing proof that the Colonies will rally to the defence of the Empire in the hour of need and the help they then gave shows clearly that though unable to lend material naval aid they will be able to give invaluable military assistance to Great Britain.

There is not a little to be said for adopting the policy of a tacit agreement that the whole burden of Naval preparation for war should be borne by Great Britain alone in exchange for the assumption by the Colonies of the obligation to train sufficient military forces to assist the British Army in the protection of Imperial interests throughout the world.

There would be little objection to such a policy if we could be sure that none of the Colonies would ever under any circumstances desire to sever their connection with the Mother Country. While there is no reason to anticipate such a contingency it is not for Great Britain to assume the impossibility of it; the strength of the Imperial bonds lies in their being elastic and voluntary and it is inconceivable that, should Canada or Australia desire to withdraw from the Empire, such a wish would ever be opposed by force of arms. The Colonies desire, and rightly desire, to preserve their autonomy, and a purely dependent position in regard to naval power is incompatible with their healthy development as semi-independent states.

Australia has already a small nucleus of Naval Militia and desires to develop the local naval defence, other Colonies may follow her example and whatever opinions may be held in this country as to the fighting value of such local navies and the economic and administrative difficulties attendant on their development it is clear that on broad Imperial grounds the Admiralty having given their advice on the specific subject of naval strategy should not seek to further oppose the legitimate Colonial aspirations to naval power . . .

Notwithstanding the many objections to the Agreement of 1902 the Admiralty have always tried to make the best of it and are prepared, if necessary, to continue to fulfil their obligations until its expiration in 1918, but the Commonwealth Government having now proposed that its provisions should be re-considered, their Lordships would offer their opinion that the Agreement of 1902, being based on incorrect principles, is incapable of satisfactory modification. The Admiralty therefore propose:

1   That the Agreement should be terminated.

2   That the obligation to maintain any particular number of ships on the Australian Station should cease. Some undertaking on the part of the Admiralty would probably be required that Australian ports would be visited at intervals and this would be unobjectionable.

3   The present subsidies should cease, no subsidies being accepted in future to which conditions limiting the authority of the Admiralty are attached.

4   While adhering to their previously expressed opinion that Imperial interests can be most efficiently and economically served by 'one navy under one authority', the Admiralty would not oppose the legitimate aspirations of the Colonies to maintain naval forces under their own control.

5   The force of Colonial seamen now enlisted for service in the Imperial Navy and Colonial Naval Reserve to be amalgamated with the local Colonial forces; the Admiralty would continue at the request of the Colonial Governments to embark Colonial Seamen for training and service in any of H.M. Ships employed on the Australian Station. These men to be paid by Colonial funds, but when embarked to receive British rates of pay only, the difference between British and Colonial rates being treated as a deferred payment.

Colonial Seamen when embarked in H.M. Ships should be liable for service in any part of the world in peace or war, but should always be discharged at Australian ports on the termination of their period of service in the Imperial Navy, when they would revert to the Colonial force and be available for service in the local flotillas.

6   The Admiralty to further in every way in their power the establishment and development of the new Colonial navies viz, – by lending Officers and men as instructors, training Colonial seamen at the request of the Colonial Governments and affording them advice and help in ordering and obtaining material.

7   To discuss with Colonial Governments the functions and status of the Colonial Naval forces, and the possibility of utilising them to assist in the Pacific Islands patrol duties under the orders of the Naval Commander-in-Chief and for the survey of Australian waters.

Although Australia may be ready to agree to these proposals as furthering the provision of a naval force of their own, it is improbable they will be equally acceptable in New Zealand, which Colony does not desire to create a local Navy and is suggesting increased contributions to Naval funds; there will undoubtedly be some difficulty in finding a solution of the question which will be equally acceptable to all parties. Still if the two larger partners to the Agreement are decided on the desirability of terminating it, it will be difficult for New Zealand to insist on its being carried out in its entirety.

[48] *The Secretary of the Admiralty to the Under-Secretary of State of the Colonial Office, draft*

20 March 1907

M[y] L[ords] C[ommissioners of the] A[dmiralty] have had under consideration your letter of the 18th January last No. 30,965/B6 forwarding copies of two resolutions proposed to be submitted to the Colonial Conference by the Governments of Australia and New Zealand respectively on the subject of defence generally and especially on the Naval Agreement.

With reference to the suggestion that a memorandum might advantageously be prepared in the Admiralty dealing generally with the whole subject, which would form a basis for discussion by the Colonial Conference, I am to request you will represent to the Earl of Elgin that after careful consideration, M[y] L[ords] are of opinion that it would not be desirable to lay before the Conference any memorandum which might be considered as expressing the definite views of the Admiralty on this question. M[y] L[ords] are before all desirous that it should be understood that their chief object is to conform in every possible way to the wishes of the Commonwealth and New Zealand Governments, and they therefore propose to instruct the Admiralty representatives to confine themselves to an attitude of attention to the views which may be formulated by the Colonial delegates, without committing themselves to any opinions on the subject.

M[y] L[ords] trust that Lord Elgin will be able to concur in this line of procedure, which they are of opinion is the one best calculated to achieve a satisfactory result.

[49] *Deakin to Admiral Sir John A. Fisher, First Sea Lord, copy,
with Fisher's annotations*

Melbourne
12 August 1907

'Personal'

The friendliness and frankness with which you were kind
enough to personally explain to me the Admiralty view[1] of Naval
Policy and its relation to the Squadron in Australian waters lends
me to take advantage of your courtesy again in this informal
fashion in order to assist towards a better mutual understanding.
It is probable that I shall have to put some questions to Lord
Tweedmouth officially at an early date but write to you now
unofficially because owing to the pressure upon me while in
London I had no opportunity of entering with him into the particulars involved in the proposal, as I was able to do with you. I shall
write him too but more briefly and from another standpoint.

1 No – it was my personal view which he asked me for.

At present I am supposed to be resting for recuperative purposes
and not to be concerned with business of any kind. This matter,
however, will not wait.

The situation is this. According to your judgment the present
subsidised Australian Squadron ought not to be continued.[2] Its
best ships should be removed and united with those of the Indian
and China Squadrons in one joint Eastern Fleet of powerful
vessels. If war broke out this would be done at once now under
the Agreement,[3] so that the concentrated naval force in those seas
might be brought to bear upon our foe wherever he might be
found; on our coasts; off Japan, or off Colombo. Consequently the
sooner our present Squadron can be merged in this joint Eastern
Squadron in time of peace, so as to be ready for war, the better.
The £240,000 subsidy paid by Australia and New Zealand does
not compensate you for its severance in time of peace from the
other two Squadrons now existing.[4] It would pay you to forego
the subsidy and get your best ships into that Squadron free from
the limitations imposed by your Agreement with us of 1903. While
we hold you to that bargain we are impairing the striking force
of the Navy in the East, instead of increasing it, as was formerly
supposed.[5] In the interests of Australia, if they can be considered
alone, the same course is necessary.[6] The best defence of this
country can be secured by a joint Eastern Squadron of powerful

ships operating wherever necessary.[7] Both the Empire and Australia are therefore losing instead of gaining by the present Agreement.[8]

2   Yes, this is my private opinion and always has been so.
3   Yes.
4   No, nothing like! Never was there such an extravagant waste of money, ships and men as this agreement entails on the Admiralty.
5   Yes.
6   Yes.
7   Yes.
8   This was my opinion expressed to Mr Deakin.

If the three existing Squadrons were consolidated so far as their best vessels were concerned the rest of the ships now on this station would be left as at present quite apart from any agreement. They would patrol the Pacific, conduct surveys, and make their present rounds as they do now. Their base would be in Sydney, where they would use all the accommodation you now possess. They would be seen there and elsewhere on our coasts as occasion required. In addition the new concentrated Squadron would visit us say once a year in order that our capitals, which are all on the seaboard, might be kept in touch with the British Navy. This would be the new order of things after the Agreement was canceled, and without any new Agreement being required.[9]

9   He grabs a lot in this paragraph, but on the whole it is what would be done.

Under these circumstances the Commonwealth would devote itself to the defence of its harbours and coasts. It would spend the sums advised by the Committee of Imperial Defence Report in protecting our harbours by shore works. It will in addition add local floating defences. You strongly urge submarines,[10] at each principal port; two at least in Sydney and Melbourne and one at each of the other capitals, together with some swift ocean going destroyers capable of patrolling our coasts.[11] Pending the building of the latter perhaps you could give us a couple of the best cruisers of the 'P' class that you are laying aside in the course of your reorganisation. We could man these for the time being with Australians now engaged in the Squadron if you thought fit to spare them for a fixed period. In any event whatever ships and men we have will be available in time of war in the event of an attack upon our coasts, in order to act with the concentrated Royal

Navy Squadron, or any part of it, in our own waters. I understand that these submersibles and destroyers would afford a very real help to your Squadron and be of great value from the point of view of Imperial Defence in these seas.[12] They and the harbour works, etc. would represent our naval contribution instead of the present subsidy.

10 Yes.
11 Yes.
12 Yes.

These submersibles and destroyers, built, manned, and maintained at the sole expense of the Commonwealth, would remain under the control of its Government.[13] Their distribution and movements would be entirely subject to that Government at all times, but its officers and men would either be engaged here under the same conditions as those of the Royal Navy or be obtained from the Royal Navy. They would serve on our local vessels for the usual term on this station, whatever it might be, and then pass into other vessels of yours to continue their training elsewhere so as to keep them, while here, up to at least an equal standard of efficiency to that required everywhere in the Royal Navy. They would remain members of that Navy in every sense, recruited and serving under its laws. They would be regularly inspected here by the Admiral or his deputies and be subject to Naval discipline with all the penalties and privileges associated with it. Australia would pay them while they were on this Station at Australian rates of pay, though of course they would accept the usual deductions necessary to continue their title to share in the Royal Navy Pension Fund. Preference would be given wherever possible in our vessels to Australian Officers and seamen as opportunity occurred. Our ships would fly the white ensign with the Southern Cross and be altogether Australian in cost and in political control as to their movements and stations. In everything else they would be part of the British Navy, the officers and men being simply seconded for fixed terms for service under our general control, but in every other respect indistinguishable from the men in the Imperial Squadrons here or elsewhere.[14] In time of war, in my opinion, they would be placed by the Commonwealth Government of the day directly under the Admiral commanding the Eastern Squadron, since he would be the highest naval authority in this part of the world. I doubt if any conditions would be imposed upon this transfer at such a time, but it must be clearly understood that any decision on these points must rest absolutely in the hands of the responsible Government of Australia when such an emergency arises.[15]

13   Yes during peace.
14   These details I told him would have to be carefully thought out, but on the whole I concur in all this.
15   Yes.

I should like to have the benefit of your closest criticism of my statement, of your own position, and upon this outline as a whole, which, but for the breakdown of my health, would have been submitted to you some weeks sooner. If pressed for time I may have to cable some questions to the Admiralty before long which this exposition of my views may help to make intelligible to you. Of course any suggestions or comments of yours will be very welcome. Although our intercourse in London was unfortunately brief your grasp of the whole position was so firm and comprehensive that I feel we cannot do better than help you to combine Australian and Imperial Defence in one.[16] They always must be one. We want the most effective ships and efficient men here with ample prospects of advancement to the latter when they merit it. We also want a flexible relation as intimate as possible between our Government and the Admiralty which shall encourage the development of our local defences to the fullest extent and in such a form as to supplement to the best advantage the Imperial Navy in our hemisphere . . .

16 I simply made no secret of my detestation of the Agreement.

The main principles laid down at the Colonial Conference were that:

(1)   Admiralty will not oppose the formation of a local flotilla of Submarines and Destroyers:
(2)   Some means of fusion must be provided between the local flotilla and the Imperial Navy. Otherwise the isolation of the former would be fatal to its existence, see especially page 475 (Mr Deakin's remarks):
(3)   The burden of disposing of the men entered under the present agreement must fall on Australia (see p. 470)
(4)   Admiralty must be unfettered in the disposal of the fleet except as regards the local Flotillas, see p. 180 and 481.

The effect of (2) is that Australia must maintain a proportion of the personnel for her local Flotilla in excess of actual requirements, a proportion which is continuously under training in the Royal Navy, the units of which will pass into the local Flotilla and so keep it at the proper pitch.

But there is no reason why the training of this proportion should be confined to the Australian Station. If this were so the Admiralty – as shewn above – would find themselves fettered in the disposition of their ships while the 'pitch' of the local Flotilla would suffer if the training was restricted to the Australian Station only.

There is also the difficulty, (which has been referred to in previous papers) that on account of colonial jealousies the patrol of the Islands should be independent of Colonial influence. This would not be the case if the ships engaged on the patrol were manned entirely by Australians.

The conclusion seems to be that Admiralty should not accept any arrangement restricting the sphere of action of the Imperial vessels in the manner proposed.

If any concession is made it might be in respect of loaning 2 P. Class Cruisers to provide for the Reserve but the Australians of the permanent service ought certainly to be liable for service anywhere.

It is quite possible that a P. Class Cruiser might be wanted at any moment in China much more urgently than in Australia, and if the present proposals were agreed to, not one of them could be moved.

[50] *Speech by Deakin*

13 December 1907

When asking you to make a far larger provision for naval development than has even been attempted here, we require to recollect at the very outset that we owe to naval power and to the British flag our freedom in, and ownership of, this territory, the power to retain it, the whole of our political liberties and social standards. The Commonwealth is governed by a policy appropriately termed that of a 'White Australia', because the 'white ensign' flies all round our coast. Withdraw that and peril would be instant . . .

The whole defence of the sea and its control is to be a matter for the British Government and the British navy. The defence of our shores is to be left to Australia, except that there may be a small flotilla of Australian vessels capable of being used by the Navy as a part of its squadron. That represents a political transformation. An equal transformation has taken place in the strategical policy of the Admiralty, which affects us most materially. The old doctrine, so far as I understand it, appeared to be that the strength of the mother country was to be asserted by the presence in every important sea of a separate fleet to patrol its waters and to maintain British interests against attack, so that whereever difficulty arose there would be a fleet in that particular portion of the globe prepared for duty. But in recent years the whole view of the Admiralty experts seems to have changed, and although there is still a certain amount of localisation of forces, the doctrine of concentration has been rapidly developed, and is now being acted upon all round the world. As honourable members are aware, the old fleets of the Empire in Europe and on the American coast – they have been withdrawn from the latter – have been massed. Second or third rate ships have been discarded to the scrap-heap. The most powerful vessels of the old squadrons have been brought together, in order, instead of having a separate fleet in every ocean, and on almost every coast, to have fleets commanding great areas, consisting of the most powerful vessels, expeditious, and heavily armed, which, when concentrated, are enabled to operate at any particular point with greater effect than was ever attempted before. That is a transformation, as I understand it, of the system of naval strategy which has a great deal of importance for us. There was foreshadowed, in 1903, a sphere of operations for the Australian Squadron enlarged by the addition of the India and China seas. We now know that the fleet usually in Australian waters would be centred in time of war in accordance with the policy of concentration. The best ships of our squadron would be united with the best ships of the India and China Squadrons, and they, operating together, would become responsible for any force anywhere in those three seas. When the first Agreement was sought to be made, in 1881, and was afterwards made in 1887, with the separate Australian Colonies, there was a demand by Australia for the protection of our local shipping by a special fleet, which, though Imperial, was to be in part paid for by us, and was allotted to our coast. That consideration largely disappeared under the Agreement of 1902, and would now, in accordance with present views, disappear altogether... While feeling that for every constitutional reason, any flotilla created and maintained by the Commonwealth must be under Commonwealth

control, I have grown more and more deeply to realise the risks
of our attempting to create a small force solely of our own, in
which the men and officers would have no hope for experience or
advancement except within its bounds. A small flotilla of that
description would remain a thing apart, not directly committed to
the high standards of the Imperial Navy. In the Imperial Navy, as
honourable members are aware, the men and the officers on every
station are changed at short periods. Elaborate provisions are
made to prevent them becoming hide-bound, sit-at-ease, indif-
ferent, or mechanical . . . I ventured, therefore, to attempt to find
a means by which we could get the whole benefit of connection
with the Admiralty and the Imperial fleet, sharing its standards,
its training, and its prizes, and yet maintain the Australian
character of our flotilla, and so made the suggestion which I now
summarise. Let our officers and men be engaged here, under the
same conditions as those of the Royal Navy, or be obtained after
they have served in the Royal Navy. Let them serve on our local
vessels for the usual term on this station, whatever it may be, and
then pass into other ships of the Royal Navy to continue their
training elsewhere. This would keep them, while here, up [to] a
standard of efficiency equal at least to that required everywhere
in the Royal Navy. They would remain members of that Navy in
every sense, recruited and serving under its laws. Their services
in our ships would count in the same fashion as upon similar
vessels in the Navy. They would be regularly inspected here by
the Admiral or his deputies, and be subject to naval discipline and
to all the penalties and privileges associated with such discipline.
Australia would pay them, while they were on this station, at
Australian rates of pay, though of course they would accept the
usual deductions necessary to continue their title to share in
the Royal Navy Pensions Fund. Preference would be given whereever possible in our vessels to Australian officers and seamen at
every opportunity that occurred. Our ships would fly the White
Ensign with the Southern Cross, and be altogether Australian in
cost and in political control as to their movements and stations.
In everything else they would be part of the British Navy, the
officers and men being simply seconded for fixed terms for service
under our general control, but in every other respect indistinguishable from the men in the Imperial squadrons here or elsewhere.
In time of war they would almost certainly be placed by the
Commonwealth Government of the day directly under the
Admiral commanding the Eastern Squadron, since he would be

the highest naval authority in this part of the world. I doubted in London, and still continue to doubt, if any conditions would be imposed upon this transfer at such a time, but it must be clearly understood that the decision on these points must rest absolutely in the hands of the responsible Government of Australia when the emergency arises ...

What I contend for is a constitutional, not a naval, principle. Even a desirable thing loses much of its desirableness when an attempt is made to force it on a self-governing community, having the right to choose its own path, and accept the consequences. It will be dangerous for the Admiralty to insist on a supremacy which, if misadventure befell, would place the whole responsibility upon them. The Government of the Commonwealth, representing the Australian people, is entitled in this, as in every other matter, to speak and act for them.

[51] *Deakin to Northcote*

Melbourne
24 December 1907

I have the honour to inform your Excellency that, in considering the new defence schemes of the Government, an outline of which was presented to Parliament on the 13th instant, your Ministers are of opinion that one of the most vital considerations is the maintenance of the personnel of an Australian flotilla in the highest attainable efficiency. Since the vessels owned by the Commonwealth must be few and small when compared with those of the Royal Navy, and the number of officers and men correspondingly limited, if our service were isolated their opportunities of promotion and for keeping themselves abreast of the latest naval development in tactics, mechanical appliances, and instruction must be comparatively very restricted.

2   The present proposals now awaiting discussion aim at securing the same efficiency as that of the Royal Navy by manning the flotilla as far as possible with Australians, who will be to all intents and purposes members of that navy. It is desirable that a preference in Commonwealth ships should be arranged for those born or reared within its borders, providing they qualify for vacancies. All those engaged would, while serving here, receive the extra allowance at present paid to Australians in the squadron on this coast. They would be relieved and replaced regularly at the end of three years or other term fixed by the Admiralty for the

crews of His Majesty's ships on the station. They would follow the same courses of training under precisely the same conditions.
3   By the end of the first commission it is hoped that a further body of seamen will have been recruited in Australia, so that the vessels of the flotilla might always contain a large proportion of Australians in their complements.
4   While in these seas they would be under officers of the same standing as those holding similar positions in the Royal Navy, and subject to disciplinary laws in all respects analogous to those to which they would be amenable while serving on His Majesty's ships in other parts of the world.
5   The crews of the Australian flotilla and the flotilla itself would therefore form in effect an additional branch of the Royal Navy, maintained without cost to the mother country, and preserving its standard of efficiency.
6   Ministers will be greatly indebted if the practical methods of giving effect to this tentative outline of an arrangement can be considered and commented upon by the Lords Commissioners of the Admiralty, in order to guide the Government of the Commonwealth in framing its naval policy of harbour and coast defence.

[52] *'War-Ships on the Great Lakes of North America',*
memorandum by the General Staff for the Committee of Imperial
Defence, signed by N.G. Lyttleton, Chief of the General Staff
17 January 1908

'Secret'
The General Staff desire to bring the question of war-ships on the Great Lakes of North America to the notice of the Committee of Imperial Defence.
Naval command of the Great Lakes, more especially of Lakes Erie and Ontario, would be of grave importance to Canada in the event of war with the United States, and recent increases in the number and power of United States' war-vessels in these waters have inspired some anxiety in the Dominion.
2   By the Rush-Bagot Agreement of 1817, the naval strength of each country is limited to:

1 armed vessel on Lake Champlain.
1 armed vessel on Lake Ontario.
2 armed vessels on the Upper Lakes.

The size of these vessels is limited to 100 tons and their armament to one 18-pr. gun. The construction of armed vessels on the lakes is forbidden.

This Agreement can be denounced by either party at six months' notice ...

9   The growth of the naval and military power of the United States, and the decision that the British Navy is absolved from responsibility for securing the command of Lake Ontario on the outbreak of war, has increased the importance of the Agreement from the British and Canadian point of view.

10   The United states appear to have three main reasons for wishing to amend or to abrogate the Agreement. In the first place they point out, and with justice, that the size of vessels is far greater now than when the Agreement was concluded ninety years ago, and that even if its principles are sound its conditions are completely out of date. This is an argument for amendment. In the second place they wish to take advantage of the ship-building facilities which they now possess on the shores of the Lakes, and to allow their builders there to construct minor war-vessels for the United States' Navy, not for use on the Lakes, but on the ocean. Thirdly, they want to develop reserves for their navy from the marine population of the Lakes. These two last proposals, if acceded to, would completely alter the Agreement in principle.

11   The balance of advantage under the Agreement of 1817 lies with the United Kingdom and Canada. It is undoubtedly out of date as regards the size of the war-vessels permitted on the Great Lakes, but any attempt to amend the Agreement in this direction will lead to requests being pressed for its modification, so as to allow the construction of war-vessels. Any alteration in this direction would be entirely in favour of the United States, with which Canada would be entirely unable to compete in ship-building. No restrictions which would be enforceable in practice could prevent the United States from having numbers of war-vessels on the stocks ready to be finished off quickly on emergency.

12   On the four Upper Lakes the preponderance of the United States in ordinary shipping is so great that command of these waters in the event of war must, at any rate at the commencement, fall to them. On Lake Ontario the Canadians predominate, and the retention of this predominance is essential to the defence of their country. So long, then, as the Canadians retain the control of the Welland Canal, between Lakes Erie and Ontario, and so long as the Erie Canal and its branches are not enlarged, additions to the strength of the United States on the Upper Lakes by

alterations in or by abrogation of the Rush-Bagot Agreement, however inconvenient, are not of vital importance. It is not likely, however, that Lake Ontario could be excluded from any modification of the Agreement, and in any case the United States could make such an exception valueless by enlarging the Erie Canal.

13 If no action is taken, and the United States persist in infringing the Agreement, the uneasiness felt in Canada must deepen, and lead to counter infringement, or refusal to permit the passage of these vessels through the Canadian canals. This could only lead to friction and to the acceleration of the construction and deepening of the United States' canals. Modification of the Agreement in the direction desired by the United States must inevitably lead to the same results. So also would abrogation, but in this case at least there would be no room for subterfuges and consequent recrimination. All three courses must lead to that competition in armaments which for all three countries it is most desirable to avoid.

14 The General Staff therefore considers that it is of great importance, in the interests of peace and in the military interests of Canada in the event of war, that the original conditions of the Agreement should be adhered to as closely as possible; that opportunity should be taken on the occasion of each addition to the naval forces of the United States on the Great Lakes to press this view on the United States' Government, while formally reserving to ourselves the right to make additions to our own forces up to the total strength of those of the United States.

The General Staff also believe that gradual modifications of the agreement are more likely to engender ill-feeling than its total abrogation. If, however, abrogation is to come, the responsibility for it should be thrown upon the United States.

[53] *Greene to the Under-Secretary of State for Colonial Affairs*
10 February 1908

I have laid before my Lords Commissioners of the Admiralty your letter of the 20th December, forwarding a further telegram, dated the 17th idem, from the Governor-General of the Commonwealth of Australia, on the subject of the proposed revision of the Australasian Naval Agreement . . .

With regard to the subject of the control of the local naval force in time of war, my Lords consider it of great importance that there should be no misunderstanding, and they had hoped that the discussion which took place at the interview between Mr Deakin and the representatives of the Admiralty on the 24th April last made the position sufficiently clear.

In time of peace the different self-governing dominions comprised in the British Empire have power to maintain and employ for harbour defence and for police purposes armed ships and vessels in their own waters; they have also a limited power to maintain and employ ships of war for wider duties, provided that an Order of His Majesty in Council is obtained as provided for in the Colonial Naval Defence Act of 1865 (28 & 29 Vict., cap. 14). In time of war the circumstances are entirely changed, and as under international law there is only one executive authority in the British Empire capable of being recognised by foreign States, Colonial ships of war cannot operate independently of the Royal Navy except to the limited extent referred to above.

The executive power of the Crown as the central authority of the British Empire must be applied as regards foreigners in the same manner and under the same conditions whereever a military or naval force is in existence, and the same responsibility for any action taken by Colonial or home ships of war will rest upon this central authority. It is essential, therefore, that officers commissioned by His Majesty should have full power and responsibility in accordance with their rank, whereever they may happen to be serving. Accordingly it follows that not only must the local force when associated with the Royal Navy recognise orders given by the Admiralty or Naval Commander-in-chief, but also that their officers must submit to the command of any senior naval officer during the time they are in company with him.

While, therefore, my Lords recognise the force of the contention as a general principle that the Government of a self-governing Colony should have power to control its own waters the movements of the local force it maintains, and that this force should not be moved away from Colonial waters without the concurrence of the responsible Government, yet in the conduct of operations of a warlike nature it would be of vital importance that the vessels should come under the general command of the Commander-in-chief on the station and be subject to his orders and directions.

It would be obviously impracticable for any defensive operations to be carried out satisfactorily without that close co-operation which unity of direction secures, and my Lords are glad to notice that Mr Deakin in his speech before referred to expressed the opinion that in the event of operations in Australian waters the Commonwealth vessels would in almost every imaginable circumstance be placed wholly under the control of the Commander-in-chief for the time being.

It was understood at the Conference that the Colonial Ministers recognised that it would be an advantage to the fleet to be able to rely upon the existence of a force of destroyers and submarines in the waters of distant parts of the Empire, but this advantage would be much diminished if the vessels are not in time of emergency to be placed under the command of the Commander-in-chief or Senior Naval Officer.

[54] *'Liability of Defended Ports abroad to Attack by Torpedo Craft', memorandum by the Colonial Defence Committee, signed by J.R. Chancellor, Secretary, Colonial Defence Committee*
June 1908

'Secret'
9   In open waters remote from the centres of naval power of belligerents ... the conditions are somewhat different. It might happen in the early days of a war that the stronger naval power in seeking to secure the ultimate command of the sea was unable to develop its naval strength in distant waters. The torpedo craft of the weaker power might therefore remain at sea by day in such waters without incurring serious risks of capture; and if they were able to find, within striking distance of a suitable objective, secluded harbours in neutral or hostile territory, the violation of which might not for some time be detected or prevented, operations at a considerable distance from their base in home territory might be undertaken by torpedo craft of the inferior naval power ...
16   After a careful review of the special circumstances affecting the position of each port, the Admiralty are of opinion that, under present conditions, the only British ports abroad at which special fixed defences against torpedo attack need be provided are Gibraltar, Malta, and Halifax.

[55] *Greene to Under-Secretary of State for the Colonies*
20 August 1908

With reference to Admiralty letters of the 10th February, and 29th May, 1908, I am commanded by my Lords Commissioners of

the Admiralty to acquaint you that they have made a careful
enquiry into the scheme proposed by the Prime Minister of the
Australian Commonwealth for the establishment of a local naval
force in substitution for the existing Naval Agreement . . .

It is estimated that the total cost of building and equipping six
destroyers, nine submarines, and two depot or parent ships will
amount to 1,277,500L, as shown in Appendix II, made up as
follows:

6 destroyers	£ 473,500
9 submarines	£ 496,000
2 depot ships	£ 308,000
Total	£1,277,500

The manner and conditions under which these vessels will be
constructed are left for future consideration, but their Lordships
will give all the advice and assistance that the Commonwealth
Government may desire, it being considered an advantage that
the details of construction and armament should correspond
with the general requirements of the Admiralty.

The annual maintenance of these vessels, including repairs,
stores, and depreciation will amount to 186,000£. It has been
assumed that the repairs will be carried out at local shipbuilding
yards, and will not involve any charges upon Sydney dockyard,
which will be reserved for Imperial purposes, as hitherto.

As regards personnel, the numbers required are estimated at
seventy-nine officers and 1,125 men, and the total annual cost,
including pay and allowances, victualling, &c, will amount to
160,000£ per annum as shown in Appendix III. It must be noted
that the cost includes half pay and retiring allowances of officers,
and pensions and gratuities of men, calculated on the assumption
that the pay will be precisely the same as in the Imperial Navy,
and particularly that gratuities on leaving and service pensions are
treated as an equivalent for the higher salaries and wages prevailing in Australia. The experience of the present Agreement has
convinced their Lordships that any attempt to combine a higher
rate of pay in Australia with the ordinary conditions of pay and
service prevailing in the Imperial Navy must be abandoned.

It is suggested that pensions to men for long service should be
awarded after twenty-two years, and disability pensions after less
service, under the conditions laid down in the King's Regulations,
and that gratuities after short periods of service in the fleet should
be payable on rules similar to those obtaining in the Royal Fleet

Reserve. Under these rules a man can obtain a gratuity of 50£ at the age of 40 after having served five years (or more) in the fleet, followed by service in the reserve consisting of five-year periods up to a total of twenty years. The former condition should be generally applicable to the skilled ratings, and the latter to the general service men, the time taken in training the skilled men required for destroyers and submarines precluding the adoption of an engagement terminable after five years only.

It will be noticed that the scheme as here developed will involve a larger charge upon the Commonwealth funds than that hitherto payable, but my Lords have reason to believe from the statements made by Mr Deakin that it will not be in excess of the amount he was prepared to pay to give effect to the scheme. The total annual charge as set forth is estimated at 346,000£, and even if this should be somewhat under the mark, it is not considered that a flotilla constituted on the lines desired by the Commonwealth Government could be provided at less cost, possessing, as it will, all the advantages of close connection with the Imperial Navy.

As previously stated, my Lords consider that the security from over-sea attack of the Empire generally[,] of which the Australian continent form an important part, is best secured by the operation of the Imperial Navy, distributed as the strategic necessities of the moment dictate. At the same time they recognise that under certain contingencies, the establishment of a local flotilla acting in conjunction with the Imperial forces would greatly assist in the operations of the latter. My Lords also recognise the importance, politically, of fostering a feeling of security among the inhabitants of the coast towns of the Commonwealth by the provision of a local force which will always be at hand. In the absence, therefore, of any direct contribution to the expenses of the Imperial Navy, my Lords will be ready to co-operate in the formation of such a flotilla, subject to a satisfactory understanding being arrived at in regard to the general administration of the force.

At the same time their Lordships cannot disguise from themselves the fact that the carrying out of the scheme will involve many difficulties, but it is hope that with a readiness of both sides to overcome them, a satisfactory arrangement may be concluded. Many more details remain still to be considered and settled if an agreement is arrived at on the general lines indicated above, such as the manner in which the scheme is to be brought into operation, the settlement of the financial details, &c.

My Lords will accordingly await a further expression of opinion from the Commonwealth Government upon the scheme generally before proceeding to consider such further details.

I am to add that their Lordships understand that the question of the position of the present local defence force and of the Royal Naval Reserves will be considered separately and independently of the scheme referred to in this letter.

[56] *Memorandum by the Judge Advocate of the Fleet, R.B.D. Acland, on the Disciplinary Provisions Contained in the Australian Defence Acts 1903–4, and Regulations made Thereunder*

November 1908

In reference to the letter of the Admiralty to the Colonial Office dated the 20th August, 1908, I venture to direct the attention of their Lordships to certain points which have occurred to me connected with the legal position of the officers and men of the Australian force with regard to discipline.

It seems to me, for the reasons stated below, that unless the disciplinary provisions of the Naval Defence Acts 1903–4 passed by the Australian Commonwealth Parliament and the regulations made thereunder are radically altered, a position of very great difficulty, and possibly even of danger, will be created with regard to the discipline of His Majesty's Navy . . .

Compare the position of the Commanding Officer of one of the Australian armed ships with that of the Commanding Officer of a vessel in the Royal Navy, who, under section 56 (2) of the Naval Defence Act, has power to try any offence not capital which is not committed by an officer and inflict a punishment up to three months' imprisonment.

These provisions make a serious inroad into the position of a Commanding Officer, and at the same time lower the position of officers generally, for they are made liable to be punished summarily by the Commandant or Commanding Officer for such offences as disobedience or neglect of duty, instead of being solemnly tried by court-martial.

. . . It is, however, worth while to direct attention to the more important matters in which the Australian Acts and Regulations made thereunder (which have the force of law) are inconsistent with those which obtain in the Royal Navy.

First and foremost the punishments which may be inflicted are very much reduced. Two instances will suffice. In the list of offences above referred to in (2) is 'causes or conspires with any other person to cause mutiny or sedition.' The maximum term of imprisonment is three months with or without hard labour; the same for No. (4) 'uses violence to superiors,' which would include knocking the captain down on the quarter deck. Next it is provided in the same section that every member before being reduced or dismissed for any alleged offence may, if he so request, be tried by court-martial. By section 95 the refusal to be sworn or answer, &c., may be punished by a fine not exceeding 100£, no power to imprison. By section 96 every person who is tried by courts-martial may be assisted in his defence by counsel, which, I suppose means counsel with the ordinary rights of his profession, e.g., cross-examining without the leave of the President, and addressing the court. If the offence charged be punishable with death, he shall be entitled to be defended by counsel at the expense of the Crown. This would include in time of war, where the force would be under the Naval Discipline Act, any case of a seaman striking a second class petty officer or marine striking a lance-corporal.

In view of the fact that the present proposal appears to be that the men of the Australian naval force shall sometimes be serving on ships of the Royal Navy, and sometimes on ships of the Australian flotilla, I submit that the provisions of the Australian Defence Acts and Regulations to which I have called attention require serious consideration.

Whether the maximum punishments provided in the Australian Code are sufficient to maintain the perfect discipline, submission to authority, and instant obedience, essential to the naval service, is a matter on which I am not competent to form an opinion. But it seems clear to me that the introduction of a considerable number of men of the Australian force accustomed to the disciplinary regulations referred to above on the lower deck of a ship of the Royal Navy could not fail to cause comparisons to be made between the powers of the captains and the maximum punishments which could be inflicted in one force or the other, to foster discontent with the existing order of things in the Royal Navy, and possibly even to lead in the long run to serious political differences.

Appendix IV: Opinion of Law Officers upon the subject of the Authorities of the Colonies of Australia arming and employing vessels for the

*service of Her Majesty's Local
Government in that Colony.*

Signed: J.D. Harding (Queen's Advocate), Richard Bethell (Attorney-General), W. Atherton (Solicitor-General), Robert Phillimore (Admiralty Advocate), and R.P. Collier (Admiralty Counsel and Judge Advocate of the Fleet)
21 December 1860

We are of opinion that the consent of the Crown to the Act of the Colony of Victoria, without an Act of Parliament authorising that consent, would confer upon vessels equipped under the Colonial Act the legal character of British vessels of war only within the limits of the territorial jurisdiction of the Colony; i.e., within 3 miles of the shores, but that beyond these limits such vessels would not be entitled to this character. Inasmuch as the Colonial Act clearly contemplates a naval force to be employed beyond the territorial waters of the Colony, we are of opinion that such an Act is ultra vires of the colonial Legislature, both because it trenches upon the prerogative of the Crown, and because the vessels of foreign States would not be bound to respect its provisions . . .

2 We think that all vessels of war in the Colonies intended to navigate beyond these territorial limits should be commanded by officers holding commissions from the Crown, and be essentially part of the Royal Navy of England. There can be no serious difficulties in the way of an arrangement between the Crown and the Colony upon this important subject.

It would not seem unreasonable that, in consideration of the Colonies supplying or contributing to the expenses of certain vessels of war, those vessels should be stationed on their coasts, and charged with the especial duty of protecting them.

And it appears to us that by some arrangement of this kind, into the details of which we do not enter, the wishes both of the colonists and of the Home Government would be best carried into effect.

We think it right also to point out another mode by which the same object might be perhaps more advantageously effected, viz., by the passing an Imperial Statute so framed as to confer on the Crown large discretionary powers of entering into such arrangements with the various Colonies as may in each particular case conduce to the end proposed, viz., the maintenance in each Colony requiring it of a naval force especially appropriated to the defence of that Colony, but forming part of the Imperial Navy of the Crown.

[57] *'The Defence of St Helena', memorandum by the Admiralty*
14 January 1909

'Secret'

1 The Admiralty have already expressed the opinion that the strategical requirements in the Southern Atlantic can only be satisfied by defending both St Helena and Ascension, as, if either remained undefended, a hostile commerce destroyer would be able to use the anchorage for coaling, either from supplies on shore or from a collier or prize. (*Vide* Appendix 2 to C.D.C. Memo No. 391 M)

2 They are not prepared to admit that the provision of defences at St Helena in order to deny the anchorage to a hostile commerce destroyer desirous of coaling there would necessarily involve the provision of defences and garrisons at many ports now undefended, as suggested by the Colonial Defence Committee in paragraph 16 of their Memorandum No. 391 M. The Admiralty would be the first to protest against the general adoption of the principle that defences should be provided for the negative purpose of denying the use of anchorages to an enemy.

They would point out that by slightly defending both St Helena and Ascension it is possible to deny to a hostile raiding cruiser the only coaling anchorages in an immense area, and so to render large tracts of important trade routes almost immune from the possibility of attack. This strategical advantage, if not absolutely peculiar to St Helena and Ascension, is at any rate not possessed in a remotely comparable degree by any other spot in the British Empire . . .

4 It is worthy of notice that both St Helena and Ascension are important telegraph stations on the Eastern Telegraph Company's line from England to the Cape via Madeira and St Vincent. Although the loss of St Helena and Ascension would not entirely isolate the Cape from England, it would cut the best and most direct route, since the alternative West Coast route touches at a great many foreign ports, and the East Coast route is longer. The defences render interruption of this important cable route far more difficult than would otherwise be the case . . .

[58] *G.B. Simpson, Officer Administering the Government of New South Wales, to the Earl of Crewe, Secretary of State for the Colonies*

26 March 1909

I have the honour to report, for your Lordship's information, that a representative Meeting of the citizens of Sydney was held in the Town Hall last night, and, according to the Press reports, the Meeting was largely attended and the utmost enthusiasm prevailed. At the Meeting the following Resolutions were unanimously passed:

1. 'That in the opinion of this Meeting of Citizens glorifying in the traditions of the British Race of which they are a part the time has arrived for the Commonwealth to take an active share in the Naval Defence of the Empire.'
2. 'That in view of the expressed determination of Britain's rivals to challenge her Naval supremacy Australia should present a Dreadnought to the British Navy as the immediate expression of her invincible resolve to stand by the Mother Country and take her place in the Empire's fighting line.'

An influential Committee was then appointed to carry into effect the above Resolutions.

2   Speeches in support of the Resolutions were delivered by leading citizens, including the Lord Mayor, the Right Hon. G.H. Reid, late leader of the Opposition in the Commonwealth House of Representatives, the Hon. J. Cook, the present leader of the Opposition in the House of Representatives, and many letters and telegrams of a patriotic character were received and read at the Meeting.

3   A diversity of opinion appears to exist amongst leading public men here as to whether it would not be more expedient to substantially increase the present Naval subsidy than to offer a Battleship. The Premier of this State is strongly in favour of renewing the Naval Agreement and increasing the present Subsidy, but considers the proposition to present a Battleship inappropriate, and many others favour this opinion.

[59] *Resolution of Canadian House of Commons*
29 March 1909

That this House fully recognises the duty of the people of Canada as they increase in numbers and wealth to assume in larger measure the responsibilities of national defence.

The House is of opinion that, under the present constitutional relations between the mother country and the self-governing Dominions, the payment of regular and periodical contributions to the Imperial Treasury for naval and military purposes would not, so far as Canada is concerned, be the most satisfactory solution of the question of defence.

The House will cordially approve of any necessary expenditure designed to promote the speedy organisation of a Canadian Naval Service in co-operation with, and in close relation to, the Imperial Conference, and in full sympathy with the view that the naval supremacy of Britain is essential to the security of commerce, the safety of the Empire, and the peace of the world. The House expresses its firm conviction that whenever the need arises the Canadian people will be found ready and willing to make any sacrifice that is required to give to the Imperial authorities the most loyal and hearty co-operation in every movement for the maintenance of the integrity and honour of the Empire.

[60] *Telegram from Simpson to the Secretary of State for the Colonies, received Colonial Office*
4 April 1909, 12.35 p.m.

With reference to offer of [a] Dreadnought [my] Minister asks me to send you following message (begins) 'Governments of New South Wales and Victoria are of opinion that inasmuch as Naval defence is under the constitution exclusively a matter of federal concern and in respect of any proposal for co-operation in Empire defence Australia should speak with one voice it is most desirable that Governments of New South Wales and Victoria should not take action till the opportunity has been afforded the Commonwealth Parliament when it meets to make a proposal to Imperial Government on behalf of all the States. Therefore if the Commonwealth Parliament resolves to make the offer of a Dreadnought to the Imperial Government[,] Governments of New South Wales and Victoria have agreed to contribute proportionately to the cost of the same. Despatch has been addressed to Prime Minister of

Commonwealth of Australia embodying above views. If however Commonwealth Parliament does not adopt this course Governments of New South Wales and Victoria have agreed immediately to take the necessary steps to obtain the authority of their respective Parliaments to share costs of a Dreadnought on a per capita basis (ends)

[61] *The Earl of Dudley, Governor-General of the Commonwealth of Australia, to the Secretary of State for the Colonies [Crewe]*

15 April 1909

Prime Minister of the Commonwealth has asked me to submit to your Lordship, for consideration of His Majesty's Government, the following memorandum of the question of Naval Defence:

'Whereas all the dominions of the British Empire ought to share in the most effective way in the burden of maintaining the permanent naval supremacy of the Empire.

And whereas this Government is of opinion that, so far as Australia is concerned, this object would be best attained by encouragement of naval development in this country so that people of [the] Commonwealth will become a people efficient at sea and thereby better able to assist [the] United Kingdom with men as well as ships to act in concert with the other sea forces of the Empire.

The views of the present Government, as a basis of co-operation and mutual understanding, are herewith submitted:

(1)   The Naval Agreement Act to continue for the term provided for;

(2)   The Commonwealth Government to continue to provide, equip, and maintain the defences of naval base for the use of the ships of the Royal Navy;

(3)   In order to place Australia in a position to undertake the responsibility of local naval defence, the Commonwealth Government to establish a Naval Force;

(4)   The Commonwealth Government to provide ships constituting the torpedo flotilla and maintain them in a state of efficiency, wages, pay, provisions, and maintenance of officers and men;

(5)   The sphere of action of the Naval Force of the Commonwealth to be primarily about the coast of [the] Commonwealth and its territories;

(6)   The administrative control of the Naval Force of the Commonwealth to rest with the Commonwealth Government. The officer commanding to take his orders from the Commonwealth Government direct, proper sequence of command by officers appointed by the Commonwealth being maintained. The forces to be under naval discipline administered in [the] same way as in the Royal Navy;

(7)   Whilst employed about the coast of [the] Commonwealth or its territories, whether within territorial limits or not, the vessels forming the Naval Force of the Commonwealth to be under the sole control of [the] Commonwealth. Should the vessels go to other places, the said vessels to come under the command of the naval officer representing the British Government, if such officer be senior in rank to the Commonwealth officer. Provided that, if it be necessary to send these vessels or any of them on training cruises outside the waters referred to, arrangements shall be made with the Lords Commissioners of the Admiralty through Naval Commander-in-Chief on the Australian station;

(8)   In time of war or emergency or upon a declaration by the Senior Naval Officer representing British Government, that a condition of emergency exists, all the vessels of the Naval Force of the Commonwealth shall be placed by the Commonwealth Government under the orders of Lords Commissioners of the Admiralty. The method by which the vessels shall come under the orders of the Senior Naval Officer would be by furnishing each Commander of an Australian vessel with sealed orders and instructions (to) the effect that upon the declaration to him by the Senior Naval Officer representing British Government that a state of war or emergency exists, such sealed orders shall thereupon be opened and, in pursuance of their provisions, he shall thereupon immediately place himself under the orders of the Senior Naval Officer representing British Government;

(9)   It is, however, to be understood that if the services of any of the Coast Defence vessels be desired in seas remote from Australia, the approval of the Commonwealth Government shall first be obtained to their removal;

(10)   To ensure the highest efficiency, the Lords Commissioners of the Admiralty to be asked to agree to the naval Commander-in-Chief on the Australian station making, at [the] request of the Commonwealth Government, periodical inspection of the vessels of the Naval Force of the Commonwealth, Naval School of Instruction, and Naval Establishment;

(11)   Lords Commissioners of the Admiralty to be asked also to approve of the service on the flotilla of such officers of the Royal Navy as may be mutually agreed to for service as instructors and specialist officers, and to receive officers of the local flotilla for instruction at the torpedo, gunnery, and other schools in the United Kingdom;

(12)   Lords Commissioners of the Admiralty to be asked to give opportunities from time to time for officers and men specially selected by the Commonwealth being attached to battle fleets or torpedo flotillas in European waters for special instruction, the expense to be borne by Commonwealth; and

(13)   For special facilities to be given, by arrangement with the Naval Commander-in-Chief on the Australian station, for the vessels of the flotilla being exercised in conjunction with the ships of the Royal Navy on the Australian station, subject to the command of such combined exercises being held by the Naval Commander-in-Chief of the Royal Navy on the Australian station.'

In concluding his memorandum, [the] Prime Minister assures me that Commonwealth Government would highly appreciate the receipt, at earliest possible moment, of the views of His Majesty's Government on the foregoing proposals.

[62] *'Memorandum, showing Certain Properties Proposed to be Transferred from the Admiralty to the Government of the Dominion of Canada and the Conditions upon which such Transfer is Proposed to be made', copy*

1909

1   Subject to confirmation by Act of the Imperial Parliament or Order in Council made thereunder the following properties as specified on the accompanying plans will be transferred from the Admiralty to the Government of the Dominion of Canada under the Conditions hereinafter stated, viz:

[Halifax]
The Royal Naval Dockyard and Hospital.
The Commander in Chief's House and Grounds.
The Recreation Ground and Cemetery.
[Esquimalt]
The Royal Naval Dockyard and Hospital.
The Naval Coal Stores at Thetis Wharf.
The Magazine Establishment on Cole Island.

The Royal Naval Recreation and Drili Ground with buildings thereon.

The Royal Naval Cemetery with Chapel building.

2   The transfer will be made subject to the following conditions, viz:

(i)   If the Dominion Government fail to maintain the above mentioned properties at Halifax and Esquimalt in a state of efficiency, or make any alteration in the buildings, wharves, jetties, &c. or in the present use of the sites: or if they fail to maintain the existing depth of water alongside the frontages of the properties, conveniences at least equal in character to those which exist at present shall be provided by the Dominion Government at the same port.

(ii)   The Dominion Government will arrange for the stocking of coal or other fuel at Halifax and Esquimalt in a suitable manner for the use of H.M. Ships and will allow their local representatives to take charge of it, the necessary arrangements being settled as occasion requires by the Admiralty and the Dominion Government.

(iii)   The Dominion Government will grant all facilities required by H.M.Navy, including user of workshops and appliances by men of the Fleet, whenever wanted, at any Government Establishments, of which the Dominion may now or in future be possessed, such facilities, with the exception of labour and materials, to be given free of cost.

(iv)   The Dominion Government will inform the Admiralty before carrying out any proposal which they may have in view to use the above-mentioned properties for other than Naval or Military purposes . . .

[63] *Paraphrase of a telegram from Dudley to the Secretary of State for the Colonies [Crewe], received Colonial Office*
19 April 1909

My telegram of 16th March transmitting the proposals of Australian Government as to Naval Defence Policy. I think it well to remind your Lordship that owing to the unsettled condition of politics here no great reliance should be placed on these proposals as a final expression of Australian opinion. The proposals have no Parliamentary sanction as yet and a change of Government might involve considerable modification of details. I am sending by mail despatch on subject.

I see that the London press have published extracts from scheme. Can you inform me by telegraph whether Imperial Government authorised publication?

[64] *Telegram from the Secretary of State for the Colonies to the Governors-General and Governors of Australia, New Zealand, Cape of Good Hope, and Newfoundland sent*

30 April 1909, 6 p.m.

The Prime Minister of the United Kingdom as President of the Imperial Conference has desired me to ask you to convey the following message to the Prime Minister of (the Commonwealth of Australia) (the Dominion of (New Zealand) (Cape Colony) (Newfoundland)).

'It will no doubt, be within your knowledge that on the 29th March the Canadian House of Commons passed a Resolution to the following effect: [see document #37 above] . . .

I understand that the Dominion Government propose that its Defence Ministers should come here at an early date to confer with the Imperial Naval and Military Authorities upon technical matters arising upon that Resolution.

His Majesty's Government have also before them recent patriotic proposals made by Australia and New Zealand, proposals most highly appreciated by the Mother Country, and demanding very cordial and careful consideration both as to principle and detail.

I desire, therefore, to commend to you the following important suggestion, namely, that a Conference of representatives of the self-governing Dominions convened under the terms of Resolution I of the Conference of 1907, which provides for such subsidiary conferences, should be held in London early in July next. The object of the Conference would be to discuss the general question of naval and Military Defence of the Empire with special reference to the Canadian Resolution, and to the proposals from New Zealand and Australia to which I have referred.

I assume that as the consultation would be generally upon technical or quasi-technical naval and military matters that the other Governments of the self-governing Dominions would elect to be represented as in the case of Canada by their Ministers of Defence, or failing them by some other members of the Government assisted by expert advice, but it is entirely for the

Government of /the Commonwealth/ New Zealand/ Cape Colony/ Newfoundland/ to decide the precise form of its representation.

The Conference would, of course, be of a purely consultative character, it would be held in private and its deliberations would be assisted by the presence of members of the Committee of Imperial Defence, or of other expert advisers of His Majesty's Government. I am addressing a similar message to the other members of the Imperial Conference.

I am strongly of opinion that an early confidential exchange of views between His Majesty's Government and the Governments of His Majesty's self-governing Dominions beyond the seas will be of the greatest mutual advantage, and I therefore trust that your Prime Minister and his colleagues will see their way to adopt the proposal.

(To Newfoundland only: At present juncture I presume your Prime Minister will suspend definite answer until the elections are over.)

(To Cape only: I recognise that at the present time the Government of Cape Colony in common with the other South African Governments which are contemplating the probability of early union may not be in a position to take an active part in such a Conference, but the absence of any representatives of the South African Dominions from its deliberations would be a serious detriment to the completeness of the Conference.

Please repeat this telegram to the Governors of Transvaal, Orange River Colony, and Natal.

I have informed them that I have communicated [a] message to you for Prime Minister of Cape Colony which is also for communication to other Prime Ministers in identical terms.)

[65] *Grey to Crewe*
Government House, Ottawa, 11 May 1909

'Secret'

I have the honour to inform your Lordship that I have had more than one long talk recently with Sir Wilfred Laurier on Naval matters.

Sir Wilfred Laurier holds that any money withdrawn at present, from the construction of much needed public works, would be a diversion from the best and most effective contribution which Canada can make to the strength of the Empire. He points to the contribution of $100,000,000 of public money to the Canadian

Pacific Railway, and asks whether the construction of this railway, which has made Canada, has not been a more effective contribution to Imperial Defence than ten Dreadnoughts would have been. There can be only one answer to this question.

Sir Wilfred's ambition is to complete the construction of two more Transcontinental Railways. He asks, will not these two new railways mean more to Imperial Defence than the number of Dreadnoughts purchasable with the public money put into their construction.

The desirability of constructing the Georgian Bay Canal without further delay is admitted, not only for development, but for strategical purposes. Sir Wilfred is unwilling at present, for prudential financial reasons, to embark upon this enterprise. The enlargement of the Welland Canal is also desired for the purpose of diverting to Montreal the Lake Traffic, which now finds its outlet through United States Ports. This work is also postponed, for similar reasons.

In the opinion of Sir Wilfred, the construction of these public works is the most effective immediate contribution Canada can make to Imperial Defence, and must rank in Importance prior to the building of Dreadnoughts. The population of Canada is upwards of 7,000,000. When the two new Transcontinental Railways, the Georgian Bay Canal, and the Welland Canal enlargement are completed, and when Canada has increased her present population of 7,000,000 to fifteen or twenty millions, Canada will be able to contribute $100,000,000 to Dreadnoughts, and will do so, if necessary. Any money diverted from the immediate task of building up the strength of Canada – even the appropriation of the small amount of $10,000 for assisting Messrs. Baldwin and McCurdy to continue their aerodrome experiments at Baddeck, Nova Scotia – is regarded by His Majesty's Canadian Government as a false move.

Sir Wilfred does not want Canada to be involved in any immediate contribution to a scheme of Defence outside such measures as may be necessary for the protection of her own shores. He is willing to do whatever may be considered necessary in this direction and with this object is sending the Ministers of Defence over to England to consult with the Imperial authorities. He does not wish to go further at present.

Although Sir Wilfred is opposed to the idea of holding a subsidiary Conference in July, his reply to your cable of April 30th is couched in terms which will enable the Canadian Ministers of

Defence to take part in it. Pressure of public opinion here, as well as in England, Australia and New Zealand, will probably make the co-operation of the Canadian Ministers of Defence with the Ministers of Defence of the other self-governing Dominions, in the July Subsidiary Conference, necessary.

Sir Wilfred, as you are aware, has as genuine and deep-seated a conviction as any member of the Peace Society on the question of armaments. He regards them with all the horror of a man who sees in them only the advancing shadow of impending National bankruptcies. He will not admit that there is any necessity for taking immediate action. He repeats that Canada will be ready to shed her last drop of blood and to spend her last dollar to maintain the naval supremacy of the Crown; but he will not do anything to prevent that supremacy from being challenged.

[66] *Memorandum for the 1909 Imperial Conference by Reginald K. McKenna, First Lord of the Admiralty*

20 July 1909

On the 16th March of this year statements were made on the growing strength of foreign navies by the Prime Minister and the First Lord of the Admiralty on the introduction of the Navy Estimates for 1909–10.

On the 22nd March the Government of New Zealand telegraphed an offer to bear the cost of the immediate construction of a battleship of the latest type and of a second of the same type if necessary. This offer was gratefully accepted by His Majesty's Government. On the 29th March the Canadian House of Commons passed a resolution recognizing the duty of Canada, as the country increased in numbers and wealth, to assume in a larger measure the responsibilities of national defence, and approving of any necessary expenditure designed to promote the speedy organisation of a Canadian naval service in cooperation with, and in close relation to, the Imperial Navy. On the 15th April Mr Fisher, the Prime Minister of the Australian Government telegraphed that, whereas all the British Dominions ought to share in the burden of maintaining the permanent naval supremacy of the Empire, so far as Australia was concerned this object would be best attained by the encouragement of naval development in that country. (On Mr Deakin succeeding Mr Fisher as Prime Minister a further telegram was sent on the 5th June, offering the Empire an Australian *Dreadnought* or such addition to its strength as may be determined after consultation in London.) . . .

2   If the problem of Imperial naval defence were considered merely as a problem of naval strategy it would be found that the greatest output of strength for a given expenditure is obtained by the maintenance of a single navy with the concomitant unity of training and unity of command. In furtherance, then, of the simple strategical ideal the maximum of power would be gained if all parts of the Empire contributed, according to their needs and resources, to the maintenance of the British Navy.

3   It has, however, long been recognised that in defining the conditions under which the Naval Forces of the Empire should be developed, other considerations than those of strategy alone must be taken into account . . .

4   The main duty of the forthcoming Conferences as regards naval defence will be, therefore, to determine the form in which the various Dominion Governments can best participate in the burden of Imperial defence with due regard to varying political and geographical conditions. Looking to the difficulties involved, it is not to be expected that the discussions with the several Defence Ministers will result in a complete and final scheme of naval defence, but it is hoped that it will be possible to formulate the broad principles upon which the growth of Colonial naval forces should be fostered. While laying the foundations of future Dominion navies to be maintained in different parts of the Empire, these forces would contribute immediately and materially to the requirements of Imperial defence.

5   In the opinion of the Admiralty, a Dominion Government desirous of creating a navy should aim at forming a distinct fleet unit; and the smallest unit is one which[,] while manageable in time of peace, is capable of being used in its component parts in time of war.

6   Under certain conditions the establishment of local defence flotillas, consisting of torpedo craft and submarines, might be of assistance in time of war to the operations of the fleet, but such flotillas cannot cooperate on the high seas in the wider duties of protection of trade and preventing attacks from hostile cruisers and squadrons. The operations of destroyers and torpedo boats are necessarily limited to the waters near the coast or to a radius of action not far distant from a base, while there are many difficulties in manning such a force and keeping it always thoroughly efficient.

A scheme limited to torpedo craft would not in itself, moreover, be a good means of gradually developing a self-contained fleet capable of both offence and defence. Unless a naval force – whatever its size – complies with this condition it can never take its proper place in the organisation of an Imperial navy distributed strategically over the whole area of British interests.

7   The fleet unit to be aimed at should, therefore, in the opinion of the Admiralty, consist at least of the following:

1   Armoured cruiser (new 'Indomitable' class), which is of the 'Dreadnought' type.
3   Unarmoured cruisers ('Bristol' class),
6   Destroyers,
3   Submarines,
    with the necessary auxiliaries, such as depot and store ships, &c., which are not here specified.

Such a fleet unit would be capable of action not only in the defence of coasts, but also of the trade routes, and would be sufficiently powerful to deal with small hostile squadrons should such ever attempt to act in its waters.

8   Simply to man such a squadron, omitting auxiliary requirements and any margin for reliefs, sickness, etc., the minimum numbers required would be about 2,300 officers and men, according to the Admiralty scheme of complements.

9   The estimated first cost of building and arming such a complete fleet unit would be approximately £3,700,000, and the cost of maintenance, including upkeep of vessels, pay, and interest and sinking fund, at British rates, approximately £600,000 per annum . . .

14   As regards shipbuilding, armaments and warlike stores, etc., on one hand, and training and discipline in peace and war, on the other, there should be one common standard. If the fleet unit maintained by a Dominion is to be treated as an integral part of the Imperial forces, with a wide range of interchangeability among its component parts with those forces, its general efficiency should be the same, and the facilities for refitting and replenishing His Majesty's ships, whether belonging to a Dominion fleet or to the fleet of the United Kingdom, should be the same. Further, as it is a sine qua non that successful action in time of war depends upon unity of command and direction, the general discipline must be the same throughout the whole Imperial service, and without this it would not be possible to arrange for that mutual cooperation

and assistance which would be indispensable in the building up and establishing of a local naval force in close connection with the Royal Navy. It has been recognized by the Colonial Governments that in time of war the local naval forces should come under the general directions of the Admiralty.

[67] *Paraphrase Telegram, Dudley to the Secretary of State for the Colonies, received Colonial Office*
22 July 1909, 7 a.m.

'Immediate and confidential'

It is desirable I think confidentially to inform you that it is my opinion that although [the] present Australian Government's attitude towards [the] forthcoming Naval Conference is one of good will, and although they are prepared to enter it with as open a mind as possible and in a friendly spirit to consider any recommendation it may make with regard to Australian action in naval defence yet I think there is a certain amount of uneasiness among Ministers with regard to its possible conclusions and especially with regard to Australian offer of the cost of a 'Dreadnought'. The establishment of a considerable flotilla of torpedo boats was proposed by the late Labour Ministry. This policy was popular among a considerable section of the people because it offered a visible result for expenditure. If, however, two million pounds of Australian money were expended in building a battleship, some fear that the result of this sacrifice would never be seen in Australian waters. I think therefore that if it was possible to formulate some scheme under which this money would be expended on a type of vessel or vessels that would be employed in South Pacific, the Australian Government would be relieved of considerable difficulty and would be able to harmonise [the] patriotic offer with local prejudice. If a cruiser squadron were inaugurated by the expenditure of the money from Australia and New Zealand for permanent employment in the South Pacific, it is possible that from time to time additions might be offered to what would be considered to a great extent to be a local force.

I give this information because I believe it represents [the] views of [my] Ministers under [the] present political circumstances, but as I have pointed out already in my despatches my opinion is that Australia and perhaps New Zealand would much better employ the money in building cruisers of their own, Imperial ships being left free and unrestricted by any conditions.

[68] *Summary of Result of Meetings of the Inter-departmental Conference on Status of Dominion Ships of War*

19 August 1909

Canada

The Canadian representatives explained in what respect they desired the advice of the Admiralty, in regard to the measures of naval defence, which might be considered consistent with the resolution adopted by the Canadian Parliament on the 29th March, 1909.

While, on naval strategical considerations, it was thought that a fleet unit on the Pacific as outlined by the Admiralty might in the future form an acceptable system of naval defence, it was recognised that Canada's double seaboard rendered the provision of such a fleet unit unsuitable for the present.

It was represented on the part of the Admiralty that it would be difficult to make any suggestions, or to formulate any plans, without knowing approximately the sum of money which Canada would spend. The Canadian representatives then suggested that two plans might be presented: one incurring an annual expenditure of 400,000£, and the other an expenditure of 600,000£, omitting in both cases the cost of the present fishery service and hydrographic surveys, but including the maintenance of Halifax and Esquimalt dockyards, and the wireless telegraph service, estimated at some 50,000£ a-year.

Taking, first, the plan for the expenditure of 600,000£, after discussion the Admiralty suggested that the Canadian Government might provide a force of cruisers and destroyers comprising four cruisers of improved 'Bristol' class, one cruiser of 'Boadicea' class, and six destroyers of improved River class. As regards submarines, it would be advisable to defer their construction because they required a highly trained and specialised complement.

The *Boadicea* and destroyers might be placed on the Atlantic side, and the 'Bristol' cruisers divided between the Atlantic and Pacific Ocean. The number of officers and men for this force of eleven ships would be 2,194, and the cost of the vessels suggested, including repairs and maintenance, interest and sinking fund on capital expenditure, and pay, &c., of personnel, at Canadian rates, would not, it was anticipated, exceed 600,000£ a-year.

If it was decided to limit the plan to an expenditure of 400,000£ a-year, the Admiralty suggested that one 'Bristol,' the *Boadicea*, and two destroyers should be omitted, in which case only 1,408 officers and men would be required. Two 'Bristols' would then be placed on the Pacific, and one 'Bristol' and four destroyers on the Atlantic coast.

Summaries are attached to this memorandum, giving the estimated details of the expenditure at British rates.

Pending the completion of the new cruisers, which should be commenced as early as possible, an arrangement might be made for the loan by the Admiralty of two cruisers of the 'Apollo' class so that the training of the new naval personnel might be proceeded with at once. The vessels would be fitted out and maintained at the expense of Canada, and the officers and men provided by volunteers from the Royal Navy, but paid by the Canadian Government. They would be lent until they could be replaced from time to time by qualified Canadian officers and men. The Admiralty would be willing also to lend certain officers for organising duties and for the instruction of seamen, stokers, &c.

Arrangements would be made to receive Canadian cadets at Osborne and Dartmouth.

In any consideration of the question of providing new docking facilities the Admiralty suggested that the docks should be designed of sufficient size to accommodate the largest ships, whether for war or commerce, as apart from the mercantile advantage such docks might be used in case of an emergency by armoured cruisers and battle-ships. Docks of this kind might be placed on the Pacific, the Atlantic, and the River St Lawrence.

The question of the flag also was discussed, and it was arranged that the Admiralty would give the matter consideration and would communicate its views at a later date to the Canadian Government.

Any necessary Acts of Parliament which would have to be passed should be considered so as to place the discipline and general regulations of the naval forces as much on Admiralty lines as possible, having due consideration to local requirements. The legislation should also provide for the formation of a naval reserve and naval volunteer force.

In order to encourage a good class of men to make the naval service their profession for life, it has been found advisable in Great Britain to provide for pensions.

Other details, such as the training of officers and men, organisation, discipline, &c., were discussed, and a general agreement was arrived at that the wishes of the Canadian Government would be met as far as possible.

Australia

The suggestions made in the Admiralty Memorandum formed the basis of the discussion and the following arrangements were provisionally adopted, viz.:

Australia should provide a Fleet unit to consist of:
One armoured cruiser (new 'Indomitable' class).
Three unarmoured cruisers ('Bristol' class).
Six destroyers (River class).
Three submarines (C. class).

These vessels should be manned as far as possible by Australian officers and seamen, and the number required to make up the full complement for immediate purposes should be lent by the Royal Navy.

In peace time and while on the Australian station this Fleet unit would be under the exclusive control of the Commonwealth Government as regards their movements and general administration, but officers and men should be governed by regulations similar to the King's Regulations, and be under naval discipline, and when with vessels of the royal Navy, the senior officer should take command of the whole. Further, when placed by the Commonwealth Government at the disposal of the Admiralty in war time, the vessels should be under the control of the Naval Commander-in-chief.

The Australian Fleet unit should form part of the Eastern Fleet of the Empire to be composed of similar units of the Royal Navy, to be known as the China and the East Indies units respectively and the Australia unit.

The initial cost of such a Fleet unit was estimated to be approximately:

One armoured cruiser (new 'Indomitable' class).	2,000,000
Three unarmoured cruisers ('Bristol' class), at 350,000£.	1,000,000
Six destroyers (River class), at 80,000£.	450,000
Three submarines (C. class), at 55,000£.	165,000
Total	3,695,000

The annual expenditure in connection with the maintenance of the Fleet unit, pay of personnel, and interest on first cost and sinking fund, was estimated to be about 600,000£, to which amount a further additional sum would have to be added in view of the higher rates of pay in Australia and the cost of training and subsidiary establishments, making an estimated total of 750,000£ a-year.

This annual cost should be disbursed by the Commonwealth, except that the Imperial Government, until such time as the Commonwealth could take over the whole cost, should assist the Commonwealth Government by an annual contribution of 250,000£ towards the maintenance of the complete Fleet unit.

The annual subsidy of 200,000£ under the existing agreement should be paid as heretofore by the Commonwealth to the Imperial Government up to the time when the existing Australian Squadron should be relieved by the new Australian Fleet unit.

When desired, officers and men of the Australian service might be sent for training and service to vessels and training schools of the Royal Navy and their places taken by officers and men of the Royal Navy who, with the approval of the Admiralty, should volunteer for service in vessels of the Australian Navy.

The dockyard, &c, at Sydney, should on the completion of the Fleet unit be handed over to the Commonwealth Government free of charge, on condition that it is to be maintained in a state of full and complete efficiency, and that it shall not be diverted from its original purpose.

The construction of the armoured cruiser should be undertaken as soon as possible, and the remaining vessels should be constructed under conditions which would ensure their completion, as nearly as possible, simultaneously with the completion and readiness for service of the armoured cruiser, which it is understood would be in about two and a-half years.

Training schools for officers and men should be established locally and arrangements made for the manufacture, supply, and replenishment of the various naval, ordnance, and victualling stores, required by the squadron. Until stores and munitions of war are manufactured in Australia the vessels of the Australian unit should be supplied as far as possible with stores, ammunition, and ordnance stores in the same manner, and at the same cost, as other vessels of His Majesty's Service.

Great stress was laid upon the maintenance of the same general standard of training, discipline, and general efficiency both in ships and officers and men.

New Zealand

The proceedings which took place at the various meetings are summed up in the form of a letter from Sir Joseph Ward to the First Lord of the Admiralty, and a reply from the latter, which are printed in full.

Sir Joseph Ward to Reginald
McKenna, 11 August 1909

At to-day's meeting you explained that the general idea underlying the Admiralty memorandum was that the present East Indies, China, and Australian squadrons should be treated strategically as one Far Eastern or, as you thought as a preferable term, Pacific station, and that each of the principal portions of this station should have a complete Fleet unit, the Commonwealth Government maintaining one unit in Australian waters in lieu of the present Australian squadron, and the Imperial Government providing the remainder, the *Dreadnought* cruiser presented by New Zealand forming the flagship of the China unit.

I think it will conduce to clearness if I state my views in writing, hence this memorandum.

I expressed myself as generally satisfied with this arrangement as a strategic plan, but I would point out that if, as I understand, Australia is providing an independent unit, it means the superseding of the present British Australian Squadron, and the fact of that being done would, on its completion, determine the Naval Agreement with Australia and New Zealand, thus creating an entirely new position.

I favour one great Imperial Navy with all the Overseas Dominions contributing, either in ships or money, and with naval stations at the self-governing dominions supplied with ships by and under the control of the Admiralty. I, however, realise the difficulties, and recognise that Australia and Canada in this important matter are doing that which their respective Governments consider to be best, but the fact remains that the alterations that will be brought about upon the establishment of an Australian unit will alter the present position with New Zealand.

New Zealand's maritime interests in her own waters and her dependent islands in the Pacific would, under the altered arrangements, be almost entirely represented by the Australian Fleet unit, and not, as at present, by the Imperial Fleet. This important fact I consider necessitates some suitable provision being made for

New Zealand, which country has the most friendly feeling in every respect for Australia and its people, and I am anxious that in the initiation of new arrangements with the Imperial Government under the altered conditions the interests of New Zealand should not be overlooked. I consider it my duty to point this out and to have the direct connection between New Zealand and the Royal Navy maintained in some concrete form.

New Zealand will supply a 'Dreadnought' for the British Navy as already offered; the ship to be under the control of and stationed whereever the Admiralty considers advisable.

I fully realise that the creation of specific units, one in the east, one in Australia, and, if possible, one in Canada, would be a great improvement upon the existing condition of affairs, and the fact that the New Zealand 'Dreadnought' was to be the flagship of the China-Pacific unit is, in my opinion, satisfactory. I however consider it is desirable that a portion of the China-Pacific unit should remain in New Zealand waters, and I would suggest that two of the new 'Bristol' cruisers, together with three destroyers and two submarines, should be detached from the China station in time of peace and stationed in New Zealand waters; and that these vessels should come under the flag of the Admiral of the China unit; that the flagship should make periodical visits to New Zealand waters; and that there should be an interchange in the service of the cruisers between New Zealand and China, under conditions to be laid down.

The ships should be manned, as far as possible, by New Zealand officers and men, and in order that New Zealanders might be attracted to serve in the fleet, local rates should be paid to these New Zealanders who enter, in the same manner as under the present Australian and New Zealand agreement, such local rates being treated as deferred pay.

The determination of the Agreement with Australia has of necessity brought up the position of New Zealand under that joint Agreement. I therefore suggest that on completion of the China unit the present Agreement with New Zealand should cease, that its contribution of 100,000£ per annum should continue and be used to pay the difference in the rates of pay to New Zealanders above what would be paid under the ordinary British rate. If the contribution for the advanced rate of pay did not amount to 100,000£ per annum, any balance to be at the disposal of the Admiralty.

The whole of this fleet unit to be taken in hand and completed before the end of 1912, and I should be glad if the squadron as a whole would then visit New Zealand on the way to China, leaving the New Zealand detachment there under its Senior Officer.

Reginald McKenna to Sir Joseph
Ward, 18 August 1909

The suggestions made by you at the meeting on the 11th and recited by you in your letter to me of the same date have been carefully considered and concurred in by the Admiralty.

The present Naval Agreement with Australia and New Zealand will not be renewed, and in view of this fact and the other special circumstances referred to by you, the part of the China Fleet unit, as set out by you, will be maintained in New Zealand waters as their head-quarters. Your wish that the ships of the fleet as a whole, or at any rate the armoured ships and the cruisers, when completed, should pay a visit to New Zealand on the Way to China shall also be carried out.

I take this opportunity on behalf of the Admiralty of repeating their sincere thanks to the New Zealand Government for taking so important a part in the inception of the present Conference. The Admiralty feel that every effort should be made to work out a scheme acceptable to the people of New Zealand, having regard to the patriotic action taken by yourself and your Ministers in March last.

[69] *Greene to McKenna*
31 December 1909

Since my conversation with you before Christmas I have been in communication with the Foreign Office and the Colonial Office, and also with the Treasury Solicitor, on the subject of the status of Colonial ships of war outside Colonial waters. Both Sir Charles Hardinge [Permanent Under-Secretary of State for Foreign Affairs] and Sir Charles Lucas [Under-Secretary of State of the Colonial Office, and first head of its Dominions Department], whom I consulted, are of opinion that a conference between representatives of the Foreign Office, Colonial Office and of the Admiralty would be very desirable as the first step towards the consideration of the question, and if you approve it is proposed to write to the Foreign Office and Colonial Office officially and invite them to nominate representatives. Sir Charles Hardinge said that he himself would probably attend and would bring with him one of their Legal Advisers; Sir Charles Lucas said also that he or one of the Assistant Secretaries would attend also with one of their Legal Advisers, and Mr Mellor, the Treasury Solicitor, said that he would come or send also one of his Assistants. I have therefore had a letter drafted of which I send you a copy and I shall be glad to know whether you approve of the letter being sent now. It is not likely that we shall be able to hold a meeting before the middle of next month, but it is hoped that in the course of one or two meetings we may be able to form some conclusions as to the course to recommend.

I may add that I have had a conversation also with Mr Acland, the Judge Advocate of the Fleet, and he said that he would gladly do what he could to assist in the consideration of the subject. At this stage, however, the question to be settled is more of a general character than one which concerns discipline: later on I have no doubt that Mr Acland's help would be valuable. [Minuted: Added as third representative of Admiralty after first meeting of Conference.]

I propose that Admiral Bethel[e] [Rear-Admiral the Hon. Alexander E. Bethell, Director of Naval Intelligence] and myself should represent the Admiralty at the conference as we have both been considering the subject since the Imperial Conference last year. Believe me, Yours very truly,

[Minuted: First Lord
signified his approval {AW?}
3/1/10]

[70] *[Greene] to the Solicitor H.M. Treasury, draft*
20 January [19]10

I am commanded by My Lords Commissioners of the Admiralty to acquaint you that they have had under consideration the question of the status of Colonial Ships of War outside Colonial waters as to which no definite conclusion was arrived at during the discussions which took place in the course of the Imperial Conference last August.

At the time when the Admiralty Memorandum laid before the Conference (see Parliamentary Paper Cd: 4948/09) was being considered it was concluded that should the Dominion Governments adopt the policy of establishing navies of their own it would be necessary to come to an understanding on the following points:

(1)   The measures requisite to give Colonial naval forces the international status of warships of a Sovereign State.

(2)   The means of employing Colonial naval forces on Imperial Services in time of peace as well as war.

As both the Canadian and Australian Governments have now taken practical steps towards the establishment of such naval forces, the former having undertaken to purchase a 2nd Class unarmoured Cruiser from His Majesty's Government, and the latter having sent instructions for the building by contract of the armoured and unarmoured cruisers of a Fleet Unit, it becomes necessary to consider these questions without delay.

It appears to my Lords that the subject might in the first place advantageously be considered by an inter-departmental Conference between representatives of the Foreign Office, Colonial Office and Admiralty, and I am to enquire whether you or your representative could arrange to take part in the Conference.

It is understood that the Foreign Office will be represented by Mr Walter Langley, C.R., and Sir W.E. Davidson, KCMG., CB., and the Colonial Office by Sir Charles Lucas, KCMG., CB., and Mr J.S. Risley.

My Lords propose to nominate as representatives of the Admiralty the Director of Naval Intelligence and the Assistant Secretary.

[71] *Memorandum, 'Pacific Fleet – provision and disposition of',*
by [Rear-Admiral the Hon. Alexander E. Bethell] Director
of Naval Intelligence

25 January 1910

It was decided at the Colonial Conference held last year that 3 fleet units consisting each of 1 Indefatigable, 3 Bristols, 6 destroyers (River class) and 3 submarines (C class) should be placed in the Pacific and known under the name of the Pacific Fleet . . .

It is understood that tenders have been asked for the construction of the Australian and New Zealand Indefatigables and that the orders will shortly be placed. These vessels will be ready for commissioning by the middle of 1912 and the remaining vessels of the units ought to be completed by the same date, though there is no absolute pledge that this will be done except in the case of the New Zealand ships (First Lord's letter to Sir Joseph Ward of 18.8.09).

It is not known whether any steps have been taken towards the provision of any of these vessels, but in view of our position towards Germany in the matter of modern destroyers and the more rapid rate at which she is completing her annual programmes than we are, it is not seen how 12 River class destroyers could be sent abroad. The question of whether one 'Indefatigable', six 'Bristols' and six submarines can be spared from Home waters without replacement also requires consideration.

[72] *G.J. Desbarats, Deputy Minister, Department of Marine and Fisheries, Ottawa, to Lord Strathcona, High Commissioner for Canada*

London
1 June 1910

I have the honour to request that you will ascertain from the Lords Commissioners of the Admiralty whether they have arrived at any decision as to the proper Flag to be flown by the ships of the Naval Service of Canada; and whether, should it be determined that the ships of the Navies of the Overseas Dominions are to fly the White Ensign, their Lordships will state what will be the distinguishing Flags to be borne by Flag Officers of the Imperial Service lent to Canada, and by Flag Officers of the Naval Service of Canada.

I also wish to be informed as to whether the Lords Commissioners of the Admiralty will allow the ships of the Naval Service of Canada to use the same flags for signalling purposes, and the same signal books, etc. as the Royal Navy: I would point out that the only ships to which it would be necessary at present to issue these books would be the *Rainbow* and *Niobe*, each of which will be commanded by an Officer of the Royal Navy, and officered by Officers of that Service.

[73] *'Canada – Halifax Defence Scheme', Colonial Defence Committee Memorandum?*

June 1910

4   The first duty of British fleets and squadrons on the outbreak of hostilities will be to seek to engage the enemy's naval forces with a view to securing the command of the sea. The vital importance of sea command is now so well understood that an enemy possessing a powerful battle fleet is unlikely to undertake organ-

ised attacks on defended ports until an attempt at least has been made to cripple our naval power by fleet actions.

5   So long as the ultimate command of the sea remains in dispute, an enemy's battleships will, as a general rule, be reserved for employment in fleet actions. The only targets on shore against which they would be likely to expend their heavy ammunition are hostile naval bases at which repairing facilities exist, and at which stores are accumulated on so large a scale that their destruction would cause appreciable embarrassment.

[74] *'Report of Interdepartmental Conference on Status of Dominion Ships of War', submitted to the Admiralty Board by Greene*

4 July 1910

'Confidential'

The Inter-departmental Conference appointed in January last to consider the status of Dominion ships of war, in continuation of the scheme of cooperation in Naval Defence discussed at the Imperial Conference, 1909, have completed their enquiry & their report is submitted herewith for their Lordships' consideration.

The conclusions of the Conference are summarized in Section VII, pages 32–341, in which the adoption of a scheme for a limited Imperial navy is recommended as giving the maximum of the advantage to be derived from the cooperation of the Dominions in the naval defence of the Empire & this scheme seems to be the most in agreement with the general policy of the Admiralty.

Both the Australian & Canadian gov[ernmen]ts are proceeding with the establishment of naval forces on the lines discussed at the Imperial Conference, & it is very desirable that a definite agreement should be arrived at with these governments on the subject of the relations of their ships of war to the Royal Navy. It is accordingly submitted that copies of the report be referred as soon as possible to the Colonial Office & Foreign Office & the views of the Secretaries of State be invited upon the scheme recommended by the Inter-departmental Conference.

The question of the best manner of approaching the Dominion gov[ernmen]ts is also for consideration. It is probable that a meeting between the representatives of the Dominion gov[ernmen]ts & HM Gov[ernmen]ts would be advantageous.

[Minuted] Propose to approve.

Copies will be issued to
members of the Board. C[hief
of] S[taff] 5.7.10 . . .

VII – General Conclusions

121   In the preceding remarks, we have set forth fully our views
on the difficult and complicated matters referred for our consideration, and it only remains now to summarise our conclusions.
From the historical account which we have given of the position
of the self-governing Dominions in the matter of naval defence,
it will be evident that the time has come when, in some cases at
any rate, it is no longer possible to treat as a practical policy the
payment by them of a money contribution, and it must be accepted
that the Dominions wish to share with the Imperial Government
the burden of maintaining a navy. It is therefore necessary to do
all that is possible to assist the Dominions to organise their naval
forces in the most efficient manner, and at the same time it is
most desirable to induce them to accept a position of such close
relationship with the Royal Navy that the two will be virtually
part of the same Imperial force and that the assistance of the
Dominion naval forces can be invoked in carrying out Imperial
services in time of peace as well as in war.
122   With this object it is essential that the Dominion ships should
have the international status of British ships of war, and that the
officers should have the international status of duly commissioned
British naval officers, with all the other honours of the Royal
Navy.
123   Unless this policy of intimate association with the Royal
Navy is pursued the Empire as a whole will gain little practical
advantage from the establishment of Dominion naval forces, while
those forces will lose both in prestige and efficiency. The
Dominions evidently contemplate such close association and the
closer this can be voluntarily made the better, in order to secure
harmonious co-operation and preclude the possibility of friction
within the Empire and international differences without. The
advantage of close connexion between a local service and the
Imperial Service was recognised clearly by Mr Deakin, the late
Prime Minister of the Australian Commonwealth. In the speech
before referred to which he delivered in the Australian House of
Representatives on the 13th December, 1907 (after his return
from the Imperial Conference) he stated 'a small flotilla of that

description would remain a thing apart, not directly committed to the high standards of the Imperial Navy. I ventured, therefore, to attempt to find a means by which we should get the whole benefit of the connexion with the Admiralty and the Imperial Fleet, sharing its standards, its training and its prizes, and yet maintain the Australian character of our flotilla.' (See Appendix I, p. 41.) It is true that these remarks applied to proposals not identical with those which led to the adoption of the scheme of a fleet unit, but they illustrate vividly what in the opinion of one of the foremost men in Australia would be the result of establishing a naval force cut off from close connexion with the Royal Navy.

124 Since the ships are to be provided at the expense of the Dominions, it is only reasonable that it should be left to those Governments to control the administration, and, in time of peace, the disposition of the ships. In war time the ships cannot be used without the consent of the Dominion concerned, but, if used, they should be under the direction of the Admiralty. In peace time some special provision must obviously be made, or else it would be within the power of the Dominion Governments to order or permit their ships to take action in relation to foreign Powers for which the Imperial Government would be responsible, but which they would not be able to prevent or control. In such an event the Imperial Government might be seriously hampered in the control of the foreign policy of the Empire and might be committed to a policy, or even to a war, of which they did not approve. This danger is not merely academical, but may easily arise.

125 It is because of the risk of grave complications with a foreign State owing to the possibility of hasty or ill-advised action on the part of a Dominion fleet, or rather of a ship or a Commanding officer, that we consider that it would not be wise for the Imperial Government to adopt a policy of laisser-aller or to legalise the establishment of a Dominion fleet with uncontrolled authority to act out of its own waters. As already pointed out, the position of a fleet is different from that of a military force or from any other administrative service whose field of action is necessarily confined to its own territory; and excellent as has been the result of leaving the British Dominions beyond the Seas [ms addition] to manage their internal affairs, it is altogether another matter to authorise a Dominion fleet to act independently of the Imperial Government on the high seas and in foreign waters.

126 Accordingly the result to be aimed at on this view is that while the Dominions should not either in peace or war be under

an absolute obligation to permit any active use of their ships, the Imperial Government should possess both in peace and war effective means of precluding such action as in their opinion would affect foreign relations; in war this would involve the acceptance by the Dominion Governments of the principle that their naval forces would not take any action whatever, except with the approval of the Imperial Government, other than measures of self-defence within their own territorial waters.

127   Such powers on the part of the Imperial Government ought, if possible, to have a legal sanction, and not to depend merely on an agreement or understanding between the two Governments; by this is meant that the Imperial Government should have the power of taking effective disciplinary action, either by means of a court-martial or otherwise, against any Officer who contravenes Admiralty orders.

128   We have referred to the alternative methods which might be adopted in securing a working union with the Royal Navy, and it is clear that if the Dominions are prepared to accept it, the alternative of a united Imperial Navy is the more satisfactory. The objections which may be raised to it are probably more apparent than real, and with full explanations it might not unreasonably be laid before the Governments of the Dominions for their acceptance, on the ground that it would best provide for:

(1)   The efficiency of the Dominion naval forces and the dignity of the flags;

(2)   The interchange of Officers and men between the Dominion and Imperial services, thus providing the possibility of a career which will attract men of first rate ability to the Dominion services;

(3)   The effective co-operation of the Dominion naval forces with the Royal Navy, whether in peace or war; and

(4)   The avoidance of dangerous international incidents.

129   Further legislation would appear to be necessary whichever scheme may be adopted; in the case of a united Imperial Navy for the purpose of removing doubts and adapting the provisions of the Naval Discipline Act to the new conditions; in the case of auxiliary Dominion fleets, for the purpose of placing the discipline of the naval forces outside the territorial limits of the Dominions on an effective legal basis.

Note [added under signatures]

The Conference was constituted as follows:

Admiralty

(Chairman) Rear-Admiral the Hon: A.E. Bethell, CMG., Director of Naval Intelligence,

Mr R.B.D. Acland, kc., Judge Advocate of the Fleet,

Mr W. Graham Greene, CB., Assistant Secretary of the Admiralty.

Colonial Office

Sir Charles Lucas, KCMG., CB., Assistant Under-Secretary of State,

Mr H.W. Just, CB., CMG., Assistant Under-Secretary of State and Secretary of the Imperial Conference,

Mr J.S. Risley, Legal Assistant.

Foreign Office

Mr W.L.F.G. Langley, CB., Assistant Under-Secretary of State,

Sir W.E. Davidson, KCMG., CB., KC., Legal Adviser to the Secretary of State.

Treasury Solicitor & King's Proctor

Mr A.H. Dennis, Assistant Solicitor.

[75] *'Principles of Imperial Defence', memorandum by the Colonial Defence Committee, s.v. 'Local Defences', et seq., signed by Chancellor*
7 July 1910

'Secret'

Local Defences

26 It is not the purpose of this Memorandum to deal in detail with the military requirements of the self-governing Dominions, which must necessarily vary in the case of each according to the local conditions and geographical position. It is of course essential that each of the component parts of the Empire should possess military forces of sufficient strength to ensure the maintenance of public confidence and to be capable of dealing effectively with oversea attack on the scale which in each case is considered to be reasonably probable. It is also necessary, as explained below, to provide fixed defences and garrisons for naval ports and for certain commercial ports where merchant shipping can, in case of need,

seek protection from molestation or capture by the enemy's commerce raiders.

Beyond this there is the important question of co-operation in Imperial defence which was discussed at the Imperial Conference on Defence in 1909. It is recognised that the Governments of the self-governing Dominions reserve to themselves the right of deciding, when the occasion arises, as to whether they will provide military forces for Imperial service outside home territory in any war in which the British Empire may be engaged, and as to the strength and composition of any force which may be provided by them for this purpose. It seems certain, however, that there will be a general desire amongst the self-governing Dominions to contribute to the common defence of the Empire in any war which seriously threatens its integrity. It may therefore, perhaps, be well to emphasise once more the fact that modern warfare will seldom admit of the hasty improvisation of military forces subsequent to the outbreak of hostilities. In a few instances hasty improvisation may be practicable, but it is probable that in the future great naval and military events will immediately follow even if they do not precede, a declaration of war. If, therefore, organisations have to be improvised, staffs created, transport and equipment provided, and plans matured, after the outbreak of war, the value of any assistance given would be greatly lessened, even if it were not altogether belated.

Any military system, therefore, to be effective must be capable of providing, when a crisis comes, for the immediate mobilisation and concentration in organised bodies of the troops required for home defence and for any force which the Government of the Dominion concerned may desire to contribute to the common cause of Imperial defence.

27   In regard to the question of the local defence of naval ports and commercial harbours, it has already been explained that the protection afforded by the navy constitutes their main security from attack from the sea in war; but it is obvious that naval action alone will not serve to guarantee absolute immunity from hasty predatory attacks by hostile warships, which may have succeeded temporarily in evading the vigilance of our fleets.

To render all ports of the Empire absolutely secure against such attacks, the provision of extra local defences at each port would no doubt be necessary. As, however, our resources are not unlimited, it is not possible to make every portion of the Empire secure against all the contingencies of war. All that can be aimed at is to provide for the defence of important and vulnerable objectives against such forms of attack as, having regard to the strategic conditions, are regarded as reasonably probable.

28   In estimating the need for providing passive defences at particular ports and localities, due regard should be had to these considerations lest an unduly large proportion of the resources available for the defence of the Empire be diverted from employment on offensive action by which alone decisive results can be obtained.

As His Majesty's ships are likely to be fully occupied in keeping touch with and engaging the enemy's fleets, the movement of troops oversea in the early stages of a maritime war is an operation attended with grave risk and inconvenience. The Admiralty are therefore unable to undertake responsibility for the safe convoy of war garrisons to foreign stations within any particular period after the outbreak of war. On the other hand, to send out troops in anticipation of war in a time of strained relations is not practicable, as the reinforcement of foreign garrisons at a time when delicate negotiations are in progress would have a bad effect and might precipitate hostilities. It has therefore for many years been the accepted policy that the Imperial garrisons of defended ports abroad should be maintained at war strength in time of peace.

The different kinds of objectives on land likely to invite attack and the scale and nature of oversea attack to which they may severally be exposed in a naval war will now be considered with a view to furnishing a basis for the calculation of the standard of local defences required for their protection.

Naval Ports

29   In order to afford facilities for the repair of our ships and their supply with munitions of war, it is necessary to maintain certain naval bases permanently in time of peace. With a view to ensuring that local defences are provided on a uniform system, naval ports have been divided into three categories, viz: (a) first-class naval bases; (b) second-class naval bases; and (c) war anchorages.

First-class naval bases are ports where naval dockyards are established capable of building or repairing His Majesty's ships in all respects, and where permanent depots of men, ammunition, and stores of all kinds are maintained. There are at present no first-class naval bases outside the United Kingdom.

Second-class naval bases are ports where naval dockyards are established capable of carrying out repairs on a lesser scale, and where permanent depots of men, ammunition, and stores of all kinds are maintained. Gibraltar, Malta, Hong Kong, Simonstown and Bermuda are classed as send-class naval bases. When the measures now being initiated by some of the self-governing Dominions for the establishment of local navies under their own control are further advanced, some ports in these Dominions will fall within the category of first-class or second-class naval bases.

War anchorages are ports situated in important strategic localities, which do not possess the qualifications of naval bases of His Majesty's ships, but afford facilities for the replenishment with men, ammunition, or stores, conveyed to them by rail or by sea. At the present time there are no places outside the United Kingdom fortified as war anchorages; but in war certain fortified commercial ports will probably be used for this purpose by His Majesty's ships.

30   Local defences are required for the protection of these naval dockyards and establishments. In view of the strategic considerations explained above, the defences at ports of the British Empire need not provide complete protection against prolonged operations. They should be capable of affording security from sudden capture or destruction by such raiding naval and military forces as might temporarily succeed in evading the vigilance of our squadrons, or, in the case of ports in seas remote from our centres of naval power, against such organised naval and expeditionary land forces as might be dispatched against them in the event of the command of the waters in their neighbourhood temporarily resting with the enemy.

31   As regards purely naval attack, the fixed defences as a rule need only be sufficiently formidable to deter attack by a fleet not supreme at sea, and therefore not in a position to risk incurring serious loss of fighting efficiency. To fulfill this object they should be capable of inflicting sufficient damage to place attacking ships at a disadvantage in a subsequent naval action, or to compel their immediate return to a base for repairs.

32   The main object of the enemy's ships in attacking our defended ports will generally be not so much to destroy the defences as to inflict the maximum of injury on the material resources that the defences are designed to protect. Attacking vessels would seldom be left unmolested for sufficient time to enable them to silence coast batteries preparatory to a bombardment of their real objective; and even if it happened that the

opportunity were offered, the amount of ammunition which would have to be expended to inflict serious injury on modern coast batteries would probably be far in excess of the supply that attacking ships would have at their disposal for such a purpose.

It must be remembered that a bombardment by battleships of objectives on shore would not be undertaken if the attacking ships were liable to encounter a hostile squadron before being able to replenish their ammunition supply. This consideration would have great weight in the case of attack on British ports in distant seas, the majority of which are remote from the naval bases of possible enemies. The knowledge that facilities for repairs and refitting could only be obtained after a long voyage to a distant base would cause a commander to hesitate before exposing his ships to the risk of incurring serious injury.

33   Until a few years ago the radius of action of sea-going torpedo craft was limited to a distance of about three hundred miles from their base. And it was assumed that naval ports more than three hundred miles distant from the nearest foreign torpedo flotilla base might be regarded as immune from attack by sea-going torpedo craft. In view of the increased size and sea keeping capacity of modern torpedo craft, the Admiralty are of opinion that this rigid rule cannot in existing circumstances properly be applied to all ports alike; and they consider that in estimating the vulnerability of ports to torpedo attack each case must be judged on its merits.

34   Attacks by torpedo craft on battleships or cruisers lying in harbour are most to be feared during the period immediately preceding the outbreak of hostilities. The chief safeguard against such attacks lies in the maintenance of so large a force of torpedo-boat destroyers in instant readiness for war that no hostile torpedo craft would be able to approach our ports without incurring great risks of capture or destruction.

To meet the possibility of evasion of our destroyer flotillas by the enemy's torpedo craft, however, special fixed defences must be provided at certain naval bases.

35   Effective defence against torpedo attack can in some cases be provided by the construction of breakwaters to contract the entrance channels of harbours. Such permanent obstruction are, however, usually very costly, and as it sometimes happens that their construction, by causing silting, would injuriously affect the depth of water in a harbour, this form of protection is not always

applicable. Special fixed defences are therefore usually provided at naval bases exposed to attack by torpedo craft.

At ports where special anti-torpedo defences are not provided His Majesty's ships will make their own arrangements for defence against attack by torpedo craft, whether sea-going or carried in the enemy's ships.

36   Fixed defences, however formidable, will not render a fortress secure against attack by an expeditionary land force. This form of attack can only be dealt with by an adequate force of mobile troops and guns capable of being rapidly moved to any threatened point.

As explained in paragraphs 23 and 24 above, the size of the expeditionary land force to be provided against, and consequently the strength of the garrison required to oppose it, will vary at different naval ports according to the proximity and resources of the nearest bases of possible enemies and the number of the troops there available and to the distribution of our fleets on the outbreak of war.

Temporary Naval Bases

37   In addition to the permanent naval bases which are required as main sources of supply, a fleet operating on the offensive – as the British fleet must always do – may require advanced bases of a temporary and auxiliary nature to enable it to keep close touch with the enemy. This necessity arises from the fact that the systematic transhipment of coal, ammunition, and other supplies constantly needed by a fleet can seldom be effected at sea, and is as a rule an operation requiring that ships shall anchor in smooth water. The distance of the permanent bases may be inconveniently great for the purpose. The essential qualifications for these temporary bases will, therefore, be suitable position, sheltered anchorage, and security from attack by the enemy.

A suitable position may, in some cases, be found in the nearest British fortified commercial port (see paragraph 38 below), which will also afford a sheltered anchorage and a certain measure of protection from attack. Such positions, however, may not be the most advantageous, and the occupation of an unfrequented natural harbour – such as that utilised at the Elliott Islands by the Japanese fleet off Port Arthur – has the advantage that secrecy can more easily be maintained. The continued proximity of the fleet itself will constitute the chief safeguard of such a base, and any local defences – especially against attack by torpedo craft – will, when required, be organised by the navy. In the case of a fortified commercial port the scale of armament to be provided is not affected by the possibility of the utilisation of the port as a temporary naval base.

Fortified Commercial Ports

38   Certain commercial ports require fixed defences, but, before discussing the standard of these defences, it is necessary to consider briefly the extent of protection, direct and indirect, which will be afforded to British commerce and commercial ports by the navy. Such protection must be regarded as the main security of these ports from attack in war, and all defensive measures on land must be regarded as merely ancillary.

An enemy possessing a powerful fleet is unlikely to undertake organised attacks on commerce or commercial ports until an attempt at least has been made to cripple our naval power, for which latter purpose his cruisers are likely to be required in the first instance to act in conjunction with his battleships. Isolated attacks on merchant vessels met during the progress of some strategic movement may indeed occur, but regular attacks on commerce in distant waters, if they take place at all at the beginning of a war, are more likely to be carried out by armed merchant vessels than by hostile cruisers, which are not likely at that stage to be available for such service. In view of the supreme value of armoured vessels in war and of their great cost and consequent small numbers, it is improbable that a squadron would undertake a subsidiary operation such as the attack on a commercial port, if the defence were of such a nature that the attackers would run the risk of losing even one of their number, or of receiving such injuries as to involve risk of capture or immediate return to a base. Of recent years foreign naval Powers have almost without exception ceased to lay down any but small unarmoured cruisers, and the armoured cruisers now under construction approximate to the battleship type. The great value of such armoured vessels as adjuncts to the battle fleet renders it improbable that they would be detached for attacks on commerce or on commercial ports until the struggle for the command of the sea had been decided. The older types of armoured cruisers may, however, become available in the future for subsidiary operations of this nature.

The British naval reply to attacks on commerce would probably involve extended operations with cruiser squadrons and single ships taking full advantage of the facilities afforded by our numerous commercial ports as coaling places and as centres for the collection and distribution of intelligence relating to the movements of the enemy. The hostile raiding vessels would in all probability resort to methods of evasion in order to inflict the maximum of damage on our trade with the minimum of risk to themselves, and, in order to avoid observation and opposition alike, they would probably frequent open water on the great trade routes rather than the vicinity of our coasts and commercial ports, where their presence would be quickly reported. It is obvious, however, that the action of British naval forces alone could not guarantee the absolute immunity of our merchant shipping from interference, or prevent the possibility of hasty predatory raids on undefended ports if the injury capable of being inflicted would justify the risks involved.

39   Certain fortified commercial ports on frequent trade routes are, therefore, advantageous as coaling stations and harbours of refuge, where merchant vessels can, in case of need, seek protection from capture or molestation, and await a favourable opportunity of continuing their voyages. Fixed defences may also be required at great commercial ports if valuable and vulnerable resources of national importance, such as docks and extensive plants of machinery, are collected within such a limited and exposed space that they might be seriously damaged by the gun fire of a raiding cruiser if no defences existed. Fortified commercial ports are classified as defended mercantile ports, coaling stations, ports of refuge. The object of fixed defences, in such cases, is to keep the enemy's cruisers at a distance.

40   The measure of defence required is that which will involve such risk of injury to the attacking cruiser as would not in the opinion of a naval commander, be justified by the possible advantages to be obtained. Thus, while the destruction of the naval resources at a primary base might exercise a considerable influence on the course of a war, the damage effected by a predatory raid on shipping and other private property at a commercial port could rarely be commensurate with the risks involved in coming under the fire of even light defences on shore. Foreign cruisers carry only a limited number of rounds of common and semi-armourpiercing projectiles for each of their heavier guns, and it is reasonable to assume that no commander would care to expend a large proportion of this small stock of ammunition in engaging forts when he might afterwards have to meet a British warship

with his magazines depleted. To a naval commander the strong probability of sustaining injury or even disablement to his ship, at a distance from a port where repairs could be effected, would constitute another power deterrent. It follows, therefore, that the strength of the armaments at fortified commercial ports should be governed, not so much by the remote contingency of having to carry on an engagement with raiding vessels, as by the necessity of deterring hostile ships from closing to positions from which they could deliver, with immunity to themselves, a deliberate and effective fire against British shipping lying in a harbour of refuge, or against docks and other similar property of value at a great commercial port.

41   Even where the prompt arrival of a British naval force is not a reasonable probability, as might, for instance, be the case if a single cruiser attempted a hasty raid on a comparatively remote commercial port, defences on a very moderate scale should be a sufficient deterrent to make the enemy turn his attention to less hazardous undertakings. A raiding cruiser has always the commerce at sea as an alternative objective, and when a choice of defended and undefended objectives presents itself the undefended objective will naturally be selected, even when the resistance capable of being offered by the defences is not formidable.

42   The risk of attack by torpedo craft on merchant-ships lying in harbour is exceedingly remote. The mere destruction of a few merchantmen would hardly be considered a sufficient inducement for the employment of these vessels, whose proper role is the attack of warships; and so long as British battleships remain at sea, the enemy's torpedo craft are extremely unlikely to be thrown away in attempting subsidiary operations of this nature. Moreover, to sink merchant-vessels without at least warning and examination would be contrary to the principles of maritime warfare: and the difficulty of distinguishing neutral ships the sinking of which might lead to diplomatic complications, would be a further deterrent consideration to a belligerent.

In these circumstances, it is not considered necessary to provide commercial harbours with special fixed defences for their protection against attack by torpedo craft; but dock and lock gates, where they exist, should be protected by any suitable contrivance which will prevent injury by a torpedo; and guards should be posted to prevent the treacherous application of charges of explosives.

43   As regards the scale and nature of oversea land attack to be provided against at commercial ports, all that is required is that the infantry garrison should be capable of dealing with raiding attacks by the small landing force that can be transported by such hostile ships as may temporarily succeed in evading our squadrons. The greater the size of such a raiding force and the greater the distance oversea it must traverse, the greater will be the risks it will incur of interception and destruction by British warships. The maximum strength of the raiding force to be provided against will thus vary for different ports according to their proximity to foreign bases and to the dispositions of our fleets. It may be assumed that the hostile landing force that could be conveyed oversea by tactics of evasion would rarely be of a greater size than could be transported by three or four cruisers, possibly accompanied by one or two armed transports.

44   This limitation on the scale of oversea attack will not, however, be applicable to the exceptional cases of fortified commercial ports in waters remote from our centres of naval power, where, as explained in paragraph 23 above, it is conceivable that the local command of the sea might be in certain eventualities temporarily rest with the enemy.

In determining the scale of probable oversea attack on such ports each case must be considered on its merits, due regard being had to the value of the objective, to the hydrographical and topographical conditions, and to considerations of distance, time, and relative naval and military strengths.

Other Commercial Ports

45   Fixed defences are not in all cases required for the security of commercial ports. The inducements to attack or to bombard the comparatively small amount of shipping likely to be found in ports remote both from the main trade routes and from hostile naval bases may easily be outweighed by considerations of ammunition supply and by the knowledge that such attacks would reveal the position of the raiding vessels, which would probably find, in open water, better opportunities and greater security for their predatory operations.

In some cases, hydrographical conditions, such as the existence of a long, narrow, and easily blocked river approach, may prove an effective deterrent to raids which, owing to the indirect influence of naval operations, must, for the ships of a Power inferior at sea, necessarily be of a hasty and fugitive character.

On the other hand, cases may occur in which the configuration of an open harbour and the depths of the water render it useless to attempt to protect shipping lying at anchor by means of coast defences. It is obvious that no secure port of refuge exists if an enemy's ship can easily destroy vessels lying in a roadstead while herself beyond the effective range of shore batteries.

46   Even in the case of commercial ports where fixed defences are not provided the presence of an alert and well-organised military force can afford effective protection to local interests. In the presence of such a force, skillfully disposed, the risks involved in landing men from raiding vessels are greatly increased. Without landing men, no substantial injury could be inflicted to property on shore, except by resorting to an expenditure of ammunition which may be regarded as prohibitive when no object of strategic importance is involved. A threat of bombardment, with a view to extorting an indemnity, might seem likely to serve the purpose of an enemy assuming that British authorities could be found to comply with such a demand; but such a requisition should unhesitatingly be rejected. Surrender would entail more moral, if not material, loss to the place than would result from the few shells which a raiding vessel could afford to expend, and by yielding the supplies that it had been incapable of defending, a Colony would assist the enemy in inflicting damage elsewhere on British vessels or neighbouring Colonies. If on the other hand the requisition were refused, landing parties would have to be sent ashore to obtain payment. If the defenders were properly disposed, the use of the ship's guns against them would be ineffective; and under modern conditions the difficulty of effecting a landing in the face of the opposition that can be offered by a force intimately acquainted with the ground would be so great, that the possible advantages to be secured could scarcely be commensurate with the risks involved.

Again, coaling could not be carried out in a port, nor could coal on shore be seized until the defending force had been driven away, and this would scarcely be possible without landing greatly superior numbers. Even boat attacks with a view to cutting out colliers or lighters lying close inshore would be difficult under the rifle fire of the garrison. To deny coal in this manner to an enemy, while preserving it for the use of British shipping, will, in time of war, be one of the chief duties of the local forces of the smaller Colonies.

Cable Landing Places

47   The maintenance of submarine cable communications throughout the world in time of war is of the highest importance to the strategic and commercial interests of every portion of the British Empire. Owing to the wide geographical dispersion of the various parts of the Empire, no Power would suffer so much as ourselves from the interruption of cable communications. Attempts by the enemy to destroy cables must therefore be regarded as a probable operation in war.

48   It is probable that cruisers or armed merchantmen will be specially allotted to this duty by the enemy. As regards the possibility of cables being cut in deep water, the Hydrographer to the Admiralty has expressed the opinion that a vessel provided with a suitable grapnel would have little difficulty in lifting and cutting a cable within the 100-fathom line, on a bottom fairly free from rocks, at any point where its position was accurately known. It is therefore manifest that an enemy might, in certain circumstances, be able to cut cables lying in shallow water out of sight of land, or at any rate without exposing himself to the fire of troops on shore.

49   It does not, however, follow that protective measures on land are useless. It is by no means certain that an enemy's raiding cruisers or armed merchantmen would be in possession of sufficient accurate information as to the route of a cable to enable them to find it without tedious grappling operations; and as time would be an all-important consideration in the operations of such vessels, which will be in constant danger of being brought to action by British warships, they might well prefer to attack the cable where it would be readily accessible, that is, at the shore ends. If no measures of defence are provided on land, it would be easy for a small party to land, destroy the cable-huts and instruments, follow the cable down to the sea from the hut, and tow the end out seaward, cutting off pieces from time to time. No technical knowledge would be required, and no tedious grappling operations would be involved. The measure of defence required on land at cable landing places is evidently that which will involve such risks to the crews and boats of an attacking vessel as will deter her from attempting an enterprise which at best would only result in a temporary interruption of the cable communications. Such a measure of defence should be provided at places where local forces exist.

Signal Stations

50   In order that all the channels at our disposal for the collection and distribution of naval intelligence throughout the Empire may be safeguarded, it is necessary that suitable measures should be taken for the protection of all signal stations which will be maintained for observing the movements of shipping in time of war.

Radio-telegraphy stations should if possible be situated some distance inland, where they would not be exposed to bombardment from the sea or to the danger of raiding attack by parties landed in boats from passing warships. Visual signal stations must necessarily be in exposed positions, and when not situated within the defended area of a defended port or locality, they should as a general rule be provided with an infantry guard sufficient to deter attack by a small armed party such as might be landed from a cruiser or armed merchantman.

[76] *Admiralty Draft, to [Captain R. Muirhead Collins RN] High Commissioner for the Commonwealth of Australia*
6 [16?] July 1910

I am &c . . . to acquaint you that they [i.e. the Admiralty Board] have had before them your letter of the 7th July No: 11,158, forwarding copy of a telegram from the Commonwealth Government enquiring as to the description of the Flag and Badge to be flown by the Australian Fleet Unit.

In reply I am desired to state that the subject of the Flag to be flown by the Naval Forces of the Dominions established in accordance with the conclusions of the Imperial Conference 1909 has been, together with other matters, under their Lordships' careful consideration, and it has become clear that the question of the Flag is so intimately connected with the status to be given to the ships with respect to Foreign Nations and other important matters that it cannot be settled until the whole question of the relations of Dominion Ships of war to the Royal Navy has been determined.

It is proposed to make this the subject of a separate communication to the Government of the Dominions, and therefore it will not be possible to give any decision as to the flag in the meanwhile.

[77] *Memorandum by the Admiralty on the 'Status of Dominion Ships of War', Committee of Imperial Defence*

August 1910

'Secret'

Since the meeting of the Imperial Defence Conference in 1909, to consider the question of the naval and military defence of the Empire, the new naval arrangements provisionally agreed to by the representatives of Canada and Australia have received the approval of the Governments of these Dominions, and certain action has been taken by them to give effect to the same. It has, therefore, become necessary to consider the manner in which the principles of these new arrangements should be applied, so that, while the administration and control of the Canadian and Australian naval forces should rest with the Dominion Governments, there should be the same standard of training and discipline in these forces as in the Royal Navy, and that they should all as far as possible develop as integral parts of one Imperial Navy. The questions to be settled are briefly:

(a)   The measures requisite to give the Naval Forces of the Dominions the international status of war-ships of a sovereign State.

(b)   The means of employing the Naval Forces of the Dominions on Imperial services, so far as the Governments of the Dominions consent to such employment, in time of peace as well as during war.

For the sake of convenience the latter question will be considered first, as the former presents no great difficulties and is partly involved in the latter. The problem presented by the proposed establishment of naval forces by the Dominions is unique, and there is no precedent in history to which an appeal can be made in determining their status; it requires, therefore, careful investigation in three respects in particular.

The first question is the legal position of ships of war of the Dominions when beyond territorial waters, and whether Imperial legislation may be required to supplement Dominion legislation or to give it validity beyond territorial waters.

The second question is the method by which and the extent to which uniform discipline should be maintained. Homogeneous discipline appears to be necessary on several important grounds. In order that the Dominion naval forces may be of full value in time of war, it is essential that they should be under similar discipline and training in time of peace. The fact that Imperial officers and men are to be lent for service in the Dominion ships and interchanged with their officers and men is another reason, for the Admiralty might find some difficulty in justifying their action except on conditions of uniformity. The need for uniformity is even greater if the ships of the Dominions are to be employed on Imperial service, and are to undergo training with ships of the Royal Navy.

The third question for consideration, apart from the legal points involved in the first question, is that, while the Dominion ships are intended primarily and mainly for local requirements, yet as a mobile force they will in any case, and more especially if employed on Imperial service, be continually passing beyond territorial waters, and will, therefore, be liable to come into contact with foreign national ships, and also with other British ships of war; accordingly their position must be regulated and defined. Questions of international relations and of peace and war are under the Constitution vested in the Crown (which with regard to such questions acts on the advice of the Imperial Government), and therefore Imperial control over the Dominion naval forces in some form and to some extent would seem to be unavoidable. Accordingly the question of the nature and extent of the control of the Admiralty, to be exercised with the consent of the Dominion Governments over their naval forces when outside territorial waters, requires to be determined.

The question of the manner and extent of the employment of Dominion ships on Imperial service turns largely upon the code of discipline that governs them, because, unless there is some sanction which the Imperial Government can enforce, there is no direct means by which the personnel of the new naval forces which are being brought into existence, capable of voyaging anywhere, can be made to comply with the orders of the Central Government. Further, if the Royal Navy and Dominion naval forces are to act together as one fleet or to be anything more than quasi-foreigners to one another when they meet, it is essential that, if possible, the same disciplinary code should be applied to each component part.

Thus it was agreed as regards Australia, when the conferences took place in 1909 between the representatives of the Admiralty and the Commonwealth, that while, in peace time, and on the Australian station, the ships maintained by Australia should be under the exclusive control of the Commonwealth Government as regards movements and general administration, the officers and men should be under naval discipline, and when with ships of the Royal Navy the senior officer should take command of the whole.

Unless some arrangement of this kind is made it must be recognised that international difficulties of a very grave nature may arise, owing to the fact that a mobile armed force has been established, over whose action the Central Government would have no control, though the ultimate responsibility would rest with them.

In considering the question, therefore, the following points must be borne in mind:

1. That the matter is one which is important in peace as well as in war. Wars arise out of acts done in times of peace;
2. That if the Dominion vessels are placed at the disposal of the Admiralty in time of war, provision must be made for rank and command in time of war; and
3. That any violent change of rank and command coming into force in the early days of the stress of war is to be deprecated.

Accordingly, if possible, the system adopted should secure the following results:

1. The ships must have the international status of British ships of war, and the officers the international status of officers duly commissioned by the authority of the British Crown.
2. The Imperial Government should be able to control any action that may be taken by the ships of a Dominion which might possibly involve the whole Empire in international difficulties.
3. It being naturally the wish of the Dominions to keep the ships of war provided by them as far as possible under their own control, it is desirable that the control – for international purposes – of the Imperial Government should be restricted within the narrowest possible limits.
4. It is important that service in the Dominion naval forces should be such as will be likely to attract good men of all ranks . . .
5. So far as varying regulations and conditions of service might permit, there should be facilities for an interchange of ships, officers, and men between the Dominion naval forces and the Royal Navy; this seems to be especially desirable from the point of view of the Dominions.
6. The Dominion ships, officers and men should attain to such a degree of efficiency and prestige, and have such a system of disci-

pline as to enable them to co-operate effectually with the Royal Navy when necessary, and as a consequence of such effectual co-operation to share in all the honours and privileges of that body.

It is not necessary, before considering the best means of securing the above results, to examine the constitutional and legal position of the Colonies at length. It is sufficient to state that serious questions might possibly arise as to the capacity of the Dominion Parliaments to legislate effectively for the discipline of their naval forces outside their territorial waters.

It should be the aim of all parties, if possible, to avoid raising such questions by coming to some arrangement which will secure the objects in view without touching upon these delicate and difficult matters.

This can be done by adopting the policy of what may be called a united Imperial Navy. Under this policy the objects to be aimed at would be that the naval forces of the Dominions should form an integral part of an Imperial Navy, subject to whatever limitations may be necessary in order to secure to the Dominion Governments as great a power of control as is compatible with the one main principle that all the naval forces within the Empire should form single whole. It is not implied that the Dominion Governments, as representing the Crown, will cease to be owners of their ships, or that they cannot refuse, if they think fit, to place them at the disposal of the Imperial Government either in peace or in war; but it is suggested that whether within or without their own waters, the Dominion naval forces should be under the same Discipline Act as the Royal Navy.

The first step necessary to carry out the policy of a united Imperial Navy is the assent of the Dominions to the proposition that the officers and men are 'persons in or belonging to His Majesty's Navy and borne on the books of one of His Majesty's ships in Commission' within the meaning of section 87 of the Naval Discipline Act: in other words, that the Naval Discipline Act should apply to them everywhere . . .

With regard also to the form of the officer's commissions, there are many advantages in the adoption, both for officers commissioned by the Dominion Governments and for officers commissioned by the Admiralty in the ordinary way, of one form of commission extending to service in all the naval forces of the Crown, including the Royal Navy and any Dominion fleet.

These proposals are recommended on the following grounds:

The moral effect will be to encourage the feeling of unity in the Imperial service, and to increase the prestige of the Dominion officers. In peace time it will assist the course of administration to dispense with the necessity of temporary commissions being granted by the Dominion Governments to officers of the Dominion fleets detached for service in the Royal Navy. In war time it may be necessary to transfer officers at any moment from or to the ships of the Royal Navy, and of the Dominions, if effective co-operation is to exist between them, and in view of this possibility it is necessary that each officer's commission should be operative in any ship of His Majesty, whether provided by the Admiralty or by the Government of a Dominion. For international reasons it is undesirable, at least in war time, to have more than one form of commission in one belligerent force.

The problem of making provision for times of war is far less difficult than that of providing for times of peace. Both Australia and Canada appear to accept the position that when the ships of the Commonwealth and of the Dominion are placed at the disposal of the Admiralty in time of war the officers and men must be subject to the Naval Discipline Act. If the scheme of a united Imperial Navy be adopted no further Imperial legislation will be required beyond that which has been already indicated . . .

The question of the actual flag to be flown by Dominion ships of war under the new conditions contemplated would be determined by the position which these vessels will occupy in relation to the Royal Navy. Under the scheme proposed the relations of the Dominion ships to the Royal Navy will be very intimate, and having regard to the convenience of peace administration and to efficient co-operation with the Royal Navy in time of war, the use of the White Ensign without any distinctive badge may well be offered in the first instance to the Dominion Governments. At the time of the Imperial Defence Conference in 1909, the question of the flag did not come up prominently, but certain enquiries were made by the representatives of Canada and Australia, and a wish was expressed by them that the White Ensign with the distinctive emblem of the Dominion in the fly might be sanctioned. The use of the White Ensign without any distinctive badge apparently did not occur to the representatives of the Dominions or to the representatives of the Admiralty; but the granting the use of the White

Ensign, without a distinctive badge, would foster and maintain a feeling of union between the Dominion naval forces and the Royal Navy, and would show that it is desired to offer the new Dominion ships a full share in the heritage of the British Navy . . .

Since the ships are to be provided at the expense of the Dominions, it must be left to those Governments to control the administration, and, in time of peace, the disposition of the ships. In war time the ships cannot be used without the consent of the Dominion concerned, but, if used, they will no doubt be placed under the direction of the Admiralty. In peace time some special provision must obviously be made, or else it would be within the power of the Dominion Governments to order or permit their ships to take action in relation to foreign Powers for which the Imperial Government would be responsible, but which they would not be able to prevent or control. In such an event the Imperial Government might be seriously hampered in the control of the foreign policy of the Empire, and might be committed to a policy, or even to a war, of which they did not approve. This danger is not merely academical, but may easily arise.

[78] *Telegram, Crewe to the Governor-General of Canada [Grey]*
15 August 1910

With reference to enclosure to my despatch No. 571 3rd August [, the] Admiralty have been engaged in consideration of various legal and international questions arising out of the creation of Dominion fleets as arranged at last year's Conference and I am sending by mail for communication to your Ministers copy of a memorandum on the subject. Admiralty consider that for full understanding and settlement of this matter personal conference with representatives of your government is indispensable and as questions involved are in part legal they suggest that advantage should be taken of [the] presence of [your] Minister and Deputy Minister of Justice in Europe who will no doubt be shortly returning to this country. In view of [the] early transfer of cruisers to your Government [the] Admiralty are earnestly desirous that [a] conference should take place as early as possible and I should be glad if your Ministers could agree and could send necessary communications to [the] Ministers here. It would be most useful if they could be authorised at all events to come to some provisional arrangement with [the] Admiralty.

[79] *Memorandum by the Assistant Admiralty Secretary [Greene]*
on the 'Status of Commonwealth Naval Force, Report of
Conference with High Commissioner', referred to [Thomas] the
Admiralty Secretary, [Admiral Sir Arthur K. Wilson] the First
Sea Lord and [McKenna] the First Lord

8 February 1911

As approved by Their Lordships [,] the bearing of [the] Admiralty Memorandum of August 1910 upon the organisation of the new Commonwealth Naval Force from the point of view of status and discipline has been discussed with the High Commissioner for Australia, Sir George Reid, and the result has been embodied in the attached report which has been signed unanimously.

The report follows the lines of the Memorandum but a slight departure from the procedure suggested originally has been made in order to apply more simply the principle of full local control combined with general control by the Admiralty. Whereas in the former paper it was proposed that in certain circumstances there should be reserved to the British Commander-in-Chief in Australian waters his power of command or discipline under the Naval Discipline Act, it is now proposed that the Australian Commander-in-Chief should in Australian waters (as defined) have an independent position similar to that of a Commander-in-Chief of a Station under the Admiralty, and further that the Commonwealth Minister of Marine should have power to issue a Commission to an Officer of the Commonwealth Force empowering him to order Courts Martial. This modification is not important provided that the Imperial Naval Discipline Act and King's Regulations are applied in all essential respects to the Dominion Naval Force, and also that the scheme is accepted as a whole.

It is understood that Sir George Reid is forwarding a copy of the report to his Government for consideration. It is desirable, therefore, that the action to be taken on the report should be considered without delay. It is unlikely that the Commonwealth Government will now express any final opinion until the meeting of the Imperial Conference, but it would be an advantage that the report, if approved by their Lordships, should be referred as soon as possible to the Colonial Office and also the Committee of Imperial Defence, who have the subject under consideration.

Referred for any remarks.

[80] *Report of a Sub-Committee of the Committee of Imperial Defence assembled to formulate questions connected with Naval and Military Defence of the Empire to be discussed at the Imperial Conference, 1911*

March 1911

'Secret'

Appendix III, 'Co-operation of the Naval Forces of the United Kingdom and of the Dominions, General Principles laid down by the Admiralty.'

The British, Canadian, and Australian fleets to be sister members of the King's Navy, hoisting the white ensign at the stern and the distinctive flag of each nation on the jack staff.

2   Each fleet to be administered by its separate Admiralty.

3   As an essential condition of this partnership and as a preliminary to its formation the Dominion Admiralties to agree with the British Admiralty upon a common system of naval discipline and training.

4   By arrangement, officers and men to be interchangeable between the three fleets.

5   Each fleet to confine its operations primarily within the area of its own station. Special arrangements to be made between the Admiralty concerned and the British Admiralty when the particular service on which a Dominion fleet is engaged takes it outside its own area. Visits by Dominion fleets to foreign countries to be subject to approval by Imperial Government.

6   In the event of war the British Admiralty to control the whole of the King's Navy, but each Admiralty to be at liberty to withdraw from the partnership before joining in hostilities. If a Dominion Government has resolved upon common action in a war, it is not to be at liberty to withdraw its fleet from the partnership so long as the war lasts.

Appendix IV 'Organisation of Australian Naval Force, Provisional arrangements agreed to by the Representatives of the Admiralty and the High Commissioner of Australia;'

Signed by R.B.D. Acland [Judge Advocate of the Fleet], A.E.

Bethell [Rear-Admiral the Hon. Alexander E. Bethell, Director of Naval Intelligence], W. Graham Greene [Assistant Secretary of the Admiralty], and G.H. Reid [Sir George Reid, Australian High Commissioner], 31 January 1911

The following proposals are put forward as a whole, as steps in the direction of carrying out the arrangements provisionally arrived at during the meetings of the Imperial Defence Conference held in 1909, which are recorded as follows in Parliamentary Paper Cd. 4948:

1 'In peace time and while on the Australian Station the (Australian) Fleet Unit would be under the exclusive control of the Commonwealth Government as regards their movements and general administration, but

2 'Officers and men should be governed by regulations similar to the King's Regulations, and be under naval discipline, and when with vessels of the Royal Navy the Senior Officer should take command of the whole ...'

The Australian Fleet Unit should form part of the Eastern Fleet of the Empire, to be composed of similar units of the Royal Navy, to be known as the China and the East Indies Units respectively and the Australian Unit.

I Control

The question of 'control as regards movements of ships and general administration' divides itself into two main heads: (1) As regards the provision of matériel and personnel, and what may be called the internal economy of the ships; (2) control of the movements of the ships: (a) in time of peace; (b) in time of war.

(1) This is to be entirely within the province of the Commonwealth. It was common ground in discussion between the Admiralty and Australian Representatives at the Conference that the training of the cadets and of the men and boys before they actually join a ship should be the same, as far as possible, as that provided for entrants into the Royal Navy.

At the present moment the question of the subsequent training of junior officers in the Australian Service is not pressing, as for some time they will be directly under officers steeped in the traditions of the Royal Navy, and the Admiralty have already expressed their readiness to arrange for officers being appointed to ships of the Royal Navy for instruction. (See Cd. 4948, p. 26.)

(2) (a) Control of Movements of Ships in Time of Peace

It having been agreed that the Australian Fleet Unit shall form part of the Eastern Fleet of the Empire, it follows from the Admiralty point of view that strictly the vessels of that unit would be liable to come under the command of the Commander-in-Chief of the Eastern Fleet if he is the Senior Officer. But on the other hand, if this were acted on to its full extent it would be an infringement of the principle which has been equally accepted that the Commonwealth Government should be supreme in the matter of directing the movements of the Commonwealth Fleet Unit whilst in Australian waters.

To avoid this difficulty the following proposal is made:

Each unit of the Eastern Fleet to have its own station, and in ordinary circumstances not to leave it, the Australian Station under the command of its own Commander-in-Chief being so defined as to include the adjacent high seas to an extent sufficient to provide for voyages between different parts of the Commonwealth and its dependencies, and for the necessary local movements of the ships incidental to manoeuvres, instruction, &c.

If the Australian Government desire to send the vessels of the Australian unit outside the limits of their station, so defined, arrangements should be made with the Imperial Government in a manner similar to that which is usual between the Admiralty and the Foreign Office. Should any of the Commonwealth ships, at the request of the Admiralty and under the orders of the Commonwealth Government, be sent on Imperial duty beyond the limits of the Australian station, they would, whilst so detached – in accordance with the routine which now prevails in the Royal Navy in similar cases – be under the command of the Commander-in-Chief of the station on which they might be temporarily serving.

Ships of any unit entering or passing through another station to report themselves to the Commander-in-Chief of that station, who would not, however, interfere with their movements or internal economy.

Arrangements to be made between the Admiralty and the Commonwealth Government for the ships of the Australian unit taking part in fleet cruises, or for any other joint training considered necessary, under the Senior Commander-in-Chief. While so employed the ships to be under the command of that officer, who would not, however, interfere in their internal economy further than the circumstances of the case necessitated.

The appointment to the Naval Board by the Governor-General, on the advice of his Ministers and with the concurrence of the Admiralty, of a specially selected Flag Officer of the Royal Navy as the Chief Naval Adviser of the Minister of Defence, appears to be highly desirable, having regard to the special arrangements proposed.

Such Flag Officer, or other officers and men as may be needed for the Australian unit, to be lent by the Admiralty to the Commonwealth under arrangements to be agreed upon. Preference to be given to officers and men of Australian birth or origin, but they should all be volunteers for the service.

Service of officers of the Royal Navy in the Australian Naval Force to count in all respects for promotion, pay, retirement, &c., as service in the Royal Navy.

Under these arrangements the Commonwealth Government would have 'the exclusive control during peace time and while on the Australian Station of their fleet unit as regards movements and general administration.' It would only be placed under the command of a Senior Officer of the Royal Navy independent of that Government when they saw fit, and even then such Senior Officer would have instructions not to interfere in matters of internal economy more than absolutely necessary.

Under the provisional arrangement made at the Imperial Defence Conference 1909 many questions of seniority must arise between officers commanding vessels of the Royal Navy and those of the Australian Fleet Unit. In order to determine these questions the names of such officers should be shown in the Navy List, their seniority being determined by the date of their commissions in either the Royal Navy or Australian service, whichever is the earlier.

2   (b)   Control of Movements of Ships in Time of War

It was provided in the arrangement above referred to that, 'when placed by the Commonwealth Government at the disposal of the Admiralty in war time, the vessels should be under the control of the Naval Commander-in-Chief.'

Accordingly, in time of war the Commonwealth Government would be free to place their vessels at the disposal of the Imperial Government or not, as they decide. If the vessels are so placed, they would become an integral part of the Imperial Fleet under the command of the Commander-in-Chief, and remain under his authority wherever they may be, within or without the 'Australian

Station,' and be liable to be sent anywhere during the continuance of the war.

In peace time the orders of the Minister of Defence would be sufficient for placing the Commonwealth ships under the command of an officer of the Royal Navy, but the transfer in time of war requires something more formal if only to emphasise the seriousness of the step.

It is desirable, therefore that the Australian Fleet unit should only come in the fullest sense under the command of a Commander-in-Chief other than its own after the making of a formal order of the Governor-General in Council placing the fleet unit, in the words of the Colonial Naval Defence, 'at the disposal of His Majesty for general service' for the period of the war.

II   Discipline

In turning to the naval discipline part of the subject, it is essential to bear in mind that, as already mentioned, it has been accepted that the officers and men should be under naval discipline, and that the Australian Fleet Unit should form part of the Eastern Fleet of the Empire.

The due carrying out of this provisional agreement involves two important points: first, that the disciplinary code in force in what may be considered for the present purpose two parts of one great fleet should be the same; second, that as a fleet exists primarily for fighting purposes, there should be as little change as possible when the time of trial comes.

There are many difficulties – constitutional, legal, and other – in devising a code of discipline for a fleet which is for practical purposes sometimes to be one whole and at others two or more separate parts, one of which is built, equipped, and manned by a sovereign Power, and the other under the authority of a Legislature which, however extensive its powers, has not the full attributes of sovereignty.

Among these difficulties is the extent to which a Dominion has power to legislate for persons outside its territorial waters; assuming that it has such power, there is the probability, amounting almost to a certainty, that statutes enacted by two or more Legislatures would not be identical. There is the question of what would be the true international position in peace and in war of ships of war equipped, and with officers commissioned, by authority of a non-sovereign Legislature; there is the difficulty which must arise of providing in all circumstances for tribunals to punish infractions of the disciplinary code; and lastly, there is the importance of securing to the Commonwealth Government the power of deciding whether or not discipline, as it is understood in the Royal Navy, shall apply to the officers and men of the ships provided by the Commonwealth.

In order to obviate all these difficulties the following proposal is made:

That the Naval Discipline Act should be so amended as to make it clear that the expressions 'His Majesty's navy' and 'His Majesty's ship' shall, for the purposes of that Act, include any fleet unit and ships composing that unit respectively, which is provided and maintained by a self-governing Dominion, where the Governor-General in Council orders that the Naval Discipline Act shall apply to the officers and men of the naval force maintained by that Dominion. In the amending Act, power should be reserved to the Governor-General in Council to prevent the application of any sections of the Naval Discipline Act, except those contained in Parts I, II, III, IV, and Section 86 to 98 inclusive of Part VI, and Section 101 of Part VII.

Further, the Act should be amended so as to provide that the expressions 'Admiralty,' 'Lords Commissioners of the Admiralty,' of 'Secretary of the Admiralty,' should mean in relation to any ship provided and maintained by a Dominion, or to any officer of man borne on her books, the Minister at the head of the Naval Department of the Government in question; and that in Section 46 the expression 'United Kingdom' should mean, in reference to any such ship, officer, or man, the Dominion providing and maintaining them. (The object of this last proposal is to maintain inviolate the jurisdiction of the Colonial Courts to try persons belonging to the local force for offences committed on shore in the Dominion against the ordinary law of the Dominion.)

No doubt much drafting detail would have to be considered hereafter to give effect to this proposal, but if the principle is accepted that the Act is to apply to the Dominion Fleet units (though only by virtue of an Order of the Governor-General in Council) all the difficulties mentioned above will be overcome.

(1) It leaves the Dominion free to decide whether it will accept the code of naval discipline for the time being in force in the Royal Navy; but, if it does accept it, it secures that the code shall be identical throughout, and avoids the difficulties which might

arise from parliamentary complications or want of accurate drafting.

(2) No change will be necessary when, in time of war, Dominion vessels are transferred.

(3) It avoids the difficult questions as to the power of the Dominions to legislate extra-territorially and yet provides for the code being in full force whether the ship is in home or foreign waters or the waters of another Dominion.

(4) It settles definitely the status of Dominion officers and men vis-à-vis foreign nations.

(5) By making the officers definitely officers of His Majesty's navy it decides all questions of naval precedence and courtesy, both among themselves and with regard to foreign officers.

(6) It avoids all the difficult questions with regard to courts-martial. The officers being officers of His Majesty's navy, and being also members of a force provided and maintained by a Dominion, can be commissioned to order courts-martial either by the Admiralty or by the Minister of Defence, or both, and they would be qualified to sit, in accordance with their seniority, on a court-martial for the trial of an offender of either branch of the service.

So much for the Naval Discipline Act. But the law of the navy is embodied not only in the Act, but also in the King's Regulations and Admiralty Instructions, and it is desirable that such parts of the Regulations as deal with the duties of the different officers, with discipline, and the powers of punishment vested in the Captains and executive officers, should also be identical in both branches of the service.

It is suggested that on the one hand the Dominions should have the power of making whatever regulations they think fit, provided that they are not inconsistent with the Naval Discipline Act; but on the other hand, the Admiralty consider that they should be free to withdraw the officers and men of the Royal Navy from the Dominion service unless, at least, all those chapters which affect either the fighting efficiency or the discipline of the ship are enacted in identical terms . . .

The scheme, as herein sketched, covers the main points dealt with in order to carry out the Provisional Agreement. Once the main principles have been accepted, other points of great but subsidiary importance will have to be considered before a complete scheme can be worked out.

Lastly, there remains the question of the flag. With regard to this, the Admiralty earnestly desire that the flag to be flown in all the Dominion fleet units should be the White Ensign together with the Dominion flag on the jackstaff, the ships of the Royal Navy flying the White Ensign with the Union Jack on the jackstaff.

P.S. [Initialled by] G.H.R. [Sir George Reid]

The question of the flag is reserved for the decision of the respective Governments.

[81] *Certified Copy of a Report of the Committee of the Privy Council of Canada approved by His Excellency the Governor-General, signed by R. Boudreau, Clerk of the Privy Council*
3 March 1911

The Committee of the Privy Council have had before them a report, dated 1st February 1911, from the Secretary of State for External Affairs, representing that, with a view to promote the dignity and importance of the Canadian Navy, and to emphasise the fact that the Command-in-Chief thereof is vested in the King, Your Excellency's Advisers humbly desire that the official title of the 'Naval Forces of Canada' shall be 'The Royal Canadian Navy.'

The Committee, on the recommendation of the Secretary of State for External Affairs [Charles Murphy, Secretary of State for Canada and ex-officio for External Affairs], advise that Your Excellency may be pleased to ask the Right Honourable the Principal Secretary of State for the Colonies to cause this request of the Government of Canada to be laid before His Majesty.

All which is respectfully submitted for approval.

[82] *Minutes of the 109th Meeting of the Committee of Imperial Defence, discussing Committee of Imperial Defence paper 70-C, 'The International Status of the Dominions During a War in which the United Kingdom is Engaged.'*
24 March 1911

MR MCKENNA [First Lord of the Admiralty] read paragraph 4 of his Memorandum printed in the paper under discussion, which he said embodied his views as to the General Policy. He also desired to add that in the particular case of Australia it had been agreed between the Commonwealth Government and the Admiralty that once the Australians had placed their naval forces at the disposal of the Admiralty they were not at liberty to withdraw them while the war lasted. If the question dealt with in this paper were now raised, this agreement would come to an end, and it was improbable that by pressing this question we should improve our position. No naval agreement had yet been made with Canada.

SIR ARTHUR NICOLSON [Under-Secretary of State for Foreign Affairs] said that his views were contained in his Memorandum in the paper under discussion. If the United Kingdom were in a state of war, her Dominions were necessarily in a state of war also. For them to proclaim their neutrality was to announce their secession from the Empire. If there was any doubt of what their attitude was likely to be, it was quite useless to communicate the report of Lord Hardinge's [Viceroy of India] Sub-Committee to them.

MR HARCOURT [Secretary of State for the Colonies] suggested that the omission of paragraph 3 of the abridged report (CID Paper 124-B) might perhaps obviate the question being raised. The difficulty was one of domestic politics in the Dominions. There was also grave risk of information, however secret, leaking out. The publication of confidential matters in the Dominion newspapers was a frequent occurrence.

SIR WILLIAM NICHOLSON [Chief of Imperial General Staff] pointed out that the Dominions were constantly enquiring how their military forces could aid us in case of war, which did not look as if they really entertained any idea of not taking part in a serious struggle. He suggested that the addition of the words 'outside their own territories' at the end of paragraph 3 might perhaps solve the difficulty.

LORD HALDANE [QC, Liberal MP] said that he thought that we should act on the assumption that the Dominions would take part in a serious war. There was very little doubt that they would, and he was sure that their statesmen were quite alive to the fact that they would really have very little choice, none unless the enemy so willed; but for reasons of internal politics, which were quite comprehensible, they declined to bind themselves beforehand. The substitution of the word 'discretion' for 'freedom' in the fourth line of paragraph 3 might get over the difficulty to some extent, or, perhaps better still, the amendment of the last two lines to read: 'it will not in any way prejudice their control over their own affairs.'

SIR ARTHUR WILSON [First Sea Lord] said that it was undesirable to communicate our secret plans to the Dominions if there was any doubt as to what their attitude was likely to be on the outbreak of war. The prompt execution of the measures proposed by Lord Hardinge's Committee in the Dominion ports was of less importance than in Home ports, in as much as there was less chance of the ultimate escape of the enemy ships.

MR CHURCHILL [Home Secretary] did not think that any of the amendments suggested for paragraph 3 of the report removed the difficulty, and he was inclined to doubt whether the measures proposed by Lord Hardinge's Sub-Committee were of sufficient importance to warrant the risk of raising this dangerous question.

MR MCKENNA said that he feared that the addition of the words proposed by Sir William Nicholson would tempt the Dominions to tie up their fleets in local waters, which was exactly what the Admiralty most wished to avoid. The omission of paragraph 3 would not suffice, as the report contained an intimation that on the outbreak of war the Home Government would telegraph a direct order to the Dominion Governments to take definite hostile action against enemy ships, and this of course raised the same question of their liberty to abstain from taking part in the war. They took the view that the enemy would not necessarily attack them; it was quite possible that he would not; and they claimed accordingly the right to decide whether they would attack him or not. He did not think that there was any great importance or urgency in communicating the report to the Dominions. It could be sent to the naval Commanders-in-chief now, but he did not think the advantages to be gained by providing for the execution of the policy laid down by Lord Hardinge's Sub-Committee in Dominion ports was worth the risks we should run of raising this very awkward question.

LORD ESHER [Liberal MP] pointed out that the report made it clear that very careful arrangements were necessary if the policy laid down was to be carried out, and that it would be impossible to communicate full details of necessary action expected from them on the outbreak of war by telegraph. These arrangements required elaborate machinery, which must be set up beforehand. Improvised measures would certainly fail.

THE PRIME MINISTER [Mr Asquith] agreed that the report

with paragraph 3 omitted would be likely to invite enquiry, while its inclusion gave sanction to a doctrine which was not tenable in law without secession.

The action to be taken by the Dominions in carrying out the policy laid down by Lord Hardinge's Sub-Committee was especially important, for, as was shewn in paragraph 31 of the original report [CID paper 120-B] in the particular case of Germany, whereas on a given date there were German steam-vessels aggregating 67,000 tons in the ports of the United Kingdom, there were on the same date German steam vessels aggregating 80,000 tons in British ports oversea. The arrangements for their detention in port as proposed was not so much a naval question as one for the custom-house officers, assisted where necessary by the police or other forces.

He was a good deal impressed by the danger of leakage of information as to our secret plans for war.

A satisfactory solution of the matter was very difficult to discover. Perhaps the best course would be to try and make some opportunity to talk the matter over with the representatives of the Dominions, when they come, before taking any further steps.

(Conclusion) It would be desirable to have a preliminary discussion of the subject with the representatives of the Dominions before taking further action.

[83] *[Admiralty Secretary?] to H.W. Just, Assistant Under-Secretary of State for the Colonies*

1 April 1911

Private draft (unsent)

We had our second meeting with Mr Smith [Canadian KC, special representative of the Laurier administration] yesterday afternoon, when we discussed more in detail the question of the status of the Canadian Fleet.

Mr Smith had evidently been impressed by our previous discussion and though he again put forward the wish of the Canadian Government that their naval force should be treated as a separate Navy yet he admitted the advantages to that force of an arrangement which provided for a close connexion with the Royal Navy. He concentrated his remarks upon the questions of the flag and free movement of the ships in any part of the world. As to the former he had Sir W. Laurier's instructions to make a strong point of the flag at the ensign staff bearing the Canadian arms, but I think we shook him in his views and he will probably be content to report that the Admiralty were still strong on the matter and leave the question to Ministers as Sir G[eorge] Reid [Australian High Commissioner] did. On the second point he seemed to think that if the Government only looked for a notification by the Canadian Government that they proposed to send a ship to some foreign or Colonial port and that the Imperial Government would thereupon simply intimate whether there was, or was not, an objection to the visit, Sir W. Laurier would probably be satisfied. He asked, however, whether the notification would have to pass in every case through the Colonial Office, as he thought it would be more palatable if the Canadian Government could notify their wishes as to the proposed visit to a foreign port direct to the Foreign Office. As to that we pointed out that a recognised channel between a Dominion and Home Government was desirable and that the point would have to be considered by the Colonial Office – but in a matter of purely formal procedure adherence to this general rule might be relaxed, if no inconvenience resulted.

We further discussed the disciplinary position of the officers of the various Services under the proposed arrangement with the Dominion Governments, and the circumstances in which the senior officer of any Service would take command – it being explained that the administrative control would in all cases rest with the Government maintaining the ships – and I think we reassured him that an agreement of such a kind would not interfere in practice with the autonomy of the Dominion Service, while it would be a very great advantage to that Service especially during the process of development.

We also discussed the area of the proposed Canadian Naval Stations on the Pacific and Atlantic sides and pointed out that within those stations the Canadian Ships would have the same freedom of action as ships of the Royal Navy on their separate stations.

[Minuted] Not sent – First L[ord] arranging to write to W Harcourt.

[84] *Memorandum by [Greene?], 'Status of Canadian Naval Force, Conference with Mr R.C. Smith K.C.,' referred to [Thomas?] Admiralty Secretary, [Wilson] First Sea Lord, and [McKenna] First Lord*

3 May 1911

A copy of the arrangement discussed with Mr R.C. Smith, who was deputed by the Canadian Government to confer with the Admiralty, is submitted. It is based on the principles approved by the Imperial Defence Committee and follows in the main the lines of the arrangement discussed with Sir G. Reid, as representing the Australian Government. It is to be observed, however, that Mr Smith expressed a doubt whether the Canadian Government would consent to the wording of paragraph 6, in regard to its having no liberty to withdraw its ships during the course of hostilities when once in co-operation with the Royal Navy. On the part of the Admiralty it was pointed out that the paragraph restricting the freedom of the Admiralty to send their ships into Canadian Stations should not be interpreted to mean more than that they would not station ships on those Stations and that they would not send ships into Canadian ports without notification to the Canadian Government, and the wording was only acquiesced in inasmuch as Mr Smith had used it in his personal telegram to his Government.

It was definitely understood that the question of the flag was reserved for negotiation between Ministers at the Conference.

Mr Smith was requested to consider the Memorandum as confidential and it was not to be made public, it being only to prepare the way for the Conference.

Naval Service of Canada

1   The Canadian Fleet to be a sister member of the King's Navy, the other members being the Fleet of the United Kingdom and the Fleet of Australia.

2   All units of the King's Navy to hoist a common ensign, the White Ensign, as the symbol of the authority of the Crown and each to hoist also the member's own distinctive flag.

3   The Canadian Fleet to be entirely under the administrative control of the Dominion Government.

4   The training and discipline of the Canadian Fleet to be uniform with that of the Fleet of the United Kingdom.

5   In time of peace:

a)   The Canadian Government to have entire control of the movements of ships in the Canadian Naval Stations, which shall be constituted so as to include in the Atlantic Station the waters North of 30° and West of 40° and in the Pacific Station the waters North of 30° and East of 180°.

b)   The Stations so defined to be treated as reserved to the Canadian Fleet, and the Admiralty not to send their ships to Canadian waters except {after notification to / with the concurrence of} the Canadian Government. United States ports, Bermuda and Newfoundland to be outside this arrangement.

c)   In the event of the Canadian Government desiring to send ships to foreign ports, or ports outside such Stations, the proposal to be {concurred in by / notified to} the Imperial Government in such time and manner as is usual between the Admiralty and the Foreign Office.

d)   Outside of such Stations, the Canadian Ships, when with ships of Royal Navy, to have the same relations to other ships of King's Navy as obtain in the Royal Navy. The Senior Officer, whether Canadian or Imperial, to have command but no power to direct the movements of the ships of the other Service, unless engaged by arrangement on joint Imperial Service, or by arrangement on joint Fleet cruises or exercises.

6   In the event of war, the Canadian Government to be at liberty to withdraw its Fleet from membership with the King's Navy before joining in hostilities. The British Admiralty to control the whole of the King's Navy engaged in hostilities and if a Dominion Government has resolved upon common action in a war it is not be at liberty to withdraw its Fleet from the membership so long as the war lasts. In accordance with the Canadian Naval Service Act, 1910, the placing of the Canadian Fleet at the disposal of the Admiralty to be the subject of a formal Order of the Governor General in Council.

7   The Naval Discipline Act to be amended so as to enable the Governor General of a Dominion by Order in Council to apply the Naval Discipline Act and any amendment thereof to the Ships, officers and men provided and maintained by that Dominion, and the Act to be also amended so as to provide that when the Act is so applied the expressions 'His Majesty's Navy' and 'His Majesty's Ships' shall include for the purposes of that Act the Fleet provided and maintained by such Dominion and the ships comprising such Fleet respectively. Power to be reserved to the Governor General

in Council to prevent the application of any sections of the Naval Discipline Act except those contained in Parts I (except section I) II, III, IV, and Sections 86 to 98, inclusive, of Part VI and Section 101 of Part VII.

The Act to be further amended so as to provide that the expressions 'Admiralty', 'Lords of the Admiralty' and 'Secretary of the Admiralty', shall mean, in relation to the Fleet and ships provided and maintained by a Dominion or any officers and men borne on the ships' books, the Ministerial authority charged with the administration of the Naval Service of the Government in question, and that in relation to such officers and men the expression 'England or the United Kingdom' whereever used should mean the Dominion providing and maintaining the Fleet and ships.

8   Should a Court Martial have to be assembled and a sufficient number of officers of the requisit rank not be available in the Canadian Naval Service at the time, arrangements to be made by the Admiralty, at the request of the Minister of Naval Service, for detailing additional officers from the nearest Station of the Royal Navy.

[85] *Imperial Conference 1911, 'Memorandum of Conferences Between the British Admiralty and Representatives of the Dominions of Canada and Australia'*
June 1911

'Confidential'

The naval services and forces of the Dominions of Canada and Australia will be exclusively under the control of their respective Governments.

2   The training and discipline of the naval forces of the Dominions will be generally uniform with the training and discipline of the fleet of the United Kingdom, and, by arrangements, officers and men of the said forces will be interchangeable with those under the control of the British Admiralty.

3   The ships of each Dominion naval force will hoist at the stern the white ensign as the symbol of the authority of the Crown, and at the jack-staff the distinctive flag of the Dominion.

4   The Canadian and Australian Governments will have their own naval stations as agreed upon and from time to time. The limits of the stations are as described in Schedule (A), Canada, and Schedule (B), Australia.

5   In the event of the Canadian or Australian Government desiring to send ships to a part of the British Empire outside of their own respective stations, they will notify the British Admiralty.

6   While the ships of the Dominions are at a foreign port a report of their proceedings will be forwarded by the officer in command to the Commander-in-Chief on the station or to the British Admiralty. The officer in command of a Dominion ship so long as he remains in the foreign port will obey any instructions he may receive from the Government of the United Kingdom as to the conduct of any international matters that may arise, the Dominion Government being informed.

7   The Commanding Officer of a Dominion ship having to put into a foreign port without previous arrangement on account of stress of weather, damage, or any unforeseen emergency will report his arrival and reason for calling to the Commander-in-Chief of the station or to the Admiralty, and will obey, so long as he remains in the foreign port, any instructions he may receive from the Government of the United Kingdom as to his relations with the authorities, the Dominion Government being informed.

8   When a ship of the British Admiralty meets a ship of the Dominions, the senior officer will have the right of command in matters of ceremony or international intercourse, or where united action is agreed upon, but will have no power to direct the movements of ships of the other service unless the ships are ordered to co-operate by mutual arrangement.

9   In foreign ports the senior officer will take command, but not so as to interfere with the orders that the junior may have received from his own Government.

10   When a court-martial has to be ordered by a Dominion and a sufficient number of officers are not available in the Dominion service at the time, the British Admiralty, if requested, will make the necessary arrangements to enable a Court to be formed. Provision will be made by order of His Majesty in Council and by the Dominion Governments respectively to define the conditions under which officers of the different services are to sit on joint courts martial.

11   The British Admiralty undertakes to lend to the Dominions during the period of development of their services, under conditions to be agreed upon, such flag officer and other officers and men as may be needed. In their selection preference will be given to officers and men coming from, or connected with, the Dominions, but they shall all be volunteers for the service.

12   The services of officers of the British fleet in the Dominion naval forces, or of officers of these forces in the British fleet, will count in all respects for promotion, pay, retirement, &c., as service in their respective forces.

13   In order to determine all questions of seniority that may arise, the names of all officers will be shown in the Navy List and their seniority determined by the date of their commissions, whichever is the earlier, in the British, Canadian, or Australian services.

14   It is desirable, in the interests of efficiency and co-operation, that arrangements should be made from time to time between the British Admiralty and the Dominions for the ships of the Dominions to take part in fleet exercises or for any other joint training considered necessary under the Senior Naval Officers. While so employed, the ships will be under the command of that officer who would not, however, interfere in the internal economy of ships of another service further than absolutely necessary.

15   In time of war, when the naval service of a Dominion, or any part thereof, has been put at the disposal of the Imperial Government by the Dominion authorities, the ships will form an integral part of the British fleet, and will remain under the control of the British Admiralty during the continuance of the war.

16   The Dominions having applied to their naval forces the King's Regulations and Admiralty Instructions and the Naval Discipline Act, the British Admiralty and Dominion Governments will communicate to each other any changes which they propose to make in those Regulations or that Act.

Naval Discipline (Dominion Naval Forces) Bill Memorandum

The Bill is intended to give effect to the agreement arrived at by the Imperial Conference with respect to the government and discipline of the naval forces of the self-governing dominions and their relations to the Royal Navy and to one another, namely, that there should be power by mutual agreement between the home Government and the dominion Government, and between one dominion Government and another, to provide that the fleets and ships of the Royal Navy and of the dominion navy and of one dominion navy an another dominion navy when co-operating together should be treated, so far as discipline is concerned, as belonging to the same service, so that an officer of one such force should be able to be treated as a superior officer in relation to an officer of another such force and to assume command over the ships belonging to such other force and that for the purposes of courts-martial officers could be taken indiscriminately from both or all such forces.

The Bill is drawn in the form of declaring what is the effect of legislation passed by a self-governing dominion applying the Naval Discipline Act to the naval forces raised by the dominion. It does not touch the question as to whether the Legislatures of the self-governing dominions have power to legislate with respect to the discipline of their naval forces outside the limits of the dominion.

The Bill will not apply to any dominion unless two conditions are satisfied:

(1)   that the Legislature of the dominion has applied the Naval Discipline Act to its naval forces;

(2)   that the dominion has made provision that the Bill should apply to the dominion.

If these two conditions are fulfilled the effect of the Bill will be to place, when so desired so far as discipline is concerned, the dominion naval forces in the same position in relation to the British naval forces and to one another as if each were an integral part of the Imperial naval forces.

Draft of a Bill for declaring the effect of the Naval Discipline Acts when applied by the legislatures of self-governing dominions to the Naval Forces raised by such Dominions.

Be it enacted by the King's most Excellent Majesty, by and with the advice of the Lords Spiritual and Temporal, and Commons, in this present Parliament assembled, and by the authority of the same, as follows:

1   (1)   Where in any self-governing dominion provision has been made (either before or after the passing of this Act) for the application to the naval forces raised by the dominion of the Naval Discipline Act, 1866, as amended by any subsequent enactment, that Act, as so amended, shall have effect as if references therein to His Majesty's Navy and His Majesty's ships included the forces and ships raised and provided by the dominion, subject, however:

(a)   in the application of the said Act to the forces and ships raised and provided by the dominion, and the trial by court-martial of officers and men belonging to those forces; to such modifications and adaptations (if any) as may have been or may be made by the law of the dominion to adapt the Act to the circumstances of the dominion, including such adaptations as may be so made for the purpose of authorising or requiring anything, which under the said Act is to be done by or to the Admiralty or the Secretary of the Admiralty, to be done by or to the Governor General or the authority charged with the administration of the naval service of the dominion; and

(b) in the application of the said Act to the forces and ships of His Majesty's Navy not raised and provided by a self-governing dominion in relation to the forces and ships so raised and provided, to such modifications and adaptations as may be made by His Majesty in Council:

Provided that where any forces and ships so raised and provided by a self-governing dominion have been placed at the disposal of the Admiralty, the said Act shall apply without any such modifications or adaptations as aforesaid.

(2) This Act shall not come into operation in relation to the forces or ships raised and provided by any self-governing dominion unless or until provision to that effect has been made in the dominion.

(3) For the purpose of this Act the expression 'self-governing dominion' means the Dominion of Canada, the Commonwealth of Australia, the Dominion of New Zealand, the Union of South Africa, and Newfoundland.

2 This Act may be cited as the Naval Discipline (Dominion Naval Forces) Act, 1911.

Notes of Meeting at the Admiralty, 20 June 1911

Australia

Present: Reginald McKenna (First Lord of the Admiralty),
Admiral of the Fleet Sir A.K. Wilson (First Sea Lord),
G.F. Pearce, (Minister of Defence of the Commonwealth of Australia),
Rear-Admiral Sir Charles Ottley (Secretary of the Imperial Defence Committee),
Rear-Admiral The Honourable A.E. Bethell (Director of Naval Intelligence),
Mr W. Graham Greene (Assistant Secretary of the Admiralty), Mr G.W. Johnson (Colonial Office),
Captain Maurice P.A. Hankey (Assistant Secretary of the Imperial Defence Committee).

The Memorandum of Agreement drawn up between the representatives of the Admiralty and the Commonwealth was discussed clause by clause and certain amendments were approved. Mr Pearce raised the question of the proposed limits of the Australian Station, which he would like to see extended to include the island of Timor on the North and the British Solomon Islands, the New Hebrides and New Caledonia on the West. It was pointed out to him that the present international obligations of H.M. Government as regards the New Hebrides and other foreign political considerations rendered it undesirable to extend the limits of the Station at present. The limits would be fixed by mutual agreement from time to time and circumstances in the future might remove the present difficulties. Sir A. Wilson also stated that the so called policing of the Pacific Islands would be done by one or two small vessels forming part of the China Squadron and would not as a matter of course devolve upon the vessels stationed in New Zealand waters, which were not of the appropriate class. On this understanding Mr Pearce concurred in the limits as stated in the Schedule.

The Bill drafted to amend the Naval Discipline Act was agreed to without modification.

It was decided that there would be no objection to the publication of the Agreement with Australia in the official report of the Conference.

Notes of a Meeting at the Admiralty, 29 June 1911

Present: Reginald McKenna (First Lord of the Admiralty),
Admiral of the Fleet Sir A.K. Wilson (First Sea Lord),
L.P. Brodeur (Minister of Naval Service, Canada).
G.F. Pearce, (Minister of Defence of the Commonwealth of Australia),
Rear-Admiral The Honourable A.E. Bethell (Director of Naval Intelligence),
Mr W. Graham Greene (Assistant Secretary of the Admiralty),
H.W. Just, (Colonial Office),
Captain Maurice P.A. Hankey (Assistant Secretary of the Imperial Defence Committee).

The draft Memorandum of the result of Conferences between the Admiralty and the representatives of Canada and Australia was discussed and certain amendments were approved.

In connexion with the draft Bill declaring the effect of the Naval Discipline Act when applied by the Dominions Mr Brodeur observed that the Canadian Government would think it necessary to legislate independently of Imperial legislation. In his opinion the Bill as drafted for the Admiralty seemed to imply a right to legislate for the Dominions. It was pointed out to him in reply that the Bill did not purport to go beyond the cases where the officers and men of the Dominion Forces might be affected by Imperial legislation or where officers and men of the Royal Navy might be affected by Dominion legislation. The Bill left the Dominions as well as the Imperial Government perfectly free to legislate as they might think fit as regards their own forces. Mr Pearce concurred in this view. Finally Mr Brodeur [,] while still maintaining that it would be necessary for the Canadian Government to deal with the relations of their Service to the Royal Navy by their own legislation[,] said he did not object to the bill from the point of view of Imperial legislation and declared that the Canadian Government had no intention but to follow closely the Naval Discipline and Regulations of the Royal Navy.

It was decided that the ships maintained by Canada and Australia should be called His Majesty's Canadian and Australian ships; and also there was no objection to the naval forces in each Dominion being called the Royal Canadian Navy and the Royal Australian Navy if it were so desired.

The questions of uniform and commissions were referred to. Sir Arthur Wilson pointed out the disadvantage of officers having to modify their uniform according to the Service in which they might happen to be serving at the time, and that it was desirable to avoid unnecessary expense. Both Mr Brodeur and Mr Pearce, however, stated that so far as they could judge there would be a wish that some slight distinction should be associated with the Dominion uniform such as the substitution of the maple leaf for the laurel leaf on the buttons of the Canadian Uniform. Mr Pearce stated a similar distinction was also contemplated by the Australian Government. The matter would be further considered.

As regards Commissions, it was agreed that it would be an advantage from the point of view of interchangeability of officers if the Commissions in the Dominion forces were worded similarly to the Admiralty Commissions, and that the question should receive further consideration.

[86] *E.P. Morris, Prime Minister of Newfoundland, to the Administrator of Newfoundland [Chief Justice Sir W.H. Horwood?]*

12 August 1911

By reference to schedule 'A' of the memorandum in question, [i.e. Status of Dominion Naval Forces] it appears that the Canadian Atlantic station includes the waters north of 30 degrees north latitude, and west of the meridian 40 west longitude. This includes the waters surrounding Newfoundland, and you will observe that, under the provisions of paragraph 4 of the memorandum, the Canadian Government may establish their stations in the waters referred to in schedule 'A' . . .

With a view therefore, of there being no misunderstanding hereafter, Ministers suggest that the attention of the Canadian Government be drawn by the Secretary of State for the Colonies to this matter, and that it be made clear that the zone in which their navy may operate in the North Atlantic waters should in no degree encroach on Newfoundland's territorial and maritime waters, and that it be pointed out to them that the memorandum referred to in no way alters the status existing previous to the drawing of that memorandum.

The Newfoundland Government, or its representatives at the Imperial Conference, were not consulted in relation to this memorandum, nor were the Newfoundland members of the conference present when the terms of the memorandum were agreed upon, nor were they given any opportunity of considering its principles or details, and the writer saw it for the first time when it was forwarded by your Excellency a few days ago.

[87] *Lord Harcourt, Secretary of State for the Colonies, to [Horwood?], the Officer Administering the Government of Newfoundland*

13 September 1911

I have the honour to acknowledge the receipt of your despatch No. 70 of the 14th August transmitting a communication from your Prime Minister on the subject of the status of Dominion Navies.

2   In reply I have to request that you will inform your Ministers that the Canadian stations were formed for the purpose of assigning to the Canadian Fleet a definite area of action in the

same manner as Commanders in Chief of Stations abroad have a definite command and in the arrangements discussed with the Canadian Ministers Newfoundland was expressly excluded as requiring special provisions. In determining the limits of the Canadian Atlantic Station there was no intention on the part of His Majesty's Government or of the Government of Canada to alter the existing position with regard to the control and jurisdiction of the Government of Newfoundland over its territorial waters, or to empower the Dominion Government to establish a station in those waters without the consent of the Government of Newfoundland.

3   The supervision of the Newfoundland fisheries is now carried out by a ship of the Imperial Navy and there is no present intention of disturbing this arrangement.

4   I trust that this statement will be satisfactory to your Ministers.

5   I am sending a copy of the correspondence to the Governor General of the Dominion of Canada for the information of his Ministers.

[88] *Winston S. Churchill, First Lord of the Admiralty, to Herbert Asquith, Prime Minister*

14 April 1912

I have of course been casting about for a naval policy for the Dominions. Canada is soon coming to ask advice, a[nd] no one can be satisfied with the present arrangements into [sic] Australia a[nd] N[ew] Z[ealand], except as the 1st stage.

Briefly, what I am coming to is this: – G[rea]t Britain will keep a Navy strong enough to deal with the strongest probable combination in the decisive theatre. This means concentration, a[nd] consequent abandonment of all seas except those in wh[ich] the supreme issue will be settled. After a decision by battle had been obtained, we [wou]ld of course spread our fleets again a[nd] restore the situation in any waters, however remote. But what is to happen before such a decision or before war comes at all? We shall be increasingly held concentrated in home waters. The war may never come, a[nd] all the time Imp[erial] a[nd] Colonial interests will lack the support of available naval force in many parts of the world. What we want in fact is a movable squadron or Fleet, not tied up by the main situation, cruising ab[ou]t the Empire giving each part a turn a[nd] going to any threatened point; its influence w[oul]d be felt everywhere, not only in the actual station where it was, but wherever it might if necessary go. Here then is the fundamental division of labour wh[ich] the Mother Country should make with her Colonies: – 'We will cope with the strongest combination in the decisive theatre, You shall patrol the Empire'. Local navies may have been the only possible first step. Can we not now take the next – namely a joint Dominion Sq[uadron]? Separately, these navies are weak a[nd] even ridiculous. One Dreadnought et practeria nihil! But combined they might make a force wh[ich] no European power [wou]ld face without dispersing its own home combination, a[nd] consequently releasing ours.

The scale of the Sq[uadro]n a[nd] the time taken to complete it w[oul]d be varied to suit moods a[nd] circ[umstance]s. Let us make the plan on sound principles of policy a[nd] strategy, a[nd] the spaces can be filled in later. The following is only my sketch:

The Imperial Squadron.

Battle Cruisers	Australia	1
	New Zealand	1
	Canada	2
		4
Light Armoured Cruisers	S. Africa	2
	India	2
		4

Control and Duties

In times of peace, to move constantly from station to station spending 3 or 4 months in rotation in the waters of each Dominion, the movement being regulated by a sub-c[ommit]tee of the C[ommittee of] I[mperial] D[efence] composed of one rep[resentati]ve from each Dominion and one or two Imp[erial] rep[resentati]ve. Also to move to any threatened point a[nd] discharge any special duty.

Discipline, Training, administration a[nd] war employment to rest with the Admiralty.

Observe – the sq[uadro]n is to be additional to the European force, so that we are not sponging on them for our own needs.

Service in this sq[uadro]n may be specially arduous a[nd] always from home. Therefore it w[oul]d be right to pay Colonial rates of pay to officers a[nd] men. As we shall have to find the bulk of the officers a[nd] men at any rate for some years. I sh[oul]d propose to let our people be selected for not more than 2 y[ea]rs at a time.

The extra pay w[oul]d thus be operative over the whole service, a[nd] not localized to particular crews in special stations. This w[oul]d avoid many difficulties wh[ich] now beset us in Australasian waters. Each w[oul]d have his turn or at least his chance.

Beyond this moving squ[adron] other necessities come into view. Suppose they are required to proceed to Australian waters or to Vancouver for a time. They will want Docks. They will want to complete their war organisation with a flotilla. Therefore each Dominion can develop its own coastal defence on a sound principle, viz of affording the means for the Imp[eria]l Sq[uadro]n to operate from their own coasts with full effect whenever the need arises. All this will take time. But the path will be clearly visible. On the arrival of the Imp[eria]l Sq[uadro]n at Sydney or Vancouver or Simonstown, it will find docking facilities, stores coal a[nd] the destroyer or submarine flotilla wh[ich] make it a Fleet in integrity. We set up a design looking far ahead, into wh[ich] all Colonial naval activities can be fitted as they arise. We show them the way; a[nd] it is the only way, to provide for their own proper naval defence, we holding off the big dog meanwhile.

If you like this general idea, the way to give effect to it is to persuade the Canadians to summon a conference, in Canada, of the other Dominions. We will go over there, it being their show primarily, to give them the Ad[miralt]y view a[nd] clinch the whole thing. We [woul]d take the 1st Battle [added in margin] Cruiser Sq[uadro]n in Aug[ust] (all being quiet here) to show them what their force w[oul]d look like when complete.

'Four when you want them instead of one all the time'. The argument is v[er]y fertile a[nd] pliant to all the needs a[nd] prejudices that have to be met.

As for S. Africa, we have only to ask Botha I am sure. N.Z. a[nd] Australia have done their part already.

The cost both of construction a[nd] maintenance to be borne by the Dominions.

Maintenance not v[er]y great. 4 B[attle] Cruisers at £120,000 p.a. a[nd] 4 small cruisers at £60,000 p.a. for the general purpose; the rest, flotillas etc, being local at discretion.

The stay of the sq[uadro]n at each station if necessary to be proportioned to the am[oun]t of contribution. But this is a detail.

Will you please talk to me ab[ou]t this when opportunity offers, a[nd] give me y[ou]r guidance! [initialed]

[89] *Private office, draft press notice, forwarded by H.C.C. Davidson, Colonial Office, to G.A. Steel [Private Secretary to the First Lord] at Admiralty*

18 May 1912

The Admiralty have recently been in communication with the Government of the Dominion of New Zealand upon the employment of the battle cruiser which is now building at the charge of the Dominion for presentation to the Royal Navy. It had been intended that this vessel should be stationed in the Far East, but the Government of New Zealand have in response to Admiralty enquiries and suggestions, expressed their wish that the Admiralty should employ this vessel wherever her services can be most useful. His Majesty's Government have gratefully accepted this intimation. As the British squadron on the China Station has recently been reinforced by the *Defence*, the Admiralty have decided that the battle cruiser *New Zealand* can best at present be employed in home waters. She will accordingly as soon as she is completed visit the Dominion of New Zealand probably in the early part of next year, after which she will join the First Battle Cruiser Squadron in the First Fleet which her arrival will complete to its full strength of five ships.

[90] *Churchill to Robert L. Borden, Prime Minister of Canada*

25 July 1912

'Confidential'

The Prime Minister greatly regrets that his duties will not allow him to be absent from England during September, but he and the Cabinet are quite agreeable for me to go to Canada for a short visit if you think it would be useful. I must of course receive a regular invitation which should perhaps be extended to the Board of Admiralty. I could start at the end of August so as to arrive shortly after your return. I should propose to visit Vancouver and Esquimalt to inspect the docking facilities there and examine the naval questions which arise. A Cruiser Squadron would visit Quebec and Montreal, and we would send one of our best Cruisers from the China Station to Vancouver. In order that we might go fully into the technical matters, on which we have touched in our Conferences here, Prince Louis of Battenberg would accompany me and one or two officers of the War Staff. I shall have to get back by the middle of October. If you have any doubts as to the utility of such a visit, don't hesitate to tell me so frankly. I shall be in London Saturday morning and it would give me great pleasure to see you at the Admiralty if you find yourself free.

[91] *Borden to Churchill*

28 August 1912

At our early interviews, and especially at that of the 16th July, you were good enough to give me two assurances which were especially encouraging and satisfactory to both myself and my colleagues:

1st   That an unanswerable case for an immediate emergency contribution by Canada could, and would be made out by the Admiralty.

2nd   That this case would be made in two-fold form; one confidential and in detail for the Cabinet, the other public and more general in character for Parliament and the people.

The first form is essential, but the second is vital. Such a case must come from the Admiralty, and it would not be useful or possible for us to undertake its preparation. In all our debates, and in all our pledges, we have definitely and positively based our 'emergency' action upon information to be supplied by the Admiralty; and we must be in a position to lay it before Parliament as coming from them and bearing their authority. Such 'emergency' or immediate action will have to be abandoned for the present unless an unanswerable or at least a sufficient case is presented in such form as can be submitted to Parliament. Upon returning to Canada I shall be obliged to admit to my colleagues that up to the present no such case has been placed before me in that available form. This I sincerely regret, as I have at some inconvenience postponed my departure from the 23rd to the 30th in order that the case in its twofold form might be considered, discussed, and settled before I left England.

[92] *Churchill to Borden*

29 August 1912

'Personal and Confidential'

The typescript Memorandum sent you on the 26th instant deals, as you rightly apprehend, solely with the 'emergency case,' and further memoranda will be sent you as soon as they have been completed on the other points mentioned in your letter, which more particularly concern the question of permanent policy.

I do not know how the case for immediate action on the part of Canada as set forth in the Memorandum can be strengthened, but, if there are any directions in which you feel it has not been sufficiently made out, and will indicate them, they shall receive careful attention.

The fact emerges with extreme clearness that, unless special measures are taken by Canada in the immediate future, it will be necessary next year for the Imperial Government to make some further provision, over and above the very great exertions they are already making, in order not merely to secure the safety of the United Kingdom, but to enable them to continue to afford to the Empire as a whole the naval protection it has so long enjoyed.

I had not wished to put this so nakedly, even in a confidential document. There cannot be any doubt at all as to the conclusions which should be drawn from it.

With regard to a further Admiralty Memorandum, which you can make public and lay before the Canadian Parliament, while I still think, for the reasons stated in my letter, that the course there proposed would have been a more convenient one, I hold strictly by the promise I made to you on the 16th July.

A Memorandum in a form which can be published and presented to your Parliament shall be at once prepared, and will follow you by an early mail.

I hope that before leaving you will return the confidential Memorandum with any improvements which may occur to you noted upon it. Both documents will then be sent to you together with as little delay as possible.

The Admiralty Memorandum for publication will necessarily deal with facts only, and it will be for you to draw the conclusions and state your policy upon them.

[93] *[Churchill? to Borden?] date added in manuscript*
14 September 1912

The following is not only Secret, but private to yourself: [illegible]

We have assumed throughout that Canada wishes her assistance to take the most direct and effective form of war power. There are several classes of vessels which Canada could build which would be of use to the British Navy or set our money free to build other vessels: for instance, small ocean-going cruisers or fleet auxiliaries of various kinds. But what we have gathered from many Canadian sources led us to believe that Canada would wish to figure in the very front fighting line, and to supply those units which will themselves directly determine the fate of a naval war. Now there are only two kinds of ships which will really do the business of inflicting capital injury, namely Dreadnoughts so called, and submarines. All the rest, important and indispensable as are their services in trade protection, scouting, screening, sweeping, etc., are appurtenances; the battleship and the submarine alone are the principals. So great is the development of the new weapon, the submarine, that on purely military grounds we would just as soon have any Canadian money spent on submarines as on Dreadnoughts. We must have both. You could not build the submarines: they are a highly specialized form of construction. Only a few of them would be wanted for your own coast. If the others were to be of any service to us, they would have to be permanently stationed in the North Sea: they would be highly localised and altogether alienated from Canada. Your people would never see them, would hardly hear of them. They have no names, only numbers. Their construction would not excite any enthusiasm, nor would it take a form which the mass of people who know very little about the Navy could understand. To the ordinary layman it would look as if large sums of Canadian money were being frittered away on a host of 'small fry' vessels for British Home waters. Although, therefore, the military argument is evenly balanced between these two types, and both are of supreme consequence at the present time, we have no hesitation, on surveying all the circumstances, in asking you not for submarines but Dreadnoughts. There is a great and legitimate sense of pride which the Canadian people feel in contributing the finest and most powerful ships in the world to the Imperial Navy, in naming them after Canadian Provinces, and in watching their progress to and fro about the world. I am sure, therefore, that the course we advise is equally good for us and far better for you. But in order that [manuscript addition: NB] there may be no misunderstanding between us at a later stage, and in view of your wish that Canadian aid should be additional to pre-arranged British programmes, I must tell you that we contemplate only building 4 battleships next year instead of 5 as announced, and spending the money for the fifth battleship in building every submarine that the British Yards at full pressure can undertake within the year. We should not propose to spend less money, but we should spend a part of it in a different way. It is essential that no reference should be made to this at all and it must be made entirely between you and me, for it is possible that we may even announce 5 big ships to Parliament, and leave our foreign friends to find out 7 or 8 months later that one of them has taken a very different form.

[94] *Borden to Churchill*

3 October 1912

'Confidential'

As I explained to you in England, the objection may be strongly urged in some parts of this country that any considerable sum to be provided by Canada for immediate aid towards increasing the naval strength of the Empire ought to be expended in this country. On the other hand, it is sufficiently manifest to me that the construction of battleships of the largest and most powerful class cannot be undertaken in this country within a reasonable period, having regard to the grave conditions which we are called upon to confront. At our interviews in London, I suggest to you a possible solution, which you will doubtless recall. If the Canadian Parliament should vote a large sum of money sufficient for the construction of two or three battleships of the largest type, and should authorise the expenditure of that money in the United Kingdom, it would seem not only practicable, but reasonable, that the great shipbuilding firms to which the contracts might be let should undertake the beginning of a shipbuilding industry in one or two parts of this country. The points to which our attention has been particularly directed are: Halifax in the Maritime Provinces, and Montreal or Quebec in the Province of Quebec. I do not suggest this in any spirit of bargaining, but you, of course, realise that conditions of a somewhat difficult character from a political standpoint will have to be encountered. It is of vital importance that any proposals which we make to Parliament shall be carried to a successful issue, as otherwise the moral effect upon the whole Empire will be disastrous. Thus it seems to be not only in Canadian, but in Imperial interests as well, that everything should be done to overcome local prejudice or sentiment of the character suggested.

Moreover, you will perhaps agree in the importance from an

Imperial standpoint of the early establishment not only of dock-yards but of naval bases provided with the necessary equipment both for building and for repairing war vessels of the smaller type at least.

I cannot too strongly emphasise the importance of the consider-ations which have been very briefly set forth in this letter. Possibly the particulars, the early arrival of which your recent cable announces, may cover this point.

[95] *Borden to Churchill*

5 October 1912

'Confidential'

In writing to you on the 3rd instant, with respect to the import-ance at this juncture of stimulating or encouraging the beginnings of a ship-building industry in Canada, I omitted to allude to the proposal or suggestion, more than once discussed between us, that war-vessels, such as destroyers or small cruisers, might be built within a very early period in Canada, and that possibly an arrange-ment might be made to divide the extra cost between the two Governments.

[96] *Minute by Oswyn Murray, Assistant Secretary to the Secretary of the Admiralty*

15 October 1912

The local conditions at Auckland which make it impossible for submarines to be berthed there unless a harbour be provided were certainly not known either to the Admiralty or to the New Zealand representatives at the time when the agreement was arrived at, & may therefore be properly regarded as a new factor to be taken into consideration. It would not be unreasonable to represent these conditions to the New Zealand Government, & to say that, in these circumstances which were not foreseen in 1909, the advan-tage of two submarines is not sufficient to justify the somewhat large expense involved, & that another destroyer would be a better & more economical disposition of force. This would prob-ably lead the New Zealand Government, if desirous of retaining the submarines for any reason, to offer to pay for the harbour.

It would be desirable, however, in the first instance to ascertain the comparative cost of upkeep (including depreciation) of

1 River class Destroyer, as against –

2 C class submarines

as it might turn out to be false economy to make the substitution. A.G. could state this: Concur that we must offer something to replace the submarines.

[97] *Memorandum by Murray to the Secretary of the Admiralty*
16 October 1912

The remarks of the C[hief] O[f] S[taff, Admiral Sir Henry Jackson] do not appear to deal with the particular aspect of the question which is probably of most interest to New Zealand.

The words in the telegram from New Zealand 'especially with regard to the question of Naval Bases' doubtless refer to the fact that under the agreement New Zealand was to have one of the subsidiary bases, and that 2 Bristols, 3 Destroyers, and 2 Submarines belonging to one of the two Fleet Units (Australia providing the third Unit) were to be based on New Zealand in peace time. In other words New Zealand, in return for her contribution, was to feel that she was really being brought into the Pacific Fleet formation in the same way as, though in a less degree than, Australia.

The C.O.S.'s statement, no doubt, shows that the Cruisers on the China and East Indies Stations can be so reckoned as to make up the numbers of Cruisers fixed for the two Fleet Units, but the unfortunate point is that none of the Cruisers included in his statement will be based on New Zealand. Instead of having in her waters two 'Bristols' forming an integral part of a Fleet Unit, New Zealand will only have for the present at all events, two 'P' class cruisers which, on C.O.S.'s own showing, are not reckoned in the Fleet Units at all. In other words she is to all appearance being excluded from the part originally assigned to her in connection with the Pacific Fleet organisation. (C.O.S. only deals with cruisers, not with T.B.Ds. and Submarines.)

It appears to me that, in our reply, either (1) we should take the line that we are carrying out the Pacific Fleet organisation, but owing to the changed position at home it will only be possible to send out 'Bristols' gradually; that consequently for the moment the *Newcastle* in China, *Highflyer, Fox, Flora* in East Indies and 2 'P' class cruisers in New Zealand waters will represent the 6 Bristols, and that as this is obviously a weaker force than 6 Bristols, the *Kent* and *Monmouth* which are outside the approved 'Unit' organisation are also being kept in Eastern waters, or (2) we should say straight out that the situation at home owing to the new German Fleet Law necessitates a postponement of the Pacific Fleet organisation, and ask their concurrence in this.

There has been so much in the Press about the help we are giving Australia &c. that one can quite understand New Zealand fearing that she is not receiving equal treatment.

Minute by Chief of Staff, Admiral Sir Henry Jackson

I think it very desirable that the Admiralty should appear to be adhering to the arrangement of 1909–11 & if so the first alternative suggested by the Ass[istan]t Sec[retar]y would be preferable. If you concur, I will ask him to prepare a draft on those lines for consideration.

[98] *Minute by Murray to the Secretary of the Admiralty*
19 October 1912

The arrangement as regards stationing in N[ew] Zealand waters a force of 2 cruisers, 3 destroyers and 2 submarines was definitely agreed to at the Imperial Conference in 1909. See passages in attached print marked A (page 5) and B and C (page 3).

The agreement was dictated mainly by political reasons. Sir J[oseph] Ward [Prime Minister of New Zealand from 1907 to March 1912] gave it to be understood that his policy of a contribution to an Imperial Navy could not be maintained unless the people in New Zealand saw the flag flying on their coasts, as they would never agree to the Australian Navy representing their interests alone. My recollection is that [Admiral of the Fleet] Sir J[ohn] Fisher [retired as First Sea Lord in January 1910] suggested the two submarines, and at the time it was not pointed out that this would involve large expenditure in order to provide accommodation for them.

In 1911 when Sir J. Ward was in London for the Conference the question of establishments on shore came up in discussion and he was informed that the Admiralty could not contemplate starting in New Zealand large establishments while they were transferring those in Australia to the Commonwealth Government. But even then it was not realised that it would be necessary to provide such a large expenditure on shore in respect of the two submarines only. Looking to what has passed since 1909 there is no doubt that the Admiralty are pledged to make New Zealand waters the head quarters of a small division of the China-Pacific Squadron, but these ships need not be tied to New Zealand and it was always considered that such ships as were suitable would be available for general service in the Pacific Islands and elsewhere.

In the circumstances it seems that the only way of dealing with this special expenditure for the submarines is either to represent that before they are sent the New Zealand Government must provide the necessary accommodation, or to argue that the advantage of 2 submarines is not sufficient to justify the incidental expenses and that, say, another destroyer would be a better disposition of force. It should be borne in mind that New Zealand contributes £100,000 a year to the Navy.

[99] *Vice-Admiral Sir George Fowler King-Hall, Commander-in-Chief Australia, to Churchill*

Admiralty House, Sydney
22 October 1912

The N[ew] Z[ealand] Gov[ernmen]t intend consulting me as to the advisability of their throwing their lot in, as regards naval matters, with the Commonwealth. Public opinion in N.Z. is decidedly veering that way; and the N.Z. Cabinet are divided, but I trust after I have seen them, that they will decide to join forces. I believe they would give £500,000 towards the R.A.N. and I suppose in this case the Home Govt. would take over the N.Z. battleship altogether – I think Admiral Henderson's programme should be expedited if possible; there is lots of money in the country and no lack of volunteers of youths and boys for the R.A.N. They are all very favourably reported on, and are certainly most intelligent, keen in their work, and full of esprit de corps. I sincerely hope that the Admiralty will send out the best officers to the R.A.N. It is so important to start them off on the right lines, and rivet them to the R.N. which at present is being done. If we fail to help them, the R.A.N. will become a sectional force, and jealousy arise and lead to a separative movement... The R.A.N. have been put under my command in all respects when in presence of my flag.

Minutes:

W.S. Churchill [First Lord]:   I am not at present convinced of the advantage of this: & in any case CinC has no authority to advocate so important a change of policy.   25.11

Admiral Beatty [Naval Secretary]:   I much regret the line the

C in C Australia is taking, & I think he sh[oul]d be told not to interfere in any questions of policy.

No one can yet say whether the Commonwealth Navy will be a success or not there are many people who predict not, it w[oul]d there[fore] be a misfortune for New Zealand's Navy to be absorbed in that of the Commonwealth. 27.11.12

Churchill: Please draft a telegram to C in C as above.   28.11

[Oswyn] Murray [Assistant Secretary]:   Your telegram of 7 Nov stated that [the] joining up of Dominion & Commonwealth forces raised many difficult questions which should not be pre-judged on the spot without full consultation with the Admiralty. In his reply Adm[ira]l King Hall said he was to meet [the] Dominion Cabinet [at the] end of this month to discuss question, but he would be most careful not to compromise the Admiralty. It should not be necessary to send a further telegram, unless the C in C is old he should take up a position of strict reserve . . .   29.11

[100] *Minute by [Rear-Admiral Herbert E.P. Cust], Hydrographer*
26 October 1912

Owing to the generally open and exposed character of the coasts of New Zealand, the frequency of gales, the limited number of safe harbours and the great distances which separate them, these waters are not considered suitable for the navigation of submarines and their employment must necessarily be restricted to the immediate vicinity of certain ports.

With regard to Auckland. An examination of the charts and sailing directions does not lead one to anticipate the difficulties mentioned by Captain Hickley, on the contrary one would have concluded the harbour and vicinity were very suitable for the operations of this class of vessel.

North easterly and easterly winds are experienced on an average on 88 days each year, the most prevalent throughout the year being those from West to South. There are numerous wharves alongside which vessels can lie and presumably submarines could find safe berthing accommodation under shelter of these even during gales.

It seems hardly credible that such a sea could get up so as to prevent submarines getting out of harbour.

[101] *Churchill to Borden*

4 November 1912

I have given careful consideration to your two letters about the encouragement of the shipbuilding industry in Canada. I recognise the importance of such a policy on general grounds not less than from the immediate Canadian standpoint, and any practical scheme for Admiralty co-operation would command my support. The main difficulty to be surmounted is to obtain that high degree of expert knowledge and experience which modern warships require for their efficient construction.

We might, however, in the first instance agree upon certain classes of vessels with which it may be considered that competent Canadian shipyards would be able to deal. The most suitable classes of vessels with which to inaugurate the system would be light cruisers, oil-tank vessels, and small craft for auxiliary services. We should, if it would meet your views, be prepared to invite tenders from approved Canadian firms for the construction of some vessels of such classes in the near future.

It would be understood that progress with this policy would have to be dependent on the prices quoted being reasonable, having regard to all the circumstances (including the fact that Canada will be prepared to share any extra cost), and also on the time required for construction not being excessive as compared with the dates fixed for completion of similar ships in England. No fixed scale or proportion of orders could be guaranteed to Canadian firms. We would begin by giving some orders at once, and further progress would depend upon the development of the industry and the extent of our programme. The Admiralty would, of course, remain wholly responsible for the design of all vessels, and for the supervision of the construction of those building in Canada. Arrangements for this could be worked out in detail later and should not present any difficulty.

[102] *Copy of cipher telegram from Churchill to King-Hall*

7 November 1912

Official reply to your telegram No. 129 about control of Royal Australia Navy in War is being sent. Joining up of Dominion and Commonwealth forces raises many difficult questions which should not be prejudged on the spot without full consultation with Admiralty. Withdrawal of New Zealand battle cruiser would be extremely inconvenient. You should temporise tactfully and urge reference to Admiralty before final decision.

[103] *Decipher telegram from King-Hall to Churchill*
Melbourne 11 November 1912

'Personal and Confidential'

Submit that your speech [of] 9th November as cabled[,] namely that [the] Commonwealth Prime Minister [Andrew Fisher] was explaining how Australian Fleet would be controlled by [the] Admiralty in war time[,] has placed [the] Prime Minister in [an] awkward position through the omission from your speech as cabled of the fact of this control emanating from the Commonwealth in accordance with Article No. 16 of the Memorandum of Conference of June 1911. It is advisable if you could take an early opportunity of bringing out the fact that the executive control must be with the Commonwealth and that it rests with them to co-operate and transfer it to the Admiralty. I have been in communication with the Governor General of Australia [Lord Denman] and Mr Fisher. The former has also telegraphed to you on this subject.

[104] *Paraphrase of telegram from the Governor-General of the Commonwealth of Australia [Lord Denman], to the Secretary of State for the Colonies, [Harcourt], received, Colonial Office*
13 November 1912, 11.45 a.m.

'Secret'

I have received from the Prime Minister [Andrew Fisher] the following telegram which he has received from the High Commissioner.

'Secret. I am desired by the Admiralty to inform the Commonwealth Government that in view of the present critical state of affairs it is necessary to expedite the completion of all ships under construction with a view to commissioning them if required in case of emergency. The Admiralty are anxious that if the Commonwealth Government concur the *Melbourne* should be expedited in common with all the other vessels. Such action may involve delay in the final completion of the *Melbourne* should the period of suspense be prolonged and a slight increase of expence which the Admiralty propose to defray. If the Commonwealth Government agree the Admiralty have taken precautionary steps so that the *Melbourne* will be ready to receive her crew on November 16th. Very urgent. The Government of the Australian Commonwealth may deem this a fitting opportunity of showing their desire to co-operate in imperial emergency.'

Would you be so good as to inform me if this is correct and if so I would suggest that the usual channel of communication should be employed by the Admiralty. (signature indecipherable)

[105] *Colonel Sir Arthur Young, Governor, Commander-in-Chief and High Commissioner to the Malay States, to Harcourt*
Singapore
14 November 1912

I have the honour to confirm my telegram of the 12th November that a resolution has been passed in the Federal Council on that day which is as follows:

'With a view to the strengthening of the British Empire and maintaining her naval supremacy the Rulers and the people of the Federated Malay States desire to offer to His Majesty's Government a first class armoured ship and it is resolved by this Council that this offer should be made.

The vessel to be constructed as soon as His Majesty's Advisers consider it desirable.

Details as to payment to be decided hereafter, but payment to be made within five years if possible.'

2.   His Highness the Sultan of Perak moved the resolution setting forth that the gift was offered to show his loyalty and his gratitude for the protection given by Great Britain to the Federated Malay States and the prosperity that these States had attained owing to that protection: he mentioned that the cost of the ship which was offered would be not less than 20 millions of dollars . . .

8.   There are suggestions made which have come to my knowledge that this Government has been informed by the Home Government that the Federated Malay States should contribute to the Navy and that this Government has practically brought pressure to bear on the Rulers of these states to pass this resolution. I therefore considered it desirable when I spoke on the resolution to make the remarks which I reproduced below from the revised shorthand report:

'There had been, continued His Excellency, some conjectures as to how this resolution came into being. To remove any misconception he wished to state the circumstances which led to this resolution. Some time ago the Chief Secretary, Mr Brockman, represented to him (His Excellency) that he thought the Federated Malay States were not perhaps assisting to an extent commensurate with their prosperity in the defence of the Empire and he stated that he knew that the rulers of these States were of the same opinion.

He agreed with the Chief Secretary on both points. Later on the Chief Secretary interviewed His Highness the Sultan of Perak on the subject. His Highness not only spoke on the subject with cordiality but received it with enthusiasm . . . The Chief Secretary then interviewed His Highness the Sultan of Selangor and His Highness received the subject with the same cordiality and enthusiasm as the Sultan of Perak. As they had already heard that day, the ruler of Negri Sembilan was of the same opinion and the representative of His Highness of Penang had also spoken in the same strain'.

9.   On the 13th November I received, whilst the Council was sitting, your telegram of the 12th November stating that His Majesty's Government accepted the offer: I read the telegram and the contents were received by the Rulers and Members of the Council with applause.

[106] *Crewe to the Governor General of India in Council [Baron Hardinge of Penshurst]*

6 December 1912

I have considered in Council your Despatch No. 5 (Marine Department), dated 1st August 1912, with which you forward a number of memorials from officers of the Executive and Engineering Branches of the Royal Indian Marine, praying for the grant of commissions, and recommend that statutory powers may be obtained for the issue of commissions under the authority of His Majesty's Royal Sign Manual, and having the signature of the Secretary of State for India.

2   You refer to the statement of Lord George Hamilton [sometime First Lord, and then Secretary of State for India] in his Despatch of 31st May 1900, No. 71, that if, after experience of the effect of the orders which the Admiralty had then issued to

the Royal Navy regarding marks of respect to be paid to officers of the Royal Indian Marine, your Government should be convinced that these officers still labour under disabilities only removable by the grant of commissions, he would address the Admiralty again on the subject. I apprehend, however, that a proposal to confer permanent commissions in the Royal Navy on all officers of the Royal Indian Marine, a service not under the control of the Lords Commissioners, would not be likely to be approved, and I gather that you are conscious of this difficulty in recommending that the proposed commissions should appoint to the Royal Indian Marine. Your desire is to secure for the officers of the Royal Indian Marine a larger measure of respect in their relations not only with the Navy, but also with the Army. Commissions, it appears to me, could not be granted merely for the purpose of securing salutes and other marks of respect. It would be necessary that they should confer certain specific powers of command.

3 I request therefore that Your Excellency's Government will consider what powers the proposed commissions should confer, and will submit drafts of the forms of commission which you would recommend for Executive officers and for Engineers. I desire, however, to point out that the grant of such commissions is an act of the Royal Prerogative, and that to invoke statutory powers for this purpose seems neither necessary nor appropriate.

[107] *'Naval Defence', memorandum on naval defence requirements, prepared by the Admiralty for the Government of Canada*

December 1912

Oversea.   It has been necessary within the past decade to concentrate the fleet mainly in Home waters.

In 1902 there were 160 British vessels on the oversea stations against 76 to-day.

7 Naval supremacy is of two kinds: general and local. General naval supremacy consists in the power to defeat in battle and drive from the seas the strongest hostile navy or combination of hostile navies wherever they may be found. Local superiority consists in the power to send in good time to, or maintain permanently in, some distant theatre forces adequate to defeat the enemy or hold him in check until the main decision has been obtained in the decisive theatre. It is the general naval supremacy of Great

Britain which is the primary safeguard of the security and interests of the great Dominions of the Crown, and which for all these years has been the deterrent upon any possible designs prejudicial to or inconsiderate of their policy and safety.

The rapid expansion of Canadian sea-borne trade, and the immense value of Canadian cargoes always afloat in British and Canadian bottoms, here require consideration. On the basis of the figures supplied by the Board of Trade to the Imperial Conference of 1911, the annual value of the overseas trade of the Dominion of Canada in 1909–10 was not less than 72,000,000£, and the tonnage of Canadian vessels was 718,000 tons, and these proportions have already increased and are still increasing. For the whole of this trade wherever it may be about the distant waters of the world as well as for the maintenance of her communications, both with Europe and Asia, Canada is dependent, and has always depended, upon the Imperial Navy, without corresponding contribution or cost.

Further, at the present time and in the immediate future Great Britain still has the power, by making special arrangements and mobilising a portion of the reserves, to send, without courting disaster at home, an effective fleet of battleships and cruisers to unite with the Royal Australian Navy and the British squadrons in China and the Pacific for the defence of British Columbia, Australia and New Zealand. And these communities are also protected and their interests safeguarded by the power and authority of Great Britain so long as her naval strength is unbroken.

8   This power, both specific and general, will be diminished with the growth not only of the German Navy, but by the simultaneous building by many Powers of great modern ships of war.

Whereas, in the present year, Great Britain possesses 18 battleships and battlecruisers of the Dreadnought class against 19 of that class possessed by the other Powers of Europe, and will possess in 1913, 21 to 24, the figures in 1914 will be 31 to 33; and in the year 1915, 35 to 51.

The existence of a number of navies all comprising ships of high quality must be considered in so far as it affects the possibilities of adverse combinations being suddenly formed. Larger margins of superiority at Home would, among other things, restore a greater freedom to the movements of the British Squadrons in every sea, and directly promote the security of the Dominions. Anything which increases our margin in the newest ships diminishes the strain, and augments our security and our chances of being left unmolested.

9   Whatever may be the decision of Canada at the present juncture, Great Britain will not in any circumstances fail in her duty to the Oversea Dominions of the Crown.

She has before now successfully made head alone and unaided against the most formidable combinations and she has not lost her capacity, by a wise policy and strenuous exertions, to watch over and preserve the vital interests of the Empire.

The Admiralty are assured that His Majesty's Government will not hesitate to ask the House of Commons for whatever provision the circumstances of each year may require. But the aid which Canada could give at the present time is not to be measured only in ships or money. Any action on the part of Canada to increase the power and mobility of the Imperial Navy, and thus widen the margin of our common safety, would be recognised everywhere as a most significant witness to the united strength of the Empire, and to the renewed resolve of the Oversea Dominions to take their part in maintaining its integrity.

10   The Prime Minister of the Dominion having enquired in what form any immediate aid that Canada might give would be most effective, we have no hesitation in answering, after a prolonged consideration of all the circumstances, that it is desirable that such aid should include the provision of a certain number of the largest and strongest ships of war which science can build or money supply.

[108] *Draft letter, Greene to the Under-Secretary of State, Colonial Office*

[January 1913]

I have laid before My Lords Commissioners of the Admiralty your letter of the 2nd October, No.30784/1912, transmitting a copy of a telegram from the Governor of New Zealand requesting on behalf of his Government information as to the future Naval policy in the Pacific Seas especially with regard to the question of Naval Bases and the proposed Commonwealth Navy.

2   In reply I am to request that you will inform the Secretary of State that My Lords have not had under consideration any change in the general policy discussed at the Imperial Conference of 1909 and 1911, by which certain Squadrons would be provided for the protection of the Pacific, the ships of the Royal Australian Navy

forming one unit, and others being provided by the Admiralty, New Zealand contributing towards the expense.

3   In pursuance of this policy arrangements are being made to provide for the maintenance in New Zealand waters of the number of vessels agreed upon (Auckland being their head quarters) when the ships of the Royal Australian Navy will arrive in Australasian waters and Sydney Yard will be transferred to the Commonwealth authorities.

4   As regards the ships to be supplied by the Admiralty – for service in the Pacific the Government of New Zealand has already generously given its consent to the Battle Cruiser *New Zealand*, which it was proposed to allocate to the China Squadron, being employed elsewhere in view of the general naval requirements, and My Lords regret that the same urgent considerations will render it necessary that the process of replacing the ships now in Far Eastern waters by vessels of the types originally contemplated should only be gradually carried out. In the meantime, however, My Lords will arrange so that the actual naval force available in the China and Pacific seas (apart from the vessels of the Royal Australian Navy) will bear due relation to the strength originally agreed upon.

5   The arrangements agreed upon in 1909 and 1911 contemplated [provided in first draft] that two Cruisers of the 'Bristol' Class, 3 River Class Destroyers, and 2 Submarines would ordinarily in peace time be maintained in New Zealand waters. It is accordingly proposed to send out the Destroyers in the course of 1913, but My Lords will be glad if the New Zealand Government will agree to the temporary retention of the two 3rd class cruisers now on the station until other cruisers can be detailed. With regard to the two submarines, My Lords have recently had under their consideration reports as to the serious limitations upon the operations of vessels of this type in New Zealand waters owing to the generally open and exposed character of the coasts, the frequency of gales, and the great distances which separate the few harbours in which they could safely be stationed. After careful consideration Their Lordships have come to the conclusion that these limitations would so seriously detract from the utility of the submarines that it would for the present be undesirable to station vessels of this type in New Zealand waters. A fourth Destroyer in those waters would, in Their Lordships' opinion, constitute a much more effective disposition of force, and they will be glad to receive, in the

circumstances, the concurrence of the Government of New Zealand in this substitution.

6   I am to request that Mr Harcourt [Secretary of State] will cause a reply to the Governor's telegram to be despatched in the foregoing sense.

[109] *Minutes of a private Admiralty committee meeting to discuss Australian and New Zealand naval co-operation*

c. 10 November 1912

A meeting was held in the First Sea Lord's Room on Friday, 10th January 1913, at which the following were present: First Sea Lord [Sir Francis Bridgeman], Additional Civil Lord [George Lambert], Secretary [Sir William Graham Greene], Chief of Staff [Admiral Sir Henry Jackson], Assistant Secretary [O.A.R. Murray].

The question was considered of the policy to be adopted in the event of an official expression of Admiralty views being requested in regard to the proposal that the New Zealand Government should co-operate with the Australian Commonwealth Government in developing an Australasian Navy.

It was decided that, whatever the views of the Admiralty might be, it would not be desirable officially to express any preference either for or against such an agreement between the two Dominions. The probability is that any such expression of opinion would hamper the Admiralty seriously at some future time. The attitude of the Admiralty should, therefore, be restricted to one of readiness to assist New Zealand in whatever Naval Policy she may decide to adopt. The following views were also expressed:

(1)   Speaking generally, the development of an Australasian Navy on the lines proposed must involve a lower standard of efficiency than if the same strength were added to the British Navy.

(2)   In the event of New Zealand co-operating with Australia, the Admiralty may possibly be asked, or may have pressure brought upon them to offer, to return the *New Zealand* for service in Australasian waters. In view of the unconditional character of the original gift, this is perhaps not very probable, but it must be borne in mind as a possible contingency.

(3)   Apart from this question, and looking at the matter from the point of view merely of finance and manning, there is probably no great balance of advantage on either side. If the combination

takes place the Admiralty, on the one hand, (a) will lose the New Zealand contribution of £100,000; (b) will probably have to provide for the manning of the additional ships for the Australasian Fleet for some years to come; (c) will have to provide with crews such ships as may be required for Pacific Island work from the China Station.

The Admiralty, on the other hand, (i) will be relieved of the necessity for maintaining 2 Bristols 3 Destroyers and 2 Submarines in New Zealand waters, and (ii) will be spared the capital expenditure and maintenance expenses due to the establishment of a base at Auckland.

The question of the reply to be made to the New Zealand Government's enquiry as to Admiralty policy in the Pacific (M.01658) was also considered, it appearing probable that this enquiry was made with the view of ascertaining what is the present alternative to the Australian scheme for co-operation.

A proposed reply to this enquiry has been drafted, but in view of the reports that have appeared to the effect that a representative of the New Zealand Government is to be deputed to visit this country in order to discuss Naval questions, it is suggested that before sending this reply the Colonial Office should be asked to ascertain by telegraph whether this visit is to take place, and if so whether the representative will be empowered to discuss the matters covered by this enquiry, as in that event it would be considered desirable to await his visit rather than deal with the matter by correspondence.

[110] *Churchill to the First Sea Lord [Battenberg] and the Chief of Naval Staff [Admiral Sir Henry Jackson]*
17 January 1913

I hear disquieting reports of the prospects of the Canadian naval policy. It is possible that a dissolution may be forced either by the Senate rejecting or by the Lower House obstructing the measure. It is by no means certain that Mr Borden would not carry the country and his policy. But in any case there would be delay, and the chance of a contrary result must always be faced. I wish you to consider what steps would be necessary in the event of this expected aid being (a) postponed for a year; (b) lost altogether; having regard to the Cabinet decision about the strength to be maintained in the Mediterranean and all those discussions earlier in the year.

It may, I think, be helpful if we are able to assign a certain special role for the Canadian ships after they were available, and announce it as soon as possible. I am told that my speech at the Shipwrights' Dinner last year is regarded in many quarters in Canada as the most hopeful line of advance: and I am therefore inclined to foreshadow the formation of an Imperial Squadron additional to the existing organization of 8 squadrons. This squadron would consist of the *New Zealand*, the *Malaya*, and the 3 Canadian vessels, the whole steaming at above 25 knots. It would be based on Gibraltar and held available for general Imperial service. The distances from Gibraltar to different parts of the world should be computed in days steaming of such a squadron, and I should like to see a Table prepared. Gibraltar is well situated as a base from which such a squadron could range widely over the North and South Atlantic, could enter the Mediterranean, could return to the North Sea, could pass round the Cape or through the Suez or Isthmian Canals. (?Could such ships use the two canals.) What also is the state of Gibraltar Harbour? Is it sufficiently deep for these 5 vessels to lie there? Could the Gibraltar Dockyard repair them? Will the rocks lying about the bottom have been removed by the time the ships are ready?

The whole of this project should be examined, and I should like to have a paper on it as soon as possible.

It may be desirable and even necessary to dwell on this when the Estimates are presented in March.

Should the Imperial Squadron be based on Gibraltar, the 4th Squadron would, I presume, have to be accommodated in Home Ports. By that time Rosyth will, however, be ready and the congestion will be greatly relieved.

Jackson; C[hief] O[f] S[taff] Concur [initial] 23 1 13.

[111] *Assistant Under-Secretary of State, Colonial Office [Sir H.W. Just], to the Secretary to the Admiralty [Greene]*
Downing Street
21 January 1913

'Immediate and Secret'

I am directed by Mr Secretary Harcourt to request you to inform the Lords Commissioners of the Admiralty that he observes from the Navy List for January that His Majesty's Ship 'Defence' has been transferred from the China Station to the Mediterranean Station and that His Majesty's Ship 'Hampshire' is under orders to take her place.

2   It will be remembered that when it was suggested to the Government of New Zealand that the battle cruiser *New Zealand* should be appointed to the Home Fleet instead of to the China Station, they were informed that 'the Admiralty will send at once the armoured cruiser "Defence" a sister ship to the "Minotaur", to the China Station, effectively securing British preponderance in those seas'.

3   It is presumed that it was by an oversight that Mr Harcourt was not informed of the intention to substitute what appears to be a weaker vessel for the 'Defence' and he will be glad to learn as soon as possible in what terms the Lords Commissioners desire . . .

[112] *Admiralty War Staff note in response to First Lord's letter [Churchill] of 17 January*

22 January 1913

With reference to First Lord's question as to the steps which would be necessary – should the Canadian contribution be lost or delayed – to give effect to the Cabinet decision as to the future position in the Mediterranean, it is submitted that hitherto the Colonial ships have been excluded from all calculations as to future requirements by express and definite orders to that effect.

Estimates as to the future composition of the British Mediter- ranean Squadron have been based on the decision to maintain at the expense of the Home Government a force equal to that of Austria, provided out of the 60% margin in armoured ships over Germany.

On this assumption the postponement or cancellation of the Canadian contribution would have no effect on these estimates. The only result – admittedly a very important one – of the stationing of three Canadian Dreadnoughts in the Mediterranean, would be to raise our force from a status of bare equality to one of unquestionable superiority as far as Austria was concerned. It would practically re-establish our former Mediterranean security if it was definitely accepted that the Colonial ships were additional units to the force maintained by the Home Government on the above basis.

With reference to the second portion of the First Lord's memorandum the attached table has been prepared based on the 1/5 N.D.P. speed of the 'New Zealand', 16 knots, at which her coal consumption is about 217 tons per day, and the radius of action about 5300 miles. The remainder are supposed to be oil burning ships of 'Queen Elizabeth' type.

There is sufficient space and depth of water at Gibraltar for the squadron he suggests, and the Yard should be able to undertake the refit and repair of the ships without undue strain.

The 'Queen Elizabeth' type should just be able to get through the Suez Canal if oil tanks are nearly empty, and designed draught is not exceeded. As they would need to be very carefully handled, 2 days have been allowed for the transit of all the squadron, and a further 2 days for taking in the 12,000 tons of oil required at Suez to enable them to proceed on their voyage.

Passage	Days
Gibraltar to Dover	3.2
Gibraltar to Halifax	6.9
Gibraltar to Panama Canal	11.2 (a)
Gibraltar to Rio Janeiro	11.1
Gibraltar to Simons Bay via St Vincent	15.3 (b)
Gibraltar to Port Said	5
Gibraltar to Singapore	22.3 (c)
Gibraltar to Auckland via Suez	34 (d)

(a)   Distance via Port Royal is 4520 miles.
(b)   Allowing 1 day for coaling at St Vincent.
(c)   Allows 2 days in Canal, and 2 days for oiling at Suez.
(d)   Allows 2 days in Canal, 2 days for oiling at Suez, and 1 day oiling at King George Sound.

[113] *Churchill to Colonel James Allen, New Zealand Minister of Finance and Defence*

14 February 1913

'Secret'

We have been thinking over the conversation which we had with you on Monday, and there are one or two points which I ought again to bring to your notice before our next meeting.

In the first place, let me preface my remarks by stateing that His Majesty's Government do not depart from the opinion they expressed in previous conferences, that it is for the responsible Governments of the Dominions to determine what they think best for their own interests, and that, while the Admiralty think it right to leave no doubt as to their views upon the strategical position, yet they will do their best to give full effect to the ultimate decision of your Government.

At the present time New Zealand and Australia are perfectly safe from all danger of external attack, because they are protected by the naval power and the alliances based on the naval power of Great Britain. The only event which could expose them to any danger would be the destruction of the British naval power in Home waters. If that occurred, there are no preparations which could be made by Australia and New Zealand jointly or separately which would, during at any rate the next fifteen or twenty years, enable those Dominions to cope with the naval power of Japan. The Japanese have at present, built and building a navy of sixteen battleships and four of the best battle cruisers in the world, together with many minor vessels, and they are maintaining and increasing their navy each year by Estimates which already exceed 9 1/2 millions sterling. You will see for yourself from these figures that any naval force which by their utmost efforts Australia and New Zealand could provide in the near future, could play only a small and subsidiary part in their defence against oversea attack. By far the greater part of the work would have to be done by the Navy maintained by the British taxpayer. This being so, the right and wise policy for New Zealand is to make the best contribution she can to the general strength of the British Navy as the only weapon which can effectually protect her.

From this point of view the gift of the 'New Zealand' for general service could not have been better. It is a valuable aid to the British Navy, and appreciably increases the margin of superiority upon which British naval supremacy depends. We have always loyally done our best to make the Australian arrangements a success. But the Admiralty cannot hesitate at all in expressing their strong preference for the method of naval assistance which has been adopted in the past by New Zealand. Useful as is the material reinforcement of this fine vessel, the moral effect, so important alike to our safety and our policy, is far greater. There is no other way in which the Dominion of New Zealand could for such a comparatively small expenditure of money have produced such a great effect on the British Empire or on the European situation.

We should regret any wish on the part of the Government of New Zealand to withdraw the battle-cruiser from the general service to which she has been dedicated. At the same time you should realise that you and your Government have only to express the desire for us to give prompt effect to it. It would be unnecessary, from the strategic point of view at the present time, to place the *New Zealand* on the China Station. She would not find there any vessel of equal strength belonging to any of the European Powers. The Japanese Navy, on the other hand, is out of all proportion to any force we should maintain in China waters during the continuance of the alliance, whether that force were strengthened by the addition of the *New Zealand* or not. If, however, you and your Government urge that the 1909 Agreement provided for the stationing of a battle-cruiser in China waters, and if this request is made to us, we will meet it as quickly as possible, and substitute the *New Zealand* for two of the armoured cruisers, the 'Minotaur' and one other now on the China Station, which are at present sufficient for the strategic needs of the situation.

I may point out that while the number of armoured ships on the China Station is at present four, the Admiralty under the Agreement of 1909 only proposed to keep there one Dreadnought battle-cruiser. The others are maintained in pursuance of the understanding with Japan that we should have on the station a force superior to that of the strongest European fleet in the Far East and for general Imperial purposes.

So far as New Zealand waters are concerned, we recognise the desirability of encouraging the development of naval sentiment and naval interests in the Dominion. We are reluctant to take any steps which should appear to commit us to favouring the principle of local navies against that previously adopted by New Zealand of one Imperial Fleet. We should, however, be prepared to afford facilities to New Zealanders for training and serving in British ships of war, and we will assist, to the best of our ability, any steps that may commend themselves to your Government for building up a New Zealand Naval Reserve. The waters of New Zealand and the harbours available do not lend themselves to the effective employment of torpedo-craft, nor is there any service which at present such vessels could render which could not be better otherwise provided for. In response to your wish, a statement will be prepared as to the adequacy of the existing military defences of New Zealand ports to ward off any casual cruiser attack that might possibly be made upon them. We are prepared, however, to maintain on the New Zealand Station for the purposes of training a New Zealand Naval Reserve, as well as for the ordinary cruising work in the islands, the two light cruisers which are at present on the Australian Station (*Cambrian* and *Psyche*, or similar vessels), and to send a third in lieu of the three destroyers and two submarines mentioned in previous discussions. These vessels are quite equal as ships of war to commerce protection and other military purposes likely to be required of them in New Zealand waters in the present period.

It is true that the Board of Admiralty offered in 1909 to detail two 'Bristol' type cruisers in New Zealand waters, but, as I pointed out, great changes have since occurred in the naval situation. The need of fast light cruisers is serious at home; and there is no military reason which at present requires them in New Zealand. In the event, however, of newer or better vessels of any foreign Power being stationed in or about New Zealand or Australasian waters, we should of course at once replace our vessels so as to maintain an effective superiority. In the meantime, if your Government is desirous of having newer ships and of further aiding the naval resources of the Empire, you could not do better than undertake the building of modern light cruisers to replace in the course of the next few years the older vessels which will be based on New Zealand ports. The cost of maintaining the three ships in New Zealand waters would considerably more than absorb the 100,000£ a-year contribution which the Dominion pays at present. There would be no difficulty in earmarking the money to that object were that desired. The cost of maintaining the 'New Zealand' battle-cruiser herself, which is considerably over the 100,000£ a-year, would, of course, continue to be borne by us so long as she is available for general Imperial service. It would no doubt be possible to arrange for some regular means of consultation between the Commodore of the three New Zealand cruisers and yourself or your successors in the Ministry of Defence in regard to the training of the New Zealand Naval Reserve, whose development would form a feature of our future policy. Were a force of trained naval reservists gradually to become available in New Zealand there would be no great difficulty in an additional ship being stationed in reserve at Auckland, which in time of war or emergency might be mobilised for service.

You will, I hope, understand that these are suggestions which arise out of our discussion, and not final or formal proposals for decision. We should be very glad if you would consider them well and let us know when you would like to have another discussion upon them.

[114] *Allen to Churchill*

18 March 1913

'Secret'

I have already acknowledged receipt of your letter of the 14th February last, and I thank you for the valuable information contained therein. I am also obliged to you for the suggestion that you would be prepared to maintain on the New Zealand Station two light cruisers, and to send a third in lieu of the three destroyers and two submarines mentioned in previous discussions. It may shorten further deliberations on the naval question in the Pacific if I write what is in my mind.

Your letter raises some controversial questions which we have already talked over, and which do not lend themselves to discussion by letter. Upon one point, however, there must be no misunderstanding. I refer to the question raised by you that you were 'reluctant to take any steps which would appear to commit us to favouring the principle of local navies against that previously adopted by New Zealand of one Imperial Fleet.' My comment upon this sentence in your letter is that local units and an Imperial Fleet are not necessarily antagonistic. Sentiment and other conditions do lead to the desire for local units but there is not the slightest reason why the Dominions interested, if they do not realise it at the first, should not at a very early date learn the strategical importance of unity in control and the necessity for directing their individual efforts in providing separate units in such a way that each unit will fit in, one with the other, whenever concentration does take place. Let me further add that in my judgment an Imperial Fleet should consist of ships, it is true, and also of personnel provided by each constituent part of the Empire.

As to the 1909 Agreement, you say, 'great changes have since occurred in the naval situation.' This may be so, but certainly New Zealand thought the situation fairly serious in 1909 and gave expression in material form to her desire to help the Mother Country. In view of what has transpired in my talks with yourself and with the Admiralty, and considering what you have written in your letter of the 14th February, I cannot ask that the Agreement of 1909 should be carried out, but I do desire to know whether New Zealand can feel assured that any policy agreed upon will have some permanency. You will no doubt understand that the New Zealand Cabinet must consider any suggestions made before a final conclusion can be arrived at.

The principles that guided me in what I now write to you are:

(a)   The utilisation of New Zealand national sentiment and local patriotism to give our people that interest in Naval Defence which, in my opinion, would not be forthcoming by the payment of a subsidy to Great Britain, to use as Great Britain may determine.

(b)   Sound Strategy – In this I include some provision for advice to Australia and New Zealand, which would so direct their efforts in acquiring or constructing ships of war that the unit of one would fit in with the unit of the other, or that the units of both would fit in with any Imperial Squadron when concentration took place. This would avoid waste. Further, I include in sound strategy a definite understanding that all units in the Southern Pacific, including the British unit in China waters, should concentrate for manoeuvre purposes in peace time: and that they should all be at the disposal of the Admiralty in case of war or if war were imminent.

Australia, in utilising the first principle, has adopted what I believe to be, both with regard to ships and men, a permanent policy, and has made such progress that the Australia unit alone will be more effective than the Australasian Squadron which it replaces; and in doing this Australia has, to a very large extent, given relief to the Mother Country.

For the next year or two New Zealand can do but little in acquiring ships of her own, but some provision must be made to enable us to carry on the policy already in existence, namely, the training of our men for the Navy. Your suggestion enables this to be done, but I should be glad of your advice as to this training. To me the present arrangement is not satisfactory, and I would like to be in a position to report to the New Zealand Cabinet what the Admiralty considers ought to be done to produce the efficient sailor, and, in process of time, the higher ratings and officers. I see no prospect of training officers ourselves for some time to come, but there is available to us the recently established Naval College in Australia, and, if the Admiralty would accept the training there for entry into the Imperial Navy, we could utilise that institution and at the same time help our Australian neighbours. Can this be agreed to, or, in case this is not possible, can you offer a larger number of naval cadetships to New Zealand?

Then, in deference to New Zealand national sentiment, which I hope to see further developed, will you extend your generous offer of the *Cambria* and the *Psyche* for 'the purpose of training a New Zealand Naval Reserve?' What I suggest is that you should place these two ships at the disposal of the New Zealand Government. This would give to New Zealanders that direct interest in naval concerns which to me seems so valuable. As to the third ship mentioned in your letter, though I do not think we can, under the circumstances, expect you to supply more than three ships, still it would have been a great satisfaction to me had I been able to report to the New Zealand Cabinet that there were two other cruisers, in addition to the *Cambrian* and *Psyche*, available for manoeuvring purposes, and which would fit in with the Australian Fleet. If this third ship, or the two if you can let us have them, were for the time being also placed at the disposal of the New Zealand Government, it would add to our interest generally; would assist us in our administration of the Cook Islands; and would enable us to say to the people of the Commonwealth that we had something with which to assist them.

The 100,000£ a-year now provided by New Zealand, and the whole of the present New Zealand Naval Reserve would be available for the three or four ships. The cost would probably exceed 100,000£ per annum, and I should be obliged if you would let me know by how much.

I have not suggested that New Zealand should acquire ships at present, partly because for the next year or two the country has sufficient financial responsibility, and partly because I believe some organisation should be created to advise Australia and New Zealand before any further ships are acquired. If Canada and South Africa were included in this organisation it would be even more effective. Do you think it impracticable that the Prime Minister and Ministers of Defence of the Commonwealth and of New Zealand, and, say, one or more representatives of the British Government or the Admiralty, should from time to time meet and discuss this question of the provision of necessary ships, and other cognate questions, from the point of view of Imperial necessities in the Southern Pacific? Indeed, I do not see why such a representative body might not be considered a Sub-Committee of the Committee of Imperial Defence, to give advice to their respective Governments on concentration in peace time for manoeuvre purposes, or on any question affecting the interests of the Navy in the Pacific.

It is not possible to foretell what may happen in the future, but it seems to me that the growth and development of Dominions like Canada, Australia, and South Africa must, in an evolutionary way, lead to an alteration in the organisation for Imperial control. This readjustment at some future date has been in my mind, and I make the suggestion in this letter in the hope that the Mother Country, Australia, and New Zealand may make an experiment in the direction of common control in the South Pacific which, if successful, may assist the Empire in working out the problem of the future, namely, readjustment to suit the new conditions which are certain to arise.

It will give me pleasure to meet you at any time up to the 18th April, on which date I leave England on my return to New Zealand. I should be much obliged if you would give me a few days' notice of any meeting.

[115] 'Representation of the Dominions on the Committee of Imperial Defence', note by the Secretary [Captain M.P.A. Hankey]

4 April 1913

'Secret'

The proposals of His Majesty's Government in connection with the representation of the Dominions on the Committee of Imperial Defence, as also the communications in this connection that have been received from the Governments of the Union of South Africa and the Commonwealth of Australia, are shown in the annexed papers.

2   The Government of the Union of South Africa 'doubt whether the idea of a Minister of the Union residing in London for the purpose of constantly representing the Union Government on the Imperial Defence Committee is practicable,' and Ministers point out that it is always open to the Union Government to seek advice from the Imperial Defence Committee in writing, or, in more important cases, to ask for a personal consultation between that Committee and the representative, or representatives, of the Union Government.

3   The Commonwealth Government confine themselves to stating that it is impracticable for a Commonwealth Minister to visit England during 'the ensuing year,' and to suggesting that, in view of the importance of co-operation in naval defence, a subsidiary conference should be convened in Australia early in 1913.

4   No replies to the Secretary of State's despatch of the 10th

December, 1912, have as yet been received from the Governments of Canada and New Zealand.

5   At the 119th Meeting of the Committee of Imperial Defence, however, Mr Borden, speaking in connection with the proposals for the representation of the Canadian Government on the Committee of Imperial Defence, said:

' . . . so far as my colleagues and I are concerned, we are entirely in sympathy with the proposals, and we have really little doubt that the same favourable consideration will be taken by Canada as a whole;'

and again in Ottawa on the 5th December, 1912, in introducing the Canadian Naval Bill, the Canadian Prime Minister said:

'I am assured by His Majesty's Government that pending a final solution of the question of voices and influence, they would welcome the presence in London of a Canadian Minister during the whole or a portion of each year. Such Minister would be regularly summoned to all meetings of the Committee of Imperial Defence, and would be regarded as one of its permanent members. No important step in foreign policy would be undertaken without consultation with such a representative of Canada. This seems a very marked advance, both from our standpoint and from that of the United Kingdom. It would give to use an opportunity of consultation, and therefore an influence which hitherto we have not possessed. The conclusions and declarations of Great Britain in respect of foreign relations could not fail to be strengthened by the knowledge that such consultation and co-operation with the Oversea Dominions had become an accomplished fact.'

[116] *Churchill to Allen*

7 April 1913

We have now considered very carefully your letter of the 18th March.

Dealing first of all with their definite proposals, I am glad to say that the Admiralty are able to station three cruisers in New Zealand waters for the local protection of New Zealand and her commerce and for general Imperial service. We are quite ready to assist you in building up a New Zealand Naval Reserve, and a scheme will be prepared in detail for that purpose. It will comprise arrangements for the training of New Zealanders in the three cruisers of the New Zealand Squadron, and also for keeping the trained men requalified after they have completed their service afloat, as well as the procedure for communication and consultation between the Officer Commanding the Squadron and the New Zealand Government on these matters. In this way a New Zealand Naval Reserve will be developed, and, as soon as this force attains the necessary strength, a fourth cruiser will be sent to New Zealand and placed in reserve at Auckland. In time of emergency the complement of this vessel could be completed from the New Zealand Reserve together with the crew of the 'Torch' now employed on patrol duty in the islands, which vessel would be laid up in time of war . . .

It will not be possible to transfer any of the ships we propose to maintain on the station to the administration of the New Zealand Government, if that is what I rightly understand you to suggest in your letter. New Zealand has at present no organisation or naval personnel for manning and maintaining such a squadron, and the vessels in addition discharge important general Imperial services quite outside those which are strictly concerned with New Zealand interests . . .

You also speak of the need that 'some organisation should be created to advise Australia and New Zealand before any further ships are acquired.' We do not see how any such organisation could possess greater knowledge of this subject than the Admiralty, and the whole of our services are and will always be at the disposal of New Zealand or Australia.

[117] *Allen to Churchill*

9 April 1913

The organization for maintaining the ship or ships may not exist in New Zealand at the present moment, or may be very imperfect; still we have in Auckland a good dock and expensive equipment for repairs and maintaining and mechanics quite competent to carry out repairs. The necessary naval personnel could surely be acquired.

In suggesting some organisation to advise Australia and New Zealand before any further ships are acquired, it never occurred to me that the organisation could possess greater knowledge of the subject than the Admiralty. My object was to bring the Mother country, Australia, and New Zealand into closer touch on naval questions; to use the organisation as a means for advice on the acquiring of ships by each, so as to avoid waste; and to insure that whatever action each country took had in view the use of the units as a whole. Further, it did seem to me a desirable ambition on the part of New Zealanders, whilst gradually assuming their fair share of the burden of Imperial Defence, to educate themselves for the day when important changes must take place in administrative control if the Empire is to hold together.

[118] *Committee of Imperial Defence, 123rd Meeting*
11 April 1913

Imperial Naval Policy, New Zealand

COLONEL ALLEN [New Zealand Minister of Finance and Defence], continuing, said . . . he wished to recommend to the Cabinet in New Zealand some policy which would give the people of that country control over their own naval arrangements. For example, he wished the Government of New Zealand to have control over the training of their own men. Under existing arrangements they knew nothing whatever about it, and were completely out of touch with their men. This was most unsatisfactory to the Government and people of New Zealand, and the only way to rectify it was for the New Zealand Government to take over the administrative control . . .

THE PRIME MINISTER [Herbert Asquith], asked whether the ideas of the New Zealand people on the naval question might not be grouped into two categories: first, the sentimental desire to see the fruits of their own efforts, and, second, the technical fact that, whatever the abstract strategical principles might be, they would feel more secure if they were protected by ships under their own control.

COLONEL ALLEN agreed that that was so, but the second feeling was stronger than the first. Formerly the British Fleet was predominant in every sea, but that was no longer the case. New Zealand was growing in wealth and importance, and her people conceived it their duty to guard the standard of civilization to which they had attained. They therefore thought it safer gradually to provide for their own protection by their own Fleet, but that Fleet would form part of the Imperial Fleet in time of war or if war were imminent . . .

COLONEL ALLEN . . . thought that they must contemplate the possibility of the [Anglo-Japanese] alliance being terminated.

It took a long time to develop a navy. Their policy in New Zealand was frankly opposed to free intercourse with Japan. They did not know what the Japanese really felt about that, or for what purpose they were developing a great fleet. If New Zealand was weak they might be forced to sacrifice something they cherished. Whether they were right or wrong, they did not feel safe. In the First Lord's letter of the 14th February he laid great stress on the inability of Australia and New Zealand to develop their naval forces. He did not think his estimates would be accepted in either country as being based on adequate statistical data . . .

MR CHURCHILL [First Lord of the Admiralty] said that . . . what was necessary was for the New Zealand Government to offer sufficient inducement to the necessary ratings to volunteer for service in the New Zealand Navy. So far as they could the Admiralty would help. He felt, however, bound to point out that the officers and men would need some assurance as regards their future prospects. In Canada they had started a local navy, but a new Government had come to power and had changed the policy. No doubt any disturbance to the prospects of volunteers would be compensated, but this fact might cause officers and men to hesitate before they volunteered for such service. It was mainly a question of the terms offered. If the personnel on board was mainly British the administrative control must be with the Admiralty; when the balance was the other way, the transfer to the New Zealand Government could be arranged. He was quite willing to discuss the details and to meet the wishes of New Zealand so far as possible.

[119] *'Imperial Naval Policy', memorandum read by Churchill at the 123rd Meeting of the Committee of Imperial Defence*
April 1913

The safety of New Zealand and Australia is secured by the naval power and the alliances based on the naval power of Great Britain. No European State would or could invade or conquer New Zealand unless the British Navy had been destroyed. The same naval power of Great Britain in European waters also protects New Zealand and Australia from any present danger from Japan. While Japan is allied to Great Britain, and while Great Britain possesses a sufficient margin of naval superiority, Japan is safe from attack by sea from the great Fleets of Europe. In no other way in the years that lie immediately before us can Japan protect herself from the danger of European interference. It would appear that the reasons which have led Japan to contract and renew the Alliance will grow stronger with time. The growth of German interests in China, the re-building of the Russian Fleets in the Baltic and the Black Sea, and the general development of European Navies on a scale far greater than Japan can afford to imitate, will require her increasingly to seek that sure protection which British naval supremacy can so easily afford.

The obligations of Great Britain to Japan under the Alliance are not limited to preventing an armada being dispatched from European waters to alter suddenly the balance of naval strength in the China Seas. We are bound to maintain in these waters a force superior to that of any other European Power, and consequently any danger to Japan arising from a gradual increase of European squadrons in the Far East is also provided against.

The Alliance has now been renewed up to the year 1921, with the full concurrence of the oversea Dominions. It is not to be expected that, even after that date, Japan will have less need of that powerful friend at the other end of the world who guarantees her from all European molestation. Quite apart from the good sense and moderation which the Japanese have shown since they became a civilized Power, and quite apart from the great services mutually rendered and advantages derived by both Powers from the Alliance, there is a strong continuing bond of self-interest. It is this that is the true and effective protection for the safety of Australia and New Zealand.

If the British Fleet were defeated in the North Sea, all the dangers which it now wards off from the Australian Dominions would be liberated . . .

From this point of view the profound wisdom of the policy hitherto adopted by New Zealand can be appreciated. In giving a splendid ship to strengthen the British Navy at the decisive point, wherever that point may be according to the best principles of naval strategy, the Dominion of New Zealand have provided in the most effectual way alike for their own and for the common security . . .

The situation in the Pacific will be absolutely regulated by the decisions in the North Sea. Two or three Australian and New Zealand Dreadnoughts, if brought into the line in the North Sea, might turn the scale and make victory not merely certain, but complete. The same two or three Dreadnoughts in Australian waters would be useless the day after the defeat of the British Navy in Home Waters . . .

It is recognised, however, that time will be required before the true principles of naval policy are comprehended in the Dominions, and that in the interval arrangements must be made to develop, so far as possible, their local naval establishments . . . The choice which is open to us in the immediate future is not one between having Australian vessels in the right strategic stations or the wrong, but between having them in the wrong or not having them at all . . . The Admiralty have, therefore, on political grounds, co-operated to the best of their ability in the development of the Australian Fleet Unit . . . It is with the object of combining sound military principles with local aspirations that the design of an Imperial Squadron has been conceived . . .

There should be developed severally in Canadian, Australasian, and South African waters a naval establishment, with docks, defences, and repairing plant, which would enable the Imperial Squadron to operate in each theatre for a prolonged period.

Side by side with this there should be developed in each of these three theatres, so far as may be necessary, and allowing for local conditions, the local defence flotillas, both destroyers and submarines, for the purpose of both defending their bases and establishments and of operating in conjunction with the Imperial Squadron when it arrives. Great ships move easily and swiftly about the world, but small craft are, by their nature, localised, and can only cross the world with difficulty and effort.

Thirdly, the Dominions should locally maintain the light cruisers necessary not for fighting battle fleets but for commerce protection in their own waters, and these cruisers would also combine with the Imperial Squadron, when it arrives, to make the fleet complete in all respects.

[120] *Allen to Churchill*

24 April 1913

'Confidential'
Your letter of the 18th April was handed to me at Euston Railway Station just before leaving on my return to New Zealand.

It will give me pleasure to recommend the New Zealand Government to adopt as closely as possible the general features of the scheme prepared for the Australian Government. If the suggestions which I shall make are agreed to, then legislation will be introduced on the lines indicated in your letter, and the kind offer of the Admiralty to 'lend an experienced officer' will be gratefully accepted.

It does not seem to me necessary in the early stages of the scheme to set up a Naval Board in New Zealand, and I should be glad to have your opinion on the suggestion made by me, I think to Sir Graham Greene [Admiralty Secretary], that the Officer in command of the training vessel should be the advisor of the New Zealand Government through the Minister who may be placed in charge of Naval Defence. This officer would be given whatever staff was deemed necessary.

The difficulty of a 'Career for officers and men' is realised by me, and I made some suggestions on this matter to Sir Graham Greene, which I now repeat, and I should be grateful to you if you will criticise them and give me the benefit of your valuable advice, or, if you do not consider them satisfactory, will you kindly make some other proposal. What I suggested was that the Royal Navy should be available for service by, and [for the?] promotion of, New Zealand men who had served on the New Zealand training ship; the two Admiralty ships in New Zealand waters to be used in the first instance, and, when no further positions could be provided on these, then the principle should be applied to ships in the China Squadron. The training ship under the administration of the New Zealand Government and the Admiralty ships in New Zealand waters to be manned as far as possible by New Zealand trained men: and these men to be recognised for service under the New Zealand Government whenever required on due notice being given.

If neither of the Admiralty ships were acquired by New Zealand, then in order to get over the difficulty of unequal pay I suggest that on the transfer of New Zealand men our Government should make up the difference in pay to those serving on the particular ships to which our men might be transferred. This same principle might be followed in the case of New Zealanders allotted to a ship or ships in the China Squadron till the time arrived when promotion in the Royal Navy would carry sufficient emolument to induce the New Zealander of higher rating to accept it without increased pay following his further promotion.

I have not at my hand material to judge of the difference in pay, deferred and ordinary, between those serving in the Royal Navy and those in the Royal Australian Navy: and, assuming that the New Zealand rates of pay will approximate to the Australian,

I should be glad if you could let me have a memorandum indicating the difference.

I am not sure whether the same difficulty with regard to pay applies to officers, and should be obliged if you would advise me on this matter. It seems clear that the opportunities for promotion for officers in New Zealand seas would be insufficient, and the Royal Navy should be open to them.

Whilst on the question of 'Officers,' I believe it is correct to say that at one interview with you and the First Sea Lord a definite statement was made that the training in the Australian Naval College would be accepted for entrance into the Royal Navy. Would you please inform me how many cadetships would be open to New Zealanders through Osborne and through the Australian College?

You will also remember that the 'special entry of naval cadets' was discussed and that I pointed out the difficulty New Zealanders would find in complying with paragraphs 3 and 5 of the Regulations on account of their distance from Great Britain. If it is considered desirable that New Zealanders should enter in 1914, 1915 and 1916 for these cadetships I fear it will be necessary to modify the regulations.

I gratefully acknowledge your kindly sympathy in placing one ship at our disposal. I do not propose to ask the New Zealand Government to purchase, because experience may show us that a different class of ship may be more suitable for our purposes, and, further, I understand that you generously offered to lend us one vessel if we paid all costs, including maintenance, repairs, &c. The estimate of the cost will, I understand, be sent with the papers which are to be forwarded to me through the High Commissioner.

Before leaving the training of personnel there is one other matter which I desire to bring under your notice. It would be wise to make our training and pay as nearly as possible uniform with the training and pay adopted by Australia. The Commonwealth has provided two methods of entry and training:

(a)   As boys, between the ages of 14 1/2 and 16. To commence with twelve months' training in H.M.A.S. *Tingira*. From this source future warrant officers, &c., will be obtained.

(b)   As men, between the ages of 17 and 25.

In New Zealand we have one training ship for boys, the *Amokura*, and it is possible we may be able to utilise this ship for the purposes of source (a). If, however, we find that the *Amokura* is not large enough or is not suitable for the work, could you lend us temporarily a larger ship to replace the *Amokura*, or one which, with the *Amokura*, would enable us to make sufficient provision for the training of boys for the Navy on the same basis as that adopted by Australia?

I should be glad if you will send me an estimate of the annual cost of the two ships that you intend to station in New Zealand waters. I ask this because it is my intention to advise the New Zealand Government to repeal the provision for the contribution of 100,000£ a-year to the Royal Navy, and in place thereof to increase the amount so as to provide for the payment of the training-ship and, for the time being, for the annual cost of the two other ships, if this is within our means.

In your letter of the 7th April you write that a scheme will be prepared for 'the procedure for communication and consultation between the Officer Commanding the Squadron and the New Zealand Government' on the matter of training. I hope that, notwithstanding some change as to the training, this scheme will be gone on with, and that it may embrace other matters than training. We living in New Zealand are deeply interested in Pacific questions, and it appears to me that some communication and consultation on subjects other than training may be mutually beneficial. For instance, a report from, or consultation with, an officer visiting the Cook Islands, now a portion of New Zealand, would be very valuable to the New Zealand Government.

Referring to the last paragraph but two in your letter of the 18th, you may rest assured that our object in New Zealand is to preserve the close connection between the New Zealand service and the Royal Navy: indeed, my earnest desire is to create in, I hope the not far distant future, one Imperial Navy adequate for Empire needs wherever these may become apparent. I have already assured you that the New Zealand Government will be advised by me to place any ships that New Zealand may have under its administration at the disposal of the Admiralty in the event of war, or if war were imminent.

In conclusion, I thank you again for your patience in listening to what I have had to say: for the time, and for the careful consideration you have given to the whole matter. I shall have much pleasure in reporting to the Prime Minister and Members of the Cabinet in New Zealand that the Admiralty has met us with every desire to make effective that great Navy which all portions of the Empire must rely on to maintain security all over the world.

Thanking you for your good wishes for the voyage, and hoping some day to have the very great pleasure of welcoming you in New Zealand.

[121] *Governor-General of Canada [The Duke of Connaught] to Harcourt, forwarding a note from Borden to Churchill, draft cable*

1 June 1913

Following from Prime Minister for Mr Churchill

Confidential, private and personal. We appreciate most thoroughly [the] unfortunate situation arising from [the] action of [the] Senate [which rejected Borden's Naval Aid bill] and assure you of our desire and intention to retrieve that situation as speedily and as effectively as possible. Please consider practicability of having construction of three ships undertaken immediately by your Government upon assurance of Canadian Government that on or before their completion we will introduce a bill authorising Canadian Government to pay for them and take them over under same conditions and for same purpose as set forth in Naval Aid Bill. I am prepared to make an early statement to Parliament that, we shall give this assurance to your government. We would in ordinary course have majority in Senate before completion of ships and thus be enabled to fulfill our engagement to take them over and pay for them. Public opinion in Canada is overwhelmingly in favour [of the] Naval Aid Bill but we cannot hold [a] general election until after [the] redistribution [of electoral boundaries] and probably not before autumn of 1915.

[122] *Harcourt to Connaught, draft*

4 June 1913

Please communicate [the] following to [the] Prime Minister [Sir Robert L. Borden] from [the] First Lord of Admiralty [Winston S. Churchill] (Begins) Private, Secret. Cabinet this morning felt that for us to begin the three proposed Canadian ships pending Canada being able to assume responsibility for their cost would be open to criticism in both countries as seeming to go behind the formal decision of [the] Canadian Parliament and that we have no right at present to assume that [the] Senate's vote could be reversed.

We felt that both you and we should remain perfectly free to deal with the future.

On the other hand [the] Admiralty have to meet the emergency referred to in my published memorandum of September last by having proper numbers of ships[,] for [the] whole world defence of [the] Empire apart from [the] needs of United Kingdom[,] available at a definite date.

If your Bill had been carried the three Canadian ships would have been ready for battle in the third quarter of 1915.

We cannot therefore delay action and orders will be issued forthwith to begin the last three ships of the British 1914–1915 programme already announced now instead of in March next as hitherto prescribed.

This acceleration imposes a substantial addition to our expenses this year but safeguards the imperial naval position effectively for another six months during which time further discussions can if desired take place.

Announcement will be made to House of Commons tomorrow by me.

Private personal, we warmly appreciate the spirit of your telegram, and will welcome further consultation. [message ends.]

[123] Borden to Churchill

Ottawa
25 June 1913

'Confidential'

I shall not take up your time in reciting the inner history of the events connected with the prolonged discussion of the Naval Aid Bill and of its rejection by the Senate. There are, however, a few facts with which you should be made acquainted and certain considerations which it is desirable to bring to your attention.

1   Laurier's amendment of the 12th of January to our resolution adopted that portion of our proposal which provided for an immediate appropriation of thirty-five million dollars for increasing the effective naval forces of the Empire. His proposal was not necessarily inconsistent with ours inasmuch as the ships were subject to recall upon reasonable notice and could have been utilized in the establishment of the two fleet units which he favoured.

2   It became evident in March that the Opposition had embarked

upon a definite policy of obstruction. Their press and a number of their friends urged them to adopt such tactics as would ensure an early dissolution. We had good reason to believe that some of their leaders were very much alarmed at any such prospect; and if dissolution had been brought about it is most certain that our majority would have been substantially increased, as we have convincing evidence that many prominent Liberals are greatly dissatisfied with the attitude and action of their party on this question.

3 You may naturally ask why under the circumstances we did not advise the Governor General to dissolve Parliament. The considerations which influenced us in taking the opposite course were as follows:

(a) The unfortunate effect of an election at the present juncture upon the general business interests of Canada in view of the financial stringency and of other relevant conditions.

(b) Unless a redistribution Bill was first passed a further election would be necessary within two years.

(c) Even if an election were held upon this issue and the Government sustained by an increased majority, there could be no assurance as to the action of the Senate which might still find some pretext for defeating the measure.

(d) Our own friends throughout the country were strongly opposed to a dissolution at the dictation of the Opposition and of their majority in the Senate.

4 When it became abundantly apparent that a policy of defiant obstruction had been adopted by the Opposition, we were forced to resort to Closure which was passed with less difficulty than we had anticipated. Very shortly afterwards the Bill was put through Committee, ample time having been allowed for any further relevant discussion. Upon the Third Reading of the Bill we were not forced to resort to the Closure, the Opposition being very much divided as to the expediency of further opposition.

5 Up to this time it was anticipated that the Bill would pass the Senate with certain amendments which we were disposed to accept. Confidential proposals were made to us by leading Liberals in the Senate and the measure as proposed to be amended would have fulfilled an important purpose, as at least two battleships or battle cruisers would have been provided for the common defence of the Empire while the balance of the appropriation would have been devoted to harbour and coast defences and auxiliary ships.

In the end, however, Sir Wilfred Laurier insisted upon the rejection of the Bill giving to his friends in the Senate a choice between that course and his resignation. Many Liberal Senators who voted against the Bill did so with reluctance and against their better judgment.

6   The action of the Senate naturally involved much disappointment throughout the country. The situation was under careful consideration by the Government for some days after the rejection of the bill; and I need not do more than allude to confidential communications which passed between us, copies of which are enclosed for convenient reference.

7   On the day of prorogation Sir Wilfred Laurier came to the House with a carefully prepared speech. A copy of the Hansard report is forwarded for your information although the important portion was telegraphed to you at the time. His question gave me the opportunity which I desired of making an announcement that eventually we expected to take over and pay for the three ships which you propose to lay down in substitution for those which Canada would have provided under our Bill. The announcement came as a surprise to Laurier and it obviously disconcerted and nettled him. On the other hand it has been received with much acceptance and approval throughout Canada.

[no paragraph 8]

9   In case you propose to lay down the three ships it is important from our standpoint that they should be of the character, class and fighting value which we proposed and announced. We understand perfectly that there cannot be any agreement whether formal or informal in this respect, as you are officially confronted with the fact that the Parliament of Canada has declined to make any provision. This, however, need not preclude informal and confidential discussion. It is not possible for me to visit England this year but one of my Colleagues, the Hon. W.T. White, Minister of Finance, will sail for the United Kingdom on the first of July. He has taken a deep interest in this question and is thoroughly familiar with all its aspects. His visit is not official but is purely for the purpose of a much needed holiday and rest. I have asked him to discuss informally and confidentially with you the whole situation for the purpose of facilitating future developments along the lines above suggested.

10   I quite realise the possibility that the situation may have so changed from your point of view during the past three or four weeks that the suggestion above set forth may be impracticable

or without value. In that case, kindly give any necessary explanations to Mr White; who is thoroughly informed as to the whole situation on this side and who has my full confidence in this and in all respects.

It would be ungrateful for me to close without expressing on behalf of my Colleagues and especially on my own behalf our very deep and warm appreciation of the most loyal and generous support which during all these trying months you have accorded us in connection with our proposals. I need hardly assure you that whatever can be done by this Government under the circumstances to retrieve or to improve the situation will be undertaken most gladly and cordially; and in this the public opinion of Canada will be entirely with us.

[124] *Churchill to Borden*

30 June 1913

'Secret and Personal', [draft?]

Many thanks for your telegram sending me a copy of your statement to the Dominion Parliament upon the prorogation.

Let me explain to you the course we are adopting. Parliament has sanctioned 5 capital ships being built by the Admiralty in the programme of 1913–14. i.e., between the 1st April 1913 and the 31st March 1914. Of these, we had proposed to begin 2 in the dockyards in August and not to start the other 3 until the end of February or the beginning of March next year. It is our practice to obtain sanction for ships from Parliament a considerable time before they are actually begun, so that we always have a certain margin in hand for emergencies, which enables us to complete the necessary numbers of ships at the dates required. It is upon this reserve power that we have drawn for the present. We have not added to the number of ships under construction, nor to the total commitments of the Admiralty; but we have begun 3 of our own ships now instead of in March next. The effect of this is to secure the position for a further six or seven months, during which there is still time for Canada to act. The effect of this acceleration can only be temporary and will have passed away by February next year. In that month we shall have to decide upon the programme of 1914–15, and if Canada is still unable to take any action, we shall have to make new proposals.

The 5 ships which are now being begun are not therefore, as you seem to suppose, ships begun by this country which Canada

can at any time take over, but ordinary ships of the British programme which have been antedated in construction. It would,
however, be open to you at any time in the next few months to
take these ships over as Canadian ships building at Canadian
expense, and there would still be time for us to lay down another
3 ships for ourselves in the ordinary course in March next. But I
do not think the Cabinet would approve of our starting 3 ships
on the chance of your being able to take them over at a much
later date in their construction. That might so easily be mis-represented as Imperial interference in Canadian affairs, or as an
attempt on our part to prejudge the issue between Canadian
Parties. If, therefore, you wish to take over these ships between
now and next February, there would appear to me to be two
courses open.

The first, which I put with great diffidence – for I confess I
do not fully understand the controversies between Parties in the
Dominion Parliament – would appear to be that you should take
an ordinary Vote in Supply either for the whole cost or for the
first installment of the cost of these ships under the existing Canadian Naval Defence Act; and subsequently, when you have a
majority in the Second Chamber, to introduce the measure regula-
ting the conditions of their employment. I notice that this course
has been not unfavourably referred to by the Opposition. Of
course I cannot judge at all over here what its bearing would be
upon Canadian politics. But after all the great thing is to get the
ships started. They will take nearly 3 years to build, and there is
therefore plenty of time to settle what is to be done with them. I
fear it may be you will tell me that it is impossible even as an
interim device to escape from the deadlock which the action of
the Senate has created. If so, you will I am sure not be vexed with
me for putting it before you in the same spirit of full and friendly
consultation in which we have been working together throughout.

The only other course which I see open is for the great contracting firms like Vickers and Armstrongs to begin the ships as a
speculation on the declaration of the Canadian Government that
they intend to take them over some time before they are completed. The risk of their being left with the ships on their hands
would not be a very great one even if by misadventure the policy
and intention of the Canadian Government was changed; because
if the ships were built or in an advanced stage of construction, the
British Government simply could not allow them to pass into any
other hands but ours or yours. The danger of having these great

weapons loose in the world would be so compulsive that we should be forced, not as a matter of agreement or contract, but by the pressure of events, to take them over ourselves. I should imagine from what I know of them, that the great firms would not hesitate, if pressed by you, to go ahead and build for stock. The process, however, cannot be considered as either economical or regular; and while I mention it as a possible course, I must not be understood as recommending it. All that could be said for it is that it would get the ships under weigh without involving the British Government in a controverted question of Canadian politics.

If both of these courses are inadmissible, and no other can be suggested, it will be necessary in the first Quarter of 1914 for us to take further decisions. The 3 ships which were the subject of your Naval Aid Bill would have been ready for battle from the middle of 1915 on, and they would have raised our margins of naval strength available for the whole-world defence of the Empire – apart altogether from the defence of the United Kingdom – to the requisite level during the whole of 1916. In fact their building would have practically cancelled out the later developments of the Australian navy. If nothing can be done to begin them effectively before March next, we shall have to take steps to fill the gap of a different character to the temporary expedient we have now adopted.

And here I must put the whole position before you. In the Admiralty Memorandum of September last we, at your request, advised Canada that the form in which your naval aid could most usefully be given was in the construction of capital ships of the greatest power. That advice was and is in our opinion absolutely sound. If Canada desired to contribute effectively towards the general naval defence of the Empire, large ships of great power, which could move all over the globe and be available for service at any threatened point, alone fulfill the requisite conditions. Flotillas of small vessels, whether light cruisers, torpedo boat destroyers, or submarines, are necessarily localised to particular theatres, and there is no way at present in which general aid could be given by Canada except by capital ships. No other form of construction was so necessary or so suitable, and none other embodied the issues in so simple and concrete a form. Our position is, however, very different from yours. The measures which have already been taken will assure us in the autumn of 1915 of 41 Dreadnought ships, without counting the 'Australia'. If we made a great development and increase in torpedo craft of all kinds in the narrow waters of the North Sea and Mediterranean, we should thereby liberate for our whole-world Imperial service a certain number of capital ships and thereby increase the mobility of the Imperial fleet just as effectually as if we built additional Dreadnoughts. My naval colleagues consider that for less money than 3 capital ships would cost, we could by a greatly increased flotilla construction in the narrow seas liberate 3 ships for general service. The scale of the capital ships and their cost are continually rising, and even since we first discussed this question, the price of these great weapons has risen by nearly half a million apiece, i.e., three complete with all their stores and ammunition would cost 8 1/2 millions instead of the 7 mentioned in your Bill. On the other hand the power of the torpedo and the efficiency of the submarine are also developing with very great rapidity. In six months time we shall know a good deal more than we do now. Left to ourselves, therefore, if there were no prospect of Canadian aid being available in time, and we were forced at the expense of the taxpayers of the United Kingdom to do the best we could for the whole-world defence of the British Empire, it is very probable that we should not attempt in March next to fill the gap by building 3 Dreadnoughts, but would instead develop other forms of naval construction of an essentially localised character to release our existing great capital ships for whole-world service. This apparent change, although perfectly explicable and arising out of a natural development of events and the differences in the circumstances of Canada and Great Britain, might be easily mis-represented, so as to prejudice the policy on which you are embarked, and to which you are with our full agreement and upon our responsibility in no small degree committed. It appears to me to be another reason why action should be prompt if it is to be taken at all.

There is a further reason which should have your serious consideration. If you will very kindly read the marked passages in my statements to Parliament this year, which I attach, you will see that one of the first objects of our policy is to procure a cessation, or at any rate a check in the building of capital ships by Germany. This idea of a 'naval holiday' is not at all visionary but quite practical, and personally I have real hopes that, with the heavy demands of army expenditure in Germany and the resolute and rapid advance of the British naval power through the simultaneous action of Great Britain, Canada, and the other Dominions, together with developments in the technical sphere, Germany may next year or the year after feel it her interest to enter into an agreement with us to suspend for a fixed period the construction of capital ships. If this hope were to be realised, we should of course endeavour to persuade other foreign countries and the self-governing Dominions to come into line, so that no great ships would be built throughout the world during the period of truce. For this reason, it seems especially important that the Canadian ships should be definitely started and, as it were, out of the way with the currency of this year. Then we should have a clear field to say to Germany: 'the year 1913–14 is done, and the ships of that year, including the Canadian ships, are settled. But next year, 1914, or the year after, 1915, are still open, and we are quite prepared to join with you in a suspension in the construction of capital ships during one or both of those years. If you do not build your two in 1914, we will not build our 4 which are the answer to them' – and so forth. If, on the other hand, the Canadian ships are still unsettled, an element of uncertainty would be introduced which would be fatal to a happy issue so greatly to the advantage of civilization. If Canada had only been able to come in now with her 3 ships, I am convinced that we should have got into a position where the Germans would be very glad to have made a clear-cut bargain of a limited character. Convince them by facts that it is useless for them to continue the rivalry and then make it well worth their while by a good, fair agreement, to pull up.

Pray treat this letter not only as confidential, but as personal to yourself, and let me have your views upon it and the situation in the same spirit. There is now plenty of time for the situation to be examined in all its aspects.

[125] *Churchill to Allen*

31 July 1913

'Confidential'

The suggestions made in your letter of the 24th April in regard to the organisation of the Proposed New Zealand Naval Force have now received careful consideration, and I enclose a Memorandum summarising briefly the chief features of the scheme as proposed by you and developed in the correspondence and conversations which we have had upon the subject, and adding such other details as in the opinion of the Admiralty are necessary for its success.

It is fully recognised, I think, that this scheme is in a great measure a preparatory one, its object being to bring into existence as soon as possible an organised system of naval training for New Zealanders.

On the question of a career for officers, the Admiralty are prepared cordially to accept your suggestion that New Zealanders should be entered and trained as officers of the Royal Navy, and not merely as officers of the New Zealand Naval Force, and also, as at present advised, are prepared to agree that the Royal Australian Naval College should be the normal channel of entry for them. This being so, the questions as to nominations for Osborne, or 'Special Entry' under the regulations recently promulgated, become of minor importance. It is, however, proposed that the two nominations for Osborne already at the disposal of New Zealand annually should be retained. Although with the Australian Cadets' College available, I presume that not many New Zealand boys would desire to come to England for their training, yet, of course, there is nothing to prevent them from presenting themselves in any numbers before the Selection Committee for entry to Osborne or for 'Special Entry.'

As pointed out in the Memorandum, the total number of entries annually is not a matter that can be settled entirely by reference to your or our wishes, but must bear some relation to the numbers that will actually be needed at a later date for the ships to be acquired.

As officers of the Royal Navy, the pay of these officers must naturally be fixed according to the ordinary Royal Naval scales, and I do not think that any difficulty as regards pay should arise in their case.

As regards the men of the New Zealand Naval Force, your proposals as regards entry, training, and subsequent service have been adopted and somewhat elaborated. On the whole the Admiralty would have preferred that the men when trained, should have been available for service in the Royal Navy on any station, instead of only in the China and East Indies Squadrons as proposed by you, because this limitation lessens the opportunities for our own men to serve on these stations for which we always have many volunteers. But although I mention this point, I do not desire in any way to press you with regard to it.

The difficult questions arising out of the differences in the rates of pay fixed for men in the Royal Navy and those that will obtain in the New Zealand Naval Force (assuming that you adopt the Australian rates) have been carefully considered. Whilst, however, we much appreciate your suggestion that the New Zealand

Government would make up the difference in pay to Royal Naval ratings serving in any of our ships to which men belonging to the New Zealand Naval Force might be transferred, we do not think it advisable to take advantage of it, either in New Zealand waters or in the China and East Indies Squadrons.

In the latter Squadrons we think the difficulty might reasonably be got over by treating the difference between the two rates as 'deferred pay' in the case of your men because the higher rates of pay are based on the higher cost of living in New Zealand and this would not affect the New Zealander whilst serving on the China or East Indies Station. In the ships of the Royal Navy in New Zealand waters, however, this solution would not be practicable, and the existence of the two rates of pay must simply be put up with for the probably brief period during which it may be necessary for your men to be trained or to serve in ships of the Royal Navy on that station.

Turning next to the question of replacing the *Amokura*, it is not possible for us to advise without some enquiry on the spot whether this vessel is suitable for the future training of boys for the Naval Force. If you desire it, we shall be pleased to arrange for a report upon this point to be made by officers of the Royal Navy on the station.

In the event of the *Amokura* being found unsuitable, I must say at once that it is doubtful whether any really suitable vessel belonging to the Royal Navy can be made available to replace her, much though the Admiralty would like to assist you in the matter. Should we be unable to provide a ship, we should strongly recommend the New Zealand Government to acquire a suitable mercantile vessel and to adapt her as necessary for training purposes, and the Admiralty would be happy to give any advice or assistance in carrying out this suggestion.

Two other points remain to be mentioned:

You referred in your letter to the statement in my letter of the 7th April that a scheme will be prepared for the procedure for communication and consultation between the Officer Commanding the Squadron in New Zealand waters and the New Zealand Government on the matter of training.

I should like to make it clear that this sentence was written before the proposals for the Naval Force had been developed by subsequent conversation and correspondence. Under the scheme as it now stands, the Admiralty will lend an officer of sufficient standing and experience in command of the *Philomel* to act as Naval Adviser to the New Zealand Government (through the Minister who may be placed in charge of Naval Defence) on all questions affecting the organisation and training of the Naval Force. In these circumstances it would not be necessary or practicable for the Senior Officer of the Squadron in New Zealand waters also to act as Adviser. It is proposed, therefore, that he should not deal in any way with questions of policy affecting the Naval Force, and that on the matters of routine connected with the drafting of men to ships of the Royal Navy for training, &c., he should communicate direct with the Senior Officer of the Naval Force in command of the *Philomel*. If, however, at any time the New Zealand Government should desire to have the assistance or advice of the Senior Officer of the Squadron on any of those larger matters affecting New Zealand interests to which you refer, the Admiralty will be pleased, on the subject being raised through the recognised channel, to do their best to meet your wishes.

Lastly, with regard to the financial proposals to which you alluded towards the conclusion of your letter, these are of course matters for the New Zealand Parliament to consider and determine, and I do not wish to make any comment upon them. It is desirable, however, to make it clear that the control of the vessels of the Royal Navy in New Zealand waters, which, as pointed out in my letter of the 7th April, discharge important Imperial services quite outside those which are strictly concerned with New Zealand interests, must in any case remain entirely with the Admiralty, whatever course New Zealand may take with regard to a naval contribution. If New Zealand therefore decides to make any monetary contribution to the Imperial Navy, in addition to the expenditure on the scheme now under consideration, it would be well that this contribution, on whatever basis its amount may be fixed, should be regarded as a general contribution, and not as having relation to any particular ships, even although these ships may be of special utility to New Zealand in relation to the training of the Naval Force and in other ways.

Enclosure: Draft Memorandum as to Organisation and Training of New Zealand Naval Force

1 *New Zealand Naval Force*   The New Zealand Government will proceed to pass the necessary legislation to enable them (a) to establish a Naval Force of officers and men for service either

in any ships that may in the future be maintained by the New Zealand Government, or in the ships of the Royal Navy as hereinafter provided, and for the application of the Naval Discipline Act, 1866, and amending Acts, the Naval Discipline (Dominion Naval Forces) Act, 1911, and the Admiralty instructions to this force; and (b) to reconstitute the New Zealand Royal Naval Reserve as suggested in paragraph 9.

2 *Naval Adviser* The Admiralty will lend an officer of suitable seniority and experience to act as Senior Officer of the New Zealand Government (through the Minister who may be placed in charge of Naval Defence) on all matters affecting its organisation. This Officer will also be in command of the seagoing training Cruiser to be lent to the New Zealand Government (see paragraph 10).

3 *Officers* As the New Zealand Naval Force will not at present be sufficiently developed to provide a career for officers, the officers required for the Naval Force will be lent from the Royal Navy, a preference being given to officers in the Royal Navy who are New Zealanders.

In order to provide for the future entry and training of New Zealanders as officers in the Royal Navy in sufficient numbers to meet the needs of the New Zealand Naval Force, the following modes of entry will be utilised:

(i)   Subject to the New Zealand Government making the necessary arrangements with the Commonwealth Government, the ordinary channel of entry and training will be through the Royal Australian Naval College, the training at which may, as at present advised, be regarded as equivalent to the training at Osborne and Dartmouth.

(ii) Two Cadetships for the Osborne and Dartmouth course of training will be annually at the disposal of New Zealand, on the nomination of the Governor under the same regulations as hitherto.

The total number of Cadets to be entered under (i) and (ii) should bear some relation to the number of officers likely to be required in the future for the New Zealand Naval Force. It is difficult at the present stage for an estimate to be formed of these requirements, but, pending further experience and information, it is suggested that the maximum to be entered, including entries both under (i) and (ii), should not be higher than 8 annually.

In addition to the above, the Admiralty will be glad to consider applications on behalf of New Zealanders either for entry at Osborne or for 'special entry' under the Regulations ordinarily governing these modes of entry. (See Appendices A and B attached.)

4 *Pay of Officers* The scales of pay of officers lent from the Royal Navy for service in the ships of the New Zealand Naval Force will be in all respects the same as those in force for the time being in the Royal Navy, until a separate New Zealand service may be established.

5 *Men* As in the case of the Royal Australian Navy, two methods of entry and training will be provided:

(a) *Entry as Boys* between the ages of 14½ and 16 in the New Zealand Naval Training Ship *Amokura* (or other vessel to be substituted for her), the course of training, &c., to be generally similar to that established in the Royal Australian Navy in H.M.A.S. *Tingira*.

(b) *Entry as Youths* and Men between the ages of 17 and 25 for training in a seagoing Training Ship to be lent by the Royal Navy, and in such other ships of the Royal Navy as may be stationed in New Zealand waters and detailed for the purpose.

The engagement will be in the case of Boys entered as a (a) to serve up till the age of 18, and for a term of seven years thereafter, and in the case of youths and men entered as at (b), for a period of seven years in the New Zealand Forces, and will also include liability to serve in any ship of the Royal Navy in New Zealand and Pacific waters, or in the China or East Indies Squadrons.

6 Service after Completion of Training After completing their training, men belonging to the New Zealand Naval Force will be available for serving in any ships that may in the future be maintained by the New Zealand Government, or in the ships of the Royal Navy in New Zealand waters or the Pacific, or belonging to the China and East Indies Squadrons. The manning of the ships in New Zealand waters will be regarded as the primary object of the New Zealand Force.

Service in the China and East Indies Squadrons will be regarded as 'Foreign Service,' and it will be understood that so far as practicable, the arrangements made will provide for alternate periods of home and foreign service, as is the practice in the Royal Navy.

7 *Pay of Men* The rates of pay, &c., in the New Zealand Naval Force will be the same as those fixed in the Royal Australian Force, which are considerably higher than those fixed in the Royal Navy (see Appendix (C), giving the rates of pay for both Services).

Men belonging to the New Zealand Naval Force serving in ships of the Royal Navy in New Zealand waters will receive these rates in the same way as if serving in ships belonging to the New Zealand Government. Men belonging to the New Zealand Force on 'foreign service' (i.e., serving in the China and East Indies Squadron) will be paid at Royal Navy rates only, the balance between such rates and those fixed for the New Zealand Force being allowed to accumulate in the same manner as deferred pay, until their return home, or to be allotted to their families. Men lent from the Royal Navy to the New Zealand Naval Force (*Philomel* and *Amokura*) will sign a special agreement to run concurrently with their existing engagement for service in the Royal Navy which would remain in force and will receive New Zealand rates of pay. Men belonging to the Royal Navy and serving in ships of the Royal Navy will receive Royal Navy rates only, notwithstanding that there may be men of the New Zealand Naval Force being trained or serving on board the same ships.

8   Transfer to Royal Navy   Facilities will be provided for men belonging to the New Zealand Naval Force to transfer to the Royal Navy at any stage during, or at the conclusion of, their term of service in the Naval Force. They will be eligible for transfer in the rating which they hold at the time, and after such transfer will receive Royal Navy rates of pay, and will be liable for service anywhere in the same way as other ratings of the Royal Navy, and subject to agreement as to the contribution towards liability for pension to be made by the New Zealand Government, their service in the New Zealand Naval Force will be allowed to count towards pension.

9   *Royal Naval Reserve*   The New Zealand Royal Naval Reserve at present consists only of 4 officers and 57 ratings, of whom only 5 seamen and 3 stokers have served in the Royal Navy. It is proposed that the Reserve as now constituted should either be disbanded or allowed to die out, and that it should be reconstituted so as to consist in future of men who have served in the New Zealand Naval Force and who volunteer on discharge to join the Reserve, and of men who have served in the Royal Navy (of whom there are understood to be many in the Dominion) their liabilities as regards drill, &c. being generally similar to those of the Royal Fleet Reserve at home, see Appendix (D), attached.

10   *Training Ships*, &c.   The New Zealand Government already have available the training ship *Amokura*. A report will be made, if desired, by competent officers of the Royal Navy in regard to

her suitability for the purpose of the new scheme of training. In the event of the *Amokura* being unsuitable, the Admiralty will consider the possibility of transferring a suitable vessel to the New Zealand Government to replace her, but should no suitable vessel be found available (as is very probable), it is recommended that the New Zealand Government should purchase a suitable merchant vessel, and the Admiralty will advise as to her adaptation for training purposes.

The Admiralty will also (subject to the sanction of the Treasury) lend to the New Zealand Government H.M.S. *Philomel* as a seagoing Training Cruiser, as well as the officers and men necessary to complete the drill complement of the vessel; and *Philomel* to be entirely under the control of the New Zealand Government in peace time, but to pass under the control of the Admiralty in the event of war for employment on any service required.

The officer lent as naval adviser to the New Zealand Government will be in command of this vessel.

The Admiralty will also give facilities for men belonging to the New Zealand Force to be trained, so far as the duties and convenience of the Royal Navy will permit, in any of the ships of the Royal Navy remaining in New Zealand waters, and also to serve in them after completion of their training in lieu of Royal Naval ratings.

It is proposed that at present the ships of the Royal Navy remaining in New Zealand waters shall be the *Psyche* and *Pyramus*. The *Torch* will also be employed on duties in the Pacific, but it is not considered that these will admit of her being brought into the scheme for training. A statement is attached (Appendix (E)), showing in a general way the manner in which it is suggested that the training, &c., shall be arranged for in the immediate future. Later arrangements would, of course, have to depend upon the development of the New Zealand Naval Force.

It is to be understood that the ships of the Royal Navy remaining in New Zealand waters will be under the control of the Admiralty, and, as forming part of that Navy, will be available for general service wherever and whenever they may be required. The Senior Officer in charge of these vessels will be instructed to give every assistance in carrying out the scheme of training and otherwise developing the organisation of the New Zealand Naval Force, but it will not come within his province to advise the New Zealand Government on matters of policy affecting the Naval Force. On matters of routine relating to the drafting of men belonging to the Naval Force for training in the ships of the Royal Navy, &c., he will communicate direct with the Senior Officer of the New Zealand Naval Force (Commanding Officer of *Philomel*).

11 *Financial Arrangements* Subject to the sanction of the Treasury, no charge to be made by the Admiralty for the loan of the *Philomel* or for her armament or any stores on board necessary to equip her for service as a training ship. As from the time of her transfer, the whole charge of her maintenance to be borne by the New Zealand Government, including the cost of any stores supplied after transfer, the same arrangement to apply to any vessel lent to replace *Amokura*. The New Zealand Government to pay the cost of all officers and men lent to the New Zealand Naval Force (*Philomel* and *Amokura*), including the usual contribution towards retired pay and pensions, and the cost of passages.

The New Zealand Government not to make any contribution towards the cost of training New Zealanders for officers for the Royal Navy if trained in England, and the Admiralty not to make any contribution towards the cost of similar training at the Royal Australian Naval College.

The New Zealand Government to pay the cost inclusive of all 'Personnel' charges, e.g., victualling, clothing, medical issues, &c., of all men belonging to the New Zealand Naval Force whether serving in ships belonging to the New Zealand Government or the Royal Navy, and also to pay the cost of the freight of men sent for service in the China and East Indies Squadrons.

The New Zealand Government will provide suitable accommodation at Auckland for a small Naval Depot for the joint use of the ships of the Royal Navy and the New Zealand Naval Force, the establishment to be for the time being maintained by and under the control of the Admiralty. Repairs and supplies for the New Zealand Naval Force to be charged for on repayment basis.

The above financial arrangements to be regarded as quite independent of any contribution which the New Zealand Government may decide to make for Royal Naval purposes, and which, although it may be based upon the cost of maintenance of the ships of the Royal Navy in New Zealand waters (see Appendix (F)), should not be regarded as made merely in respect of these.

[126] *Churchill to Allen*

25 August 1913

'Confidential'

Your letter of the 27th June has crossed a long one from me dated the 31st July, which would have been sent off earlier but for my absence in the Mediterranean and some serious preoccupations since my return.

I notice that you revert to the question of the 'Bristol' cruisers which we discussed when you were here. I then fully explained our position to you, and I understood that you concurred in our view, and that the new arrangements which we are suggesting to meet your wishes were to be taken as definitely superseding those which were outlined in 1909. We are proposing, in deference to your wishes, to maintain on the New Zealand station three cruisers instead of two; and to assist you in the development and training of a New Zealand naval personnel, one of our cruisers being specially assigned for this duty. You, on the other hand, propose for the future to utilise the New Zealand contribution primarily for the purpose of developing the New Zealand naval force, and only to devote any surplus that may be available thereafter to the general maintenance of the Imperial Navy. Although I believe you have it in contemplation to increase the total of your contribution from 100,000£ to 150,000£ per annum, it will be evident from the figures enclosed with the Memorandum recently forwarded that this amount will be considerably less than the actual cost of maintaining the squadron (*Pyramus, Psyche, Philomel,* and *Torch*) in New Zealand waters.

We have also most gratefully and publicly acknowledged the patriotic and far-seeing action of the New Zealand Government in confirming the free gift made by New Zealand of the battle-cruiser of that name for the general service of Imperial defence. We hope that the visit of this vessel to New Zealand and her mission around the Empire may make every New Zealander feel how very effective has been the service which the Dominion has rendered to the British Navy. We are very proud and glad to man and maintain this fine ship at a cost of 125,000£ a-year so long as she is available for general Imperial service. We accept in the fullest sense our duties and responsibilities for the effective defence of New Zealand and for shielding that Dominion from all menace and danger to her interests. If we are to discharge this task, especially in times like these, it is essential that we should be left free to distribute and dispose of the naval forces at the disposal of the Admiralty in what is judged to be the most effectual manner according to military and strategic needs. If and when those needs require the presence of the *New Zealand* or of other vessels of her type in the Pacific, they will certainly be sent. But we are clear and understand that you fully agree with us that to send a vessel where it is not needed, and where it could play no part in decisive events, would not be a policy on which the Admiralty would be justified in making a heavy annual outlay.

The same considerations apply to the 'Bristol' cruisers. The events which have occurred since 1909 have markedly altered the naval situation. In the work of defending the British Empire, which falls almost entirely upon the British taxpayer, we cannot afford not to make the best use, for the common defence, of the naval resources which are available. There is no class of ship with which we are less well supplied at the present time than these light, fast, modern cruisers. Great efforts are being made by the Admiralty to increase the margin of these vessels at our disposal. No less than 16 are being constructed in the programmes of the last two years; and it is probable that a further large programme will be begun in 1914–15. None of these will, however, be completed for at least a year; meanwhile, until a very large proportion of them have been commissioned, we have actually not got any 'Bristol' cruisers which can be spared from definite military duties of serious importance. We are at this present moment sending two 'Bristol' cruisers, one to the East Indies and the other to the China station, available immediately for service in the Pacific or Indian Oceans in order to cover specific German vessels in those waters. Neither of these requirements could be foreseen in 1909; and we cannot spare two more to cruise in New Zealand waters where no military need at present exists for vessels of this speed. There are apparently no foreign cruisers of an equal type which require to be met within thousands of miles of New Zealand. If any were sent from European waters we could immediately follow and match them by similar vessels. I am sure you will agree with me that it would not be justifiable from any point of view for us to remove two of these ships from stations where they are urgently and immediately needed, in order to proceed to New Zealand waters in relief of vessels which are adequate and suitable for every existing military requirement. If, therefore, the New Zealand Government desire to have two 'Bristol' cruisers immediately provided for the New Zealand station, it is important that they should realise that these vessels will have to be specially built additional to the existing British programmes, at a cost of about 700,000£. I propose, therefore, that the new proposals which you put before the Admiralty should be considered and decided in the first instance.

With regard to the drafting of a Bill to give effect to such scheme as may be approved by the New Zealand Government, I will arrange that every assistance you may require shall be afforded. It will be perfectly easy to draft a measure for the consideration of your legal advisers, to whom the New Zealand Government will naturally look for the final opinion. Our Parliamentary Counsel are, of course, not closely acquainted with the statutory procedure obtaining in the Dominion. In this matter it is legal constitutional advice rather than naval which is required, especially as the proposed legislation will be, presumably, of the same general character as that passed by the Commonwealth Government.

[127] *Allen to Churchill*
27 June 1913

The Prime Minister and some other members of Cabinet have rather forcibly impressed upon me that the arrangement to send the *Cambrian* and the *Psyche*, in addition to the New Zealand training ship under our control, is so much less than Sir Joseph Ward brought back to New Zealand ... that it lays the present Government open to political attack, and I am quite sure that you, as a politician, will realise the situation.

[128] *Paraphrase telegram from the Governor of New Zealand, Lord Liverpool, to the Secretary of State for the Colonies, received Colonial Office*
24 September 1913, 11.27 a.m.

I am desired by my Prime Minister [F.W. Massey] to telegraph to you as follows with reference to my cypher telegram of September 16th and to express his hope that an early reply may be received. 'Naval policy. First Lord of the Admiralty's Confidential communication of July 31st has been received. My Government desires to know definitely whether the Imperial Government is prepared now or within eighteen months to carry out the part of the 1909 Agreement under which two 'Bristol' cruisers were to be stationed in New Zealand waters. (2) New Zealand Government understands that the Lord Commissioners of the Admiralty advised Colonel Allen that destroyers and submarines were unsuitable for New Zealand waters, and that the Admiralty desired to abandon that part of the 1909 Agreement. (3) New Zealand accepts the training ship proposal as part of New Zealand policy but having regard to the trade interests involved deems the proposed arrangement as to the *Pyramus* and *Psyche* unsatisfactory.'

Minuted by Oswyn Murray,
25 September

It is clear from this telegram that the N.Z. Government do not in the least appreciate the Admiralty position, & that we must no longer proceed on the assumption that Col. Allen has laid before them the situation as explained to him when he was here. This being so it is for consideration whether a telegram should not be sent summarising all the main facts, before the misunderstanding proceeds any further. A draft telegram is enclosed.
Draft Telegram

It was confidently anticipated in 1909 that stationing of 'Bristols' in New Zealand waters could be arranged for without interfering with general strategic disposition of naval strength necessary in interest of Empire as a whole, but German naval expansion referred to has changed situation entirely.

[129] *Churchill to Borden*
19 December 1913

'Personal and Secret'
We have learnt to rely on each other during the last eighteen months to a degree which enables, and indeed compels me, to tell you about the situation wh[ich] is developing here. Although the preparation of armaments in Europe has undergone no relaxation, there is a powerful & widespread reaction against the expense among the populations affected. This general movement has produced its repercussion in the Cabinet, culminating in the demand that the quota of four ships appointed for the years 1914–15 should be reduced to two. The Admiralty position is unchanged from that described in the Secret & publishable memoranda.

[130] *Memorandum, marked 'A', on Canadian attitudes to the British Empire, by Lional Curtis [founder of the 'Round Table' review], sent to J.E. Masterman Smith [Private Secretary to the First Lord of the Admiralty]*

9 January 1914

The following view of the present situation in Canada is based on the writer's conviction.

1 That the Canadian people must sooner or later assume a control over foreign policy (i.e. over the issues of peace and war) no less effective than that now exercised by the people of Britain or by the U.S.A.

2 That they can only do so

(a) By separating their own foreign policy from that of the Empire and by controlling it through their own Dominion Government.

(b) or by insisting that the foreign policy of the Empire be separated from the domestic affairs of Britain and entrusted to a Government responsible no less to Canadian than to British voters.

[131] *Hardinge of Penshurst to Crewe*

29 January 1914

The Royal Indian Marine have, we consider, a strong claim to be regarded as a complement to the regular sea and land forces; with one of which they may be incorporated in time of war, and with both of which they are liable to come in contact in the course of their duty, and whilst undergoing special courses of instruction.

We are of opinion that until they receive this hall mark they will continue to labour under grave disabilities.

[132] *Harcourt to Churchill*

14 July 1914

I have spoken to the Prime Minister about the suggested visit of Admiral Jellicoe to Canada and though we see no objection to it in principle, and though it may in many ways prove useful, we feel that it is essential that Admiral Jellicoe should have very careful and detailed instructions as to what he may say to the Canadian Government and that these instructions should be previously approved by the Prime Minister and myself. You will, of course, realise that anything the Admiral says in Canada is likely hereafter to be used by Sir Robert Borden and his Government as being the official expression of the views of the British Government. Can you arrange for this.

[Minute in manuscript
by Winston Churchill]

ans[were]d
My dear Harcourt, I quite agree with what you say ab[ou]t Jellicoe's visit to Canada. He will go with full knowledge of the views held by the gov[ernmen]t as well as by the Adm[iralt]y. I shall of course impress upon him carefully that he must keep strictly to his instructions. Within these limits, he must be allowed a certain flexibility in discussion, without which he [wou]ld not fully ascertain the views of the Dominion Gov[ernmen]t. But you may depend upon him not to commit us to anything wh[ich] we shall not have authorized.

[133] *General Sir Beauchamp Duff, Aide-de-Camp General, to*
Crewe

4 February 1915

We have the honour to refer to your Lordship's Military Despatch No. 96, dated the 9th October 1914, on the subject of the difficulty experienced in finding suitable candidates for appointment to the Royal Indian Marine.
2   In paragraph 2 of that Despatch your Lordship informed us that the proposal to grant commissions to officers of the Royal Indian Marine had received your consideration, and that you had addressed the Admiralty and the War Office on the subject. Your Lordship added that the question was not one that could in any case be finally settled at an early date and that it must under present circumstances, be indefinitely postponed.
3   With paragraph 24 of our Army Despatch No. 57, dated the 7th May 1914, we had forwarded a copy of a letter from the Director, Royal Indian Marine, in which he had expressed the opinion that the grant of a commission would be the most satisfactory and economical way of filling vacancies in the service, and we stated that we concurred in that opinion. We still hold this view most strongly, in fact we consider that our only chance of attracting suitable candidates during the war lies in the grant of a commission

in some form or other. Royal Indian Marine officers are at present, as far as we are aware, the only persons having the status of officers in His Majesty's Service actively engaged in operations against the enemy, who have not the privilege and honour of holding His Majesty's Commission. It has been brought to our notice by the Director, Royal Indian Marine, that 72 officers are serving under the orders of Royal Navy officers afloat; 14 are serving under Naval Officers' orders in Marseilles, Suez, Mombassa, and the Persian Gulf; and 11 are working under the Director at operations in connection with the war in Bombay. At each Indian port, Royal Indian Marine officers are working with the defence of the port operations in co-operation with the Military and Volunteer garrisons, and in the case of the *Emden*'s attack on Madras the direction and control of the guns was undertaken by Commander W.B. Huddleston, R.I.M., the Marine Transport Officer there. It has been pointed out to us that a certain number of Royal Indian Marine officers on the retired list are serving under the Admiralty, and have presumably been granted some form of commission in the rank they held previous to retirement. It naturally causes a feeling of discontent in the Royal Indian Marine service among officers on the active list, when they realise that the only way to obtain the commission they have striven and waited for, for so many years, would be to leave the active list, were it possible for them to do so, and volunteer for service under the Admiralty.

[134] *Murray to the Under-Secretary of State for India [Sir Thomas W. Holderness]*

2 May 1915

My Lords Commissioners of the Admiralty have had under their careful consideration your letter of the 14th July last, No. M.2446, relative to the grant of commissions to Officers of the Royal Indian Marine.

2   In reply I am commanded by Their Lordships to acquaint you, for the information of the Secretary of State for India in Council, that they fully concur in his view that Officers of the Royal Indian Marine should now be granted commissions.

3   My Lords observe that the draft commission enclosed with your letter is in the form of a military commission. As, however, the Royal Indian Marine is a Naval Force they are of opinion that the commissions should preferably be Naval commissions on

the lines of the commissions recently established for Officers of the Naval Forces of the Commonwealth of Australia by Order in Council of the 9th February 1914 (a copy of which and of the form of commission is enclosed) and they would suggest for the consideration of the Secretary of state that authority to this end should be obtained by Order in Council at an early date.

4   My Lords would be glad to learn whether Lord Crewe concurs in this suggestion.

[135] *Greene to Holderness*
2 December 1915

With reference to your letters of the 3rd September, M. 17954 and 11th September, M. 27586, as to the grant of commissions to Officers of the Royal Indian Marine, I am commanded by My Lords Commissioners of the Admiralty to request that you will inform the Secretary of State for India in Council that they concur in his proposal that Temporary Commissions in the Royal Navy should be given, during the war, to all Royal Indian Marine Officers serving in H.M. Ships, or in Indian Marine ships in the Navy, and I am to request that a list of these Officers may be furnished, giving their full names, rank and date of seniority, in order that their commissions may be prepared.

[136] *Minutes of the Imperial War Conference, 4th day*
28 March 1917

SIR JOSEPH WARD [New Zealand Minister of Finance] Mr Chairman, on submitting the motion that I gave notice of some days ago, I propose, in the comparatively brief statement I will make regarding it, to avoid as far as possible any reference to the important matter of which Sir Robert Borden [Prime Minister of Canada] has given Notice of Motion to-day. To some extent the question of the future constitution of the Empire is very closely related to the future system of Empire Naval Defence, but I will keep the latter separate on this occasion, for the all-important reason that I think it is of paramount necessity that, in the event of the Admiralty, on account of pressure of work due to the war, not being able to furnish a scheme for the consideration of this Conference, one should come before a succeeding Conference. It is, in my opinion, of the very deepest concern to the people, at all events of the country I represent, that we should have the benefit of the practical experience of the members of the Admiralty who are responsible for the protection of the Empire as a whole. Up to now we have never had the benefit of the Admiralty's view definitely as to what is the best system to adopt. It is quite correct that in 1909 a Memorandum was furnished by the Admiralty dealing with these matters, in which they gave their view upon the question, really leaving it to the right (and properly so, I admit) of the individual portions of the Empire to do whatever they considered to be best. But we have never had from the Admiralty, as far as I know, any scheme to provide for the better protection of the vast and growing interests of the Empire as a whole, which will meet not only the requirements of the mother-land, but the aspirations of the individual over-sea countries as well . . . The question as to whether it is to be one Navy in the future or whether it is to be five separate navies is one upon which I think the Admiralty should be invited to express an opinion. It does not follow that the opinion so expressed may or will be assented to by the representatives of the different Over-seas Dominions; but the Admiralty should be asked to furnish what they believe to be the most effective scheme . . . After fully considering this matter, I came to the conclusion that the present is not a time to attempt to arrive at fixed decisions upon anything in regard to any Empire Navy . . . The Resolution I now submit for consideration is:

> That this Conference, while fully recognising the importance of maintaining the unity of the strategic direction of the whole Navy of the Empire, wishes at the same time to place on record the desire of the Overseas Dominions to retain, as far as may be consistent with that object, the administrative control of the forces they provide and of the expenditure which they contribute. That for the purpose of giving effect to this resolution the Admiralty be requested to work out the details of a scheme for the consideration of this Conference, with any recommendations they consider necessary for the future effective naval defence of the British Empire.

SIR ROBERT BORDEN . . . I agree also in his view that, for the purpose of obtaining expert advice, there is no authority that can be consulted with greater advantage than the Admiralty. That was my own view when the present Administration in Canada was first constituted, and for that reason at the earliest possible moment I consulted the Admiralty as to what could best be done by our Dominion for the common defence of the Empire. As to what subsequently happened, it is not desirable for me to enter into it to-day, but it would have been a great satisfaction to those who represent Canada, and, I am sure, to the Canadian people, if ships provided by Canada had taken part in the warfare upon the ocean which has been waged during the past two-and-a-half years or more . . .

Having said so much, and thus expressed myself in thorough concurrence with the purpose which Sir Joseph Ward has in view, I should like to suggest to him that the terms of this Resolution, if a Resolution is to be passed, might be modified with advantage. In the first place, I do not think it necessary that we should place on record what is contained in the first sentence of the Resolution. As I understand it, no question has arisen as to administrative control, and during the present war no question has arisen as to unity of direction. It would seem to me that unless that sentence is called for by some incident or some experience of which I am ignorant, it might as well be omitted . . .

I agree, however, that, so far as the future naval security of the Empire is concerned, it is desirable that the Admiralty should take the whole subject into consideration, and should give to us, as soon as war conditions will reasonably permit, a statement of what their views are.

[137] *Holderness to Greene*

3 May 1917

'Confidential and Immediate'

The type of commission which the Government of India now propose is that of a commission appointing to the Royal Indian Marine to be issued by the Secretary of State for India running in the Secretary of State's name (worded very similarly to Royal Navy commissions running in the name of the Lords Commissioners of the Admiralty) bearing the King's counter-signature. This type of commission had suggested itself to the Secretary of State as avoiding some of the difficulties mentioned in the letter of 3rd September, 1915.

If the Lords Commissioners of the Admiralty see no objection to the proposed grant of commissions in this form to Royal Indian Marine officers of vessels employed in mine-sweeping and consider that such commissions will regularise the position of these officers and of the vessels under their charge which may [be] armed for offence, Mr Chamberlain proposes, after taking His Majesty's pleasure, to arrange for the early issue of the commissions. They would be temporary only, pending a decision on the question of extending similar but permanent commissions to the service generally.

If the Lords Commissioners think that the grant of the commissions would not suffice for the purpose of regularising the position of the mine sweepers and officers and men on their I am to enquire whether they can suggest any other means by which this object could be attained.

[138] *Murray to Holderness*
21 November 1917

2   In reply I am to state that as the proposed commissions will not be issued by the Admiralty the precise form that they should take is not a matter concerning which this Department can advise. It should however be understood that such commissions will not confer on the recipients any rights or powers in relation to the Naval Service not at present enjoyed and in particular the mere fact that the Officers of a Royal Indian Marine ship hold such commissions does not entitle the ship to the status of a public Ship of War.

[139] *Charles Walker, Admiralty Principal Clerk to Holderness*
25 January 1918

I am commanded by My Lords Commissioners of the Admiralty to acquaint you, for the information of the Secretary of State for India in Council, that Officers of the Royal Indian Marine to whom Royal Indian Marine Commissions are granted will receive from the Navy the usual marks of respect according to their relative rank, but this will only apply to Officers on full pay.

[140] *Minutes of the Imperial War Cabinet*
27 June 1918

Future Naval Co-ordination

15   The First Sea Lord [Acting-Admiral Sir Rosslyn Wemyss] also drew the attention of the Imperial War Cabinet to the Admiralty

memorandum on the Naval Defence of the British Empire (Paper G.T.-4571). He said that the Admiralty had approached the subject with some trepidation in so far as it touched upon political ground. While fully realising the need of decentralisation of administration, as executive officers they were strongly impressed by the ideal of a single Navy and unity of command. The importance of co-ordination and unity of command had been one of the chief lessons of the present war, and was true in the naval as well as in the military sphere. In dealing with this matter the Admiralty had attempted to approach it from the wider point of view of the partnership of the nations of the Empire, and had to a certain extent taken the constitution of the Imperial War Cabinet as their guide. In any case, they had not put forward their views in any spirit of dogmatism, or with the idea of expecting a definite answer at once, but with the hope of getting their views considered, and bringing about as soon as possible an agreement on general principles which could afterwards be elaborated into a working system. The points on which he wished more particularly to enlarge were, in the first place, the need of central control in peace. Naval peace arrangements, owing to the far greater mobility of naval warfare, had necessarily to be in much closer accordance with war plans than military peace arrangements. If each part of the Empire developed a separate naval strategy in peace, it would be impossible to secure effective co-operation in war. If our naval arrangements had not been completely ready in July 1914, the fortunes of the war might have been very different. Unity of Naval strategy could only be secured by a central authority. The same applied with regard to the need for uniformity of training. He instanced a report from Vice-Admiral Sims [United States Navy], who pointed out that the American ships, although in all respects efficient when they came to join the British naval forces, were no real addition of strength – possibly the reverse – until they had learned to adopt the same methods and work on precisely the same system as the British ships. The Admiralty considered that, from this point of view, the ships of the Empire Navy should be available in any waters, and the officers of that Navy available to serve in any ship. Similarly, it was desirable to have uniformity of material, especially from the point of view of prompt replenish-ment of reserves, and, above all, unity of thought and idea, and the maintenance intact of the great traditions on which the Navy had been built up. He again emphasised that the need for unity

in these respects did not in the least apply to questions of administration.

The Prime Minister [David Lloyd George] suggested that this
very important question of naval co-ordination should be considered direct between the Admiralty and the Dominion Prime
Ministers, and, if general questions of principle arose, the matter
might be brought back to the Imperial War Cabinet for discussion.

Sir Robert Borden [Prime Minister of Canada] stated that he
had no objection to a general discussion of this subject between
the Dominion representatives and the Admiralty, although there
might be some matters of detail that Canada would wish to discuss
with the Admiralty separately.

It was generally agreed that, although action in this matter
could not be taken until after the war, it would be of great use to
the Admiralty, and advance matters, if a preliminary discussion
could take place while the Dominion representatives were in this
country, either at the Imperial War Conference or at the
Admiralty.

Owing to its secrecy, the text of Admiral Wemyss' statement
will not be circulated, but a copy is on record in the personal
custody of the Secretary of the War Cabinet [Colonal Maurice
Hankey], and can be read at the War Cabinet Offices on application to the Secretary.

[141] *Minute on 'Naval Defence of the Empire', Paymaster-in-
Chief, Charles F. Pollard [Additional Naval Assistant to the First
Sea Lord] to the First Sea Lord [Admiral Sir Rosslyn E. Wemyss]*
23 September 1918

It may not be advisable to take the rejection of the Admiralty
proposals by the Prime Ministers who attended the Imperial War
Cabinet this year, as a final rejection of the scheme. Doubt as to
the future policy of the Dominions in Imperial affairs must continue until the political atmosphere has had time to stabilise itself
after the war, and the tenure of office of the Ministers now in
power in both Australia and Canada is, to say the least, of uncertain duration. There are many who think the Laurier (or Liberal)
party is likely to have more to do with the development of Canada
after the war than the party at present in power, and the former
is said to be against the policy of a separate Canadian Fleet.

In regard to the attitude of Australia, there are already signs
of the importance of the financial aspect of the question, e.g., the

cost of developing the Henderson bases and of prospecting for oil in New Guinea: under the Admiralty scheme for the Naval Organisation of the Empire the former would presumably become an Imperial charge if it is decided to proceed with the scheme as a Naval Base, and possibly the latter also.

[142] *'Naval Defence of the British Empire', Admiralty memorandum for the War Cabinet, draft*

September 1918

It is regretted that the Prime Ministers of Canada, Australia, New Zealand, and South Africa, who attended the Imperial War Cabinet this year, should have formed the opinion that the Admiralty proposals for the Naval Organisation of the Empire are not practicable.

The Admiralty conclude that at present constitutional considerations prevent the realisation of their ideal of a single Navy for the Empire: but trust that in the future development of the Naval Forces of the Dominions this ideal, to which the Naval Staff attach the greatest importance, will be borne in mind, and that, as far as circumstances permit, the lines of development will approach as nearly as practicable the principles on which the scheme is based.

2 The Admiralty note the suggestion that they should send a representative to visit the Dominions to advise the local Naval Authorities, and will be glad to do so as early as convenient after the war. Since the views of the respective Governments, especially those which have no naval organisation at present, would be almost a necessary preface to such a visit, the Admiralty look forward to receiving, at no distant date, a communication on the subject from the Governments concerned.

3 To avoid the possibility of any misunderstanding and to facilitate further consideration of the subject, the Admiralty now propose to ask the Secretary of State for the Colonies [Viscount Milner] to inform the Dominion Governments as above, and to transmit to them at the same time copies of the paper laid before the Imperial War Cabinet and the Memorandum by the Prime Minister of Canada.

[143] *Draft memorandum, 'Commissions for Officers of the Royal Indian Marine'*

October 1918

The Royal Indian Marine is a Service under the control of the Government of India, constituted under 47 & 48 Vict. c. 38 and Indian Act XIV of 1887.

In peace time its main functions are the transport of troops and military stores, marine survey, conservancy of light-houses, co-operation with the Royal Navy in the checking of gun-running in the Persian Gulf, provision of officers for port duties at most of the important ports of India, control of the Bombay and Kidderpore Dockyards, and, if occasion should arise, suppression of piracy. The constitution of the service also contemplates that in time of war some or all of its vessels should be transferred to the Admiralty, becoming for the time being vessels of the Royal Navy. The Service had before the war 187 officers, executive and engineer.

The Government of India had submitted a proposal before the war that officers of this service should have the honour of receiving on appointment Your Majesty's full Commission instead of a letter of appointment from the Viceroy, as at present. Although the service is not ordinarily a combatant one, the Government of India considered that its officers had a strong claim to receive commissions, both from their social status and from the nature of the duties they were called upon to perform, the service being a complement to the regular sea and land forces, in one of which its officers might be incorporated in time of war, and with both of which they come into official relations in the course of their duty. This proposal was still under consideration when war broke out.

During the war several of the larger ships of the Royal Indian Marine have been made over to the Royal Navy and many officers have been lent to the Admiralty. Such officers have received temporary commissions in the Royal Navy. Apart from this the Royal Indian Marine has co-operated with the Royal Navy in the defence of Indian waters by means of its own vessels and of improvised patrol and mine sweeping craft, it has played a conspicuous part in solving the difficult problem of the transport of the forces operating in Mesopotamia, and its officers and men have shared in active operations in that area of war and elsewhere. Its personnel has been considerably increased by the addition of 106 temporary officers to its permanent cadre.

Its officers have had the following honours bestowed on them in recognition of services rendered during the present war:

D.S.O. ———————————————————— 11
D.S.C. ———————————————————— 11
C.I.E. ————————————————————— 6
C.M.G. ———————————————————— 4
C.B. —————————————————————— 1

The officers of the Royal Indian Marine are at present the only persons having the status of officers in His Majesty's service actively engaged in operations against the enemy, who have not the privilege and honour of holding His Majesty's Commission, and the Government of India urge that the circumstances of the war have shown very clearly the strength of the claim of the Service, which, before the war broke out, had already been put forward for consideration, to the concession of the honour now proposed.

The War Office and the Admiralty have given their approval to the proposal. A specimen of the commission which, subject to Your Majesty's approval, it is proposed to issue, is annexed. It will be observed that it is very similar in form to that issued by the Lords Commissioners of the Admiralty to officers of the Royal Navy. The grant of the commissions would be preceded by the issue of an Order in Council authorising this step. No change will be required in any British Act of Parliament and only a slight verbal amendment of the Indian Act of 1887 will be necessary. The Law Officers of the Crown have advised that there is no legal or constitutional objection to the course proposed.

[144] *Edwin Samuel Montague, Secretary of State for India, to the King Emperor*

18 October 1918

Mr Montague presents his humble duty to your Majesty and has the honour to submit for your Majesty's approval that present and future officers of the Royal Indian Marine should be granted Commissions from the Secretary of State for India in Council, bearing Your Majesty's Royal Sign Manual in the form of the specimen attached.

A brief account of the R.I.M. Service and of the reasons for which this proposal is submitted to your Majesty is given in the annexed Memorandum.

If your Majesty is graciously pleased to approve the proposal, an order authorising the issue of such Commissions will be submitted for approval by your Majesty in Council.

MS addition: App[roved] G[eorge V] R[ex et] I[mperator]

[145] *'Proposals for Oiling Stations for British Men-of-war and Merchant Ships Abroad', memorandum by Director of Operations Division (F) [Captain Charles P.B. Goode or Captain Alfred D.P.R. Pound], Director of Trade Division [Captain Alan G. Hotham], and Director of Supply [G.H. Ashdown], appended to minutes of Admiralty Board for*

31 October 1918

No accurate forecast can be given of the actual requirements of oil fuel for warships which may be stationed in Foreign waters after the war. The number of vessels employed on the different stations will depend on political conditions and will vary from time to time.

2   It may, however, be taken that where coal now is [?stored,] a proportion, and an increasing proportion, of oil will be required in the future; coaling stations will become more and more oiling stations as well.

3   Different Classes of Oil Bases – The oil stations may be divided into three categories:

(a)   Those in the self-governing Dominions.
(b)   Those in the Crown Colonies.
(c)   Those in foreign countries.

4   Self-governing Dominions – The oil bases in the self-governing Dominions (Canada, Australia, Union of South Africa, and New Zealand) would best be under the control of their respective Governments, and by arrangement with the Imperial Government should be kept up to a fixed standard of reserve. It might be combined with commercial supplies provided the standard of reserve fixed for warships was always maintained.

5   Crown Colonies – As regards (b), where there are dockyards and Government Establishments, such as Gibraltar, Malta, Hong Kong, and the like, the oil supply for warships will be in Government hands, and reserves would be kept up to a fixed standard; this would also be the case at places where it would not pay to keep oil for commercial purposes, such as Ascension, St Helena, and Falkland Islands.

6   Foreign Countries – For the supplies in Foreign countries (c), British firms should be encouraged to establish themselves at foreign ports both for commercial and man-of-war supply. In the case of places which are important strategically, but at which an oil depot would not be a success commercially, the British Government might assist with a subsidy or guarantee for money expended in establishing and upkeeping these bases.

7   The bases should be distributed in areas where there are numerous minor States, such as South America; in the event of trouble with one country our ships could obtain supplies from the bases in another, or at any rate our Fleet attendance oilers might, if the laws of neutrality did not allow our warships to oil there.

8   Basis of Reserves – In deciding the question of the reserves to be held on foreign stations, it is necessary to take the following data into consideration:

(a)   Basis of ship expenditure
(b)   Number of ships which may normally be expected to operate from each centre.
(c)   Extent to which it may prove necessary to reinforce the local squadron to deal with probable contingencies; e.g., despatch of a battle cruiser squadron to South American waters, a battle squadron to Canadian waters, battleships again stationed in China, &ca.

9   As regards (a), it will be necessary to compute the expenditure on the basis of the probable war consumption. This will naturally vary with the station; e.g., the Pacific, owing to absence of British bases and great extent of waters requiring patrol, will normally involve heavier expenditure than the Atlantic.

10   Unit Ship Expenditure – For the purpose, however, of this review it is thought proper to consider the provision, so far as conditions permit, of a reserve equivalent to three months' war expenditure for each station calculated on the following basis:

Oil-burning battleship	5,000 tons monthly
Oil-burning battle cruiser	5,000 tons monthly
Oil-burning light cruiser	2,000 tons monthly
Oil-burning destroyer	1,000 tons monthly

11   Force to be provided for. It is proposed to assume that a maximum force of 2 battle cruisers and 6 light cruisers may have to be provided for in emergency on the following stations:

S.E. Coast of America and West Coast of Africa
Cape
East Indies
Pacific

Three months for this force would be 72,000 tons (say 70,000 tons).
12   It is considered that this quantity should be regarded as constituting the standard of reserve to be aimed at on each station,
although varying conditions may necessitate certain departures
from this figure. The provision of this reserve, which should be
held on British territory, will not eliminate the need for suitable
bases in certain foreign territories. These latter bases should,
however, be regarded as supplementary to the main reserve, in
view of their uncertain value during hostilities.

The quantity of oil to be maintained at any one base may be
put at a minimum of 5,000 tons.

[146] *Order in Council, at the Court at Buckingham Palace*
8 November 1918

Present, the King's Most Excellent Majesty, Lord Privy Seal
[the Earl of Crawford], Lord Steward [Lord Farquhar], Lord
Somerleyton, Mr Chancellor of the Duchy of Lancaster, Sir Auckland Geddes.

Whereas there was this day read at the Board a Memorial from
the Right Honourable the Secretary of State for India in Council
[Edward Montagu], dated the 4th day of November, 1918, in the
words following, viz:

'Whereas by Orders in Council dated the 20th March, 1891,
and the 30th July, 1891, Her late Majesty Queen Victoria was
graciously pleased to approve certain titles and ranks for the
Officers of the Royal Indian Marine':

'And whereas by Order in Council dated the 13th June, 1917,
Your Majesty was pleased to approve of the designation of the
said Officers being changed in order to assimilate their titles to
those of Officers of the Royal Navy':

'And whereas it is desirable that the Officers of the Royal
Indian Marine should have the full status of Commissioned
Officers':

'I beg leave, therefore, humbly to recommend that Your
Majesty may be graciously pleased, by Your Order in Council, to
approve the following proposal':

'Officers of the Royal Indian Marine shall receive from the Secretary of State for India in Council Commissions appointing to the Royal Indian Marine Service which may, at Your Majesty's pleasure, bear Your Sign Manual.'

'The Lords Commissioners of the Admiralty and the Army Council have signified their assent to this proposal'.

His Majesty, having taken the said Memorial into consideration, was pleased, by and with the advice of His Privy Council, to approve of what is therein proposed.

And the Right Honourable the Secretary of State for India in Council is to give the necessary directions herein accordingly. Almeric FitzRoy.

[147] *Sir Joseph Cook, Commonwealth of Australia Minister for the Navy, to Sir Eric Geddes, First Lord of the Admiralty*
16 November 1918

Replying to your letter of the 13th instant I should say that the duties required of Lord Jellicoe would be to view the whole naval position in the light of the circumstances of the war, and more particularly with respect to the composition of the building scheme suggested by Sir Reginald Henderson [? in 1918 Naval Assistant to the Assistant Chief of Naval Staff, Admiral Sir A.L. Duff] some seven or eight years ago.

As you know a very elaborate scheme was propounded including a building programme to be completed about the year 1933 together with an extensive scheme of naval bases to meet these Fleet developments of the future.

The war has modified our view of many things and most certainly the navy will not have escaped the influence of the changed conditions. It is to look at the whole of the arrangements and prospects in the light of the day that we should like Lord Jellicoe's services to be made available to us. What we require is the opinion of a man whose authority and status will secure for it unquestioned respect throughout Australia. I think Lord Jellicoe such a man, hence my anxiety that you should let us have him for the purpose named.

With regard to the other appointment I hope to send you particulars early next week after receipt of a cablegram which I am expecting from Australia upon the subject.

Please accept my best thanks for your kind promise to help us in these various directions and, Believe me to be, Yours sincerely,

[148] *Murray to Geddes, first draft, sent to Wemyss*
30 November 1918

In accordance with your directions, I have, in consultation with Lord Jellicoe, drawn up the attached draft remit to him on his proposed visit to the Dominions.

Paragraph 1 directly refers to the Memorandum which Sir Robert Borden [Prime Minister of Canada] on behalf of the Dominion Prime Ministers sent to you. It is suggested that it is important to make it clear that the Admiralty are arranging this visit at the request of the Dominions, and not merely on their own initiative.

Paragraph 2: This raises an important question of policy which may require further consideration. I think that the Dominion Prime Ministers' Memorandum clearly contemplated that the object of the visits would be to advise them how to get the greatest possible efficiency, homogeneity and co-operation between all the Naval Forces of the Empire, on the lines which they have decided to follow in preference to the scheme recommended in the Admiralty Memorandum. If this be the case, it seems desirable to make it clear in the remit that the visit is arranged on this basis, for the following reasons:

(1)   In order that the Dominion Authorities may not be able to say later on that the Admiralty, by arranging this visit, withdrew implicitly the scheme recommended in the Admiralty Memorandum, inasmuch as the instructions to Lord Jellicoe contained no reservation as to the principle of an Imperial Admiralty and Navy.

(2)   In order that Lord Jellicoe himself may be quite clear that he is intended to take any proposals made by the Dominions seriously, and not merely to use them as a means of carrying on a crusade on behalf of the Admiralty scheme.

P.S. If India is to be included a separate remit will be required & will need to be prepared in consultation with India Office.

[149] *'Proposed visit of Lord Jellicoe to the Dominions and India to Advise on Naval Matters', Admiralty memorandum for the War Cabinet by Geddes*

17 December 1918

It will be remembered that on the 17th May last the Admiralty placed before the War Cabinet, for consideration by the recent Imperial War Conference, a Memorandum outlining a scheme for the Naval Defence of the British Empire, based upon the principle of a single Imperial Navy under the control of a single Imperial Naval Authority, with the local Naval Boards to manage local Naval Establishments and maintenance Services. This Memorandum was prepared in pursuance of the resolution passed at the Imperial War Conference of 1917, requesting the Admiralty to work out what they consider the most effective scheme of naval defence of the Empire.

The Memorandum was put down for discussion at the Imperial War Conference but at the meeting of 24th July it appeared that the Dominion representatives would prefer to discuss it directly with the Admiralty. The outcome was a Memorandum by the Dominion Ministers (with the exception of the Prime Minister of Newfoundland) forwarded to the Admiralty on the 15th August last, stating that 'the proposals for a single Navy at all times under a central authority are not considered practicable,' but adding:

It is thoroughly recognised that the character of construction, armament and equipment, and the methods and principles of training, administration and organisation should proceed upon the same lines in all the Navies of the Empire.

For this purpose the Dominions would welcome visits from a highly qualified representative of the Admiralty who, by reason of his ability and experience, would be thoroughly competent to advise the naval authorities of the Dominions in such matters.

As naval forces come to be developed upon a considerable scale by the Dominions it may be necessary hereafter to consider the establishment for war purposes of some supreme naval authority upon which each of the Dominions would be adequately represented.

Whilst the Admiralty do not depart from their declared views as to what would be the most effective scheme of naval defence of the Empire, they welcome this expression of the desire on the part of the Dominions to secure the greatest measure of uniformity that is possible in naval organisation, training and types of materiel throughout the Empire. The pursuit of uniformity in all these important directions cannot fail to assist the realisation of many of the objects underlying the Admiralty Memorandum.

I propose, therefore, if the War Cabinet approve, to accept the above mentioned invitation of the Dominion Premiers by appointing Viscount Jellicoe to proceed early in the New Year to visit such parts of the Empire as express a desire to be included in his Mission, in order to confer with and advise the Dominion Authorities on Naval matters.

Mr Hughes and Sir Robert Borden [Prime Ministers], on behalf of Australia and Canada, have already expressed enthusiastic concurrence in this proposal. Sir Joseph Cook [Australian Navy Minister] presses for the Australian visit to be announced as early as possible, as it would be very helpful to have it known in Australia.

Mr Long [appointed First Lord as of 16 January 1919] has cabled to New Zealand asking whether that Dominion desired to be included, and proposes to consult General Botha [Prime Minister of the Union of South Africa] on his arrival as regards South Africa.

It is thought that it would also be desirable to include India, and Mr Montagu [Secretary of State for India] has telegraphed to the Viceroy to ascertain his views.

[150] *Minutes of the Imperial War Cabinet*
18 December 1918

2   The Imperial War Cabinet gave their approval to a proposal of the Admiralty that Lord Jellicoe [former First Sea Lord] should visit, in an advisory capacity, Australia, New Zealand, Canada, and India, subject in each case to the consent of the respective Governments. This consent has already been given.

[151] *Proposed 'Telegraph Communications Board', Admiralty memorandum for the War Cabinet*
31 December 1918

The Admiralty have for some time been considering the problem of Telegraphic communications within the Empire which subject was raised at Imperial War Cabinet 42 (Minute 3) and have arrived at the conclusion that in order to provide a rapid, reliable and economical system of telegraphic communications, the requirements of cables, land-lines and wireless telegraphy need careful co-ordination.

2   Apart from its purely warlike use, Wireless Telegraphy must play a large part in the development of trade, safety of shipping, and aerial navigation and should form the basis of an Imperial News Service by which all parts of the Empire will be bound together in peace as it has been during the war.

3   The Admiralty have therefore drawn up a scheme for the formation of a 'Telegraph Communications Board', whose foundations and functions it is proposed shall follow the general lines shown in paragraphs 5, 6 and 7.

4   In drawing up this scheme the Admiralty have received the fullest co-operation from the Army and Royal Air Force who agree entirely with the proposals. The scheme has also been concurred in by the Postmaster General except in one particular, viz: as to the Department responsible for the Board, which will be referred to later in this Memorandum.

5   The general proposals are:

(a)   That a Central body for the control of the W/T interests of the British Empire should be set up as soon as possible. This body should be advisory and not executive. That is to say, it will have no telegraphic apparatus or personnel under its orders and its duties will be to advise the Government as to the line of action to be taken by the executive departments concerned . . .

[152] *'Proposed Gift of Destroyers and Submarines to the Australian Navy', memorandum by Wemyss for the War Cabinet*
3 January 1919

Sir Joseph Cook, Australian Minister for the Navy, has enquired whether it is possible that on the declaration of peace there may be some Destroyers and Submarines available which would be suitable for the Australian Navy. He points out that Australia lost two Submarines early in the war, and was unable owing to the war to complete a third which should have been ready before this, and hopes that it may be possible 'to make them good' out of our numbers, and also to spare some Destroyers.

The manner in which the war has ended makes it quite practicable to spare both Destroyers and Submarines, and in view of the fine part which Australia has played in the war, and particularly of the readiness with which she has accepted every Admiralty request as to the employment of her ships, the Board of Admiralty suggest that it would be desirable to make a free gift to her Navy of 6 Modern Destroyers and a flotilla of 6 submarines. The sanction of the War Cabinet is invited to this proposal.

[153] *William Ferguson Massey, Prime Minister of New Zealand, to Walter H. Long, First Lord of the Admiralty*

Hotel Majestic, Paris,
10 February 1919

'Strictly Confidential'

As there is likely to be some considerable delay before any report can be received from Admiral Lord Jellicoe with regard to naval matters in the Pacific, I think it is my duty to call your attention, after what has taken place during the war period, to the urgency for better naval protection of British interests in that quarter of the globe. The lesson of the battle of Coronel – with the details of which I am sure you are acquainted – will not soon be forgotten, and I noticed a day or two ago in the newspapers that one of the great Powers, with its headquarters in the Southern Hemisphere, had just launched two very powerful battleships and that the same nation had other war vessels on the stocks. These are very strong indications that the world has not yet reached the Millennium stage.

So far as the larger ships are concerned, I take it that it will be the duty of the Admiralty to make recommendations as to whether they should be controlled by the Imperial Government or the Governments of the Dominions, or partly one and partly the other; but in any case there must be a Pacific Squadron strong enough to hold its own in the years to come against anything which is likely to be pitted against it, though on that point I can speak only as a public man and not as an expert.

With regard to New Zealand's immediate interests, may I remind you of the promise made years ago to establish a naval station in New Zealand waters and sent out two light cruisers with some submarines and destroyers? I am not particularly anxious about vessels of the submarine – or even destroyer – type as I doubt whether under existing conditions they would be worth the necessary expenditure in the South Pacific, but I have no doubt about the cruisers.

New Zealand badly wants a suitable ship for training purposes. It is true that we have in New Zealand waters a gunboat – formally known as H.M.S. *Sparrow* – but if we are to encourage the maritime instincts of the race in that Dominion of the Empire, something very much better is urgently required. We should also have, with its headquarters in New Zealand, at least one cruiser of the 'Sydney' type. The *Philomel* and the *Torch* are both in Wellington Harbour, and have been there for some time, but they are obsolete and fit only for the scrap-heap.

The question will of course arise whether ships in New Zealand waters shall be controlled by the New Zealand Government under the New Zealand Defence Act or remain part of the Royal Navy.

Personally, if financial difficulties do not stand in the way, I am in favour of what is spoken of as local control and, in case of war, as provided under the Act above quoted, the ships should automatically pass to the Imperial Government to serve wherever they would be most usefully employed. This is a question which of course can be settled later on, but I am writing to you now because I do not wish the present opportunity to pass without something satisfactory being done. With its insular position, its connection with the other islands of the Pacific, its long coastline and numerous harbours, New Zealand must in future be a maritime nation. It cannot be otherwise; the younger generation of New Zealanders have the salt in their blood, and the sooner these facts are recognised and the necessary provision made, the better it will be for all concerned. I want to see proper opportunity given to young New Zealanders – and there are many who are desirous of joining the Navy as a profession – to enable them to do so.

Both the Imperial Government and the Governments of British countries in the Pacific have very serious responsibilities so far as the naval outlook is concerned, and I am strongly of opinion that we should face these responsibilities and deal with them without unnecessary delay.

[154] *Memorandum, initialled M.N. (?) to Long*
25 February 1919

First Lord

The state of affairs is as follows:

By present arrangements the British Naval Force in the Far East will consist of:

1 battle cruiser.
5 light cruisers of the latest type.
2 flotilla leaders of the latest type.
16 destroyers of the latest type.
12 submarines of the latest type.

There will also be the Royal Australian Navy, consisting of:

1 battle cruiser.
4 light cruisers.
1 flotilla leader.
5 destroyers of the latest type.
6 older destroyers.
6 submarines of the latest type.

The above constitutes a formidable force and should be fully sufficient to safeguard the interests of the Empire in those waters in any circumstances that are likely to arise in the near future.

There will also be British light cruisers on the Pacific side of both North and South America.

With the destruction of German Naval Power, we are secure in the Pacific so long as the Alliance with Japan exists, as the possibility of war with the United States in the near future need not be seriously considered.

If the Japanese Alliance is not renewed it will then become necessary to reconsider our Naval dispositions, and both Australia and New Zealand would have to be consulted in the matter. Decisions on this question are not required until after Lord Jellicoe [former First Sea Lord, on official Empire tour] has reported.

We propose to maintain on the New Zealand station at present two of the newest sloops for necessary police work among the Islands of the South Seas, for which work they are more suitable than light cruisers. These two ships will remain part of the Royal Navy.

It is quite true that *Philomel, Torch* and *Sparrow* are now fit for nothing but the scrap heap.

One modern sloop or more could well be spared as a training ship if so desired by the Government of New Zealand.

A light cruiser could also be spared, but at present we could not spare the men to man her, though we might the officers.

It might be desirable to get Cabinet approval for one modern light cruiser and two sloops to be presented to the New Zealand

Government, and they could take possession of them as soon as they are ready to do so. These ships would, of course, be entirely under the control of the New Zealand Government and would not remain part of the Royal Navy. We should, no doubt, assist by lending the necessary instructional staff etc.

It is most desirable that New Zealand should be encouraged to take to the sea, and if she desires to develop a Navy on the same lines as Australia, a logical conclusion would be that the New Zealand Station should eventually be completely handed over to her in the same way as the Australian Station has been handed over to the Royal Australian Navy.

The Admiralty would be only too glad to assist in any way they can, but I do not think that any definite steps should be taken until we get a report from Lord Jellicoe.

[155] *Massey to Long*

1 April 1919

'Secret'

Referring to our conversation this morning, my suggestions are as follows:

(1)   That the Admiralty should supply New Zealand with a light cruiser suitable for training purposes and to accommodate not less than 150 boys; this ship to take the place of the training vessel at present in New Zealand waters known as the *Amokura*.

(2)   That New Zealand should be supplied with a modern defence cruiser of a suitable type, such as the *Sydney* or *Melbourne* belonging to the Australian branch of the Royal Navy; this ship to be manned in the first instance by the Admiralty, the annual contribution of £100,000 now paid by the New Zealand Government for naval defence purposes to go towards salaries and other expenses of the cruiser.

The British ratings to be gradually withdrawn and their places to be taken by New Zealanders as the latter are available, and as financial conditions permit, until the officers and crew become practically New Zealanders and the ship a New Zealand ship with all expenses borne by the New Zealand Government but subject to the control of the Admiralty, as provided by the Naval Defence Act (New Zealand).

The present facilities for young New Zealanders to be trained for commissions in the Royal Navy to be extended if necessary.

The Commander of the cruiser, who is to have the rank of Commodore, to be the Naval Adviser to the New Zealand Government.

It is to be understood, of course that these suggestions are only tentative and subject to approval by the New Zealand Government.

I shall be glad, however, if you can give me particulars of the training ship as I understand she is urgently required to take the place of the *Amokura* which is long since obsolete and badly in need of extensive repairs.

[156] *Memorandum by the Fourth Sea Lord [Rear-Admiral Sir Hugh H.D. Tothill], 'Establishment of training ships at Indian Ports for training of Natives'*

8 May 1919

With reference to the telegram within from Lord Jellicoe [former First Sea Lord, on official Empire tour], the suggestion to establish training ships at Bombay, Karachi and Calcutta for the training of native ratings causes me some concern.

The establishment of three separate training establishments can only be put forward on the assumption that many thousand native sailors will be trained, and this implies that the fighting ships will be manned by native ratings.

I understand that the sea-faring population of India belong exclusively to the non-fighting races and my opinion is confirmed by an interview which I have had with an officer of the Indian Army now serving at the War Office.

If it is the intention to employ native ratings, it may be taken as certain that they will have to belong to men who not only have no connection with the sea, but live remote from it and regard it with horror and apprehension.

The greatest difficulty in manning the Fleet with natives would be that of the officers. There are no Englishmen resident in India from which officers could be drawn and I am informed that it would be hopeless to expect the natives of India to themselves provide the officers. Their intelligence and brain power would be quite unequal to this.

Further, it would be out of the question for these ships to be officered by men who only serve for a few years on the East Indies Station. It is universally recognised that for at least two years Englishmen are of little use in the handling of natives, and if this proposition is seriously taken up, it will apparently mean having a separate staff of officers for the Indian Navy as is now done for the Indian Staff Corps.

There would also be many difficulties in the question of race and caste. In the Army the universal practice is to have several different races in each regiment; indeed it is considered essential to do so to prevent sedition.

It may be anticipated that such a course would present well nigh insuperable difficulties in the manning of ships. It would lead to such a thing as the foretopmen being Pathans and the quarter deck men Sikhs and neither would be interchangeable with the other.

All considerations convince me that this question should be approached with caution and a most thorough investigation made before we commit ourselves to the principle.

Lord Jellicoe's instructions were drawn very wide indeed, but certainly there was no specific suggestion that fighting ships could be manned by native ratings.

[157] *National Policy – Memorandum by Hankey to the Committee of Imperial Defence*

July 1919

4   The fundamental factor, which has been the key to British policy, both internal and external, for many decades, remains unchanged, namely, our dependence on imports. But the War has considerably altered the means of paying for our imports to our disadvantage ... para 12 'the following general principles should form the basis of our policy:- i) Non-productive employment of man-power and expenditure, such as is involved by naval, military, and air effort, must be reduced within the narrowist limits ...

22   The United States of America is the most powerful nation in the world ... It is quite out of the question that we could make successful war against her, and it is doubtful whether all Europe combined could do so. America is intangible. No blockade could seriously damage her ... If Canada joined herself to the Mother-country in a War against the United States (which is doubtful) she would be overborne.

23   Fortunately, such a war is almost unthinkable. In the main,

American ideals are our ideals. If there is antagonism to the British Empire among important elements in the United States, this is not generally realised in Great Britain, where there is no corresponding antagonism . . .

24   In view of these considerations, there seems little ground for basing our naval and military establishment on the possibility of war with the United States. To this rule, however, there should be one exception, namely, that, even though some other pretext for its maintenance is found, our fleet should not be allowed to sink below the level of the United States fleet. We must always be able to protect our home waters, the approaches to our shores, and our main trade routes, other than those contiguous to the United States, which are not defensible without immense forces.

[158] *S.H.Phillips, [Principal Admiralty Clerk] for the Admiralty Secretary, to [Lieutenant General Sir H.V. Cox] Under-Secretary of State for India, Military Department, India Office*
21 August 1919

With reference to your letter of the 11th instant, No. M.26380, relative to the preparation of commissions which are to be granted to Officers of the Royal Indian Marine, I am commanded by My Lords Commissioners of the Admiralty to acquaint you, for the information of the Secretary of State for India in Council, that the seal used on Naval Commissions is the ordinary Admiralty seal, and this would not appear to be suitable for impressing on the Commissions granted to Officers of the Royal Indian Marine.

[159] *Telegram, Admiralty to Admiral of the Fleet, Viscount Jellicoe [on official Empire tour]*
20 September 1919

Colonial Office has been asked to send telegram to Dominions of which the following is a paraphrase;

Steps are now being started by Admiralty to complete reduction of fleet to post war standard. This involves considerable reduction of personnel and disposal of surplus vessels. Such vessels include battle cruisers, light cruisers, destroyers, submarines, sloops, minesweepers, patrol gun-boats, motor launches, coastal motor boats.

This involves question of how many vessels will be required by Dominions.

It is realised that decision as to total numbers required cannot be given until Lord Jellicoe's recommendations have been fully considered, but Admiralty would be greatly assisted if such requirements as can be foreseen could be communicated at earliest date.

It is anticipated, if the Dominions so desire, both officers and men of the Royal Navy would volunteer in considerable numbers to man these vessels. Such notification is desired specially as regards sloops, minesweepers, patrol gun-boats and motor launches, since the sale for commercial purposes of vessels of these classes which are surplus to Imperial requirements is a matter of urgency owing to their rapid deterioration and depreciation of present market value if retained much longer. (Ends)

India Office are being asked to send similar message to Government of India.

Information will give minimum requirements only. You should report whether you can estimate what the maximum total requirements of all the Dominions and of the Government of India are likely to be in order that the Admiralty may be able to proceed freely with disposal of surplus.

[160] *Memorandum, 'Imperial Naval Defence'*
October 1919

1   The co-operation of the Dominions and Colonies in the naval defence* of the Empire may be conveniently considered under three headings:

(a)   Nature of Co-operation.
(b)   The System of Command and Direction in time of war.
(c)   The Selection and Maintenance of Bases.

[*The word 'defence' necessarily includes every kind of offensive action which may be required for the defence of Imperial interests.]

[Part I]
Nature of Co-Operation

2   The following forms of co-operation are open to the self-governing Dominions:

(a)   The provision, manning and maintenance of Dominion units or squadrons, e.g., the Royal Australian Navy.
(b)   Payment for ships, e.g. the battle cruiser *New Zealand* by New Zealand, and the battleship *Malaya* by the Federated Malay States.
(c)   The provision of men, e.g., the Newfoundland Naval Reserve, and the South African Naval Volunteer Reserve.
(d)   Direct financial co-operation, e.g., New Zealand and South Africa.

3   The past history of Dominion co-operation in Naval defence is given in the Appendix, but the Admiralty can no longer be bound by its pre-war recommendations, and the subject must now be viewed in the light of war experience. In the years immediately preceding 1914, most Imperial defence questions were largely governed by the conception of a war in the North Sea and the immediate menace of the German Navy, with the result that Admiralty policy and Dominion sentiment were to some extent opposed to each other.
4   In May 1918, the Admiralty prepared a short memorandum on naval defence for the Imperial War Cabinet. In order to obtain unity of direction and command, the creation of a single Navy in which Dominion and British ships and men would be interchangeable was suggested in this memorandum. The Dominion Premiers rejected these proposals, and the immediate problem is the combination of independent Dominion Navies and a unified system of direction in time of war.
5   It must be remembered that the Dominions are quite free to determine the extent and nature of their contribution, and are naturally jealous of their constitutional rights. The concordance of imperial requirements with Dominion sentiment and the development of local interest in naval affairs are therefore essential factors in any scheme of defence, whilst any system which attempts to dictate the naval policy of the Dominions, or which tends to overcentralisation in London, will meet with grave difficulties in practice.
6   It now seems to be clear that co-operation on the model of the Australian Navy is likely to be the most satisfactory system in these respects, and greater financial obligations are likely to be undertaken when ships are maintained by the Dominions, manned by their own men, and based on their own ports.
7   Discussion as to the best form of co-operation is, in fact, some-

what academic because the statements of Dominion statesmen make it clear that future co-operation from Canada and New Zealand will eventually follow the Australian model. Quite irrespectively, then, of the advantages of this system, the Admiralty will be required to assist in the development of these navies, and to find a place for them in a comprehensive system of Imperial defence.

The Strategic Role of Dominion Navies

8   The control of sea communications, which underlies every problem of Imperial defence, extends to every sea. In nearly all cases the Dominions are well situated for exercising this control in certain important maritime areas, while the protection of their coastal trade and their commercial and naval ports is also a factor of considerable military importance. In these two respects, viz., the control of communications in seas distant from the United Kingdom, and the protection of their own coastal trade and bases, Dominion Navies can fulfill a very necessary and important strategical function.

9   In a European war, Australia, New Zealand, South Africa and India might undertake this work either wholly or in part in the Pacific and Indian Oceans by providing the necessary cruisers and anti-submarine craft, whilst Canada might contribute a similar share in the Atlantic. If the Pacific Ocean were the main theatre of operations, the Dominions would also be well situated for co-operating with other forces that would eventually have to concentrate in that area. The influence which Dominion cruisers might exercise, even in the face of superior forces, is well illustrated by the operations of the *Dresden, Karlsruhe, Emden* and the German Asiatic squadron.

10   As regards capital ships, a superior fighting force must be maintained in the main theatre of operations, and any dispersion which neglects this principle is opposed to sound strategy. The eventual inclusion of battleships or battle cruisers in Dominion Navies is not, however, necessarily opposed to sound strategy. If the main theatre were the Pacific or Western Atlantic, there would obviously be no objection to the maintenance of Dominion capital ships in these areas either before or after the arrival of British reinforcements. Isolated battle cruisers may also be required in distant waters to deal with similar detachments by a European Power; for example, in the late war, the presence of two battle

cruisers in the Pacific would have been justified as a counter-measure to the enemy squadron in these waters.

11   What may be called the 'initial strategical deployment' of the whole Imperial Fleet must, however, vary with the conditions of each particular war, and generalisations on the subject of strategic dispositions may therefore be misleading. The principle which emphasises the importance of a sound strategical distribution will be fully met if the Dominions agree that, subject to their decision to co-operate in a war, their ships and squadrons shall be available for service in any part of the world and to the general plan of campaign being directed by one central authority.

12   The actual application of this principle will depend on the preparation in broad outline of plans suitable for different wars, assigning definite and appropriate duties and stations to Dominions ships and squadrons, and on developing a sound system of direction and control in time of war. In peace time the necessity of moving Dominions ships from their own waters would only arise if the Empire were seriously threatened, and under these circumstances British squadrons might have to be concentrated in the Pacific or Dominion ships in European waters. It would appear, however, that a proportion of the latter would always be required to operate from their own bases.

13   The policy of forming Dominion Navies has the great advantage of stimulating national pride and effort in naval affairs, and it is therefore recommended that Canada, New Zealand and South Africa should gradually build up navies of their own on the Australian model. Considerable economy would be obtained if the Australian and New Zealand Navies combined in training, supply, &c, but the practicability of this step is a matter for discussion between their respective Governments. With regard to Newfoundland, it is suggested that she should continue to supply Naval reserve men from her seafaring population. Their numbers and the incidence of cost might, however, be revised as a result of war experience.

Size and Composition of Dominion Navies

14   As finance is the principal factor governing the size of a Navy, this question must be left for each Dominion Government to decide for itself. It would be futile to suggest any cut-and-dried scale of expenditure, for everything will depend on the future grouping of potential enemies, the strength of their armaments, and their threat to the Empire and the Dominions concerned. The

total naval force required by the Empire should, however, be discussed periodically with the Dominions, and the Admiralty should make suggestions as to the extent and nature of their co-operation through the Imperial War Cabinet or Committee of Imperial Defence.

The first charge on their estimates should be to provide self-supporting bases with the necessary docking and repair facilities.

15 The type and design of vessels may be influenced by the particular war for which they are most likely to be required. For example, Australia's contribution to the present war of a battle cruiser, light cruisers and destroyers could hardly have been improved upon, but large submarines would be particularly valuable in a war against Japan. Generally speaking, however, neither Australia, Canada, New Zealand nor South Africa will go far wrong if they build light cruisers for a European war and light cruisers and submarines for the Pacific. There will also be a place for battle cruisers in a scheme of Imperial defence if the Dominions wish to possess them.

16 Destroyers as designed at present are essentially a fleet weapon, and are uneconomical both for escorting trade, the control of coastal waters, and the defence of harbours. For the first-named work a special design of vessel is required, and for the other two, submarines are more suitable for the purposes of the Dominions. There can, however, be no finality in the design, types or numbers of ships built for the Dominion Navies, and their particular requirements should be under the continual consideration of the Admiralty Staff in close consultation with Dominion representatives.

17 Certain British destroyers and submarines have been turned over to Australia since the armistice, and the question of Canada taking over a battle cruiser, a few light cruisers, destroyers, and submarines is under consideration. No doubt, similar arrangements could be made for New Zealand and South Africa. In all cases the cruisers should be suitable for convoy duty and the submarines for service in the Pacific.

18 The officers and the majority of the personnel for manning these ships would at first have to be lent from the British Navy, but training establishments for officers and men should be started as soon as possible in the Dominions.

19 None of the Crown Colonies or Protectorates being in a position to maintain a navy, it is suggested that their contributions should be limited to harbour defence, minesweeping, and mainten-

ance of bases, with the possible exception of the Federated Malay States, which might in addition contribute a yearly sum towards the maintenance of ships in Far Eastern Waters.

Summarised Conclusions

20   The foregoing may now be summarised by saying that:

(a)   In view of their decision on the Admiralty Memorandum of May 1918, the Dominions can now best contribute to the naval defence of the Empire by building up their own Navy.

(b)   The primary role of Dominion Navies may be defined as to assist in the control of imperial communications in distant seas and the protection of their own coastal trade. The requirements of sound strategy can, however, only be met if every ship is available for war service in any part of the world, and the general plan of campaign is directed by one central authority. This question will be discussed more fully in Part II.

(c)   Each Dominion must decide its naval programme on its own responsibility. Initial financial difficulties and delay in starting may be overcome by taking over British ships and temporarily enlisting British personnel; but the root of the problem is to organise the entry and training of officers and men as soon as possible.

(d)   The particular requirements of Dominions in numbers and types of ships should be under the continual consideration of the Admiralty Naval Staff in consultation with their own representatives. A start should be made in all cases with light cruisers and submarines.

[Part II]
System of Command and Direction in War

1   The principles on which the efficient co-operation of naval or military forces depend are matters of common knowledge; and, if applied to the Dominion Navies, there will be no difficulty in co-ordinating their efforts with the British and with each other in time of war. Four different conceptions of command and direction require to be distinguished from each other:

(a)   Direction of Imperial Policy.
(b)   Higher Strategic Direction.
(c)   Executive Command.
(d)   Administrative Command.

Direction of Imperial Policy

2   The first step towards organising the defence of the Empire on a real co-operative basis is the creation of an Imperial Council or Cabinet. One of the resolutions of the Imperial War Cabinet in 1917 was to the effect that meetings of an Imperial Cabinet should be held annually, or at any intermediate time when matters of urgent Imperial concern required to be settled; and, that it should consist of the Prime Minister of the United Kingdom and such of his colleagues as deal specially with Imperial affairs; of the Prime Minister of each of the Dominions or some specially accredited Deputy possessed of equal authority; and, of a representative of the Indian people to be appointed by the Government of India.

3   The Imperial Council would have to consider questions of general policy, such as possible conflicts with other Nations, in which the Dominions would be materially interested, and would then be a convenient machine for discussing the extent of the Dominion co-operation and the naval programmes necessitated thereby.

Higher Strategical Direction

4   The policy of the Empire having been discussed, and definite decisions having been reached before the emergency has arisen, the next step is the consideration by the naval and military staffs of the ways and means of executing it. The general nature of the war, its strategical objects, the theatre of operations, and the intensity of the consequent effort, require to be investigated at the principal centre of authority and intelligence, and it is considered that this duty should be carried out by the Naval, Military and Air Force staffs, working in close co-operation with each other, assisted by Dominion officers, who should be consulted as to the functions of their own forces in the general strategical plan.

5   The attachment of Dominion naval officers to the Admiralty staff is therefore the second stage in a scheme of Imperial co-operation. They might be appointed to work in the various divisions of the staff or to advise the Chiefs of the Naval Staff. The former proposal is recommended, at any rate until the Dominion navies are more fully developed, but this question requires discussion with the Dominions. In either case, the Dominion representatives should be responsible to their own Chiefs of Staff whilst acting as advisers and liaison officers on Dominion questions to the Admiralty Naval Staff.

6   The foregoing proposal is in opposition to a memorandum by the General Staff which suggests the amalgamation of the Navy, Army, and Air Service Staffs, with Dominion representatives, into an Imperial War Staff, and the eventual co-ordination of the three services under a single Minister of Defence . . .

Executive Command

13   The third conception of command is that of execution as distinct from general direction. Once a force has been allocated to a particular task it must be controlled locally, and this especially applies to forces in waters distant from Great Britain, such as the Pacific, where considerations of time and space necessitate the execution of all plans being placed unreservedly in the hands of the local authorities, e.g., British or Australian light cruisers convoying transports from Sydney to Colombo should, so far as operations are concerned, come under the Australian Naval Staff and the local Commander-in-Chief, who might be an officer in the Australian Navy. Similarly, in the event of war with Japan, all forces in the Pacific and East Indies would be placed under one Commander-in-Chief, who would have been instructed beforehand as to the general plan of campaign, strategical objects, &c. Their execution, however, would have to be left in his hands, and any attempt at control by the Admiralty would only lead to confusion.
14   Intelligent co-operation between different ships and squadrons can only be obtained by common principles of command and staff work, and a common interpretation of tactical and strategical problems. These can only be developed by a uniform system of staff training and text books based on sound doctrines of command. The creation of a Naval Staff College in which British and Dominion naval officers are trained together is therefore the third and not the least important stage in a scheme of Imperial naval defence.

Administrative Command

15   The fourth question that requires consideration is that of the administrative command and supply organisation of Dominion navies. It is considered that the development of the Dominion navies and the successful federation of the Empire on questions of naval defence largely depend on placing the administrative command and supply of the Dominion navies entirely under their own officers, the necessary measures being taken to ensure interchangeability of material and stores. That this is quite practical

was shown by the presence of a United States battle squadron in the Grand Fleet under the administrative and disciplinary command of its own Admiral. The relationship between the Dominion and British armies in France, or between the Prussian and Bavarian armies, also illustrates this point. In the latter case, the only constitutional connection was subordination to the Imperial Command in time of war, co-operation being obtained by a uniform system of staff and technical training.

Summarised Conclusions

16   This chapter may be summarised by saying that the following means of obtaining co-operation between British and Dominion Navies are recommended:

(a)   The creation of an Imperial Council to consider questions of policy.
(b)   The representation of the Dominions on the Admiralty Naval Staff, with a view to co-ordinating the higher strategy of Imperial Naval Defence, and, the institution of a definite system of co-operation between the Naval, General and Air Staffs.
(c)   The appointment of Dominion Naval officers to the Naval Staff College so that selected British and Dominion officers may be given uniform staff training; also, the adoption of common operational and technical text-books.
(d)   Dominion officers to exercise complete disciplinary and administrative command over their own ships and squadrons, but to carry out the instructions of the Commander-in-Chief or Senior Naval Officer in all operational matters.

[Part III]
Selection and Maintenance of Naval Bases

NOTE.   The following proposals are not necessarily put forward for immediate adoption. Strategical considerations must be subordinate to policy, and the extent to which it is desirable to spend money on naval bases will depend very largely on the foreign policy of the Empire and International relations in the future.
1   It is proposed that the development and defence of naval bases should be undertaken by the self-governing Dominions on a scale commensurate with the probable requirements of Imperial

defence, the same policy being followed by India and the Crown Colonies so far as financial resources permit. If distant maritime areas are liable to become primary theatres of operations, suitable bases in adjacent Dominions should be designed to accommodate and maintain the Imperial forces that are likely to be concentrated in these areas.

2   Considerations of economy will force the Dominions to confine their construction of naval bases to the lowest margin of strategical necessity, for money spent on bases will mean less money available for ships and men. In some cases economy may be effected by combining the naval demands for docking, repair and fuel with those for commercial development. For example, Fremantle, the nearest Australian port to Europe, appears to be well situated as an entrepot for ocean freighters and smaller coasting vessels. If Cockburn Sound, near Fremantle, were developed for this purpose, it would probably meet essential naval requirements on the Western Coast of Australia and be able to pay for its upkeep.

3   In considering the location of future bases, war anchorages must be distinguished from the centres of shipbuilding, docking, and repair. The former are selected mainly for strategical reasons, whilst the position of the latter may have to be subordinated to industrial requirements. Shipbuilding may also be considered separately from facilities for repair, but the former is an important attribute of sea power, and it is suggested that the development of the Australian and Canadian shipbuilding industries should be viewed as an integral part of their naval programme . . .

War against a European Power

7   The main theatre of operations would be the North Sea, Eastern Atlantic, or Mediterranean, and Dominion naval units would probably reinforce the British Fleet in these waters, cruisers being retained abroad for convoy work, &c. The presence of the main fleet between the enemy bases and the Dominions would render the latter immune from invasion and blockade, and the necessity for bases in Dominion waters would therefore hardly arise in a European War. Minimum stocks of coal should, however, be maintained at the principal commercial ports for the use of escort cruisers, and commercial docks and repair shops should be at the disposal of the Government in time of war. If possible, these ports should also be protected from submarine attack, and arrangements made for sweeping their approaches clear of mines.

War against Japan

16   If it is decided that the Empire should be prepared for war with Japan, requirements in naval bases may be summarised as follows:

(a)   Singapore and Sydney should be considered the two primary Imperial bases in the Pacific, their docking, supply and repair services being gradually developed accordingly. With the exception of submarines, the sea-going fleet should not be based on Hong-Kong, for its use as a fleet base may lead to a fundamentally faulty strategic plan.

(b)   In the present stage of Canadian development there is no necessity for a great naval base on the Canadian Pacific coast. A fortified anchorage is required, however, and a submarine flotilla should be assembled there as an insurance against possible coastal raids or for operations further afield. So far as Japan alone is concerned, Esquimalt would serve the purpose, but it is too near the United States frontier,* and Malcolm Island anchorage (Lat. 50° 38′ N., Long. 126° 50′ W.) or one of the inlets further North, such as Bela Kula (Lat. 52° 38′ N., Long. 126° 48′ W.), is more suitable.

*Port Angelos, only 18 miles from Esquimalt, is being developed as a United States submarine and aviation base.

(c)   The fleet would frequently have to operate at great distances from its primary bases, and a mobile organisation for the equipment of temporary fleet anchorages is essential, as well as a generous allowance of sea-going fuel and depot ships.

War against the U.S.A.

17   The chances of such a war are remote, and so opposed to the fundamental interests of both parties, that the wisdom of constructing bases merely with that end in view is open to doubt. At the same time, it is well to consider the question before any new works are undertaken . . .

Summarised Conclusions

31   The foregoing may be summarised by saying that if it is necessary to be prepared for war with the U.S.A.:

(a)   A first class naval base capable of accommodating and maintaining a large fleet should be developed in Canadian Atlantic or Newfoundland waters.

(b)   Bermuda should be developed as a war anchorage and supply base, and adequately defended.

(c)   Port Castries in Saint Lucia, or some other harbour in the more isolated West Indian Islands, should be developed as a base for raiding operations on United States trade.

(d)   Canadian naval resources would be best utilised in the Atlantic, but the development of a fortified harbour on the Pacific Coast as a base for submarine operations would be fully justified. Bela Kula or some similar inlet would be suitable.

(e)   Singapore and Sydney should be developed as first-class docking and repair bases in the Western Pacific, Hong-Kong only being suitable as a base for submarine operations.

(f)   An organisation is also required so as to enable any suitable harbour to be quickly equipped as a temporary base with the necessary defences, supply, and depot ships.

Fuelling Bases and Harbours of Refuge

32   War anchorages and bases for the docking and repair of fleets and ships have only been considered in the foregoing investigation. There is a third type of base which has not been dealt with, viz. bases at focal, terminal, and intermediate points of trade and transport routes, of which Simon's Town, Bombay, Colombo, and Kilindini may be quoted as examples. They require facilities for minesweeping and fuelling, as well as protection from the attack of submarines and raiding vessels. The provision of such bases must also be considered with reference to the despatch of ships and fleets to distant areas; for example, if it were necessary to send a squadron of modern battleships to the Far East via the Cape, it would almost certainly have to use Diego Garcia as a fuelling station.

[161] *'Naval Situation in the Far East', memorandum by the Admiralty for the War Cabinet*

21 October 1919

1   The Admiralty have been considering the question of naval bases in the light of after-war events, and more particularly the situation with regard to Hong Kong.

2   It would be of great assistance to the Board if information

could be afforded as to what will be the probable attitude of the Government with regard to the renewal of the Anglo-Japanese Alliance in 1921. It is understood that the position is briefly as follows.

3   Japanese official attitude will probably be in favour of renewal but the British Government may not desire to prolong an Alliance which might cause them to be embroiled with the United States. If this presumption is correct the Treaty is not likely to be renewed, and in this event the Naval situation in the Far East will undergo complete alteration . . .

6   It is unlikely . . . in view of the days of economy ahead of us and the rapidly increasing strength of Japan, viz. an estimated 9 Dreadnoughts and 8 Battle-Cruisers by 1924, that we shall be able to maintain a Fleet equal to hers in the Far East in peace time.

7   It follows therefore, that at the commencement of a war with Japan, we should find ourselves in Naval inferiority for a time, the length of this period depending upon the amount of warning of impending trouble which we are given. It must always be remembered that great reluctance is shown by the Government to moving the Fleet during periods of strained relations, as being likely to precipitate hostilities, and hence that the despatch of reinforcements may be delayed until the last moment.

8   Taking the worst situation with our squadrons scattered, and with certain ships in reserve commission, and bearing in mind the difficulty of moving a considerable fleet with all its supply ships and auxiliaries, it is possible that a period of three months might elapse before our Naval supremacy in the Far East could be established. During this period, Japan would have had practically a free hand at sea.

9   In these circumstances, it would be faulty strategy to allow the naval forces in far Eastern Waters to be locked up at Hong Kong. They would be withdrawn, with the exception of the local defence Vessels, and based upon the probable port of Assembly of the Main Fleet, which may be assumed to be Singapore. This port may be considered sufficiently far from any Japanese possessions to make an attack upon it in force improbable during the period available before the Fleet arrives.

Hong Kong would thus be left to its own resources, until it could be relieved, open to attack by sea and probably by land. The period before relief could be afforded should not be assumed to be less than from three to four months.

10   In these circumstances, the whole nature and extent of the defence scheme for Hong Kong would require revision.

[162] *'Lord Jellicoe's Report on Australian Navy'* General Precis
not dated, [1919]

'Summary of Recommendations'
Administration

Australian Navy to be administered by a Naval Board consisting of the Minister, three naval members, (Flag Officer and two Captains), and one civil member; duties framed on Admiralty lines.

The Secretary not to be a Member of the Board.

The Minister to have an executive Naval Officer as Naval Member.

There should be direct communication between Admiralty and Naval Board on questions of Intelligence and Operations; questions of policy to be communicated through official channels.

Intelligence branch to be properly developed and an experienced officer from Admiralty sent to organise it.

Bases and Dockyards

Dock at Singapore is the only British dock existing in the Far East capable of docking 'Iron Duke' or 'Tiger' classes, and this is without bulges. Japan and U.S.A. are better provided for. 1st Class docks which are projected for Esquimalt and Vancouver too far away for eventualities under consideration. There is no dock on West of Australia between Colombo and Adelaide. A dock at Singapore capable of taking all vessels on the List is necessary, and it is recommended to make two primary dockyards in Australia; one on West Coast and one on East Coast, both partly mercantile.

Plans for dockyard at Cockburn Sound (Fremantle) on West Coast exist and should be continued; though floating docks, and not graving docks, should be considered.

Sydney does not lend itself to expansion to a primary base, and Port Stephens, north of Sydney, is recommended to take its place. No more works to be undertaken at Cockatoo or Garden Islands.

In addition to these two Primary Bases, a third at Bynoe on the North Coast is required, owing to the large distances involved. This will not be practicable till the Northern Territory is more developed, but Bynoe should be temporarily fitted as a fuel and store base.

A second advance base in the North for light cruisers and small craft is recommended at Harbour 'B'.

Port Curtis on N.E. Coast might be considered in the future as a repair base.

Harbour 'A' or 'C' should be earmarked as Fleet Assembly base against Japan, and mining defence set aside for it.

Suva, (Fiji) must not be allowed to fall into an enemy hand and should be fortified by Imperial Government.

Other bases for fueling stations are recommended, and also Jervis Bay and Frederick Henry Bay, near Hobart, as exercising bases.

The Fleet Hospital to be at one of the Main Bases.

Personnel and Training

Entry, Training, Promotions, Appointments of Officers, to be generally similar to that in force in Royal Navy.

Some entries to be allowed from boys' training ships, Midshipmen to be trained afloat.

Courses for sub-lieutenants and specialists to be held in England.

All Officers should be on a general list.

Australian Officers should serve in ships of Royal Navy to qualify for promotion to higher ranks.

Promotion to Mate recommended.

Marines Corps not required at present.

Service in Royal Australian Navy should count as good service for Officers of Royal Navy.

Ratings to engage for 12 years.

Coastguard Service to be introduced.

R.A.N.R. should be developed for ratings as well as officers.

Federal training establishment for Mercantile Marine Officers to be established.

Seagoing training ships for Cadets and boys to be provided.

An additional training ship for boys required.

Mechanical training establishment to be provided.

Australian target service to be organised.

Representative for Australia on staff of D.N.A. & T. at Admiralty.

Mining and Torpedo Training to be same as in Royal Navy.

Minesweeping training in trawlers at 4 Ports.

Submarine training at a submarine depot at Williamstown.
Anti-submarine training to be carried out in England.
Engineering College to be started.
Engineroom and Artificer ratings to be trained, as in England.
An organisation for air training capable of rapid expansion.
Discipline in Australian Navy to be improved.

Air

Air liaison office to be established in London, to keep Australia in touch with development.

Air depot ship of 4000 tons to be provided by 1921, to enable establishment and maintenance of temporary bases.

Manufacture of aircraft to be undertaken in Australia.

Types of aircraft should standardise with English aircraft as much as possible.

Limeburners Creek already requisitioned for training base.

Another station to be decided on for training in war time.

A meteorological service for flying to be established.

Aerial survey of Islands to be undertaken and Admiralty asked to lend *Argus* or *Ark Royal* for one year for this service.

Civilian flying to be developed and personnel to join reserve.

In War advanced air bases for advanced reconnaissance patrols are required . . . 12 patrols are specified, of which some would be Imperial Commitments, and also a striking force of a torpedo plane squadron is recommended. Reports are made on possible Air Sites in islands North East of Australia.

Seaplanes and flying boats appear more feasible than airplanes. Port Stephens would be central base for advanced patrols and a re-fueling depot will be needed on N.E. Coast of Australia . . .

Fuel

Main source of naval coal is Westport, New Zealand.

No suitable coal is available so far in more Northern parts.

Source of oil is, however, American or Dutch East Indian and not in British territory.

Development of shale fields in Tasmania to be pressed on with.

Investigations for all to be carried out in German New Guinea.

Potentialities of the oilfields in Portuguese Timor to be enquired into.

Reserves of oil fuel and coal to be provided at certain ports.

Trials should be carried out with Australian coals at once to try for a good Australian steaming coal.

Naval stocks of coal should be maintained from Welsh or Westport coal for present.

Stores

A Committee to be appointed to enquire into manufacture of Naval stores with view to making Australia self-contained.

Munitions

Similarly with munitions. No independent inspection of explosives during manufacture takes place in Australia. Inspection staff to be formed at once.

No explosives that have been manufactured without inspection should be allowed on board any ship. Immediate action necessary to safeguard R.A.N. ships in this respect.

Torpedo

Establishment of a Torpedo factory not yet warranted.

A range should be made at Port Phillip.

Mines

A depot for reserves near Eastern base.

Drawings and specifications of each type to be obtained from England in readiness for manufacture in Australia in war.

Wireless

Only three purely naval stations recommended in peace: Sydney, Melbourne, Townsville.

Additional war requirements to be met by commercial stations and special establishment of stations for which gear must be ready.

Commercial stations to be developed under F.M.G. and kept separate from Navy. (This is not the present plan).

Commercial stations to be taken over in war, and personnel then trained.

An experienced W/T Officer from R.N. to be obtained to organise W/T Service.

D.A.M.S.

Full provision to be made for arming merchant vessels defensively.

Stowage of equipment to be at Sydney, Melbourne, Adelaide, Fremantle.

Convoys

Australia to obtain full information of Convoy System from Admiralty.

Forces required for convoy duty in Pacific and Indian Oceans. 14 Light Cruisers, 37 Armed Escort ships.

Australia to provide: 4 light Cruisers, 6 Escort ships, to work between Fremantle and Colombo, and between Sydney and Singapore. Storage for armament of escort ships to be provided as for D.A.M.S.

[163] *Admiralty Minute, 'Lord Jellicoe's Report to the Commonwealth Government of Australia'*
31 October 1919

A preliminary survey of Lord Jellicoe's report shows that he has entered into a sphere never contemplated by the Admiralty and far beyond his terms of reference.

It appears that he has taken up the strategical question at the request of the Commonwealth Government. This is most unfortunate, since inevitably that Government will regard his views as those of the Admiralty.

In point of fact, this aspect of the question has not yet been maturely considered by the Board, though the broad policy laid down in the recent Admiralty Memorandum on Imperial Naval Defence (Board Minute 958) has been approved in principle by the Board.

It is unnecessary here to go into detail but I think the Government of Australia should be warned that Lord Jellicoe's terms of reference did not cover the strategic problem of the Pacific and that consequently his views on this point must be regarded as personal and not necessarily as those of the Admiralty.

Further, Lord Jellicoe should be warned by telegram that before informing the Governments of South Africa and Canada of his views on the strategical requirements of the future, his remarks should be submitted to the Admiralty for approval.

A copy of the Admiralty Memorandum on Imperial Naval Defence should also be sent at once to Lord Jellicoe who should be informed in the telegram proposed above that this is being done, that the memorandum contains a brief survey of the Admiralty views on the broader aspects of the subject, and that the policy outlined in it should form the basis of any recommendations he may make to the Canadian and South African Governments.

[164] *Precis of 'Lord Jellicoe's Report on New Zealand Naval Forces'*
[*c*. November 1919]

Enclosure I, 'General Precis'

This report should be considered in conjunction with the report on Australia as the two reports are largely inter-dependent and considerable portions are common to both reports. Like the Australian report, the proposals are based on the requirements of a strong Far Eastern Fleet, due to a Japanese menace.

New Zealand's Contribution   Lord Jellicoe proposes that New Zealand should contribute to the naval defence of the Empire by providing and maintaining a force to be known as the New Zealand Naval Division, to consist of:

Ships

For Far Eastern Fleet
3 Light Cruisers, ('Cassandra' class)
6 submarines, (L. Class)
1 submarine parent, ('Platypus')
   To be provided by 1926.

Harbour Defence
8 old destroyers or P. Boats.
18 Minesweepers (Trawlers)
4 Boom Defence Vessels.
   To be provided by 1923.

Defence of Trade.
7 Armed Escort ships.

It is suggested that the ships should, in the first case, be presented to New Zealand and all new ships should be oil burning only. At present there is no storage for oil, it is proposed that a coal burning light cruiser should be acquired forthwith.

The New Zealand Division would come under the control of the Admiralty in war.

Personnel   Recruiting should be organised and commenced forthwith, and the *Philomel* fitted out as a training ship.

Until the New Zealand Division is at its full strength, the officers and men should be lent from the Royal Navy.

Docks and Bases   No special Naval dockyards are required for New Zealand, as it is considered that the commercial ports meet present requirements.

Local defences for commercial ports and fuelling stations are proposed.

Administration   Proposals are made for the Administration of the New Zealand Division.

(a)   A Navy Board to be established and a Commodore appointed as Chief of the Naval Staff.

(b)   Training of officers and men to be identical with the training in Royal Navy.

(c)   Arrangements to be made for supply of munitions, stores and fuel.

(d)   Naval coal to be obtained from Westport.

(e)   Oil should be prospected for in New Zealand.

(f)   Reserves of fuel, stores, etc., to be maintained.

(g)   Trawler fishing industry to be established.

Air   No aircraft carrier is recommended but the institution of a Naval Air School.

Finance   Estimates for Naval expenditure in New Zealand based on above proposals are given in Volume I, Table VI, (amended in Table VI, attached to the covering letter).

Enclosure II, 'Summary of Recommendations'

Administration

New Zealand Division to be commanded by a Commodore.

An Officer (with rank of Commodore) senior to Commodore New Zealand Division, to be lent as Chief of Naval Staff to administer all the Naval forces and establishments in New Zealand waters.

Naval Board to consist of Minister of Defence and Chief of Naval Staff assisted by Staff Officers.

A Naval Officer of Military Branch to be liaison officer at Admiralty.

New Zealand Division to pass under orders of Admiralty under certain conditions.

Air

Air training to be same as in England.

Aircraft carrier not to be provided on account of expense of upkeep.

Small Naval flying school recommended at Auckland.

Civil flying to be encouraged and personnel to belong to reserve.

Dockyards and Bases

No special Naval dockyards required for New Zealand.

Commercial facilities considered sufficient. If new commercial docks are constructed in New Zealand they should be big enough to accommodate the largest vessels.

Auckland	to be a defended port.
Wellington	,, ,, ,, ,, commercial port.
Lyttleton	,, ,, ,, ,, ,, ,,
Otago Harbour	,, ,, ,, ,, ,, ,,

Possible anchorages and fueling bases to be defended if used, – Schemes to be prepared but no expense incurred.

Amoara,	Bay of Islands
Pelorus Sound	Queen Charlotte Sound
Port Underwood	Paterson Sound

Coal Shipping Ports:

Westport, defence recommended.

Graymouth (alternative only).

Attention is called to the possible use by an enemy of Vauua Harbour as an advanced base.

Fuel

Naval coal from Westport.

Shipping Companies should be encouraged to maintain stock of oil in New Zealand and some Pacific Islands.

Oil should be prospected for with help from Imperial Government.

Coal inspector to be appointed for New Zealand.

Reserves of fuel to be established.

Stress is laid on fact that provision of fuel supply for the fleet demands immediate action.

Personnel and Training

New Zealand Division to consist of:

(a) Regular Naval Forces (N.Z. Division of the R.N.)
(b) Reserve Naval Force (N.Z. Division of the R.N.R.)

Until New Zealanders are available, officers and men from the Royal Navy to be lent for Service under the New Zealand Government and paid at New Zealand rates of pay.

Officers of the N.Z. Division to be available for service in any of H.M. Ships.

Men to serve in New Zealand or Far Eastern Fleet.

New Zealand cadets to enter at option through either,

(a) Osborne and Dartmouth.
(b) Royal Australian Naval College,

at present not more than 8 annually.

Specialist officers to qualify in England.

Officers to be promoted on the general list of R.N.

No Marine Corps to be instituted. Marines to be lent by Admiralty.

Regulations for training advancement, re-engagements, etc., of ratings to be identical with Royal Navy.

Pay to be based on pay of Royal Navy.

New Zealand to be responsible for all pay of New Zealand Division.

Boys to be trained in *Philomel* and enter for 12 years continuous service.

Stokers to be trained in *Philomel* and enter for 12 years continuous service.

Artificers to be entered as boys.

Artisans, etc., by direct entry.

Coastguard Service to be introduced.

Eventual Numbers Required

Seaman class	1300	Annual entry	100
Stokers	650	,, ,,	50

Reserves to be enlisted and trained on same lines as the R.N.R. for:

(a) Seagoing duties, (escort ships, patrols, etc.)
(b) Shore duties, (Was Signal Stations, Wireless)

Higher G. & T. ratings to be trained in England. Remainder in England or Australia.

Submarine training in England
Higher signal ratings in Australia
Anti-submarine training:
  Officers in England
  Ratings in Australia.

Stores

New Zealand to eventually become self supporting.
A Naval Store officer to be lent to New Zealand for 6 months
to organise stores.

Munitions

New Zealand should become self-supporting as regards manufacture and have own Inspecting Staff.
No munitions to be allowed on board that have not been
inspected.

Torpedo

No factory required.
Torpedo range at Firth of Thames.

Minesweeping

Fishing industry to be developed to supply Reserve of trawlers.
2 Trawlers in full commission for training.

Wireless

High Power Stations	Nil.
Medium Power Stations	2
L.P. Stations	9
Directional	5

Philomel to be fitted with Type 16 CW for
training and as a Naval station.
D.E.M.S.
Stowage to be provided for equipment.

[165] *'Naval Mission to India and the Dominions, 1919–20',*
Jellicoe to Murray
3 February 1920

'Secret'
Be pleased to lay before the Lords Commissioners of the
Admiralty the attached general remarks on my mission to India

and the Dominions of Australia, New Zealand, and Canada, under the following general terms of reference: . . .

2   In submitting these remarks, it is necessary to point out to Their Lordships the unfavourable conditions under which the visits to the Dominions took place.

3   Apart from the uncertainty regarding naval policy at home, it was very unfortunate that neither the Prime Minister nor the Minister for the Navy was present in Australia during my visit, and there was consequently no opportunity of conferring with them and ascertaining their views. Shortly after their return to Australia a general election took place, and presumably for this reason no decision was taken on my report during 1919.

4   My arrival in New Zealand almost coincided with the break up of the Coalition Government, and with the shadow of a general election before the Cabinet, again no decision was reached during 1919.

5   In Canada the Prime Minister and the Minister for Naval Affairs had both been absent from official work for a long period owing to ill health, and although both Sir Robert Borden and Mr Ballantyne were in Ottawa during my visit, neither were in good health, whilst, for the last two weeks of my stay, the greater part of the time of the Cabinet was occupied in discussing the situation arising from the fact that the Prime Minister had been ordered to take a prolonged rest on medical grounds. Sir Robert Borden left Canada before any decision was reached in regard to a naval policy.

6   In spite of these adverse conditions, I am of opinion that both India and the Dominions, and especially the Dominions are willing and indeed anxious to bear their share in the naval defence of the Empire, if a lead is given and if they are encouraged to develop on their own lines, particularly if a sympathetic and generous spirit is shewn by the Home Government in giving ships to form the basis of their naval forces, and in lending suitable officers of high attainments with which to commence the organisation.

7   Finally I venture to urge that early Dominion representation at the Admiralty is very desirable as it would give that sense of the importance of Dominion contribution to naval defence which is at present required, and which would go far towards stimulating interest in the effort which I hope will soon be put forward by the Dominions.

[166] *Admiralty Secretary to Under-Secretary of State, Colonial Office draft*

1 July 1920

'Immediate and Confidential'

With reference to your letter of the 3rd June 1920, No. 26964/1920, on the subject of the Officers whose services the New Zealand Government desire to obtain in connection with the Naval Defence policy of the Dominion, I am commanded by My Lords Commissioners of the Admiralty to request that you will move the Secretary of State for the Colonies to inform the Governor General of New Zealand that They have been hitherto acting under the impression that, as Lord Jellicoe's report has not been adopted in its entirety, and some time must elapse before the New Zealand Naval Force can attain the dimensions recommended to be worked to in Lord Jellicoe's report, the performance by one Officer of the three functions of Naval Advisor, Commodore and Captain of the *Chatham* would for the time being provide adequately for the administration of the New Zealand Naval Force.

2 The Officer selected for the appointment is Captain Alan G. Hotham, C.M.G., R.N., who has accordingly been appointed in command of the *Chatham* and as Commodore Second Class with effect from 24th May last, on the assumption that he would act also as Naval Adviser to the New Zealand Government.

3 My Lords desire, however, that it may be made clear to the New Zealand Government that, while they are ready to meet their wishes, they are of opinion that a Commodore on shore would be superfluous until the Naval policy of the Dominion has been further developed, which will depend to some extent on the Imperial Conference which it is anticipated will be held next year.*

[* Paragraph 3 is tentatively reworded in ms.]

[167] *Paraphrase Telegram from the officer administrating the Government of New Zealand [Chief Justice the Hon. Sir Robert Stout] to the Secretary of State for the Colonies [Liverpool], received*

6 August 1920

Referring to your telegram 6th July. After careful consideration of question of appointment of the Commodore of the New Zealand division as Naval Adviser, my Government is of opinion that it is essential that the Naval Adviser should be stationed on shore. From experience of the late Naval Adviser, Captain Hall Thompson, it was found to be impracticable for one officer to be in command of a ship and also to perform duties of Naval Adviser. Although desiring that effect should be given to proposals contained in my telegram 1st June New Zealand Government does not directly wish to depart from Admiralty proposals and as an alternative suggests an officer on Commodore's staff be appointed with office on shore to deal with administrative routine work and to act as a connecting link between Commodore and Government in the absence of the former at sea.

[168] *'Empire Naval Policy and Co-operation, Outline of Empire Naval Policy' prepared by the Admiralty to form a basis for discussion at the Imperial Cabinet, s.v. Naval Measures in support of an Economic Blockade*
February 1921

Blockade Control Staffs of the Naval Control Service

1   As a result of experience in the late war, it is evident that any future wars waged by the British Empire will be fought largely with economic weapons. The close Blockade as understood in pre-war days was found to be impossible in the conditions of modern warfare. Moreover, in the early stages of hostilities the restrictions imposed by International agreements as to contraband seriously limited the value of our sea power as a belligerent. Thus the only really effective Blockade in the future will be a long-distant and absolute one, involving the stoppage of all imports and exports of the enemy country – it having been proved that all imports into a belligerent's territory do in fact assist him in the prosecution of a war.

The function of the Admiralty in such an Economic Blockade will consist in exercising control over seaborne commerce, with a view to restricting the enemy's supplies by sea, both directly, and indirectly through adjacent neutral countries.

2   This naval control inter alia entails the visit and search of all ships possibly engaged in carrying supplies to the enemy, either directly, or through contiguous neutral ports, unless these ships' cargoes have already been inspected and passed by a representative of the British Government at the port of loading. During the late war, owing to the impracticability of boarding and searching

ships at sea, it was found necessary to establish bases on shore, to which Allied and Neutral ships, suspected of carrying contraband, were diverted for examination.

3 The boarding and searching of the ships, examination of papers, etc., is carried out by naval personnel under direct control of the Admiralty, acting in conjunction with the central Blockade and Contraband authorities, who decide as to the disposal of ship and cargo – the latter being eventually dealt with by the Customs.

4 In any future war, these Blockade Control Staffs (as the organizations for boarding and search, etc., will henceforth be called) will have to be established at the commencement of hostilities. The necessary arrangements must, therefore, be made in peacetime so that the machinery can be put into motion as soon as the emergency arises.

5 The ports to be manned (both in the United Kingdom and Overseas) will depend on the country or countries with which the Empire is at war, and who, if any, are her allies. An outline scheme has, therefore, been drawn up in which the possibility of war with any nation has been considered, and the recommendations are world-wide.

Details of this scheme have already been communicated to the Dominion, Indian and Colonial Governments, through the Colonial and India Offices, while the Naval Commanders-in-Chief abroad have also been informed.

6 Europe having been the centre of operations in the recent war, the Dominions, with the exception perhaps of Canada, were not in very close touch with Naval Control, but in a future war they may be vitally affected. It must also be borne in mind that in the event of war, say in the Far East, the Admiralty would be unable to exercise complete direction at such a distance, thus necessitating the institution of a Headquarters in or near the immediate war zone.

The successful prosecution of any future Blockade must therefore depend on the closest co-operation from all parts of the Empire.

7 Detailed organisation (routine, duties, etc.), of the Control Staffs will be worked out by the Naval Staff as the post-war situation develops and the policy is defined. The question of Trading, Blockade and Enemy Shipping is under the consideration of the Committee of Imperial Defence.

8 In the meantime the scheme of Blockade Control, as outlined in the communications already referred to, will serve as a guide

to the general principles on which the Admiralty is working in preparation for the attack on Enemy Seaborne Trade.

[169] *Report of Conference held on board HMS* Hawkins *at Penang, between the Commanders-in-Chief of the China, East Indies and Australasian stations [Duff, Tothill, and Rear-Admiral Sir Edmund Percy F.G. Grant]. Addressed to the Secretary of the Admiralty*

from 7 March 1921, onwards

Appendix A, The Higher Command of the Pacific and its Head-quarters

The Conference has given very careful consideration to the question of the Higher Command of the Pacific, as it is realised that the decision on this point is probably the most important decision to be given in connection with the Pacific naval problem.

2   War experience has shown that within a definite prescribed area in which combined operations are to be carried out by various forces successfully, economical and efficient control can only be attained if it is centred in one authority. This was the case in connection with the Higher Command in France, during the latter part of the land warfare, and was also the case in connection with the control centred in the Admiralty at London of all Allied Naval Forces operating for the protection of merchant shipping in the Atlantic, and in the Channel and North Sea. Similarly, control of the Allied naval forces in the Mediterranean never achieved full success or its maximum efficiency because it was never fully centred, as desired, at Malta.

3   The area for which central control is required is dependent upon the area over which the enemy will probably exercise direct control from his headquarters; in the case under consideration, therefore, it appears that the area to be controlled by the British Higher Command would be the whole of the Western Pacific Ocean, China Seas, and Indian Ocean. The Conference therefore, considers it necessary that the Higher Command to control this enlarged 'Pacific Station' should be set up in peace and operate in war.

4   If the Higher Command could be located in London at the Admiralty, many points would undoubtably be simplified, as the Admiralty must necessarily be the ultimate authority for all naval operations, and it should have control of all the naval resources of the Empire in time of war.

Hitherto, the main sphere of naval operations has been in the vicinity of the British Isles or in the Mediterranean, and thus London, apart from its being the capital, was a convenient centre for such control. In the case of the Pacific there are two main objections to any attempt to carry out direct control from London.

(a)   The difficulty of communications, not only on account of the distance and time, but owing to the fact that communication to and from London is necessarily congested in war time owing to the enormous amount of traffic that passes to and from London on account of its position as the centre of an Empire with interests in all parts of the Globe.

(b)   The fact of the Naval Staff at the Admiralty being out of touch to a great extent, with the actual situation, and being entirely dependent for their knowledge on telegraphic communication. The Conference is therefore of the opinion that the Higher Command for the 'Pacific' must be located at some place in the British Dominions in the 'Pacific' area.

5   There appear to the Conference to be only four possible Dominions or Colonies in which to select a site, namely Malaya, Hongkong, Australia or New Zealand. Hongkong is too advanced, and for reasons known to Their Lordships, is unsuitable. New Zealand on the other hand is too remote and too isolated. The choice therefore is confined to Malaya or Australia. As, however, the direction of attack must lie in the Northern Pacific and the China Seas, the Conference considers that Malaya is the strategic centre of our defence and offence, and that Singapore is the ONLY site suitable for the location of the 'Higher Command' as well as for the main fleet base.

The loss of Singapore would be (as stated in the Admiralty war Memorandum) 'disastrous to the British Empire' and the Conference therefore has assumed (when selecting Singapore as the main fleet base and for the Higher Command) that it will be made impregnable. No alternative to Singapore is seen from which the fleet could operate successfully.

On the establishment of the Higher Command, the Conference recommend that for peace administration the East Indies, China, Australia and New Zealand Stations should remain as at present each independently under a separate command, but that the function of the Higher Command, which it is intended shall be exercised from his shore base only, should be as follows:

I In Peace

(a) Preparation of all plans and war orders for the 'Pacific'.

(b) Preparation of plans for war exercises or manoeuvres in which one or more of the four commands is engaged, and directing of such on behalf of the Admiralty from his Headquarters on shore, when one or more of these commands is placed under his orders by superior authority.

(c) Organisation and control of Pacific Naval Intelligence Centre and system, and control of secret agents.

(d) Preparation of plans for co-ordination of R/T and other means of communication in the 'Pacific'.

(e) 'Visit such places in the Pacific and Indian Oceans as were of importance for naval purposes in war, in order that he may have intimate knowledge of the general situation' (Lord Jellicoe's Report)

II In War

In war (or in period of 'strained relations,' if so directed by the Admiralty, or on receipt of the warning telegram). To be regarded as the advanced Headquarters of the Admiralty, and in that capacity from his headquarters on shore (a) direct and control all British Naval forces in the Pacific except such local forces as are reserved for local defence of ports in India or the British Dominions; (b) as the operations develop, to propose to the Admiralty the measures necessary for the successful conduct of the war.

Proposals as to the staff and accommodation of the Higher Command will form the subject of a separate report by the Commander-in-chief, China, observing that the Royal Australian, and New Zealand Navies should be represented on the staff.

[170] *'Mr Hughes on Naval Defence', memorandum by Churchill*
28 April 1921

I should like my colleagues to read the following extract from a speech by the Prime Minister of Australia [William M. Hughes], delivered in the House of Representatives, Melbourne, April 8, 1921 . . .

'Australia's existence depends upon adequate naval defence. The navy is what the people of Britain make it. It is vital to us that it should remain a great navy. The Conference is therefore important because we as Dominions may there express our opinion as to what it means to us. The question then of a satisfactory scheme of Imperial Naval Defence is one literally of life and death to Australia.

The bearing of the Japanese Treaty upon the naval defence of the Empire is obvious. As we have seen, there has lately been much talk of strained relations between the United States and Japan. Now in that lie the germs of great trouble and possibilities of infinite disaster to this world. What is the hope of the world? This is no party question. I hope that every honourable member will express himself freely, remembering only that every word said here in relation to a foreign nation is one which must be weighed before it is uttered. Words of counsel I welcome, words of warning too, but not words which, lightly uttered, make a task that is now sufficiently difficult almost impossible. What is the hope of the world? As I see it, it is the alliance and an understanding between the two great branches of the English-speaking peoples. Now here is our dilemma. Our safety lies in a renewal of the Anglo-Japanese Treaty. Yet that treaty is anathema to the Americans. America has said that she must have the greatest navy in the world; that she must have a navy sufficiently strong to defend herself. To defend herself against whom? She has left the world in no doubt, or in very little, as to whom. We not only have no quarrel with America, we have no quarrel with Japan. We have our ideals; Japan has hers. There is room in the world for both of us. We want to live in terms of amity with all nations of the earth, and we could not afford to do anything else, no matter what we might think, even if we desired otherwise . . .'

That is our dilemma. While making every effort to retain the friendship of Japan, we cannot make an enemy of the United States. We must steer our barque between Scylla and Charybdis. In some way we must attain the calm waters of port. That is the mission which the representative of Australia has to fulfill.

I need not elaborate on the advantages of a renewal of the Japanese Treaty, nor on the consequences of its abrogation. They are very obvious. It seems that our expenditure on naval defence will be, to a large degree, determined by whatever steps are taken by means of this treaty to ensure us peace or, in its absence, by what steps we take in co-operation with Great Britain to attempt to ensure it. We cannot shut our eyes to the fact that there is now great danger of such naval rivalry in the Pacific as will not only be a heavy drain on the nations directly affected, but which will have its reflex influences upon the whole world.

[171] *'Oil Fuel Reserves', memorandum by Vice-Admiral Sir Osmond de B. Brock, Deputy Chief of Naval Staff*
24 May 1921

'Confidential'
General Strategical Considerations

1   At the present time the only Powers possessing naval forces comparable to those of the British Empire are the U.S.A. and Japan.

With a one power standard it is impossible to divide the fleet between the Far East and European waters so as to be capable of meeting a threat from Japan and yet leave sufficient force in European waters to deal with a hostile America or even a combination of European Powers.

2   Mobility of the Fleet must therefore be the keystone of British naval strategy, and this can only be ensured by providing adequate fueling arrangements on the main lines of communication and supplies in the probable theatres of operations.

The situation as regards oil fuel reserves in the Atlantic is steadily improving, the Government having approved a reserve of 4,500,000 tons in the United Kingdom.

On the route to the Far East and in the Western Pacific however the position is not at all satisfactory, and until the reserves recommended in this paper are well under way we should be unable to send a fleet to the Far East at short notice or to maintain it there on a war footing after arrival.

4   Although it is to be hoped that a war between the British Empire and Japan will never arise, it is probable that the knowledge that the British Fleet can make its presence felt in the Western Pacific at short notice will cause Japan to hesitate before embarking on a course of action running counter to British interests . . .

20   Taking the case of a war with the United States in 1930. The fleet would consist approximately of:

Capital ships	29	T.B.D.'s and leaders	126
Light cruisers	28	Various (aircraft	
Submarines	30	carriers, depot	
		ships, etc.)	18

and the twelve months war consumption of this fleet with its 59 attendant oilers is estimated as 5,112,000 tons.
21   The arrangements for supply of this quantity to the Fleet in the W. Atlantic are shown in enclosure 4 from which it will be seen that after making allowance for the Trinidad fields and transporting some 750,000 tons from Persia, the whole of the United Kingdom reserve will be required, in addition to the reserves proposed for Bermuda and Halifax and the available surplus from the Mediterranean reserve.
21   [*sic*]   The probable composition of our Eastern fleet in the event of a war with Japan in 1930 would be:

Capital Ships	20	Submarines	30
Light cruisers	20	Various (aircraft	
T.B.D.'s and		carriers, depot	
leaders	80	ships, etc.)	10

The twelve months' war consumption of this fleet with 58 attendant oilers is computed to be 3,430,000 tons. Detailed arrangements for supply of this quantity to Singapore are shown in enclosure 3.

It will also be seen that for supply to an advanced base, which may be upwards of 2,500 miles from Singapore, a further 48 oilers using 460,000 tons in the year would be necessary.

Owing to the inevitable lack of transport facilities and the waste entailed when large quantities have to be taken such distances as 10,000 miles, it is not considered practicable to utilise the United Kingdom reserve in this case, and reserves must therefore be laid down at ports in the Indian Ocean, Singapore and Australasian waters.

Passage of Fleet to Singapore

22   The importance of being able to despatch the fleet at short notice and of completing the passage in the least possible time has been pointed out. The conclusions arrived at as to ports to be used and amounts required at each are shown in enclosure 5.
23   Full consideration has also been given to the passage via the Cape, but it appears to be a difficult question at present to contemplate the erection of shore reserves for such an exceptional contingency, and it is proposed that reliance should be placed on supply by tank vessels, though it is realised that delay will in that event be unavoidable. Even with large fuel reserves on this route the passage of the Fleet to the East would be slow and difficult

and everything points to the extreme importance of safeguarding the Suez Canal route.

War with a European Power

24 The main theatre of operations in this case may be in Home or Mediterranean waters.

In view of the large reserve already recommended for the United Kingdom and the proximity of the Persian oil fields to the Mediterranean, it is considered that war requirements will be met by the provision of the 300,000 tons recommended for the Mediterranean *vide* enclosure 6 . . .

Dominion Cooperation

27 The question of providing adequate fuel reserves has already been brought to the notice of the Dominion Governments and they have been invited to consider the building up of these reserves as one of the most important methods of co-operation in the naval defence of the Empire.

Included in enclosure 6 are the amounts that it is recommended should be stored at ports in Dominion and Colonial territory. Approval of the proposals now put forward will go far towards awakening in the Dominions and Colonies a sense of the importance of the fuel question . . .

Enclosure 5. Passage of Fleet to Singapore

1 Thorough examination has been made of the possible ways of carrying out this operation, and it is evident that if a speedy passage is to be ensured, certain extra fueling facilities must be provided at some of the ports, and anchorages suitable for refueling the fleet must be found in the southern part of the Red Sea and in the vicinity of Colombo . . .

4 The total time of passage from the U.K. to Singapore via Suez has been estimated at 40 days, subject to the following assumptions:

(i) Three days 'tension' before hostilities commence.
(ii) Fleet fully fueled when war is declared.
(iii) Fleet proceeds as a single unit at 16 knots (Suez to Aden capital ships 15 knots).
(iv) No delays other than waiting for fueling.
(v) Additional facilities provided as in column 4.
(vi) Eastbound oilers have precedence over H.M. Ships for canal.
(vii) Weather admits of continuous fueling day and night.

Enclosure 6, May 1921, Oil Fuel, Naval Storage Position Abroad

Location/ Reserve of Oil Fuel Proposed Tons	Oil Fuel Stock Proposed Distribution. Accommodation Base	Quantity Tons	Existing 31/3/21 Tons	Expected to Complete by 21/9/22
Mediterranean				
300,000	Gibraltar	80,000	11,200	-
	Malta	180,000	22,400	25,000
	Port Said	20,000	-	16,000
	Suez	20,000	-	-
East Indies				
730,000	Aden	100,000	-	-
	Colombo	225,000	-	-
	Kilindini	5,000	-	-
	Rangoon	400,000	-	-
Africa				
48,000	Cape	24,000	-	-
	Sierre Leone	24,000	-	-
North America & West Indies				
600,000*	Bermuda	500,000	16,000	-
(660,000)	Halifax	100,000	-	-
	Jamaica	10,000	-	16,000
	Canada (Pacific)	50,000	-	-
South America				
24,000	Falkland Islands	24,000	-	-
Pacific (China)				
1,215,000	Singapore	1,200,000	-	-
	Hong Kong	15,000	-	32,000
Australia				
400,000	To be determined later.	400,000	-	-
New Zealand				
26,000	Wellington	10,000	-	-
	Fiji	16,000	-	-
Working stock				
97,000	Bases abroad	97,000	-	-
TOTAL 3,500,000			49,000	89,000
	Cumulative Total			138,600

[172] *'Empire Naval Policy and Co-operation', letter from Walker to Hankey, Admiralty*

25 May 1921

I am commanded by My Lords Commissioners of the Admiralty to transmit herewith copy of a letter which has been received rom the Treasury, dated the 14th May, No.S 7077, respecting the contribution at present paid by the Government of India towards the maintenance of His Majesty's ships in Indian waters, in which the Treasury ask for an assurance that the Admiralty will at the Imperial Council press strongly for an increased contribution in preference to any other demands on the Indian Government.

2   The Admiralty proposals (CID 131-C) mention the following alternative methods by which India could co-operate in the Naval Defence of the Empire:

1) By forming an Indian Navy;
2) By increasing her annual subsidy;
3) By providing means for trade protection, mine-sweeping, dockyard facilities, &c.

3   Until the proposals of the Indian Government have been seen, it is not possible to determine whether it would be better for the Admiralty to accept assistance from the Indian Government in cash or in kind, and, from the point of view of naval policy, the Admiralty therefore deem it inadvisable to commit themselves in the matter as the Treasury ask. They concur, however, that there is a strong ground for pressing for an increased contribution on account of the present cost of maintaining the East Indies Squadron, unless some alternative proposal for co-operation is agreed to whereby equal relief will be afforded to this country in respect of naval expenditure.

4   The question is not so much one for Departmental decision as for the decision of the Government as a whole, and I am to represent the desirability of the Admiralty proposals contained in CID 131-C being considered by the Cabinet or by the Committee of Imperial Defence as may be considered most suitable, the views of the Treasury being presented by that Department direct.

5   A copy of this letter has been sent to the Treasury.

[173] *'Singapore, Development of as Naval Base, Memorandum',*
by the Oversea Sub-Committee of the Committee of Imperial
Defence, summary, signed by Samuel Herbert Wilson, Assistant
Secretary of CID

7 June 1921

50   To sum up, it would appear to the Oversea Defence Committee:

(a)   That the fact that a large British fleet cannot be maintained in time of peace in the Pacific must not be interpreted to mean that there is no necessity to develop one or more of the British naval bases situated there so as to ensure the docking and repairing facilities available in those waters keeping pace with the advances made in ship construction; and that unless this is done it will not, in an emergency, be possible for the British fleet to operate in the Pacific.

(b)   That, unless His Majesty's Government take the view that the safe guarding of British interests in the Pacific may be allowed to depend on the maintenance of friendly relations in all circumstances with Japan, it is essential that the existing situation should not be allowed to continue, and that steps should be taken forthwith to develop at least one naval base in those waters suitable for use by the main British battle fleet in an emergency, as a base for concentration and for repairs and supplies.

(c)   That, of the British naval bases available for use in the Far East, Singapore and Hong Kong are those most essential to the conduct by the British fleet of naval operations in the Pacific; and that, although it will be necessary to develop certain naval bases in Australia in order to make provision for meeting the needs of the Royal Australian navy, it is very unlikely that Sydney will, in normal circumstances, have to be used as a main base for a British battle fleet operating in the Pacific.

(d)   That it is impracticable, under existing conditions, to maintain in time of peace at Hong Kong a sufficient military garrison to secure for certain its safety in the early stages of a war with Japan; and that, consequently, of Hong Kong and Singapore the latter must be regarded as the most suitable for use as the rendezvous and main repair and supply base of any British fleet which may be sent to operate in the Pacific.

(e)   That, unless the defences of Singapore are strengthened, there would, under existing conditions, be every inducement to a Japanese commander in the early stages of a war to attack the fortress, with a view to denying its use to the British fleet in subsequent operations, a result which would have the effect of increasing incalculably the difficulties of the British naval commander in operating in the Pacific.

(f) That the selection of Singapore as the principal base of concentration in the Far East should not be interpreted to mean that the security of Hong Kong is not regarded as of importance, and that, since this latter port is strategically the most favourably situated of the British ports in the Far East for the carrying out of naval operations in the China seas, local naval defences, including a submarine flotilla, should be provided at the port if this can be done without any undue expenditure, since such defences would act as a powerful deterrent to a Japanese commander who contemplated an attack on Hong Kong in the early stages of a war.

(g) That, in coming to a decision as to the development or otherwise of a British port in the Far East, it is for consideration whether it is necessary for His Majesty's Government to take into account the resolution in connection with the reduction of Armaments adopted in December 1920 by the Assembly of the League of Nations, or to take into account the effect which the development of such a base would be likely to have on the relations between Great Britain and Japan . . .

Reservation by the Treasury
Representative, E.W.H. Millar

It is understood that the question of renewing the Anglo-Japanese Alliance is to be considered this year and until this question is decided the formulation of defence plans for the Far East appears somewhat premature. But in my judgment the danger of a Japanese attack must be regarded as too remote to justify incurring in existing circumstances the heavy expenditure involved in this scheme.

The hypothesis is that within the near future Japan, who is at present our ally, may consider it worth her while to challenge the British Empire, with its overwhelmingly superior naval and financial resources. To enter upon such a struggle with the remotest prospect of success she would have to be able to count at least on the benevolent neutrality of the United States and, to ensure the success of a military attack on Hong Kong, probably of China also. I submit that such a situation is almost incredible.

At the present moment it is more than ever necessary that our defence plans should be based on practical hypotheses.

His Majesty's Government has just announced that next year supply expenditure must be reduced by 20 per cent. If the Admiralty are to contribute their share of this reduction – and without the co-operation of all the Departments it will be impossible to obtain the required reduction – Navy Votes for 1922–23 must be reduced by 16,000,000£. Before the Admiralty can take any steps in this direction they must, unless the scheme for the construction of capital ships is abandoned, find savings to meet an increased expenditure next year under this head of 11,000,000£. If therefore the fresh commitments involved in this scheme are undertaken the financial problem to be faced by the Admiralty will become almost insuperable.

In short, the financial position is such that in my opinion there is no alternative but to face such risks as may be involved in the present condition of our defences in the Far East and to rely upon diplomacy to obviate them.

[174] *Appendix to the minutes of the 141st Meeting, Committee of Imperial Defence*

13 June 1921

Conclusion The Committee recognise that the question of incurring expenditure under existing conditions on the development of Singapore as a naval base is one which involves political and financial considerations outside their purview: and that accordingly a final decision in the matter must rest with the Cabinet. In view, however, of the vital importance from a strategical and Imperial point of view of developing the port of Singapore on such lines as will make it possible for the British fleet to operate in the Pacific if called upon to do so, the Committee are of opinion that at the forthcoming meeting of Prime Ministers and representatives of India, the policy favoured by His Majesty's Government should be stated as follows:

(a) His Majesty's Government fully recognise that the basis of any system of Imperial defence against attack from oversea, whether upon the United Kingdom, Australasia or elsewhere, must be, as it has always been, the maintenance of our sea power;

(b) The most pressing question in this connection at the present time is that of the measures to be taken for the protection of Imperial interests in the Pacific;

(c) His Majesty's Government are advised that for this purpose it is essential that Singapore should be available as a base of concentration, repair, and supply for the British fleet and auxiliary services, and they are accordingly prepared to take the lead in developing that port as a naval base;

(d) Owing to existing financial conditions it will not be practicable to incur a large expenditure for this purpose in the immediate future; but it is the intention of His Majesty's Government to develop the base as funds become available, and the greater the assistance that can be rendered by the oversea Governments in this connection the sooner will the necessary programme be completed.

[175] *'Reserves of Oil Fuel', memorandum by Lord Lee of Fareham, First Lord of the Admiralty*

21 June 1921

1   The complete displacement of coal by oil fuel for modern warships and the prospect of a large increase in its use by the vessels of the mercantile marine have introduced a problem with most far-reaching effects.

2   The advantages of oil over coal are so overwhelming that, in the case of warships at all events, the change cannot be regarded as being other than desirable and inevitable. As compared with an oil-burning ship, a coal-burning ship has a reduced speed, and reduced endurance at a given speed. Her radius of action is practically halved; she must return twice as often to refuel; the process of refueling is far slower, and more laborious to her crew. Coal causes smoke, oil causes none. Coal contains less heat energy than oil, and requires more stowage room. The use of oil admits of a saving of weight, a saving of space, and to immense saving in complement. In all these directions, the savings realised, and to be realised, through the transfer from coal to oil for warships are very great indeed.

3   It must be recognised, however, that this inevitable replacement of coal by oil has entirely altered the specially favourable situation in which the Navy formally stood in respect of its fuel supply, owing to the great resources of the British coal-fields, and the existence of the coaling facilities that had been built up abroad.

4   It is not necessary to labour this point beyond saying that, in order that our naval forces may retain their capacity to operate in defence of any part of the Empire, there must be an adequate

supply of oil fuel for their use, not only as a central reserve at home, but also at strategical points in different parts of the world.
5   The Government have already approved a reserve of 4,500,000 tons in the United Kingdom, and the completion of this reserve will make the situation in Home waters satisfactory; the storage abroad, however, is at present inappreciable, and the fleet is practically immobilised so far as operations outside Home and Mediterranean waters are concerned.
6   So long as there is no threat of war the situation cannot be described as dangerous, but it should be clearly realised that the fleet is unready in this respect, and it is a matter of anxiety to the Admiralty to feel that they could not respond to an urgent call for the transfer of the fleet to a base abroad, such as Singapore, for it is in the East and on the route thither that our resources are weakest.
The safety of the Empire overseas is largely dependent upon the mobility of the British main fleet, and it is well within the bounds of possibility that a foreign Power would take advantage of our weakness in this respect, where a proper provision of fuel reserves would have caused him to adopt a more cautious attitude.
7   The principles that should underlie our policy in regard to the establishment of fuel reserves are:

a)   That a year's supply of fuel for the fleet should be available within practicable transport distance of the bases from which the fleet may be required to act in the event of war.
b)   That fuelling arrangements should be made on the route to the Far East to enable the fleet to be despatched at short notice in the event of war, or of a threat of war, and to ensure completion of the passage in the least possible time.
c)   That in areas remote from the main theatres of operations small reserves should be available for the use of detached squadrons which may be expected to operate in these localities.

8   Enclosure 1 gives the quantity and distribution of naval reserves abroad which are considered necessary, and Enclosure 2 shows the requirements and the supply arrangements that could be made to meet them when these reserves are available.
It will be seen that in addition to the 4 1/2 million tons in the United Kingdom, a total of 3 1/2 million tons is required abroad, of which it is proposed that the following amounts should be in the territory of the Dominions and India:

India (Rangoon)	400,000 tons
South Africa	24,000
Canada	150,000
Australia	400,000
New Zealand	10,000
(A further 16,000 tons at Fiji)	984,000 tons

9   The Admiralty have recommended that the Dominions should be urged to look upon the provision of oil fuel reserves as one of their principal naval commitments for some years to come, for on the mobility of the main fleet rests the guarantee for their safety.

It is also suggested that the Colonies should be invited to contribute towards their defence by the provision of oil fuel depots for the use of the fleet.

Enclosure 1 includes the provision recommended for the Colonies.

10   There are, however, many factors affecting the quantities recommended which may render the completion of the programme in its entirety unnecessary, and among these may be mentioned:

a)   The introduction of the internal combustion engine into His Majesty's ships.

b)   International agreements as to the limitation or reduction of armaments, whereby the estimate of fleets to be used in 1930 could be reduced.

c)   Increase of output (beyond the allowance already made) from the oilfields that are within British spheres of influence.

d)   Discovery of new sources of fuel supply in Imperial territory or where British control can be assured in time of war.

11   The situation will be reviewed annually with a view to taking advantage of any developments which may tend to modify the ultimate requirements, but for the present the Admiralty consider it essential to proceed at the rate of progress proposed in paragraph 18 in order to ensure the mobility of the fleet as soon as possible.

12   In addition to purely naval requirements, allowance must also be made for the fact that in time of war oil-burning merchant ships employed on essential national services will undoubtedly become dependent on supply from these naval oil depots.

13   The joint memorandum (Enclosure 3), presented to the Cabinet by the Admiralty and the Petroleum Executive (5th Feb-

ruary, 1920), points out the necessity for a commercial reserve of fuel, and the danger of limitation of fleet movements that would ensue if naval supplies were drawn upon to any great extent for commercial requirements in time of war. It must be remembered that the Navy will require all the available tanker tonnage in time of war, and this makes it all the more necessary that oil-burning merchant ships should be able to bunker without having recourse to tankers.

14   It is not suggested that it is desirable, in present circumstances, to take any overt step that might give encouragement to a more rapid transfer of British merchant ships from coal to oil fuel burning than will take place in the natural order of things. It is advisable, therefore, that the commercial reserve should take the form of an additional quantity added to the naval reserves at certain bases, but earmarked to meet the requirements of essential merchant shipping in time of war.

15   To deal successfully with the problem of oil supplies in war time, it is imperative that the control and distribution shall be in the hands of one central authority, and to meet the various contingencies that may arise the Admiralty desire to have complete freedom as to the manner in which various portions of the reserves shall be used, and the ports from which supply to ships, naval or commercial, shall be drawn.

The whole of the reserves, moreover, should be compressed of that quality of oil which can be burnt by His Majesty's ships.

16   The Admiralty estimate that over and above the naval reserves a million tons is necessary for the existing oil-burning merchant ships running under the British flag and likely to require supply in time of war. This amount should be located, as shown in Enclosure 4.

17   To sum up, the reserves required are:

	tons
Naval reserve in the United Kingdom (already authorised to be completed by 1930)	4,500,000
Naval reserved abroad	3,500,000
Margin for merchant ship requirements	1,000,000

It is anticipated that about 2 million tons reserve stock will have been provided by the 31st March, 1922, leaving 7 million tons to be provided.

This total is subject, however, to the remarks in paragraphs 10 and 11 as to possible modifications in requirements in the future.

18   The Admiralty recognise that in the present financial situation it will be necessary for the reserves to be provided gradually, and I therefore propose:

That, subject to paragraphs 10 and 11, the reserves shown above be approved in principle, to be completed in fourteen years, or earlier if financial conditions admit.

The amount to be added each year to be not less than 500,000 tons, and more when financial conditions admit: this annual addition to include any quantities for which the Dominions, India and the Colonies make themselves responsible.

Enclosure 1, Oil Fuel, Naval Storage Position Abroad (Identical to Enclosure 6 in Memorandum of 24 May 1921 in ADM 116/ 3102)

Enclosure 2, Oil Fuel, Requirements and Proposed Supply Arrangements

In the East

Class of Ship	No.	Estimated Annual Consumption Oil Fuel	Source or Base	Quantity	Equivalent at Singapore	Requiring Oilers
Battleships	12	720,000	Singapore	1,200,000	1,200,000	-
Battle Cruisers	8	480,000	Rangoon	400,000	468,000	3
Light Cruisers	26	624,000	Burma	100,000		
Flotilla Leaders	80	960,000	Persia	1,530,000	1,224,000	30
T.B.D.'s						
Submarines	30	36,000	Aden	100,000	123,000*	2
Various	10	240,000	Colombo	225,000		
Attendant oilers	58	370,000	Hong Kong	15,000	15,000	–
			Australia	400,000	400,000+	–
Add for repairs						7
TOTAL	224	3,430,000		3,970,000	3,430,000	42

In the West

Class of Ship	No.	Estimated Annual Consumption Oil Fuel	Source or Base	Quantity	Equivalent at Bermuda	Requiring Oilers
Battleships	22	1,320,000	UK	4,500,000	3,779,000	66
Battle Cruisers	7	420,000	Trinidad	72,000	72,000	–
Light Cruisers	28	672,000	Persia	720,000&	525,000	21

Flotilla Leaders	126	1,512,000	Halifax	100,000	100,000	–
T.B.D.'s						
Submarines	30	36,000	Bermuda	500,000	500,000	–
Various	18	648,000	Med	170,000%	136,000	3
Attendant oilers	59	504,000				
Add for repairs						18
TOTAL	290	5,112,000		6,062,000	5,112,000	108

* After allowing for quantities consumed by fleet on passage and requirements for secondary operations.
+ Assumes that Australia stock will be positioned so as to be approximately equivalent to Singapore in respect of distance from any advanced naval anchorage.
& All commercial requirements in the East to be met from naval reserves provided in that area.
# Assumes consumption on the spot.
% After allowing for requirements for secondary operations in the Mediterranean.

[176] *Stenographic notes of the 14th Meeting of Representatives of the United Kingdom, the Dominions and India*

4 July 1921

LORD LEE [First Lord of the Admiralty]   In the first place, I would remind the Conference that before the War our naval policy was determined by and based mainly upon German development of sea power. That was the immediate and the ever-growing menace with which we had to deal, and practically everything was determined by it. In consequence nearly the whole of our Naval strength was concentrated in Home waters, within easy reach of the North Sea. At that time the Naval strength of America or Japan hardly entered at all into our calculations, although as a matter of fact we maintained then stronger squadrons than we do now on both the North Atlantic and the China Stations. Today the situation is entirely changed. The German Fleet, which was then our great pre-occupation, is now either at the bottom of the sea or its remaining units dispersed amongst the victorious Allies, and at the present moment there is no Naval Power in Europe which need cause us a moment's uneasiness. Yet, in spite of that, our historic position as the greatest Sea Power; still more our undisputed command of the sea, are both challenged as never before. It is no longer a question of our maintaining as we used to do a two-power standard, that is to say, a Navy able to deal with any two other Powers, or even the later one-Power standard plus 60 per cent. which was, I think, the final authorised standard before the War. We have now, admittedly under the stress of financial circumstances, had to retreat from both those advanced positions, and we have publicly announced to the world, that, in present circumstances, we are prepared to content ourselves with what is called a one-power standard . . .

I do not want to suggest that our future needs are only for more ships; what is quite as vital, or almost equally so, in the present situation is that we should have properly distributed at strategic points throughout the Empire adequate storage and reserves of that oil fuel upon which now the mobility of modern Fleets absolutely depends . . . The other essential requirement, in addition to the oil storage, is the provision of suitable bases for the Fleet in the waters in which they might be called upon to operate, and particularly in the Pacific. The Home Government have therefore come to the decision that Singapore must be developed as the main base for the British Fleet in the Pacific, for strategic reasons which will be explained to the Conference by the First Sea Lord and which, I think, have been also set out in papers circulated to the Dominion Ministers . . .

Prime Minister, no doubt it may be said – and no doubt it will be said by many people – 'Why cannot we avoid this intolerable competition, with its consequent financial burdens, by some kind of negotiation and agreement with the principal Powers concerned,' and I have heard it said on the platform, 'Are we to prepare Navies and Armies to fight America, who is our nearest relations, or Japan, who is our Ally?' Well, as regards the desirability, if possible, of limiting armaments by mutual agreement, we are in entire agreement, and I have expressed that view in the strongest language quite recently. But we have gone much further than that. We, alone of the three greatest Naval Powers in the world today, have already given a practical expression to our belief in the reduction of armaments. While the others have increased, and are still increasing their armaments and their expenditure, we have been reducing ours . . .

EARL BEATTY [First Sea Lord] . . . One of the essential qualities of the Fleet must therefore be, again I repeat, its mobility or its power of rapid transfer to any threatened portion of the Empire. In the event of a war with Japan, it is all-important that the Fleet should be able to reach the Far East in the shortest possible time, and provision to enable them to do so must be made today. Also provision must be made to maintain them in bases fully equipped with fuel and repair facilities when they get there. Japan may be expected to strike quickly directly hostilities commence. Hongkong, our furthest position primarily, and Borneo and Singapore to a lesser degree, are exposed to attack, and their safety will depend upon the rapidity with which the Fleet can reach the Far East. On its arrival there the main object of our Fleet is the destruction of the Japanese Fleet, with a view to holding the sea communications. Japan depends on other countries for many essential commodities, and being an Island Power, these commodities have to arrive by the sea. Hongkong is essential to the conduct of offensive operations against Japan, and every endeavour must be made to put it in such a state of defence as to make it reasonably probable that it will be able to hold out until relieved by the Fleet. An element of doubt, however, with regard to the security of Hongkong on account of its proximity to Japanese territory precludes its development as our main naval base, and it is essential that the Fleet should be provided with a secure base further to the southward, and for this purpose Singapore possesses far greater facilities than any other possessions in the Islands (position explained on the map). Singapore stands at the western gateway of the Pacific. It flanks the route from Eastern Asia to Australasia and covers the main entrance to the Indian Ocean from the eastward . . . With Singapore firmly held there is no danger of a Japanese attack in force upon Australia or New Zealand. Even with a Fleet in home waters at the outbreak of hostilities, the limited time available to the enemy during which his sea communications would be safe would render it folly for him to attempt a sea attack, conveying military Forces across a distance of 4,000 and more miles of water; and the Japanese, being what they are, would not attempt to snatch an initial advantage at such great distance from their main base. Hongkong is obviously indicated as the most probable object of attack. Considerations of time and distance render an attack upon Singapore a difficult undertaking in the interval available before the Fleet arrives on the scene, and automatically affords immunity. This is on the assumption that its defences are made sufficiently formidable as advocated by both the Admiralty and the War Office. It cannot be too strongly emphasised that without Hongkong we are not in a position to bring reasonably effective economic pressure to bear upon Japan, and its safety is imperative to the conduct of offensive naval operations . . .

MR CHURCHILL [Secretary of State for the Colonies]   This

discussion is really a very formidable discussion and deals with the most secret matters of the British Empire, and I do think that not only should the most profound secrecy be observed about it, but I do not think there ought to be any circulation of the discussion. I think it would be sufficient if the stenographers kept a record at the Committee of Imperial Defence with Colonel Hankey [Secretary to C.I.D.] which anybody could look at to refresh their memory, but that there should only be one copy in existence. It is not merely the strategic facts which are to be discussed, but if a fragment of the discussion gets out it might imperil our relations with Japan or with the United States. When you talk over the case in a hypothetical way even a fragment of it is taken to mean that we are making plans to go to war with them, and it causes the strongest possible feeling. I urge that no record except the single copy should be available.

[177] *'Empire Naval Policy. Brief Summary of the Recommendations by the Admiralty,' copy, Admiralty*
11 July 1921

'Secret'
These may be divided under the headings

I.   Imperial.
II.   Individual.

I Imperial

Apart from the urgent necessity of replacing the obsolescent units of the Empire's Battle Fleet with up to date Capital Ships – without which the 'One Power Standard' cannot possibly be maintained – the most pressing requirements in which it is hoped to secure the co-operation of the Dominions, are the development of Singapore and the establishment of strategic reserves of oil fuel.

In the Admiralty opinion Australia, New Zealand and India should be asked to contribute a considerable portion of the expenditure necessary to develop Singapore as a fleet base, since they are intimately concerned therein. Great Britain is already involved in considerable expenditure in modernising Malta as a naval base and thus enabling the fleet to be sent there, which is all a part of the general strategic plan for the safety of the Empire in the East, as the fleet at Malta is considerably nearer Singapore than in home waters.

The establishment of oil fuel reserves touches all Dominions and India and detailed recommendations are contained in 'Empire Naval Policy and Co-operation' (Chapter II).

In regard to the recommendations for Canada after discussion with the Canadian authorities, the Admiralty wish to emphasise the importance of the oil fuel reserves on the Pacific Coast, which should take priority over those on the Atlantic side, as they will be required for squadrons operating in those waters in the event of a war in the Pacific. Further, in regard to the amounts to be provided, the Admiralty wish to amend their previous recommendations of 50,000 tons on the Pacific side and 100,000 on the Atlantic side to 75,000 tons on each side.

The Imperial needs mentioned above should take precedence of the individual requirements recommended below as far as commencing work is concerned.

II Individual

Each Dominion should maintain such naval forces as their resources admit.

These forces require adequate bases, docks, ordnance depots and reserves of stores and fuel.

The Dominions should arrange for the local protection of trade in their own waters, and for the storage of guns necessary for armed escort vessels and defensively armed merchantmen and the provision of trained personnel to form their gun crews.

These and other recommendations such as a Mobile Naval Defence Organisation, a Minesweeping Organisation and Hydrographic Services are all discussed in 'Empire Naval Policy and Co-operation' and the Admiralty are prepared to advise on the relative importance and urgency, in each individual case, to be attached to these measures when they are aware of the funds available.

In the case of South Africa a programme of co-operation has already been prepared.

[178] *Stenographic Notes of the 26th Meeting of Representatives of the United Kingdom, the Dominions and India*

19 July 1921

MR MASSEY [Prime Minister of New Zealand] This question is not so easy as it appears, Mr Churchill [Colonial Secretary]. I was impressed by an opinion expressed by Mr Meighen [Prime Minister of Canada] that Australia and New Zealand are more directly concerned – more interested, as a matter of fact – than the other Dominions in Naval Defence. That may be from the point of view that we are more liable to attack, and perhaps, would be more easily defeated than the other countries of the Empire, but there is another way of looking at it which appeals to me. I am not inclined to shirk the responsibility of the country I represent in the very slightest. It is just this, that if Australia, for instance, or New Zealand, or both of them happened to be attacked and taken possession of by a foreign power, the Empire is done, because it would in such case be incapable of defending its territory, and, from that point of view, I think every country of the Empire is just as much interested as we are ...

MR E.S. MONTAGU [Secretary of State for India] I have nothing much to say that I can hope will be palatable to the First Lord. The real fact of the matter is that India suffers from poverty equally with the home country and with the Dominions and I am afraid even to a more marked degree. I should like the conference to bear in mind three things. India's net military expenditure this year is 62 1/2 crores of rupees; for the sake of convenience we can take it as sixty and a quarter million sterling at the orthodox rate of exchange ... I do not see that we can undertake on behalf of our country any further expenditure, but there are things which I should like to say to the Conference. The first is in my opinion that the development of Singapore as a base is of great interest not only to Australia and to New Zealand but also to India; and the second is, I know that I speak on behalf of my colleagues when I say we want to see an Indian Fleet upon the water. I do not believe we get our value out of the existing Royal Indian Marine and I do think that our aspirations to nationhood within the Empire would be infinitely stronger if we had an Indian Navy manned and equipped for the defence of Indian Shores against invasion.

MR CHURCHILL How much is your Royal Indian Marine?

MR MONTAGU   I am afraid I cannot give you the figure at the moment; it is quite small.

MR CHURCHILL   Three or four millions sterling a year?

MR MONTAGU   I was going to say two, but I can get the figures in a few minutes. I find on referring to the Budget figures that the estimated cost for the current year is about three quarters of a million. I do not think it produces its value. There are one or two facts I should like to point out to the First Lord which are also important in considering India's position. The first is we do at present contribute a small sum, a hundred thousand or so towards the British Navy. I do not know whether there is a comparable contribution from the Dominions.

LORD LEE   From South Africa?

MR MONTAGU   In the second place we do share with the War Office the total cost of Aden which costs us something between three and four hundred thousands a year on the present footing. Aden is a link in the communications with the Dominions which is of importance not only to India but also to Australia and New Zealand. I had hoped we should get rid of Aden but we are not getting rid of the financial liability for it. I think if the Home countries and India are responsible for all the expenditure between the Home country and India then any contribution we may make for Singapore ought to be lessened from the point of view of the contributions we make west of India.

MR CHURCHILL   Hitherto you have kept up Indian interests in the Arabian Peninsula including Aden. Now the idea is you hand over Aden to the Colonial Office paying a certain sum – less than you now pay?

MR MONTAGU   I hope it will prove to be less.

MR CHURCHILL   That was my proposal. At any rate no worse.

MR MONTAGU   The point I was trying to make is Aden is of importance to the whole Empire and the expenditure should be between India and the Home Country.

MR CHURCHILL   It is a link in the chain.

MR MONTAGU   The only other observation I would make is that when the time comes for us, as a consequence of our reduction in military expenditure, or a consequence of a reduction in our liability due to an improved position of affairs, or because we have a surplus, we make a further contribution, it will be necessary for us to negotiate with the Admiralty as to the most appropriate form it will take. At the present moment the nucleus of the Indian Navy consists of a sloop the Admiralty were good enough to give us – the '*Flower*' – which we are using for training, two boats for Harbour defence and a quantity of boom defence gear. It is a question which is highly debatable whether it would be more valuable to the Admiralty that we should contribute to Singapore or whether we should further develop Bombay or some other Port or Harbour at present undeveloped within the Indian Empire, because naturally defence begins at home, before we go abroad. Speaking as to the wishes of India what we should like to do would be to train and man ships to take the seas for the defence of India, but as regards the financial position of the Indian Empire, I am afraid we are talking of the future and not of this financial year.

MR CHURCHILL   Are you contemplating, when your finances are in a better condition, having a certain number of ships, or building vessels which would be officered by British and Indian Officers?

MR MONTAGU   Yes.

MR CHURCHILL   And manned by British and Indian ratings mixed?

MR MONTAGU   What we should like to do is to copy the example of Australia and New Zealand and have a Navy manned by Indian Officers; that will not be achieved in my opinion until we have had the assistance of British Officers to help in the training.

LORD LEE   In the meantime you recognise that the question of Singapore is of immediate urgency and indeed takes prior place.

MR MONTAGU   That is what I want to know. Do you think it would take prior place over expenditure on Bombay?

LORD BEATTY   Singapore comes first . . .

MR MEIGHEN   If New Zealand, say, is torn by violence from the family, why of course the rest of us cannot expect long to hold together. Similarly if any portion is, if we fall to the rank where we cannot protect any part of the Empire, well we are not going to last much longer. Nevertheless it is a practical impossibility to distribute the [defence] burden without having regard to the immediate imminence of the peril to the respective parts. What I mean is this. Mr Massey comes to this Conference with public opinion in his country in a certain condition. Mr Hughes [Prime Minister of Australia] comes with public opinion in his country as to this subject in a certain condition. What are the facts that account for that condition? Well, they are only too easily discernible. Their people sit under a shadow and they know just what it is. Now talk for the rest of this year would never convince the people of Canada that they sit under and as near the same shadow; you cannot do it; they do not feel that they do. If you tell them 'You are resting under the Monroe Doctrine' they dispute it. It would be idle to contend that the fact that the Monroe Doctrine does not produce a certain restful influence, though I would never lean upon it myself. I tried to show, and I tried to convince those who are open to conviction, that you cannot carry a country into a large expenditure for defence unless under such circumstances of discernible peril, as lead them to a conviction that it has to be undertaken.

[179] *'Naval Defence of the Empire – Imperial Naval Policy',*
Summary Abstract of Proceedings of 1921 Imperial Conference

Extract of notes of a meeting of Representatives of the UK, the Dominions and India, held at Downing Street on August 5, 1921, at 10.30 a.m.

Resolution adopted: 'That, while recognising the necessity of co-operation among the various portions of the Empire to provide such Naval Defence as may prove to be essential for security, and, while holding that equality with the naval strength of any other Power is a minimum standard for that purpose, this Conference is of opinion that the method and expense of such co-operation are matters for the final determination of the several Parliaments concerned, and that any recommendations thereon should be deferred until after the coming Conference on Disarmament'.

Imperial Naval Defence   The representatives of the United Kingdom, the Dominions and India, at their meeting on July 4th, 1921, decided that the Naval experts should meet at the Admiralty under the Chairmanship of Lord Lee, with a view to putting forward to a meeting of the Prime Ministers such proposals as they may think fit on the question of Empire Naval Cooperation.

The Admiralty have already, in a paper submitted to the Standing Defence Sub-Committee, summarised their recommendations in regard to the Naval policy of the self-governing Dominions, and briefly the steps which they advocate are: a) The development and coordination of independent naval forces on the part of the Dominions and India; but at the same time it is pointed out that this is not considered the ideal policy, which would be 'a unified navy under a single command.' b) Oil stocks throughout the empire to ensure mobility c) Singapore base . . .

Although the centre of naval interest has passed from the North Sea and European waters to the Pacific and West Atlantic, it will still be necessary to keep the main British fleet in a central position whence it can be moved rapidly to the East or to the West as strategic necessity may dictate. It is further pointed out that the most suitable central position must, for the time being, be in home or European waters, in close proximity to our great naval dockyards, shipbuilding yards, etc . . .

It is suggested that . . . the opportunity should be taken to impress on the oversea Prime Ministers that His Majesty's Government can no longer afford to bear the whole cost of providing for the naval defence of the Empire; and that it should be seen in what way if any the oversea Prime Ministers are prepared to recommend to their respective Governments their sharing the burden.

From a strategic point of view it would naturally be best if the Oversea Governments were prepared to give a contribution in money or in kind to the main British Fleet, thus providing a unified Empire Fleet under one command, but as they are unlikely to do this they should be invited to bear their share by: a) Maintaining up to date ships in their own local waters, the personnel being trained on the same lines as that of the main fleet so that in time of war all the British naval forces could if desired be at once put under one command. b) Arranging for reserves of oil fuel to be provided as recommended by the Admiralty. c) Australia, New Zealand and India might possibly be prepared to share in the cost of developing Singapore as a Naval Base . . .

[Treasury Memo dated in ms 23/7/1921.]

If it is essential that money should be spent on new Naval construction, surely a plan can be found for securing a suitable contribution by the component parts of the British Empire which will be stable and effective, will be based on some reasonable criterion of capacity to pay or benefit received, which will not violate the ordinary canons of British public finance . . .

[180] *Minute sheet, initiated by Naval Personnel branch of Admiralty Secretary's Department*

15 September 1921

2   The history of the arrangements for calling up the R.N.R. men in the Colonies may be summed up as follows:

3   On 29/5/03 the Admiralty wrote to the Colonial Office (N.878) proposing, in order that R.N.R. men on board merchant ships trading with the Colonies might be promptly informed of mobilisation, that copies of the Special Admiralty Order (now R.V.5 3 C) should be supplied to Colonial Officers with instructions to sign, date and promulgate it on receipt of a preconcerted signal, and requesting, if the Colonial Office concurred, that a list might be furnished of the Officers to be supplied and the number of copies required for each.

4   The Colonial Office replied (11/6/03) that in their view the signing and issuing of the Order should be wholly in the hands of the Governor, and that two or three copies would be sufficient for each Colony, the necessary additional copies being printed locally as required.

5   150 copies were accordingly sent to Colonial Office on 8/8/03(N.9241) and despatched by that Department on 22/10/03 to the Governors concerned, with a request that on receipt from the Colonial Office of a preconcerted telegram it should be signed, dated, printed and posted to the post of the Colony and advertised in the newspapers. 100 copies were also sent to India Office for distribution in India.

6   Arrangements were proposed on N.10705/03 and N.13849/03 that the Newfoundland R.N.R. should be called out by the Commanding Officer of the drill ship, and not by the Governor of Newfoundland, the necessary copies of the Special Admiralty Order being printed in advance and placed in the custody of the Commanding Officer. The Newfoundland Government took the view, however, that the Order should emanate from the Governor, and the Colonial Office concurred. The Admiralty then proposed (N.6266 of 13/6/04) that in the case of Newfoundland, Malta, Australia and New Zealand, where local Naval Reserves existed, the Governor of the Colony on receipt of information from the Colonial Office that the Proclamation calling out the Naval Reserves had been issued, should issue a similar proclamation applying the Royal Proclamation to the local Reserves, and that

the Admiralty Order should then be issued by the Naval Officers responsible.

7   In Australia and New Zealand it was pointed out by the Colonial Office that an Order in Council was required to call out the Local R.N.R. in view of Article VII of the Naval Agreement with those Dominions. The Order in Council it was proposed should be obtained by the Colonial Office with the concurrence of the Dominion Government and notified to the Dominion Government by telegram, whereupon the Governor General would notify its issue by Proclamation, and the Naval Officers responsible would issue the Special Admiralty Order calling out the R.N.R. (Imperial and local) in the Colony. To save delay the Admiralty suggested (N. 9972 of 20/8/04) that on receipt of the 'warning telegram' the Governor-General should advise the Colonial Government of the state of emergency and obtain their consent to the calling out of the local Reserves, the necessary Order in Council then being obtained, and the Imperial and local reserves being called out together.

8   The Colonial Office suggested the desirability of a uniform practice in all Colonies. It was eventually agreed however, that the Admiralty Order should be issued by the Governor in Colonies where there were no Naval Officers to issue it, and by the local Naval Officers in other cases, viz: at Gibraltar, Hong Kong, Bermuda, Malta, Newfoundland, Australia and New Zealand. It was decided that 100 copies of the Order would be enough for the smaller and 250 for the larger Colonies (the necessity for signing and dating being dispensed with). (A.L. N.12555 of 29/10/04). The total number of copies required by the Colonial Office was 4,000 (letter of 8/11/04) in N. 13733/04).

9   This arrangement held good at any rate down to early in 1914 (see Paper A.W.S.M.D. 18/14), containing a proof copy of Addenda to Mobilisation Instructions, which continues the 1904 arrangement. In 1914, however, it was apparently decided that the Special Admiralty Order should not be issued in Australia or New Zealand. (See Instructions for Mobilisation 1920, para. 98). The reason is not known, but it was perhaps considered that the notices issued by the Commonwealth and New Zealand Governments to their own R.N.R. would be sufficient warning to any Imperial R.N.R. men in that part of the world. The papers on which the new edition of the Instructions for Mobilisation was prepared might throw some light on the matter.

10   Otherwise the position remains as in 1904: i.e. the Order is

issued by the Naval Authorities in Newfoundland, Malta, Gibraltar, Hong Kong and Bermuda, and by the Officer administering the government in other Colonies and Dominions, except Australia and New Zealand.

11 On the occasion of the re-print of the Form, the Colonial Office were asked (letters N. of 16/12/20 and 12/1/21) how many copies they required to distribute to the Governors of those Colonies and Dominions where the responsibility for it rested with the Colonial authorities. They appear to think that the number of copies required is a matter for the Admiralty. Their position is not understood, as the matter primarily affects the Colonial administrations; but in the absence of any expression of opinion from them, it is considered that the previous basis of distribution, viz: 100 copies to the smaller and 250 to the larger Colonies, might be adhered to.

12 The moment seems opportune, however, to put the whole procedure for warning Imperial R.N.R. men in colonial ports on a more consistent basis.

13 If (see however, para 15 below) the warning of Imperial R.N.R. men in Australia and New Zealand is held to be sufficiently provided for by the arrangements of the local Governments for calling up their own Reserves the same would appear to hold good of Canada which has a local 'Canadian Naval Reserve' as well as the R.N.C.V.R. Moreover, in view of the principle accepted in 1904 that the duty of issuing this notice should only devolve on the Colonial Governor where there was no Naval Officer competent to issue it, there seems to be no adequate reason why in South Africa the C-in-C should not discharge this duty instead of the Governor General. The same applies, mutatis mutandis, to Singapore, Wei-hai-wei, and any other Colonies or Protectorates if any where there is a S.N.O., Intelligence Officer, or the like.

14 This would eliminate the self-governing Dominions altogether from the list of places to which R.V.53 C has to be distributed through the Colonial Office, while the list of Crown Colonies etc., furnished by that Department in N.2138/21 attached, could be reduced so as to leave only those places where there was no Naval representation.

15 It is suggested, however, that R.V.53C should be distributed in all Dominions and Colonies without exception in order that the Admiralty may be able to collect as many of the Imperial R.N.R. as are required. The Australian and New Zealand R.N.R. were at one time raised under the Imperial Act of 1859 and were called

out in time of emergency by a Proclamation under that Act subject to the issue of an Order in Council in accordance with Article VII of the Naval Agreement of 1903. But these Forces are now raised under Dominion legislation which (1) empowers the Governor General of Australia to call out the Australian R.N.R. 'in time of war' by proclamation & (2) renders the New Zealand R.N.R. liable to Active Service in time of war and empowers the Governor General in time of emergency by Proclamation to place the whole of the New Zealand Naval Forces at the disposition of the Government of Great Britain.

The case is very similar with the Canadian R.N.R. The important point to notice is that in none of these three cases can the procedure for calling out the Dominion R.N.R. suffice for notifying men of the Imperial R.N.R. that they have been called out by Royal Proclamation under the Act of 1859.

16   It is further suggested that it would be an improvement if the arrangements were altered whereby since 1904 the Admiralty Order R.V. 53 C is issued by Naval Officers in places where such officers are stationed and by Colonial Officers in other cases. No doubt there will always be Naval Officers permanently located at places like Malta, but the list of ports at which there are Naval Officers available to issue R.V. 53 C is bound to require alteration from time to time whereas there must always be a Dominion or Colonial Officer administering the Government who might quite as well take on the duty. For simplicity's sake it is considered that the Dominion or Colonial Officers should assume the duty in all cases without exception.

17   The delay in dealing with this paper is regretted but was unavoidable. [signature indecipherable]

[181] *Minute by Captain Barry Domvile, Director of Plans Division, 'War in the Far East', for the Deputy Chief of Naval Staff [Vice-Admiral Sir Roger Keyes]*

3 November 1921

Should we ever find ourselves at war with Japan it would be the first time in the history of the British Navy that the Main Fleet would be permanently based thousands of miles from Home Waters.

2   In this eventuality we should be faced by conditions entirely different from those under which we fought in the late war. As it is the accepted policy that we must be prepared for a war with

Japan it is necessary to try and ascertain how far existing arrangements would fail to meet the requirements of a fleet working under such novel conditions, and to endeavour to reconstruct our war preparation schemes accordingly.

3   It is all the more necessary to commence this investigation now as many post-war schemes are in the melting pot, and as the tendency after every war is to prepare for a future war by too close a study of the mistakes of the previous war. Many of the lessons of the late war would be entirely inapplicable to a war with Japan.

4   The mobility of the Fleet is the chief consideration. The first step towards assuring this has been taken by the initiation of a scheme for the provision of oil fuel reserves in and on the route to the East.

5   Two other important decisions relating to measures that will eventually assist the Fleet to operate in Far Eastern waters have been made. These are (a) the creation of a base at Singapore and (b) the commencement of the M.N.B. [?Mobile Naval Base?] organisation.

6   But a good deal more remains to be done, though this does not mean that preparations involving the expenditure of money are necessary at the present juncture. What is entailed is the preparation of organisations that will make it possible to provide at short notice for the repair supply and general maintenance of the Main Fleet operating in the China Seas and Pacific, as well as the laying down of the sound principle governing the system of command in the East And of the Higher Direction of the Naval War.

Repair of the Fleet

7   Singapore will not possess a first-class Dockyard for many years to come, and even when it does it will not be able to meet the demands made on it by a large Fleet. Hong Kong if still in our hands would be of great assistance but even then the output of the two Dockyards would fall far below the needs of the Fleet.

8   It is only necessary to turn one's mind to the late war to confirm this assertion. Even with the many well equipped naval and commercial yards in the United Kingdom at our disposal it was a matter of the greatest difficulty to satisfy the calls made by the Fleet.

9   There are few commercial yards within reasonable distance of

Singapore and Hong Kong, and what there are are but poorly equipped. There are practically no Docks.

10 The only way in which it would appear possible to surmount this difficulty and ensure freedom of action to naval forces (operating in Eastern Waters would be to prepare) a comprehensive organisation which would make it possible in time of emergency to expand very largely certain convenient ports and to earmark in peace time such machinery equipment and personnel at home, in the Dominions and in the Colonies as would be necessary for their war equipment.

Supply

11 Two main points are involved.

(i) The question of the ultimate peace equipment as regards armament, naval and victualling stores at Singapore and Hong Kong.

(ii) (a) The replenishment during war of the above ports, drawing, to as great an extent as possible, on sources of supply contiguous to the main theatre of operations.

   (b) The supply of a Fleet operating at an distance from them, i.e. at advanced bases.

Of the above (i) has the first call on our attention as (ii) has already been practically considered in connection with the revised edition of the 'Handbook for M.F.A.'s' [Merchant Fleet Auxiliaries].

12 Under the heading 'General Maintenance' we must include the manning requirements of the Fleet. Here again arrangements suitable to the late war would be in many respects quite unsuitable both as regards initial mobilisation and the flow of drafts to the fleet during the course of the war. For instance, it would probably be found advisable to concentrate a large part of those Reservists who happened to be East of the Mediterranean at the outbreak of war at Singapore, and not to send them home. The desirability of providing in war a Naval Depot with the necessary training establishments in a healthy situation near Singapore is also a question for consideration.

Command and Staff at Singapore and the Higher Direction of the War

13   It has been decided in principle that the Admiral in Command in the East should have his Headquarters at Singapore and that his Headquarters Staff should gradually be formed there in peace time. A start has in fact been made in laying the foundations of such a Staff.

14   We are however only at the beginning of the problem. Upon whom would the command of the naval forces in Eastern waters fall? Is a Senior Flag Officer to be appointed as C. in C. at Singapore in peace time? If so he is hardly likely to be senior to the C. in C. of the Atlantic Fleet. A solution that presents itself to the mind is that C.N.S. [Chief of Naval Staff] should move to Singapore on the outbreak of war.

15   A very large naval and administrative Staff would be essential at Singapore in war as it would be impossible for officers naval or civil in Whitehall to keep in close touch with the operational and administrative requirements of the Fleet in the East. This being the case it is obvious that, to avoid overlapping and waste of effort, a relatively small staff at the Admiralty would suffice, and that the best way to provide a skilled and an adequate war staff at Singapore would be to transfer a large part of the Admiralty to Singapore.

16   If you approve, it is proposed to take up the consideration of these questions. A series of preliminary discussions will be necessary before detailed schemes can be worked out and embodied in War Plans.

[182] *Minute by Vice-Admiral Sir Henry F. Oliver, Second Sea Lord, concerning a draft of CID paper 131-C on 'Washington Conference in its Effect on Empire Naval Policy and Co-operation', and Charles C. Ballantyne's [Canadian Minister of Marine and Fisheries] proposal that there be a common naval officers list*

21 April 1922

2   Canada has at present more Officers that [*sic*] they can employ and we train the Canadian Midshipmen in our ships to the detriment of our own Midshipmen, as the seagoing ships are overcrowded with Midshipmen.

3   Australia has too many junior Officers and is discharging some, but in three or four years they will have more Senior Lieutenants than they can emply or promote, and if all Officers are on one list, considerable pressure will be exercised to promote Australians.
4   The question of leave and of giving Colonial Officers special consideration as regards appointments is always present and there have already been difficulties with South African Officers. Political jobbery in the Dominions is brought to bear on their Navies to a far greater extent than obtains here. The relatives want Officers to serve near their homes and they stir up members who approach Defence Ministers who stir up Agents General and our influential Politicians in this country. The result is that in order to keep the peace with Dominion officials it is very difficult to refuse, and special consideration can only be given to Colonial Officers to the detriment of our own.
5   The only satisfactory arrangement would be as laid down in para 23, the Admiralty being the central authority.
6   I concur that the matter might very well come up for discussion at the Imperial Conference, but am afraid that a satisfactory decision for all parties will be almost impossible.
7   The Australian young Officers compare very favourably with ours in the Sub-Lieutenants examinations and are generally more self reliant and wide awake. The Canadian Midshipmen are a rougher type than ours or the Australians, but as they do not yet do the same examinations as our Sub-Lieutenants, like the Australians do, it is difficult to compare them educationally or professionally.

[183] *Memorandum by the Admiralty for the Committee of Imperial Defence, 'Scales of Seaborne Attack against which Provision should be made at British Ports at Home and Abroad'*
June 1922

Appendix   Table showing Scales of Attack against which it is considered that provision should be made at British ports at home and abroad in a war with Japan, France, United States of America or Italy.

Key to Formulae in 'Scales of Attack' Column in the following Tables. (Derived from 'Classification of Forms of Attack' in C.B.–0911, 'Forms of Attack from the Sea.')

Group	Section	Sub-Section	
A	-	-	Long-range bombardment
	1	-	Bombardment by armoured ships of the line
		a	Of forts
		b	Of naval and commercial dock-yards, works and shipping.
	2		Bombardment by ships other than ships of the line
		a	Of forts
		b	Of naval and commercial dock-yards, works and shipping.
B	-	-	Attacks other than by long-range bombardment, including bombardment by direct observation.
	1	-	Bombardment by armoured ships
		a	Of forts
		b	Running past forts or harbour works, shipping, &ca, at moderate ranges.
		c	Of interior positions by day after having run past forts by night.
	2	-	Blocking attacks
		a	By unarmoured ships or merchant ships.
		b	By armoured ships.
		c	By submarines.
	3	-	Attack by raider
		a	By gunfire
		b	By minelaying
	4	-	Attack by
		a	Light cruisers
		b	Destroyers
		c	Coastal motor-boats
		d	Boom smashers
		e	Minelayers
	5		Gas attack
C	-	-	Attack by submarine

1		Attack by gunfire at moderate or close ranges.
2		Attack by either mine or torpedo on shipping in the approaches to a port.
3		Attack by torpedo on ships in harbour.

Note: Blocking attacks by submarines are dealt with under B2c.

D	-	-	Attack by aircraft
1	-		Attack by aircraft carrying bombs or torpedoes.
2	-		Attack by explosive motor-boats controlled by wireless from aircraft.
E	-	-	Bombardment by super-long-range guns of the (so-called) 'Bertha' type.
F	-	-	Attack by raiding parties landed from war or merchant vessels.

Alphabetical List of Ports [not copied]

Naval Ports. Ports marked* are of such outstanding importance to the navy that it is considered that the maximum scale should be taken into consideration in making defensive arrangements, irrespective of the political situation obtaining now.

Port	Possible War	Scale of Attack against which Provision should be made.
1 Portsmouth*	France	B1ab:B2ab:B3:B4:C23:F
	U.S.A.	B3ab:C
	Italy	B3ab:C
	Japan	B3ab
2 Plymouth*	France	B1ab:B2ab:B3:B4:C23:F
	U.S.A.	B3ab:C
	Italy	B3ab:C
	Japan	B3ab
3 Rosyth*	France	B1ab:B2ab:B3:B4:C:F
	U.S.A.	B3ab
	Italy	B3ab
	Japan	B3ab
4 Singapore*	Japan	A;B;C;F
	U.S.A.	B3ab:B4a:C:F

	France	B3ab:B4a:F
	Italy	Nil
5 Malta*	France	B1ab:B2ab:B3ab:B4acd:C:F
	Italy	B1ab:B2ab:B3ab:B4acd:C:F
	U.S.A.	B3ab
	Japan	B3ab
6 Gibraltar	France	B1ab:B2ab:B3ab:B4acd:C:F
	Italy	B1ab:B2ab:B3ab:B4acd:C:F
	U.S.A.	B3ab:C123
	Japan	B3ab
7 The Medway	France	B3b:B4acde:C23:F
(Sheerness)	U.S.A.	B3ab:C23
	Italy	B3ab
	Japan	B3ab
8 Portland	France	B3b:B4abcd:C2:F
	U.S.A.	B3ab
	Italy	Nil
	Japan	Nil
9 Milford Haven	France	B1ab:B3ab:C23
	U.S.A.	B3ab
	Italy	B3ab
	Japan	B3ab
10 Invergordon	France	B3ab:C12
	U.S.A.	Nil
	Italy	Nil
	Japan	Nil
11 Berehaven	France	B1ab:B3ab:B4ab:C23:F
	U.S.A.	B3ab:C23
	Japan	B3a
	Italy	B3a
12 Bermuda	U.S.A.	A:B1ab:B2:B3:B4abde:C12:F
	France	B3ab:C12
	Italy	B3a
	Japan	Nil
13 Esquimalt	U.S.A.	B4:C:F
	Japan	B3ab:B4a:C2:F
	France	Nil
	Italy	Nil
14 Simon's Bay	U.S.A.	B3ab
	France	B3ab:B4a:C12
	Italy	B3a
	Japan	B3ab:B4a:F

15 Bombay	Japan	B3ab:B4a:C12
	U.S.A.	B3ab
	France	B3ab
	Italy	Nil
16 Hong Kong		Status quo applies
17 Sydney	Japan	B3ab:B4a:C12
(N.S.W.)	U.S.A.	B3ab:B4a
	France	B3ab:B4a
	Italy	Nil
18 Halifax (NS)	U.S.A.	B1ab:B2ab:B4:C23
	France	B3ab:C12
	Italy	Nil
	Japan	Nil
19 Suez Canal	France	B1ab:B2ab:B4:C23
and approaches	Japan	B2a:B3ab
	U.S.A.	B2a
	Italy	B2a:B3ab:C23

Note: In view of the strategic importance of the
Suez Canal to the Empire, its defence merits special consideration.

[184] *'The Washington Conference and its effect upon Empire
Naval Policy and Co-operation', memorandum by the Naval
Staff for the Committee of Imperial Defence*
28 July 1922

1   Subsequent to the Imperial Conference of 1921, three Agreements have been entered into which directly affect the Naval
Defence of the Empire.

a)   The Treaty for the Limitation of Naval Armaments.
b)   The Quadruple Treaty.
c)   The Root Resolutions regarding Submarine Warfare.

2   The effect of each of the above Agreements will be dealt with
separately.

The Treaty for the Limitation of Naval Armaments

3   The Treaty for the Limitation of Naval Armaments mainly
affects three classes of vessels – Capital Ships, Aircraft Carriers,
and to a lesser degree, Light Cruisers.

4   The Admiralty have not recommended that ships of either of these first two classes should be included in Dominion building programmes for the next few years, although it is to be hoped that by the end of the ten years' holiday in Capital Ship construction, which forms a part of the Washington programme, the Dominions will be in a better position to assist in this most expensive but important portion of the burden of naval defence.

The only Capital Ship now included in Dominion Navies is H.M.A.S. *Australia*, which has to be scrapped under the Washington Agreement.

5   Agreement was reached that Light Cruisers should not exceed 10,000 tons displacement nor carry a gun with a calibre in excess of 8 inches.

No limitation was, however, placed on the number of vessels of this class which may be constructed, and in view of the world-wide commitments of the Empire, the provision of this class of vessel is still a matter of the first importance.

6   In classes of ships other than the above no agreement was reached, and reduction in expenditure on these classes of vessels can only be made to such a degree as the requirements of the naval defence of the Empire will permit.

The strategic position in the Western Pacific has been adversely affected for the following reasons:

a)   At Hong Kong neither the existing naval facilities for the repair and maintenance of Naval Forces nor the coast defences can be increased.

b)   The United States have agreed not to develop their Naval Bases to the westward of Hawaii. In effect this rules them out, so far as effective interference with Japan in the Western Basin of the Pacific Ocean is concerned, and leaves the British Empire the sole Power to counter, with Naval Forces, any aggressive tendencies on the part of Japan.

The Quadruple Treaty

8   The Agreement of the four Powers (British Empire, United States of America, France and Japan) to hold a joint conference should a controversy arise out of any Pacific question is all to the good, as the probability of war suddenly developing has undoubtedly been reduced.

9   It must be remembered, however, that the warning which the summoning of such a conference would give, would, in all prob-

ability, only be such as to permit of existing services being expanded, and would not permit of new services being organised. 10 In addition, the diplomatic situation would be strained, and it would probably be inexpedient, for fear of jeopardising the negotiations, to make any visible preparations . . .

Dominion Co-operation still a Vital Necessity

27 In the light of the above remarks it will be seen that the world situation still calls for the maintenance by the British Empire of strong naval forces, and, it being impossible for the Mother Country adequately to maintain these unaided, the need for co-operation on the part of the Dominions in Imperial Naval Defence is as paramount as it was at the time of the Imperial Conference.

Forms of Co-operation considered of more Immediate Importance

28 Since money contributed to an Imperial Navy do not commend themselves to the Dominion Governments the following are recommended:

1) During this period of financial stringency, maintenance, by the Dominions which have hitherto possessed navies, of a healthy nucleus of a seagoing squadron which, when times are better, can be rapidly expanded.
2) Assistance by all Dominions and certain colonies in the provision of world-wide oil fuel supplies.
3) Assistance by certain Dominions and India in the development of Singapore as a naval base.

29 As regards (1).   The Canadian Government, when explaining their decision as to the recent abolition of their seagoing fleet, stated that, had the latter been maintained, overhead charges would have been out of all proportion to the defence value obtained.
30 The Admiralty cannot help feeling that this bears out the view so often expressed by them that the ideal form of Dominion co-operation lies in a unified navy with quota of men and ships supplied by the Dominions and India.
31 The Admiralty can only express the hope that where Dominions find, as in the case of Canada, that the system of separate navies shows a poor return for the money expended, they may be induced to reconsider the position and act upon the foregoing principle.
32 As regards (2).   The mobility of the fleet, on which the

Dominions and colonies depend almost entirely for their safety, can only be assured by a world-wide system of oil-fuel reserves. A fraction only of these reserves exists, and it is of the most vital importance to the empire as a whole, as well as to individual Dominions, that this state of affairs should be altered as soon as possible.

If the Mother Country is to bear this burden unaided it will be many years before all parts of the Empire can rely on the certain protection of the Fleet.

33   As regards (3). Should Japan at any time declare war on the British Empire the position of Australia, New Zealand, India, and our Eastern colonies will be one of grave danger until the arrival of the Main Fleet in the East.

For a rapid concentration of the Fleet in the East it is essential that a secure base at Singapore can be counted on.

The development of Singapore must take many years, even if substantial assistance is provided by the Dominions. The matter is, therefore, one of great urgency.

Recommendations to Individual Dominions and Colonies

34   Australia – The assistance which Australia could most usefully give at the present time is as follows:

a)   Continue the maintenance of a sea-going Fleet.
b)   Assist in the development of Singapore.
c)   Commence the provision of oil-fuel reserves.

35   As regards (b). Should Australia decide to assist the Imperial Government in this direction, she might perhaps derive greater satisfaction in providing material manufactured in Australia than in making a financial contribution. Co-operation on such lines would be welcomed.

36   As regards (c). It will be remembered that the Admiralty recommended that Australia should ultimately possess reserves for the Main Fleet to the amount of 400,000 tons as well as one year's reserve for Local Defence craft and vessels employed on trade protection in Australasian waters.

37   New Zealand – It is recognised that New Zealand is already co-operating in naval defence by maintaining the nucleus of a sea-going Fleet, but should she find herself in a position to do so, any assistance, either financial or in material manufactured in New Zealand, towards the equipment of Singapore would be welcomed.

38   The installation of a reserve of 10,000 tons of oil-fuel in New Zealand, as recommended at the Imperial Conference, is also a matter of urgency.

This provision would assist to improve the general strategic position in the Pacific, and, in view of the fact that New Zealand will probably acquire oil-burning vessels at no very distant date, it would also ensure a large measure of freedom of action for her naval forces, which otherwise they would not enjoy.

39   Canada – The Admiralty must frankly confess to their great disappointment at Canada's decision to abolish her sea-going Squadron and to confine her naval endeavour to such secondary forms of co-operation as can be left to a force maintained on a reserve basis. For reasons already pointed out, these can be of no real assistance in the naval defence of the Empire.

40   Should Canada not see her way to reconsider her recent decision and maintain at least a healthy nucleus of a sea-going force, it is hoped that consideration may be given to the policy outlined in paragraph 28 of this Memorandum.

41   It is in any case hoped that Canada will find it possible to commence the provision of oil-fuel reserves in accordance with the recommendations made at the Imperial Conference. The total amount of reserves proposed for Canada was 150,000 tons.

42   India – India, depending so much for her safety and prosperity on the Fleet, and being particularly concerned in the ability of the Fleet to reach the East quickly, can most usefully subscribe to the naval defence of the Empire by:

a)   Assisting in the development of Singapore, either financially or by providing material manufactured in India.

b)   The provision of oil-fuel reserves at Aden and Rangoon, the amounts suggested being 48,000 tons and 216,000 tons respectively.

43   The question of any further assistance which India can give in the direction of maintaining sea-going forces is ultimately connected with the future of the R.I.M., which, it is understood, is now being reorganised. Until the Admiralty have been given an opportunity of examining the details of this scheme of reorganisation, they do not feel able to make any recommendations on this subject.

44   South Africa – The best form of immediate assistance would be to provide the reserve of 24,000 tons of oil-fuel at Simonstown, which was the subject of discussion at the Imperial Conference.

45   Newfoundland – It is hoped that Newfoundland will continue to train, and, if possible, increase the strength of her Royal Naval Reserve.

Dependencies, Protectorates and Colonies

46   Malay States and Straits Settlements – The Federated Malay States have already made a substantial contribution to the sea-going forces of the Empire. Should they be in a position again to assist in meeting Naval defence requirements, and, should the Straits Settlements States outside the Federation be also prepared to co-operate to this end, the following measures would be of the utmost value:

i)   Subsidise the commercial development of Penang and Port Swettenham in such a way that those places could afford the fleet good repair and docking facilities in war, the docks to be of such size as to take the largest men-of-war that can enter those ports.
ii)   Assist towards the development of the Naval Base in the Johore Strait.
iii)   Formation of a R.N.V.R. Force (See p. 21, 'Empire Naval Policy and Co-operation.')

47   Ceylon – (i) Subsidise the commercial development of Trinco-mali, with a view to affording docking and repair facilities for the fleet in war. (ii) Formation of a R.N.V.R. Force (See p. 21, 'Empire Naval Policy and Co-operation.')
48   Hong Kong and Fiji – These Colonies, being east of longitude 110° east, and therefore affected by the status quo clause of the Four-Power Naval Treaty, are not in a position to undertake local measures. The only form of co-operation open to them is therefore a monetary contribution to the fleet.
49   Other Colonies and Protectorates – It is not possible for the Admiralty to lay down the lines on which each of the remaining Colonies and Protectorates should be asked to co-operate. For such of them as can in any way contribute towards Imperial naval defence, the following methods are open:

i)   By the organisation of minesweeping or local defence craft for use in the event of war.
ii)   By the organisation of local branches of the Royal Naval Reserve or Royal Naval Volunteer Reserve for manning the vessels mentioned in (i); for the manning of advanced bases; to

assist in the Examination Service or in a Blockade Organisation, as well as for service in the fleet.

iii)   By the organisation of Volunteer Air Forces for co-operation with the Navy.

iv)   Direct financial assistance.

v)   The provision and maintenance of Naval bases, fuel depots, & ca, necessary for the fleet in the event of war.

50   The method of co-operation must necessarily depend upon the resources and geographical situation of the Colony in question, but the active co-operation of the colonies in assisting to support the Naval burden of the Empire is essential, and would be very warmly welcomed.

[185] *Jellicoe to the Secretary of State for the Colonies [Churchill], paraphrase*

3 August 1922

'Secret'

Following message from my Prime Minister [W.F. Massey]: Begins; In a few days I am making a statement in Parliament on naval policy, and if I could give in general terms the Admiralty views on Dominion Naval policy it would be of great assistance to me. The Admiralty views as set forth in secret paper E32 of 11th July, 1921, and E.4 of February, 1921, were before the conference last year. It is not my intention, of course, to quote those papers, but I should be glad to know whether there would be any objection to my stating that Admiralty at present considers that New Zealand can best give assistance in following ways: (1) co-operation in providing money for development of naval base and reserve supplies of fuel oil for ships in Pacific waters; (2) New Zealand maintaining such naval forces as her resources permit, the first requirements being submarines and light cruisers; (3) providing depots, docks, bases and reserves of fuel and stores in New Zealand for these vessels; (4) providing trained personnel and storage of guns for merchant ships and their escorts, and also necessary organisation for local protection of trade, (5) providing mine-sweeping organisation and mobile defence organisation for ports. Ends. Jellicoe

[186] *Telegram, Churchill to Jellicoe, paraphrase*
10 August 1922

'Secret and Urgent'

Your telegram 3rd August [No. 39008/S]; Admiralty view is that only adequate Naval forces can assure Naval Defence. They must be capable of offensive action and endowed with full freedom of action. This can only be attained by adequate base and fueling facilities. Admiralty consider that however perfect local Defence measures may be these cannot in war protect any particular portion of the Empire adequately, unless a fleet capable of controlling maritime communications is possessed by the Empire. Admiralty concur generally in methods of assistance which your Prime Minister outlines, but they suggest that he should lay special emphasis on following points, which are given in order of relative importance:

(1)   Maintenance of a healthy nucleus of seagoing squadron during period of financial stringency. When times are better, this can be expanded rapidly. Expansion to consist of ocean-going submarines and light cruisers.

(2)   Provision of oil reserves in New Zealand.

(3)   Financial contribution, or supply of material manufactured in New Zealand, in order to assist in equipping naval bases. Then point 3, 4 and 5 as in your telegram under reference. Lords Commissioners have no objection to communication of these views in proposed statement on Naval policy to New Zealand Parliament. Ends.

[187] *Minutes of the 26th Meeting of the Standing Defence Sub-Committee, the Committee of Imperial Defence*
30 November 1922

THE FIRST SEA LORD [Earl Beatty] stated that as this was the first meeting of the committee which he had attended since it had been re-established under the new Government, he wished to place on record a warning as regards the naval situation in the Pacific. He wished it to be clearly understood that since the Washington Conference the situation from the naval point of view had altered in the Pacific. The United States were now incapable of naval action in the Western Pacific, thus leaving the British Empire the sole Power to counter with naval forces any aggressive tendencies on the part of Japan. We exist in the Far East on sufferance of another Power. This had already been pointed out to the Committee of Imperial Defence, and a decision had been taken that the naval base at Singapore should be proceeded with. Without Singapore we should be swept out of the Western Pacific and have no means of countering a naval offensive by Japan. There was reason to believe that if the war had taken a definite turn against us, Japan would have thrown over the Allies and associated herself with Germany, and that even during the war Japanese agents were in touch with Indian agitators. Japan might fall to a similar temptation in the future, when, by encouraging a revolt in India and raising the banner of Asia for the Asiatics, it would be no exaggeration to say that Japan would be able to wrest from us our position in India. The British Navy, under existing conditions without a base at Singapore, would be able to do nothing to counter the activities of Japan in the East. The Committee of Imperial Defence had agreed to the construction of a naval base at Singapore.

It had been laid down by the Cabinet that a great war was unlikely to arise before 1929. So far as the provision of oil storage on the eastern route, and particularly at Singapore, was concerned, the date had been extended to 1931 owing to the financial situation. However, since this decision we had been within an ace of a declaration of war, the limits of which no man could have foretold. Information had lately been received that the United States and Japan were not reducing their naval forces in accordance with the Washington Treaty until the Treaty had been ratified by France and Italy. The situation was, therefore, entirely altered. We had scrapped our capital ships in accordance with the Treaty. We were unable to send a force to the Far East to compete with Japan without a base.

It was of importance that C.I.D. Paper No. 176-C should be sent to the Dominions without delay, as in this paper were set forth the reasons why the Dominions should assist in the development of naval bases in the Far East.

[188] *Report of the Indian Retrenchment Committee 1922–23:*
'The Inchcape Report'

Constitution of the Committee

The Right Hon'ble Lord Inchcape, G.C.M.G., K.C.S.I., K.C.I.E.
(Chairman)
Sir Thomas Sivewright Catto, Bart. C.I.E.
Mr Dadira Merwanjee Dalal, C.I.E.
Sir Rajendra Math Mookerjee, K.C.I.E., K.C.V.O.
The Hon'ble Sir Alexander Robertson Murray, Kt. C.B.E.
The Hon'ble Mr Purshotamdas Thakurdas, C.I.E., M.B.E.
Mr H.F. Howard, C.S.I., C.I.E. (Secretary)
Mr J. Milne (Attached Officer)

Terms of Reference

To make recommendations to the Government of India for effecting forthwith all possible reductions in the expenditure of the Central Government, having regard especially to the present financial position and outlook. In so far as questions of policy are involved in the expenditure under discussion, these will be left for the exclusive consideration of the Government, but it will be open to the Committee to review the expenditure and to indicate the economies which might be effected if particular policies were either adopted, abandoned or modified.

Royal Indian Marine

The estimate for 1922–23 compares with the revised estimate for 1921–22 and the actual expenditure in 1913–14 as follows:

	Expenditure in Indian Rs	in England £	Total Sterling converted at Rs 15 = £1
1913–14, Actual Expenditure	24,93,000	257,100	63,49,000
1921–22, Revised Estimate	44,97,000	392,900	1,03,91,000
1922–23, Budget Estimate	67,21,000	465,200	1,36,99,000

2   The functions of the Royal Indian Marine hitherto have been:

(1)   The conveyance of troops in 'Indian waters.'
(2)   The maintenance of station ships, the tending of lighthouses

in the Red Sea, the Persian Gulf and Burma, and the Marine Survey of India.

(3)   The maintenance of the Bombay Dockyard and of all military launches.

A list of vessels in the Royal Indian Marine with details of the cost of maintaining the more important vessels is given in Addenda 1-A and 1-B.

3   Troopships.   We are informed that it is proposed to dispose of the three troopships, Dufferin, Hardinge, and Northbrook. These vessels, which were originally intended to convey troops in 'Indian Waters,' have been engaged mainly on trooping service for His Majesty's Government from whom the expenditure incurred was recovered. The total cost of maintaining these vessels for 1922–23 was Rs. 34,93,869, excluding Rs. 8,17,485 for interest on capital at 5 per cent and depreciation charges at 4 per cent. The sale of the vessels will therefore effect an automatic reduction in the provision required for 1923–24 of Rs. 34,93,869 against which must be set off the loss of receipts from the hire of the vessels to His Majesty's Government, viz. Rs 3,18,000, and the expenditure which must be incurred for care and maintenance parties for each vessel until sold, which is estimated at a total of Rs. 90,000 annually. The net reduction in the provision required for 1923–24 compared with the current year is consequently Rs. 30,86,000.

4   Station Ships, etc.   We discussed with the Director, Royal Indian Marine the possibility of dispensing with some of the other Royal Indian Marine vessels now maintained and understand that he has suggested to the Government of India that the Lawrence should be fitted up to perform her own duties and those of the Nearchus. The annual cost of maintaining the Nearchus, which is borne on the Political Estimates, is Rs. 4,86,000, of which one-half is paid for by the British Government. The Lawrence is at present utilised as a despatch vessel in the Persian Gulf and we consider that her cost, also, i.e., Rs. 4,62,000, exclusive of interest on capital and depreciation, should be borne on the Political estimates.

It should also be possible to arrange for the tending of certain of the Burma lights by trawlers and for the Clive to combine the Andaman and Burma station-ship duties and thereby release the Minto for disposal. The cost of maintaining the Minto, exclusive of interest on capital and depreciation if Rs. 2,77,000 and the disposal of this vessel less the cost of maintaining the trawlers would therefore effect some saving and the disposal of this vessel.

The *Cornwallis*, which is employed as a station-ship at Aden, attends to the lights in the southern end of the Red Sea and, as we are informed that the Board of Trade is being asked to defray the cost of attending to these lights, a small saving in the Marine estimate should result therefrom. The Dalhousie is used as a receiving ship at Bombay at an annual cost of Rs. 1,30,000, but, in view of the decision to dispose of the 3 troopships and other Marine vessels, we consider that this vessel will no longer be required.

5 Sloop and Patrol Craft. The sloop *Elphinstone* and the two Patrol Craft boats, *Baluchi* and *Pathan* were presented to India by the Admiralty after the war. The sloop is used as a relief station ship to replace vessels as required during the period necessary for their annual refit or for repairs. It is intended to keep one of the Patrol Craft boats in commission for use as a training ship and the other Patrol Craft boat in reserve for use when the training ship is not available. The annual cost of maintaining the sloop is estimated at Rs. 3,21,000 and the two Patrol Craft boats at Rs. 1,50,000. Special provision for reconditioning these 3 vessels was made in the 1922–23 Budget, the total amount provided for their maintenance being Rs. 5,86,000. There should therefore be an automatic saving of Rs. 1,05,000 for 1923–24.

6 Trawlers. Of the nine trawlers now maintained, one has been leased to the Bombay Port Trust, who bear all working expenses and the cost of repairs, etc. We are informed that the second trawler has recently been hired to the Iraq Administration of a monthly hire of Rs. 2,000 and arrangements are being made to hire the third trawler to a private individual for experimental fishing. Two of the remaining six trawlers may possibly be used as training ships in addition to the Patrol Craft boat, one will be utilised as a light-ship in the Persian Gulf in lieu of the existing light-ship and it is proposed to convert another into a water barge. The remaining 3 trawlers are at present unemployed.

We consider that one Patrol Craft boat and at least four trawlers should be dispensed with immediately.

7 Yard Craft & Military Launches. The number of yard craft in Bombay is 10, which is excessive and should be reduced. We have reviewed the use made of the 24 military launches and it has been agreed that 6, or possibly 7, of these launches can be dispensed with and a saving of at least Rs. 60,000 per annum effected.

8 River Steamers. The two river steamers *Bhamo* and *Sladen* are being replaced by two vessels which the Government of Burma

has purchased from Iraq and a saving of Rs. 1,70,000 is anticipated for 1923–24.

9   The expenditure in India is subdivided under main heads as shown below and it will be convenient to deal with each item seriatim:

Head	1913–14 Accounts Rs.	1921–22 Revised Estimate Rs.	1922–23 Budget Estimate Rs.
General Supervision & Accounts	2,04,000	4,03,000	4,20,000
Marine Survey	58,000	96,000	1,08,000
Misc. Shore Estab.	27,000	24,000	73,000
Dockyards	7,16,000	18,66,000	17,27,000
Salaries & Allowances of Officers and Men Afloat	8,23,000	18,69,000	20,99,000
Victualing of Officers and Men Afloat	3,24,000	8,58,000	7,57,000
Marine Stores and Coal for R.I.M. Vessels	12,86,000	27,71,000	28,15,000
Purchase and Hire of Ships and Vessels	5,000	9,000	9,000
Miscellaneous	3,12,000	11,68,000	5,86,000
Pensions	81,000	1,00,000	1,50,000
Total Expenditure	38,36,000	91,64,000	87,44,000
Less Receipts in India	14,43,000	46,67,000	20,23,000
Total Net Expenditure	24,93,000	44,97,000	67,21,000

10   General Supervision & Accounts – Rs. 4,20,000. The large increase in the expenditure under this heading compared with 1913–14 is mainly due to the expansion of the Accounts section consequent upon the growth of Dockyard work. The cost of this section has risen from Rs. 88,000 in 1913–14 to Rs. 2,48,000 required for 1922–23, but the percentage cost of the accounting staff to the total expenditure dealt with has decreased from 1.08 per cent to .92 per cent, so that the additional staff employed was apparently fully justified. The work in the Dockyards has recently fallen off by about 15 per cent and a pro-rata reduction should therefore be immediately possible in the Accounts section and a

further reduction when our recommendations regarding the Dock-yards take effect.

We ascertained that, although roughly 25 per cent of the cost of Supervision and Accounts is incurred in connection with the Dockyards, it has not been the practice to make a full charge for these services in the case of works carried out for the Royal Navy, Local Governments or outside bodies. Recently, however, it was decided that an appropriate share of all indirect charges should be added to the cost of work executed in the Dockyard and recoveries to the extent of Rs. 2,00,000 are anticipated for the current year. In view of the proposed reductions in the activities of the Royal Indian Marine, we consider there will be no justifi-cation for retaining the post of Deputy Director and we recommend this appointment be abolished and that the house now being built for the Deputy Director in the Dockyard at a cost of over Rs. 1 lakh should be occupied by the Director of the Royal Indian Marine for whom a house is now rented, the provision made for 1922–23 being Rs. 12,000.

It has been represented to use that considerable difficulty has been experienced by the frequent changing of the Controller of Marine Accounts. We suggest that, if possible, an officer should be retained permanently in the appointment.

We recommend that the provision for General Supervision and Accounts in 1923–24 should be limited to Rs. 3,50,000 of which Rs. 2,00,000 should be recovered from the Royal Navy and other bodies, in connection with work carried out in the Dockyard.

11   Marine Survey – Rs. 1,08,000.   The expenditure included under this head represents only the pay of Scientific Officers, Surveyors and Assistant Surveyors and their office establishment. The total cost of maintenance of the two survey ships is estimated at Rs. 7,54,000 for 1922–23 (vide Addendum 1-B). We ascertained that the complements of the two marine survey vessels have been increased as follows:

	Gross Tonnage	1913–14 Complement	1922–23 Complement
Investigator	1,185	109	118
Palinurus	444	46	54

The additional ratings comprise 12 Signalmen, 2 Sick Berth attend-ants, 2 Engine drivers temporarily employed and 1 Bhandary. We discussed the necessity for employing these additional ratings with the Director, who stated that he hoped to be able to effect a reduction in the number of signalers and in the complements of all Marine vessels for 1923–24.

A considerable portion of the cost of marine survey has in recent years been recovered from Port Trusts, Local Governments and the Anglo-Persian Oil Company, but we ascertained that no charge has been included for interest on capital, depreciation and pension charges and we recommend that in future the charges made should cover the full cost involved.

Survey work is an essential function of the Royal Indian Marine and apart from minor economies resulting from a curtailment of the complements of the survey vessels we make no recommendation.

12 Miscellaneous Shore Establishments – Rs. 73,000. The expenditure on miscellaneous shore establishments is subdivided as follows:

	1913–14	1922–23
	Rs.	Rs.
Perim Lighthouse	9,000	10,000
Coal Depots in Persian Gulf	16,000	60,000
Miscellaneous	2,000	3,000
Total	27,000	73,000

We recommend that the cost of working the Perim Lighthouse should be recovered by light dues, saving Rs. 10,000.

The expenditure under 'Coal Depots in the Persian Gulf' is mainly in connection with India's liability for the supply, free of charge, of coal to subsidised vessels of the Royal Navy employed in the Persian Gulf. The large increase since 1913–14 is due to the retention of the Coal Depot at Henjam which was opened during the war to supply coal to Government transports plying between India and Mesopotamia. Prior to the war, coal was kept only at Bushire, Muscat and Basrah and we are of the opinion that, unless it can be shown that the retention of Henjam, which we understand is situated in a very advantageous position, can be justified financially, it should be dispensed with. We are informed however that, even if Henjam is retained, a saving of Rs. 32,000 will be effected for 1923–24 by the employment of an outside contractor.

We recommend that the total provision for miscellaneous share establishments in 1923–24 should be limited to Rs. 31,000, a reduction of Rs. 42,000.

13 Dockyards. Since 1913–14 the Kidderpore Dockyard has

been closed down as a Government concern, but Rs. 25,000 was provided in 1922–23 for a care and maintenance party to look after the buildings and machinery until they are taken over by the Calcutta Port Trust or otherwise disposed of. We understand that no provision will be required for 1923–24 and there will therefore be an automatic saving of Rs. 25,000.

The expenditure incurred at Dockyards chargeable to the Royal Indian Marine is subdivided as follows:

	1913–14 Rs.	1921–22 Rs.	1922–23 Rs.
a) Administrative & Supervising Staff	2,99,000	3,30,000	4,89,000
b) Wages of Artificers & Laborers	4,09,000	15,04,000	10,63,000
c) Miscellaneous Expenditure	8,000	32,000	1,75,000

Hitherto, no commercial accounts of the working of the Dockyards have been kept and it is not therefore possible to ascertain whether the Dockyards are being economically worked or otherwise. We are informed, however, that a new system of accounts of a commercial basis has been introduced for the current year.

a) Administrative and Supervising Staff. We understand that, apart from the general increase in salaries and wages, the increased expenditure compared with 1913–14 is due to the employment of additional staff consequent upon the increase of work at the Dockyard. We are informed that a number of foremen and inspectors have been obtained from the Admiralty Dockyards in England on a three years' contract with a view to increasing efficiency and that the additional supervision provided has already enabled a considerable reduction to be effected in the cost of labor.

Ultimately it is proposed that these men shall be replaced by Indian mechanics, when sufficiently experienced, to take charge, but no immediate reduction can be effected.

b) Wages of Artificers and Laborers. There will be a large saving in the cost of wages chargeable against the Marine Service in 1923–24, owing to the reduction in the number of vessels to be maintained, but we have already taken credit for these savings.

c) Miscellaneous Expenditure. The provision for 1922–23 included Rs. 1,15,000 for underground mains for electric power and, as this expenditure is non-recurring, a reduction of Rs. 1,15,000 may be anticipated for 1923–24.

When the three troopships, the *Hardinge*, the *Northbrook* and the *Dufferin* and the station-ship *Minto* have been disposed of and the number of launches reduced, there will only remain about 8 small vessels and 18 launches, apart from yard craft, to be docked and overhauled in the Bombay Dockyards in addition to the overhauls to be done to His Majesty's ships on the East India Station. There are several other dry docks in Bombay which could be utilised and a number of up-to-date marine workshops which could be made use of in the event of some vessel having to be repaired urgently. It is self-evident, therefore, that there is no occasion to maintain in the dockyard the present large establishment which is as follows:

	Europeans		Indians		Total	
	No.	Monthly cost Rs.	No.	Monthly cost Rs.	No.	Monthly cost Rs.
Permanent Establishment	54	43,120	128	17,370	182	60,490
Temporary Establishment	2	1,650	453	24,937	455	26,587
Artificers on daily rates of pay	14	1,272	3,694	1,41,150	3,708	1,42,422
Total	70	46,042	4,275	1,83,457	4,345	2,29,499

We are of the opinion that a drastic reduction should be made and only sufficient personnel maintained to provide for current work and to leave the dockyard capable of expansion in an emergency.

14   Salaries and Allowances of Officers and Men Afloat

1913–14, Actual Expenditure	Rs. 8,23,000
1921–22, Revised Estimate	Rs. 18,69,000
1922–23, Budget Estimate	Rs. 20,99,000

A reduction of Rs. 5,43,047 will automatically be effected under this heading in 1923–24 by the sale of the three troopships and further savings by the abolition of other vessels. Apart from these savings, the Director of the Royal Indian Marine anticipates that it will be possible to effect a reduction in the complements of marine vessels and we consider that a saving of additional Rs. 50,000 should be possible for 1923–24.

We are informed that it was represented by the Director of The Royal Indian Marine, when the budget for 1922–23 was being proposed, that the rates of pay and allowances for Indian ratings compared unfavourably with those paid by the Mercantile marine and that, in consequence, difficulty was being experienced in securing and retaining a suitable class of men. The pay of Mercantile marine ratings has, however, been reduced recently and a similar reduction should therefore be possible in the case of the Royal Indian Marine ratings.

15   Victualing of Officers and Men Afloat – Rs. 7,57,000. Apart from the saving which will accrue under this heading in 1923–24 by the reduction in the number of Marine vessels, some saving should be possible in the cost of rations, owing to the fall in prices which has taken place and we consider that an additional saving of at least Rs. 30,000 should be possible.

16   Marine Stores and Coal for Royal Indian Marine Vessels. The provision under this heading for 1922–23 included Rs. 2 lakhs for the payment of arrears claims. A saving of Rs. 85,000 is also anticipated on the cost of marine stores owing to the fall in prices. It is admitted that, until recently, the arrangements for the purchase, issue and control of stores were unsatisfactory and that there was a large accumulation of certain classes of oil and cordage. There was no properly trained store officer in charge or any system of verification of stocks and indents were based on the average of the previous three years without regard to the stocks on hand. We refer to this matter later in dealing with imported stores.

With regard to coal, we are informed that the budget provision was based upon the amount to be purchased, which was considerably less than the amount actually consumed, the difference representing a reduction in stocks. Full credit for the coal consumed by the troopships, etc., which it is proposed to dispense with, has been taken in estimating the savings in respect of these vessels and some allowance must therefore be made for depletion of stocks.

17   Miscellaneous – Rs. 5,86,000. The analysis of this expenditure is as follows:

	1913–14	1921–22	1922–23
	Rs.	Rs.	Rs.
Medical Stores	3,000	9,000	3,000
Piloting and Towing	32,000	74,000	54,000
Passage and	31,000	62,000	1,06,000
Conveyance			
Repairs to R.I.M. Vessels			
at Outstations	16,000	86,000	1,09,000
Miscellaneous	10,000	1,99,000	2,64,000
Freight	14,000	50,000	50,000
Total	1,06,000	4,80,000	5,86,000

We discussed in detail the expenditure under this heading with the Director of the Royal Indian Marine, who stated that the large increase in the expenditure on repairs to the Royal Indian Marine vessels at outstations was due to reconditioning the troopers after the war and that a considerable saving might be anticipated for 1923–24. The expenditure under 'Miscellaneous' had been largely overestimated and the actual expenditure in the first six months only amounted to Rs. 27,000. We recommend, therefore, that the total provision for miscellaneous expenditure be reduced from Rs. 5,86,000 to Rs. 3,00,000, a saving of Rs. 2,86,000.

18   Receipts in India:

1913–14, Actual Receipts	Rs. 13,43,000
1921–22, Revised Estimate	Rs. 46,67,000
1922–23, Budget Estimate	Rs. 20,23,000

The receipts under this head may be subdivided as follows:

	1913–14	1921–22	1923–24
Recoveries from H.M.G.	Rs.	Rs.	Rs.
and from other Depts. in	9,76,000	13,94,000	9,96,000
India in connection with			
work carried out by the			
dockyards, etc			
Hire of Vessels	2,46,000	27,74,000	3,18,000
Sales proceeds of stores	58,000	4,00,000	4,00,000
Miscellaneous	41,000	79,000	2,87,000
Recoveries on account of			
Indian service family	22,000	20,000	22,000
pensions			
Total	13,43,000	46,00,000	20,23,000

It was represented to us that it was difficult to frame any reasonable estimate of the work which the Admiralty require to be carried out at Bombay Dockyard. The class of ships stationed in the East Indies varies from time to time and, whereas the Admiralty furnished a statement of the ships which will refit in the ensuing year, it is not possible to estimate beforehand what repairs and alterations will be required. All expenditure incurred is recoverable and, as the staff employed at the dockyards is regulated to meet the current requirements, any reduction in receipts will be offset by decreased expenditure. The receipts from the hire of vessels relate to troopships and the loss of these receipts in 1923–24, which will result from the disposal of the three troopships, has already been taken into consideration. The miscellaneous receipts comprise recoveries from the supply of coal in the Persian Gulf to vessels of the Royal Navy other than the subsidised ships.

Expenditure in England

	£	Rs. (£1=Rs.15)
1913–14, Actual Expenditure	257,100	38,56,500
1921–22, Revised Estimate	392,900	58,93,500
1922–23, Budget Estimate	465,200	69,78,000

19 This expenditure is subdivided under main heads as shown below:

	1913–14	1921–22	1922–23
Contribution toward expenses of Imperial Navy	100,000	100,000	100,000
R.I.M. furlough pay	10,000	30,000	30,000
Stores for India	117,000	216,000	194,400
Coal supplied to subsidized vessels of the R.N. in the Persian Gulf	-	-	46,600
Purchase of vessels	-	-	51,000
Sundry items	1,700	3,000	2,500
Pensions	27,700	43,900	40,700
Total	257,100	392,900	465,200

The first two items in the above table are not susceptible of reduction, except insofar as the furlough pay will be reduced by the reduction in the number of officers employed in the Royal Indian Marine. With regard to stores we are informed that, owing to the existence of large stocks, the expenditure for the current year has been restricted to £72,500. Particulars of the stocks of principal commodities on hand on April 1, 1922 and the normal consumption per annum are given in Addendum II. We understand that until recently it has been the practice to maintain – years' reserve stock and that it has been decided that in future only 2 years' supply of imported stores should be kept in reserve. The total value of imported and Indian stores on hand on March 31, 1922 was Rs. 75,11,000 and of the issues during 1921–22, Rs. 39,52,000. We consider that the reserve should be restricted to one year's requirements and we recommend that the provision for 1923–24 be limited to £75,000, a reduction of £119,000.

Coal Supply to Subsidized Vessels of the Royal Navy and the Persian Gulf. Although £46,600 was provided under this heading for 1922–23, the expenditure is now estimated at only £18,000. No restriction exists as to the amount of coal to be supplied free to Royal Navy subsidised vessels, the arrangement being that ships enter the Persian Gulf with full bunkers at Admiralty expense and leave the Gulf with bunkers full at Indian Government expense. It appears, however, that the estimate for 1922–23 was framed on a basis far in excess of the actual consumption during recent years and also prewar years and we recommend that the estimate for 1923–24 be limited to £25,000, a reduction of £21,600.

Purchase of Vessels – £51,000. The expenditure under this heading was for the purchase of a new station ship for Aden in replacement of the Dalhousie which has become unserviceable. No similar provision will be required for 1923–24.

Sundry Items and Pensions. The expenditure under these headings does not call for any special comment and we make no recommendation.

Conclusions

Having reviewed the expenditure of the Royal Indian Marine we recommend that:

(1)   the Service be drastically curtailed and reorganised on the lines suggested;

(2)   the Dockyard be worked as a separate entity on a commercial basis;

(3)   the three troopships, Dufferin, Hardinge and Northbrook be laid up forthwith and placed under care and maintenance parties until sold;

(4)   only such vessels as are necessary for the essential duties of

the Royal Indian Marine and for use as training ships for Indians be retained;
(5)   the Dalhousie, Minto, Nearchus, one patrol boat and four trawlers be dispensed with immediately and the number of military launches and Bombay yard-craft be reduced;
(6)   the cost of maintaining the Lawrence be transferred to the Political estimates;
(7)   the recoveries from Port Trusts etc., for Marine Survey work include a charge for interest on capital, depreciation and pension allowances;
(8)   the budget estimate for 1923–24 be limited to Rs. 62 lakhs, a reduction of Rs. 75 lakhs, including Rs. 4,62,000 transferred to the Political Estimates, the actual saving thus being Rs. 70,38,000.

[189]   'Naval Expenditure, Great Britain (& Ireland to 1920–21) and the Dominions', Admiralty Statistics Department
12 March 1923

Great Britain (and Ireland to 1920–21)

Financial Year	Population	Naval Expenditure (net)o £	Per Capita. L.s.d
1901–02	41,538,211	30,981,315	–.14.11
1902–03	41,592,680	31,003,977	–.14.91/2
1903–04	42,246,591	35,709,477	–.16.103/4
1904–05	42,611,375	36,859,681	–.17.31/2
1905–06	42,980,788	33,151,841	–.15.51/4
1906–07	43,361,077	31,472,087	–.14.6
1907–08	43,737,834	31,251,156	–.14.31/4
1908–09	44,123,819	32,181,309	–.14.7
1909–10	44,519,454	35,734,015	–.16.01/2
1910–11	44,915,934	40,419,336	–.17.113/4
1911–12	45,324,425	42,414,257	–.18.81/2
1912–13	45,508,243	44,933,169	–.19.83/4
1913–14	45,713,370	48,732,621	1.1.33/4
1914–15	46,089,249	103,301,86	22.4.93/4
1915–16	43,735,300a	205,735,59	74.14.03/4
1916–17	43,059,100a	209,877,21	84.17.53/4
1917–18	42,279,100a	227,388,89	15.7.63/4
1918–19	42,041,700a	334,091,22	77.18.11
1919–20	46,082,077	154,084,04	43.6.101/4

Table—*contd.*

1920–21	46,673,396	92,505,290	1.19.51/2
1921–22	42,767,530b	75,986,141	1.15.61/4
1922–23	43,062,242c	64,883,700e	1.10.11/2
1923–24	43,358,940d	58,000,000e	1.6.9

o including war expenditure
a Civilian population only
b Great Britain only, excluding Northern Ireland (1,250,531 –
Census 1911) and Southern Ireland (3,139,688 – Census 1911)
where no Census was taken in 1921
c Great Britain only – Census 1921
d estimated population
e Navy Estimates

Australia Financial Year	Population f	Naval Expenditure (net)o £	Per Capita. £.s.d
1901–02	3,773,801g	178,819	–.–.111/4
1902–03	3,841,921	149,701	–.–.91/4
1903–04	3,910,041	240,091	–.1.21/2
1904–05	3,978,161	206,036	–.1.01/4
1905–06	4,046,281	252,016	–.1.23/4
1906–07	4,114,401	256,066	–.1.23/4
1907–08	4,182,521	510,205	–.2.5
1908–09	4,250,642	267,262	–.1.3
1909–10	4,318,763	329,739	–.1.61/4
1910–11	4,374,138	1,465,034	–.6.81/4
1911–12	4,455,005g	1,634,466	–.7.4
1912–13	4,733,359	1,660,416	–.7.–
1913–14	4,872,059	1,987,101	–.8.13/4
1914–15	4,940,952	6,821,091	1.7.71/4
1915–16	4,931,988	8,470,036	1.14.4
1916–17	4,875,325	6,641,249	1.7.23/4
1917–18	4,935,311	3,766,174	–.15.3
1918–19	5,030,479	9,435,658	1.5.7
1919–20	5,247,019	5,645,374	1.1.6
1920–21	5,367,000	3,658,589	–.13.71/2
1921–22	5,436,794g	3,167,736	–.11.73/4
1922–23	5,507,472h	2,713,409i	–.9.101/4
1923–24	5,579,069h		

o including War Expenditure
f excluding full-blood aboriginals
g Census
h estimated population
i Navy Estimates

New Zealand Financial Year	Population j	Naval Expenditure (net) £	Per Capita. £.s.d
1901–02	772,719k	x	x
1902–03	795,890	21,452	–.–.61/2
1903–04	819,062	21,523	–.–.61/4
1904–05	842,234	40,742	–.–.111/2
1905–06	865,406	42,280	–.–.113/4
1906–07	888,578k	40,000	–.–.103/4
1907–08	912,556	40,000	–.–.101/4
1908–09	936,534	40,000	–.–.101/4
1909–10	960,512	100,000	–.2.03/4
1910–11	984,490	100,000	–.1.11
1911–12	1,008,468k	100,000	–.1.113/4
1912–13	1,052,627	100,000	–.1.13/4
1913–14	1,084,662	50,000	–.–.11
1914–15	1,095,994	65,874	–.1.21/2
1915–16	1,102,794	26,184	–.–.53/4
1916–17	1,099,449k	31,775	–.–.7
1917–18	1,097,672	42,002	–.–.91/4
1918–19	1,108,373	31,952	–.–.7
1919–20	1,152,602	20,075	–.–.41/4
1920–21	1,186,986	260,075m	–.4.41/2
1921–22	1,218,913k	303,517n	–.4.113/4
1922–23	1,251,701 l	256,656n	–.4.1
1923–24	1,289,371 l		

War Expenditure probably excluded during period 1914–15 to 1919–20.

x information not available
j excluding aborigines
k Census
l estimated population
m Estimates
n Expenditure

Note: The above Naval Expenditure does not include the cost of H.M.S. *New Zealand*.

Canada

Financial Year	Population	Naval Expenditure (net) £	Per Capita. £.s.d
1901–02	5,371,315P	x	x
1902–03	5,532,000	24,000s	–.–.11/4
1903–04	5,673,000	26,000s	–.–.1
1904–05	5,825,000	26,000s	–.–.1
1905–06	5,992,000	44,000s	–.–.13/4
1906–07	6,171,000	44,000s	–.–.11/2
1907–08	6,302,000	90,000s	–.–.31/4
1908–09	6,491,000	99,000s	–.–.31/2
1909–10	6,695,000	107,100s	–.–.33/4
1910–11	6,917,000	656,300s	–.1.103/4
1911–12	7,206,643P	656,300s	–.1.93/4
1912–13	7,379,500	429,227	–.1.13/4
1913–14	7,725,000	412,788	–.1.03/4
1914–15	7,928,000	265,230	–.–.81/4
1915–16	8,140,000	237,536	–.–.7
1916–17	8,361,000	254,574	–.–.71/4
1917–18	8,593,000	220,077	–.–.6
1918–19	8,835,000	163,000	–.–.41/4
1919–20	9,057,000	240,417	–.–.61/4
1920–21	9,279,000	802,024t	–.1.81/2
1921–22	8,769,489P	810,097t	–.1.10
1922–23	9,006,265r	807,768t	–.1.91/2
1923–24	9,249,434r		

War Expenditure probably excluded from 1914–15 to 1919–20.

x information not available
P Census
r estimated population
s estimated expenditure
t Estimates

Note: Exchange calculated at $4.86 = £

South Africa

Financial Year	Population u	Naval Expenditure (net) £	Per Capita. £.s.d
1901–02	598,190Y	35,000v	–.1.2.
1902–03	617,230Y	47,000v	–.1.61/4
1903–04	636,880Y	88,128v	–.2.9

1904–05	657,217Yz	85,000v	–.2.7
1905–06	656,860Y	85,000v	–.2.7
1906–07	656,490Y	72,500v	–.2.21/2
1907–08	656,130Y	85,000v	–.2.7
1908–09	655,770Y	84,782v	–.2.7
1909–10	655,410Y	83,804v	–.2.61/2
1910–11	655,040Y	98,914v	–.3.–
1911–12	1,276,242az	85,000v	–.1.61/2
1912–13	1,305,217a	85,000v	–.1.31/4
1913–14	1,330,053	85,000w	–.1.31/4
1914–15	1,354,889	85,000w	–.1.3
1915–16	1,379,725	85,000w	–.1.2
1916–17	1,404,561	85,000w	–.1.21/2
1917–18	1,429,397	85,000w	–.1.21/2
1918–19	1,454,232	85,000w	–.1.2
1919–20	1,479,068	85,000w	–.1.13/4
1920–21	1,503,904	85,000w	–.1.11/2
1921–22	1,522,924z	94,545x	–.1.23/4
1922–23	1,695,792	60,690x	–.–.81/2
1923–24	1,565,550	71,949	–.–.11

u White population only (estimated for years other than of Census).

v Contributions only received from Cape of Good Hope & Natal.

w Contribution from Union of South Africa. This ceased after 1920–21

x Estimates

Y Population of Cape of Good Hope and Natal only

z Census

a Population of the Union of S. Africa. This was formed from the self-governing Colonies (Cape of Good Hope, Natal, Transvaal and Orange River Colony) on 31st May 1910, and although the amounts contributed in 1911–12 and 1912–13 are shewn in the Navy Estimates and Appropriation Accounts as coming from the Cape of Good Hope and Natal only, it is here assumed that they are derived from the Union.

India

Financial Year	Population aa £	Naval Expenditure (net) bb
1901–02	231,605,940	x
1902–03	232,800,000	354,843
1903–04	233,010,000	348,418
1904–05	235,230,000	416,998
1905–06	236,460,000	407,300
1906–07	237,680,000	514,193
1907–08	238,920,000	396,338
1908–09	240,170,000	351,509
1909–10	241,410,000	377,697
1910–11	242,670,000	354,080
1911–12	243,933,178	484,500
1912–13	246,716,323	392,259
1913–14	247,949,904	423,303
1914–15	249,189,654	265,255
1915–16	250,435,602	522,250
1916–17	251,687,780	328,653
1917–18	252,946,218	485,669
1918–19	254,210,949	721,874
1919–20	255,482,004	1,100,855
1920–21	256,759,414	597,715
1921–22	247,138,396	692,700cc
1922–23	247,520,000	913,266dd
1923–24	247,840,000	

uncertain whether War Expenditure is included.
x information not available.
aa approximate figures, excluding native states.
bb From 1900–01 to April 1920 accounts were kept on a basis of £1 = 15 Rupees, and since April 1920 at the rate of £1 = 10 Rupees. For comparison with the earlier periods the amounts since April 1920 have been recalculated at £1 + 15 Rupees.
cc Revised Estimate
dd Budget Estimate

*   *   *

II   Monetary Contributions &C. Towards Expenditure on the Royal Navy, Admiralty Statistics Department, 15 March 1923

Notes

(1)   The amounts shewn in this Statement are included in the Naval Expenditure given in Table I (except as regards Newfoundland).

(2)   The contributions shewn are the actual amounts paid according to the Navy Appropriation Accounts. These amounts are brought to account below in the Financial Years in which they were received.

(3)   The Dominions also make payments in respect of liability for Retired Pay of Officers and Pensions of Men lent from the Royal Navy.

(4)   Payments to the Admiralty for services during the recent War are not included.

| Financial Year | AUSTRALASIA | | | CANADA | NEW-FOUNDLAND |
	Australia	New Zealand £	Total £	£	Maintenance of a branch of the RNR
1901–02	75,361		75,361	–	
1902–03	63,839 3,270a	42,827*	109,936	–	–
1903–04	209,419	13,366	222,785	–	–
1904–05	85,812	40,000	125,812	–	–
1905–06	212,885	40,000	252,885	–	–
1906–07	200,000	40,000	240,000	–	5,750†
1907–08	200,000	40,000	240,000	–	2,965
1908–09	200,000	40,000	240,000	–	2,920
1909–10	200,000	100,000	300,000	–	2,995
1910–11	200,000 14,375b	100,000	314,375	–	2,980
1911–12	200,000 7,500b	100,000	307,500	–	2,865
1912–13	175,000c	100,000			
1913–14	7,500b		282,500	–	2,970
1914–15	41,600 7,500b	50,000	99,100	nil e	3,000
1915–16	– 5,625b	50,000	55,625	–	3,000

1916–17 –			
	1,875d	1,875 –	3,000
1917–18 –	–	–	
1918–19 –		–	–	6,000
1919–20 –		–	–	1,230
1920–21 –		–	–	–
1921–22 –		–	–	–
1922–23 –		–	–	–
1923–24 –		–	–	–

a Contribution towards Survey of coast of Queensland.
b Contribution towards Survey of N.W. Coast of Australia (1910–11 includes amount due for 1909–10).
c Upkeep of *Encounter* taken over temporarily by Australia 1/7/12.
d Balance outstanding from previous year.
* includes 1901–02
e payments are however made in respect liability for Retired Pay & Pension of personnel lent from the Royal Navy.
† two years

Australia Naval Defence Act – 1887 – A sum not exceeding £126,000 a year was to be contributed by the Australian Colonies and New Zealand towards the establishment and maintenance of a British Naval Force in Australasian waters.
Naval Agreement Act – 1903. – A sum not exceeding £240,000 a year was to be contributed (5/6ths by Australia and 1/6th by New Zealand for the maintenance of an Australian Squadron and the establishment of branches of the Royal Naval Reserve.
Naval Subsidy Act – 1908 – New Zealand to contribute a sum of £100,000 annually.

SOUTH AFRICA

Financial Year	Cape of Good Hope	Natal	Union of S. Africa	Total
	£	£	£	£
1901–02	30,000	5,000g	–	35,000
1902–03	40,000	7,000g	–	47,000
1903–04	62,500	25,628	–	88,128
1904–05	50,000	35,000	–	85,000
1905–06	50,000	35,000	–	85,000

1906–07	37,500	35,000	–	72,500
1907–08	50,000	35,000	–	85,000
1908–09	49,782	35,000	–	84,782
1909–10	48,804	35,000	–	83,804
1910–11	63,914f	35,000	–	98,914
1911–12	50,000	35,000	–	85,000
1912–13	50,000	35,000	–	85,000
1913–14			85,000	85,000
1914–15			85,000	85,000
1915–16			85,000	85,000
1916–17			85,000	85,000
1917–18			85,000	85,000
1918–19			85,000	85,000
1919–20			85,000	85,000
1920–21			85,000	85,000
1921–22				–
1922–23				–
1923–24				–

g in lieu of an annual gift of 12,000
tons of Coal for the use of H.M. Ships &c.
f   increase due to adjustments on previous years.

Naval Contribution Act – 1898. – Cape of Good Hope Legislature contributed £30,000 a year towards the annual expenditure of the Royal Navy.

India
1   Maintenance of H.M. Ships in Indian Waters
2   Manning Floating Defences in Indian Harbours
3   Indian Troop Service (on account of work performed by Admiralty).
4   Repayments for services rendered by H.M. Ships in the suppression of Arms traffic in the Persian Gulf.

Financial	(1) £	(2) £	(3) £	(4) £	TOTAL £
Year	100,000	61,600	3,400	–	165,000
1901–02	100,000	61,600	3,400	–	165,000
1902–03	100,000	–	3,400	–	103,400
1903–04	100,000	–	3,400	–	103,400
1904–05	96,111h	–	3,400	–	99,511
1905–06	100,000	–	3,400	–	103,400
1906–07	100,000	–	3,400	–	103,400
1907–08	100,000	–	3,400	–	103,400

1908–09	100,000	–	3,400	–	103,400
1909–10	100,000	–	3,400	–	103,400
1910–11	100,000	–	3,400	–	103,400
1911–12	100,000	–	3,400	108,500	211,900
1912–13	100,000	–	3,400	64,000	167,400
1913–14	100,000	–	3,400	59,000	163,400
1914–15	100,000	–	3,400	46,000	149,400
1915–16	100,000	–	3,400	9,000	112,400
1916–17	100,000	–	3,400	–	103,400
1917–18	100,000	–	3,400	–	103,400
1918–19	100,000	–	3,400	–	103,400
1919–20	100,000	–	3,400	–	103,400
1920–21	100,000	–	3,400	–	103,400
1921–22	100,000	–	–	–	100,000
1922–23	100,000	–	–	–	100,000
1923–24	100,000	–	–	–	100,000

x Navy Estimates.
h L100,000 less £3,889 paid on account of
Floating Defences of Indian Harbours, the agreement in respect
of which expired in 1903–04.

Note: Since 1869 India has paid contributions of varying amounts
to the Imperial Government in consideration of services performed by the Royal Navy. Under existing arrangements which
date from 1896–7 (now under revision) a subsidy of £100,000 a
year is paid for the upkeep of certain ships of the East India
Squadron, which may not be employed beyond prescribed limits,
except with the consent of the Government of India.

III   Other Contributions, Admiralty Statistics Department,
17 March 1923

Notes.
(1)   The cost of ships purchased from the Royal
Navy is excluded.
(2)   These amounts are additional to those in

Table III

NEW ZEALAND.	cost. £
1 Battleship *New Zealand* (Presented)	£1,795,166

Note. The cost of this ship is not included
under Naval Expenditure in Table I.

AUSTRALIA.
1 Battle Cruiser *Australia*   1,705,000
Under the agreement arrived at at the Imperial Conference,
1909, the following expenditure was also incurred:
3 Light Cruisers *Adelaide, Melbourne, Sydney* 1,400,000
6 Destroyers *Huon, Parramatta, Swan, Torrens, Warrengo,*
Yarra   653,000
2 Submarines A.E.1, A.E.2,   233,500
1 Submarine Depot Ship PLATYPUS   160,000
1 Oil Tank vessel   120,000
1 Oil Fuel storage vessel   75,766
4 Hulks   25,000
                                              Total   2,667,266
                                        Grand Total   4,372,266
In addition, the following amount had been
approved for new construction outside the original
fleet unit, and for increase in cost over estimate
(up to 1920–21 inclusive)   2,135,952
FEDERATED MALAY STATES.
1 Battleship *Malaya* (Presented)   2,647,000x

x   The Total Cost of *Malaya* was £3,026,579. The contribution
of the Federated Malay States was however limited to the estimate finally submitted to the Malay Government, the balance
of £208,215 being charged to Navy Estimates 1922–23.

[190] *Devonshire to the five Dominions, draft*
27 March 1923

My Lord/Sir, With reference to my Secret despatch Dominions
No. 431 of the 23rd of December, transmitting a copy of a secret
memorandum (No. 176C) on the subject of Empire Naval Policy
and Cooperation, I have the honour to request Your Excellency/
Your Royal Highness/you to inform your Ministers, with reference
to paragraph 33 of that memorandum, that His Majesty's Government have decided to start work upon the establishment of a naval
base at Singapore, on the understanding that no considerable
expenditure will be incurred during the financial year 1923–24 and
only a moderate expenditure during 1924–25. The provision made
in the estimates for the financial year 1923/24 is £160,000, and the
total estimate for the work is £11,000,000.

[191] *Admiralty memorandum, 'Admiralty Policy Vis a Vis the Dominions, Outstanding Points up to 1923', by Captain Dudley Pound, Director of Plans*

4 May 1923

PART I

Section II

1862.   Resolution of House of Commons that Colonies should assist in internal defence.

Up to 1887 the Colonies paid contributions to Imperial Navy and in some cases maintained their own local defence forces.

1887.   The increasing dislike of the Australians to pay for a Navy which they did not see, resulted in a 10 years naval agreement, whereby an Imperial Squadron should be based in Australian waters in return for the contribution.

1898.   South Africa commenced to contribute to British Navy.

1900.   Commonwealth of Australia formed.

1902.   At the Colonial Conference of this year, the then First Lord pointed out the vital importance to the British Empire of sea power and laid stress on the necessity for central control in war.

It was found that no general resolution could be passed representing the opinion of all the Dominions.

The Dominions agreed to increase their contribution, except in the case of Canada who wished to establish local forces, and refused to make any contribution in money. In 1905 Mr Deakin, for Australia, proposed that the naval agreement should be modified as it aroused no enthusiasm and little interest was taken in the Imperial Squadron.

Admiralty deprecated the idea of a local defence force or Australian Navy.

During this period the C.I.D. [Committee of Imperial Defence] were asked by the Government of Australia to prepare a general scheme for the defence of Commonwealth Ports.

In their report stress was laid on the points that sea superiority must be maintained and that our ships must be able to go anywhere and not be tied to local defence.

[Addendum]   Imperial General Staff formed – 1907.

The Imperial Conference 1909

In 1909 the Admiralty policy was again under consideration. The original Admiralty memorandum recommended contributions based proportionally on sea-borne trade. This view was not accepted by the Sub-Committee appointed to consider the Agenda for the 1909 Conference.

The Admiralty memorandum was then rewritten (Doms. No. 16. p. 29), acknowledging that in developing naval strength, considerations other than those of strategy must be taken into account.

In short, the Dominions had to be free to bear their share in the burden as they thought best; local sentiment had to be given free play.

[Addendum]   3   Fleet Units suggested for Pacific – 1909.

Mention was also made of the need for unity of command in war, and a common standard of administration both in peace and war.

At the Imperial Conference 1909, these proposals were discussed, and in general met the views of the Canadian and Australian representatives. New Zealand, heretofore favouring contributions to an Imperial Navy, hinted that they did not desire to be represented in the Pacific by an Australian Squadron and desired to maintain touch with the Imperial Navy. (Doms. No. 17 p. 15 et seq.)

Thus were separate Australian and Canadian Navies conceived.

Dominions and Co-operation in War

The attitude in 1909 of the Mother Country and Dominions towards the broad question of co-operation in war is well represented in the words of the then First Lord of the Admiralty.

'At the Defence Conference in 1909, the representatives of the Dominions claimed that any naval or military forces which they might maintain should not be obliged to co-operate in any war in which this country was involved, unless such co-operation was approved by the respective Dominion Governments, C.I.D. 70c)'.

As a corollary to this was the feeling that co-operation could not be really satisfactory unless the Dominions had some voice in the factors which might lead to war, namely in the foreign policy of the Empire.

As a means of providing this channel for co-operation it was suggested by New Zealand at the Imperial Conference 1911 that the Dominions should be represented on the C.I.D. when any matter of overseas defence was discussed. After much discussion this proposal and one which recommended the establishment of a council of defence in each Dominion was agreed upon (C.I.D. 94c). Up to 1913 no great steps had been taken to give effect to these resolutions (C.I.D. 101c). It was probably felt by the Dominions that representation on an advisory board whose composition is entirely at the disposal of the British Prime Minister would not be consistent with their semi-independent status.

It was felt rather that more frequent meetings of the Imperial Conference should be held. At the same time the right of the Dominions to be consulted in matters of foreign policy had been conceded.

Formation of Dominion Navies, 1909–1911

The new Admiralty policy was embodied in a memorandum by the C.D.C. (Sub-Committee of C.I.D.) in a paper of July 1910 (C.I.D. 62c – para. 5). After much correspondence the Admiralty and Dominion representatives (not Prime Ministers) met early in 1911 to give effect to the decisions of the 1909 Conference.

The result, contained in C.I.D. paper 89c, may be called the charter of the Australian and Canadian Navies and is to this day (1923) regarded in Australia as the basis for their naval organisations. The points touched on are:

1. The white ensign and Dominion Jack to be flown.
2. Each fleet to be exclusively administered by its own Government.
3. Discipline and training to be standardised. Personnel to be interchangeable.
4. Control. Detailed arrangements for control in peace were laid down and cannot well be summarised.

The right of the Dominions to control, in peace, their own ships was recognised, subject to stipulations resulting from the fact that the foreign relations of the Dominions were not in their hands, but in those of the Foreign Office.

It was agreed that 'in time of war when the Dominion fleets, in whole or part, have been placed under the control of the Imperial Governments the ships were to form an integral part of the Imperial Fleet and remain under the control of the Admiralty of the U.K., and liable to be sent anywhere during the war.'

Dominion views on Naval Agreement

There is little doubt that this agreement, in the minds of Australian and Canadian delegates, represented the only workable method of co-operating in the Naval Defence of the Empire (C.I.D. meetings 111, 112, 113).

Mr R.L. Borden, M.P., the Canadian Premier, in a speech delivered in the Canadian House of Commons of [12th] January 1910, said:

'It has been suggested that instead of the organisation of a Canadian Naval Force, there should be a system of annual contributions from this country to the Mother Country; and I am free to admit that, from the strategical point of view, I would be inclined to agree with the view of the Admiralty that this would [be] the best way for the great self-governering Dominions of the Empire to make their contributions. But, Sir, from a constitutional and political standpoint, I am opposed to it from many reasons. In the first place I do not believe it would endure. In the second place it would [be] a source of friction. It would become a bone of partisan contention. It would be the subject of criticism as to the character and the amount of the contribution in both parliaments. It would not be permanent or continuous. It would conduce, if anything could conduce, to severing the present connection between Canada and the Empire. Let us remember that the British Empire as it now exists is of recent constitution. We are apt to consider it as a very old empire. The present relations of the great self-governing Dominions to the Mother Country are of recent growth and have not yet received their full development.'

On December 13th, 1907, in the Commonwealth House of Representatives, Mr Deakin, Prime Minister, said:

'At the very outset of the recent Colonial Conference in London the Prime Minister of Great Britain met us with the frank avowal that the British Government preferred no claim for money in relation to Naval Defence, and went on to add the extremely pregnant statement that the control of naval defence and foreign affairs must always go together. If honourable members appreciate the force of that axiom, they will see that it implies much, both now and in the future. It implies, also with equal clearness, that when we do take a part in Naval defence, we shall be entitled to a share in the direction of foreign affairs ...

The Admiralty were not satisfied with the contribution made;

and a section of our people were not satisfied with the bargain from our side. Consequently, when in 1905, Admiral Fanshawe delivered several speeches, and one in particular in which he pointed out the insufficiency of our contribution, I took occasion in August of that year to write to the British Government a despatch challenging his contentions. I pointed out that there was nothing distinctively Australian in the Naval Squadron maintained in these waters, that our support to it had been given in default of better means of co-operation, and that, being in no sense specifically associated with us, it roused no patriotic feeling. No exception was taken to the existence of a naval agreement between the British Government and ourselves on the contrary, that was postulated. But exception was taken to the fact that our contribution was made in money, and only indirectly in men when they entered the squadron. In no other way were they connected with us or representative of us. My objection was that Australia's part in this agreement was simply to find a certain contribution in money, and my suggestions were that we ought to substitute some active co-operation for this cash payment.'

Change of Admiralty Policy, December 1911 – August 1914

Within a year of the Imperial Conference of 1911, the following changes took place.

Mr Borden succeeded Sir Wilfred Laurier as Premier of Canada.

The Board of Admiralty was reconstituted.

Mr Massey succeeded Sir Joseph Ward as Premier of New Zealand.

The German Navy Law was passed in 1912

This latter event increased the stringency in home waters, and accordingly, when Mr Borden and his ministers visited the United Kingdom in 1912, Canada was invited to contribute 3 Battleships in lieu of establishing the Canadian Navy as previously arranged.

Mr Borden agreed, and a bill to this effect passed the Canadian House of Commons, but party feeling ran high; eventually both the scheme for a contribution and that for a Canadian Navy were dropped.

Thus the attempt to return to the system of contribution was a failure in the case of Canada.

The representatives of both New Zealand and the Union of South Africa pronounced against contributions (C.I.D. 102c and

S.A. debates), the former on the grounds that contributions made little appeal to local sentiment and aroused no enthusiasm.

Colonel Allen, New Zealand Minister of Defence, visited England in 1913, and in correspondence with the First Lord of the Admiralty definitely stated that New Zealand desired to have her own Navy. (It is open to doubt whether his views accurately represented those of his Government at that time).

After lengthy discussion the Admiralty agreed to facilitate the growth of a local navy, and as a result of this visit, *Philomel* was sent to New Zealand to train local personnel, her Captain being also 'Naval Advisor to the New Zealand Government'. She was relieved by the *Chatham* after the war, and became a stationary depot and training ship at Auckland, the two ships being known as the New Zealand Division of the Royal Navy.

Australian Naval Policy remained unchanged
Co-operation between Admiralty and Dominion Navy Boards

After the establishment of Dominion Navies, it became necessary in 1912 for machinery to be devised to enable the Admiralty to communicate War Orders to Dominion Naval Boards, and to ensure that the training of Dominion Navies was co-ordinated with methods adopted in the Royal Navy, so as to make it possible for Dominion Squadrons to co-operate in war. The Admiralty wished to have direct communication, but the views of the Colonial Office and Dominion Governments prevailed, whereby Admiralty in peace could only communicate with the Naval Boards via the Colonial Office, except in certain routine matters. Direct communication was possible in war, once the Dominion Navy had been placed at the disposal of the Imperial Government. The same rules apply at the present day. (See section III following).

Admiralty Policy, 1914 onwards

During the war the general question of Dominion co-operation did not arise until the Imperial Conference of 1917, when Naval Defence was debated at some length in a discussion arising from a resolution that the Dominions should be accorded an adequate voice in foreign policy and foreign relations. (Doms. No. 62. pp 53 to 81). At the Imperial Conference 1917, by resolution the Admiralty were asked to prepare a scheme of Empire Naval Defence, immediately after the War. (Doms. No. 62. p. xii). Imperial Conference 1918. Admiralty Memorandum of May 17th

1918 was prepared accordingly. In framing the policy it was recognised that the future might bring a closer co-operation between component parts of the Empire on questions of Imperial policy.

Contributions were acknowledged to be unpopular. The difficulties the Dominions would experience in providing adequate separate Navies were mentioned.

Stress was laid on the need for central control in war, and some higher strategical direction in peace.

To fulfill these requirements the formation of an Imperial Navy, under an Imperial Naval authority both in peace and war, was suggested. Local Navy Boards working under and responsible to Dominion Parliaments were to co-operate with the Imperial Admiralty and administer in their respective spheres all matters of training, supply and maintenance.

Representation of the Dominions on the Imperial Navy Board was to be secured by the periodical attendance of Dominion Naval Ministers as members of the Imperial Board of Admiralty

In their absence they should be in direct communication with the First Lord of the Admiralty and be represented by him.

The necessity for a Naval staff, common to all Dominions, was indicated.

Results of 1918 Proposals

The Dominion Premiers, after discussion, remarked as follows:

'The single navy at all times under one authority was not considered practicable; the war has shewn that a Dominion Navy could co-operate as part of a United Navy under a central command established after the outbreak of war.

It was admitted that the development on a large scale of Dominion Navies might render necessary the establishment for war purposes of some supreme naval authority upon which the Dominions should be adequately represented.'

It is worthy of note that the necessity for co-ordinated planning and preparation in peace does not appear to have been recognised by the Dominion Premiers.

Lord Jellicoe's visit to the Dominions

The Dominion Premiers put forward a request at the 1918 Conference that a high Naval authority should visit the Dominions in order to advise the respective Governments how they could best take part in the naval defence of the Empire.

Lord Jellicoe in 1919 visited India, Australia, New Zealand and Canada.

He presented to each Government a detailed report of their naval requirements.

The following is a very brief summary.

He advocated the principle of Dominion Navies, and recommended the formation of a Pacific Fleet to which Great Britain and the Dominions bordering on the Western Pacific and Indian Oceans would provide units in approximate proportion to their populations and their sea-borne trade, the Pacific Fleet to be able to cope with [the] Japanese Navy, and to be placed in war under a higher command at Singapore. Capital ship repair ports to be provided at the S.E. and S.W. corners of Australia, with a Main Fleet war base at an island North of Australia. (He assumed that the Japanese objective would be the invasion of Australia).

In his report to the Canadian Government, Lord Jellicoe stated that Canada had two courses of action open, either to provide a squadron for her own safety consisting of three Light Cruisers, 10 Torpedo Craft and 8 Submarines, or, alternatively, to provide herself with a much larger Fleet and thereby assist in Empire Naval Defence. Halifax & Esquimalt to be developed as Light Cruiser bases. A site to be found for a future Capital ship base on the West Coast.

Imperial Conference 1921

As expressed in E.N.P.A.C. (C.I.D.E.4) the Admiralty policy was to foster, perhaps not wholeheartedly, the formation of Dominion Navies, manned from and based on the Dominions. (E.N.P.A.C.p.3. para. 17, p. 1. para 4.)

The system of Command and direction in war was discussed and the following summary of conclusions was given.

'This chapter may be summarised by saying that the following means of obtaining co-operation between British and Dominion Navies are recommended:

(a) The creation of an Imperial Council to consider questions of policy.

(b) The representation of the Dominions on the Admiralty Naval Staff, with a view to co-ordinating the higher strategy of Imperial Naval Defence.

(c) The appointment of Dominion Naval Officers to the Naval Staff College so that selected British and Dominion Officers may be given uniform staff training; also, the adoption of common operational and technical text-books.

(d) Dominion Officers to exercise complete disciplinary and administrative command over their own ships and squadrons, but to carry out the instructions of the Commander-in-Chief or Senior Naval Officer in all operational matters.'

The Admiralty proposals were discussed by the Conference at meetings of a very secret nature of which no printed report was made.

The Resolution made public recognised the necessity for co-operation in Naval Defence and that a one-power standard was necessary. It also stated the opinion that the methods and expense of such co-operation were matters for the several Parliaments concerned but that appropriate action should be deferred pending the results of the Washington Conference.

Post Washington

C.I.D. Paper 176-C, dated August 1922, gives the Post Washington Admiralty Policy, the words used in referring to Dominion Co-operation being:

'28 Since money contributions to an Imperial Navy do not commend themselves to the Dominion Governments the following are recommended: [see document 188 above] . . .

29   As regards (1). The Canadian Government,* when explaining their decisions as to the recent abolition of their sea-going fleet, stated that, had the latter been maintained, overhead charges would have been out of all proportion to the defence value obtained.

*This statement was actually made by the Director of the Canadian Naval Service and not by the Canadian Government . . .

1923.   1st Lord's Speech at Royal Academy Banquet, 5.5.23.

'But even more do we need the forward view when we consider the ultimate foundations of our naval power. The defence of an Empire, spread over all the seas of the earth, cannot be maintained indefinitely by one small island in the North Sea. It must depend on the co-operation of all the partner states of our Oceanic Commonwealth and upon the full development of the resources and man-power of that Commonwealth as a whole. It is to the younger navies of the Empire, based on a strong and vigorous national life of great and growing communities, linked with ours in the most intimate unity of tradition and thought and training, that we must look in the future, as well as to that old parent Navy, of imperishable past renown, but also of unquenchable youth and vigour, for which I have the honour to reply.'

PART I
Section III

The following summarises the present means of co-ordination

(1) Periodical Conferences.
   Examples: The Imperial Conferences.
             The Conference of Commanders-in-Chief at Penang in 1921.

These are most valuable forms of co-ordination; for example, as a result of the Penang Conference, co-ordinated War Orders were prepared for the China and Australian Squadrons. Conferences are, however, necessarily few and far between, and normally reliance must be placed on liaison officers or written communications.

(2) Liaison Officers.
   Example: Australian Liaison Officer in London.

The term 'Liaison' is a little unfortunate, as it is generally accepted to denote a means of communication between Allies, whereas Dominion Naval forces should be regarded not as Allies but as part of one Imperial Navy.

A liason officer is only too readily regarded as something outside the ordinary organisation; he is a supernumerary and not a 'working hand', and therefore he keeps in touch only by his own efforts or by the efforts of others and not as a matter of course. It cannot be denied that a liaison officer is regarded with a certain amount of suspicion, and any information is carefully scrutinised, and perhaps 'watered down' before it is placed at his disposal.

(3) Written Communications. All Admiralty printed publications are placed freely at the disposal of Dominion Navies. There are three channels for written communications between the Admiralty and Dominion Navies.

(a) Through the Colonial Office, to the Governor General of the Dominion, and through his Prime Minister to the local Navy Board.

(b) This is a very lengthy method, and incidentally cases have occurred where Admiralty documents intended for the guidance of Dominion Navy Boards have not reached them.

(b) [sic] Through the High Commissioner. This method is quicker.

(c) In certain routine matters, direct communication obtains between the Admiralty and Dominion Navy Boards.

PART I

Section IV

Three methods of Naval Co-operation have been tried

(A) Separate Dominion Navies, administered and strategically controlled in peace by Dominion Governments, but with unity of command in war vested in the British Admiralty subject to the discretion of the Dominion Government at the time.

Examples: The Royal Australian Navy and the Royal Canadian Navy.

(B) Dominion Divisions, administered locally, but strategically controlled by Great Britain in peace or war.

The ships being either lent by Great Britain or purchased by the Dominions are maintained and manned by them wholly or in part.

Examples: The New Zealand Division of the Royal Navy.

The Union of S.A. Naval Forces.

The Newfoundland Reserve (Personnel only) (now disbanded)

(C) One Navy for the Empire, the support of which is shared by the Dominions by direct contributions of money.

Examples: Australia prior to 1913.

New Zealand prior to 1915.

S. Africa prior to 1922.

India.

Gift of £1,697,206 by New Zealand for purchase of Battle Cruiser.

Gift of £2,847,000 by Malay States for part

purchase of a Battleship.

* [Footnote.]

The essential differences between the war organisation in New Zealand and that in Australia and Canada are:

a) New Zealand Naval Forces come under Admiralty control automatically on the outbreak of war, whereas Australian and

Canadian Forces are placed at Admiralty disposal if the Dominion Parliaments so decide.

b)   All officers for the New Zealand Division are lent from the Royal Navy, whereas Australia and Canada train their own.

PART II

Introduction

Before proceeding to consider in detail the future Admiralty policy towards Dominion Naval effort and the means by which this effort may be made effective, the main principles upon which success is thought to depend will be stated.

The late war has shown that wars are no longer waged by Navies and Armies alone, but by nations in arms. The success or failure of any measure of defence rests ultimately on the sanction of the people; in war the necessary public interest is always present but in peace it is difficult to arouse.

It is for this reason that Public Interest is placed first in the list of requirements to be fulfilled.

The General requirements of Empire Naval Policy are:

(1)   Public Interest.   If Public interest is lacking, reduced naval estimates will result. This can only be corrected by the education of public opinion to a true estimate of foreign affairs as they exist now and may exist in the future together with the resultant Naval needs both for the Empire and each Dominion.

(2)   Efficient use of Forces available.   The Washington agreement, relegating us to a One Power Standard, makes it necessary that the British Empire should make the best possible use of the Forces available.

This can only be done if:

(a)   War Plans are common to the whole Empire.

(b)   The Peace Dispositions are in accordance with the War Plans.

(c)   Unity of Command in War is insured.

(3)   Provision of Personnel and Uniform training to render cooperation effective.

(4)   Provision of material.

     i.e.   (a)   Ships   number, type, location, etc.

          (b)   Bases   being guided, for the Empire as a

(c) Fuel	whole, by the strength and disposition of the potential enemy or enemies.
(d) Munitions	Dependent on (a) and (b)
(e) Auxiliaries.	

The methods by which the above requirements can be met will be dealt with to a certain extent in the proposed Memoranda to the Dominions referred to hereafter.

Full details will be prepared for the Conference.

Admiralty Policy towards Dominion Co-operation

The idea of an Imperial Navy recruited and maintained by all parts of the Empire was rejected by the Dominions in 1918.

Putting this aside there appear to be three main courses that the Admiralty might advocate. All these Courses have been tried in the past:

(1) Contributions in money or kind to be paid by the Dominions towards the British Navy.

(2) Dominions to maintain a local division of the Royal Navy (e.g. New Zealand).

(3) Separate Dominion Navies (e.g. Australia and Canada) in conjunction with a system to ensure co-ordination and co-operation.

Under the first alternative, the Admiralty would have to advocate a contribution of money or its equivalent, at the same time advising those Dominions who are already committed to separate navies to reconsider their decision.

This method is undoubtedly the most favourable from the point of view of the British tax-payer of the present day. There is, however, no doubt that, except when a case of grave emergency (as in 1909) causes them to pocket their national pride, it is so repellent to the Dominions, that no Dominion Government would dare to suggest such a thing to its electors.

An example of this is the attempt made by Sir R. Borden in 1912 to persuade Canada to contribute to the British Navy. It is hardly too much to say that this incident was largely, if not wholly, responsible for the deplorable state of Canada's Navy at the present day.

The large self governing Dominions have risen slowly but surely from Colonies to Nations, and the latter status was firmly established during the war. They regard any attempt to guide their naval policy in the direction of tribute, as an attempt to lower the status which they have gained.

For these reasons it is considered totally impossible to adopt this policy.

The second alternative is that now adopted by New Zealand and on the surface appears to have advantages not possessed by the others.

It appears to satisfy local sentiment in providing a navy under local jurisdiction; at the same time the ships are at the immediate disposal of the Admiralty on outbreak of war. In practice, however, it is to be doubted whether the arrangement would persist if the Admiralty exercised its control in war in a manner which did not satisfy local feeling (e.g. Convoy of N.Z. troops at the beginning of the late war).

This being the case, it is not very different from the third alternative.

The third alternative stated in full is:

Local Dominion Navies locally administered in peace and war, controlled by the Dominions in peace with a system to ensure co-ordinated war plans; control by a central authority in war to be effective when the Dominion Government have placed the force at the disposal of the Home Government.

This policy appeals strongly to the national sentiment of the Dominions, and history shows that it is the only form of assistance agreeable to Canada and Australia.

Looking to the future, there is little doubt that national sentiment will increase rather than diminish. It is very necessary that Imperial and national aims should not clash.

This alternative has the apparent disadvantage that money is liable to be wasted through excessive overhead charges; this is offset by the probability that the Dominions will vote more if they retain control of the expenditure, and can afford more if most of the money is spent locally.

A difficulty arises in the case of the smaller Dominions at present unable to maintain a fleet of sufficient size to be regarded as an independent unit. Even in these cases, the national feeling against contributions may be just as strong as in the larger Dominions, but the difficulties do not appear insuperable.

It is as well to recall that the navies of Australia and Canada entered into a separate existence during the somewhat exceptional period just before the war. The conditions then and during the war are hardly the most favourable for the growth of a young service on a sound footing.

Policy recommended for adoption

For these reasons it is recommended that the policy of Dominion Navies should be adopted, but it is urged most strongly that, in doing so, the Admiralty should be prepared to encourage this Policy wholeheartedly and without reservation.

A half hearted recommendation may do more harm than good.

For example, the fact that the Admiralty has not in the past viewed Dominion Navies with any great favour, is well known in Australia, and has done the Royal Australian Navy incalculable harm. Not only have a small minority of the people been convinced that they are 'backing the wrong horse' but a section of the Press has always been openly hostile, and the Australian Branch of the Navy League has loyally supported the Admiralty view and sternly opposed the Local Navy.

The most unsatisfactory effect, however, is upon R.A.N. personnel, who feel that their position is a precarious one so long as the Admiralty holds such views.

Measures for carrying out the Policy recommended

Assuming that the principle of Dominion Navies is accepted it remains to consider the means necessary to ensure 'public interest' and the 'efficient use of the forces available' without which any policy must fail.

Public Interest

As has already been stated, it is of first importance to foster a keen public interest in Naval affairs.

All Dominions have reduced their Naval efforts considerably during the past two years.

This was of course to be expected, for the same reasons which have occasioned the reduction of the British Navy.

The Dominions, however, owing to their isolation and their intense interest in their own development are less well informed of current foreign Policy and have taken a very exaggerated view of the possible benefits which may arise out of the League of Nations and the Washington Conference. They are also inclined to think that the Navy is an obsolescent arm of defence.

The only way to make the Dominions take Naval defence seriously is to persuade them that it really is vital, not only to the

Empire as a whole but to each Dominion, that the people of Great Britain cannot bear its burden single-handed.

This cannot be done by means of Admiralty Memoranda, as these will not be read in the Dominions except by a few who are associated with the Navy.

It is doubtful whether any permanent good will result even if the visiting Dominion Politicians are fully persuaded of the necessity for an adequate Navy, as, however willing they may be personally, it is unlikely that they would be able successfully to adopt a strong Naval Policy in face of the existing Naval apathy in their Countries.

Various means of educating public opinion can be suggested such as:

The inauguration of a political and press campaign like that adopted in 1909 with favourable effect on the Dominions. One of the principle features of such a campaign would be the whole-hearted encouragement of locally administered Dominion Naval Forces. Cruises of British Squadrons, the interchange of Ships, Officers and men, and use of the Navy League are also methods which suggest themselves.

Development of Dominion Navies

In giving effect to a policy of separate Dominion Navies there are several important factors to be taken into consideration.

Development of Navies the primary consideration

If this is accepted the Dominions must preserve at every stage in the process a correct balance between expenditure on (a) Effective Units (Seagoing and Naval Defence) and (b) Bases, fuel supplies and establishments.

The imperial policy up to the present, which has been to ask the Dominion to develop bases and provide fuel supplies, etc. on a scale necessary for the use of the Imperial Navy, must be so regulated in practice that it does not lead to Dominion expenditure on such items out of all correct proportion. The result may be that Dominions will accumulate bricks and mortar instead of men and ships; their naval forces in effect may be starved.

Outside assistance necessary during development

Until a Dominion Navy reaches a certain size, it will be unable to maintain a healthy growth and efficiency without outside assistance in matters such as training in higher ranks and ratings, circulation of personnel and opportunities for ships to work together, etc. The method will differ in each case and must depend on local feeling. It must also be taken into account that it is the rate of expansion of the Dominion and its resources with which the development of the Navy must correspond. Thus the navy of South Africa, where the rate of expansion is slow, will be longer in the embryo stage than, and will require different treatment from that of Australia or Canada whose population and resources are likely to increase rapidly.

There is yet another class such as Newfoundland who can hardly be expected to pass beyond the first stage of development, i.e., the provision of personnel.

Process of Development

If a systematic process of development on these lines is followed, Dominion Navies will pass through four distinct phases which should therefore be the basis of Admiralty advice.

Phase 1.   Local Defence

The Dominions should provide and assume responsibility for their own local defence services. At the same time they should prepare for the next phase by training personnel. Canada and S. Africa are now in phase 1.

Phase 2.   Provision of Ships for oversea work

When sufficient personnel is available, Dominions should take steps to obtain ships. These may be either:

(a)   Lent by British Navy.
(b)   Purchased from British Navy.
(c)   Built specially for the Dominion.

In this phase training should be continued, the British Navy should lend personnel, and a proportion of the Dominion personnel should serve in British ships on the station so as to ensure circulation.

Dominion ships should also work with British ships and utilise our naval bases, etc.

New Zealand is not in this phase.

Phase 3.   Formation of Dominion Squadrons

As soon as the Dominions possess navies sufficiently developed, the control of the station should be handed over to them.

Their squadrons would be locally administered and controlled in peace, but interchange and co-operation with British Units should be arranged to prevent stagnation.

Australia is at this stage.

Phase 4.   Dominion Squadrons to form part of the Main Fleet.

In this final stage the Dominions should be capable of providing capital ship units for the main fleet in addition to squadrons for service on their own station.

Conclusion.

Should the foregoing proposals be approved it is suggested a Memorandum be sent to each Dominion before the Imperial Conference.

(a)   Advocating the principle of Dominion Navies.

(b)   Pointing out the various phases which have to be passed through and in which category those Dominions are now in.

(c)   General recommendations as to development of present phase or progress towards the next.

(d)   Reference to any special questions such as Singapore.

[192] *Minute 203 from Jan C. Smuts, Prime Minister of the Union of South Africa, to Connaught*

Capetown
28 March 1923

With reference to His Royal Highness the Governor-General's Minute No. 2/1832 (secret) of the 18th January last, forwarding copy of Secret Despatch, Dominions No. 431, dated 23rd December 1922, from the Right Honourable the Secretary of State for the Colonies covering a Secret Memorandum prepared for the Committee of Imperial Defence (No. 176-C) on the subject of Empire Naval Policy and Co-operation, MINISTERS have the honour to inform His Royal Highness as follows:

(1)   Paragraph 44, Secret Memorandum (No. 176-C)

Tanks for the storage of oil-fuel at Simonstown are being built and it is probable that the work will be completed during the coming financial year 1923–24. As soon as the tanks are erected, arrangements will be made for a supply of oil-fuel being obtained.

(2)   Summary of further steps taken by the Union Government in connection with Naval Services.

It may serve a useful purpose to state here shortly what further steps have been and are being taken by the Union Government in the matter of Naval Services generally.

(a) S.A. Naval Service.

The Survey Sloop and two Minesweeping Trawlers obtained by the Union Government from His Majesty's Government after the Imperial Conference in 1921, are being used as intended, viz:

(i) The role of the Survey Sloop is the hydrographic survey of South African waters under Admiralty direction and when not so employed, it will carry out Fishery Research work.

(ii) The role of the Trawlers is to visit the coast ports and train the personnel of the Minesweeping Section of the R.N.V.R. (S.A. Division) referred to below.

At present these Trawlers are engaged in training their Permanent Ships personnel.

(b) R.N.V.R. (S.A. Division)

This Division of the R.N.V.R. consists of three Sections, a General Service, a Minesweeping and a War Reserve Section.

(i) The General Section at present consists of five Companies comprising volunteers who have entered for training in the S.A. Division. The personnel is intended primarily to supplement H.M. Ships on the Africa Station in time of war by supplying tradesmen such as Fitters, Carpenters, etc. They will also be available for duty in connection with the Naval Control Service, e.g., as crews for any Auxiliary Cruisers that may be put into commission, or as gun crews for merchantmen armed for self protection. Further they will man the small craft necessary for harbour control and examination services and undertake whatever Naval Transportation duties may be required ashore.

These Companies embark annually for training on H.M. Ships.

(ii) The Minesweeping Section will consist of four Flotillas and will comprise the personnel to man Auxiliary Trawlers which in time of war will have assigned to them the task of maintaining the free navigation of the Union's sea trade route. At present three Flotillas have been authorised and sixteen vessels, after having been 'sighted', have been noted as suitable for Minesweeping duties.

(iii) The War Reserve Section, Arrangements are being made to organize a War Reserve Section to which will be posted all personnel who have completed their R.N.V.R. training and also all ex-members of the Royal Navy or Mercantile Marine resident in the Union and available in case of emergency to supplement the R.N.V.R. (S.A. Division).

(c) Provision of Store and Workshop Accommodation – East Dockyard Simonstown.

The Union Government has agreed to provide new Store and Workshop Accommodation in the East (or new) Dockyard, in order to make that Dockyard self-contained and to release accommodation which might be required for the South African Naval Services in the West (or old) Dockyard.

The Union Parliament has approved of the following Services, viz:

(i) Plumbers Shop and Galvanizing Shop	£3,000
(ii) Extension of Fitting Shop with Pattern Shop	£20,000
(iii) Electric Generating Station and Battery Room	£40,000
(iv) Machinery for (i), (ii) and (iii) above	£80,000
(v) Store for Guns and Gun Mountings	£19,000
TOTAL	£162,000

An amount of £25,000 was included in the expenditure from Loan funds voted by the Union Parliament for the year ending 31st March, 1923, in order that a beginning might be made with these works.

[193] *Minutes of the 33rd Meeting of CID Standing Defence Sub-Committee*

19 June 1923

'Secret'

1   The Committee had under consideration memoranda by the Naval Staff on Empire Naval Policy and Co-operation (C.I.D. Papers 194-C, 195-C, 196-C).

THE FIRST LORD OF THE ADMIRALTY [Leopold Amery] explained that these memoranda had been prepared in two parts. C.I.D. Paper 194-C (the first part) set out the general principles of naval co-operation. C.I.D. Papers 195-C and 196-C (the second part) set out specific details applicable to Australia and New Zealand respectively. These two latter papers contained material which it would be inadvisable to publish. He explained that further memoranda were being prepared giving the detailed questions applicable to South Africa, Canada and Newfoundland. He stated that when the question of naval co-operation by the Dominions had been examined prior to 1909 the Naval Staff had preferred a contribution to Admiralty funds by the Dominions rather than the building up of separate naval units by each Dominion. This view, however, had been unacceptable to the Dominions, and in 1909 the Admiralty had accepted the principle of separate naval units for each Dominion. One object of the present paper was to show that the Admiralty were now not merely accepting as a second best, but whole-heartedly endorsing the idea of separate Dominion naval units. The Naval Staff wished to encourage the Dominions to look upon their navies no longer as small local units but rather as seagoing navies capable of forming an integral part of the main British Fleet, and desired that the Dominion naval units should reach a stage when they would be interchangeable with units of the British Navy. The Admiralty paper provided for a progressive development of the Dominion Navies through four phases. Australia was the only Dominion which had so far reached phase 3. The immediate necessity was the building by the Dominions of light cruisers, submarines and other small craft.

THE SECRETARY OF STATE FOR WAR [the Earl of Derby] enquired whether, when phase 4 was reached, the Dominion capital ships would be included in the quota allowed to the British Empire under the Washington Treaty?

THE FIRST LORD OF THE ADMIRALTY stated that Dominion capital ships were included in the British quota, but the question of the Dominion capital ships could not arise until the time arrived, as laid down in the Washington Treaty, for replacing obsolete capital ships.

LORD SALISBURY [Chairman] enquired whether the Dominions would wish to receive the Admiralty proposals in quite such a definite form unless they had been previously consulted. He suggested that some form of covering note or preliminary paragraph should be inserted in order to make it clear to the Dominions that the papers had only been prepared in the form of a Staff paper for the sake of brevity, and not in order to lay down definite demands from them.

THE FIRST LORD OF THE ADMIRALTY stated that Lord Jellicoe had informed the Admiralty that a clear definition of the Admiralty views would be of great assistance to the Dominions. This view had also been confirmed from other sources. He pointed out that Papers on similar lines had been prepared for the Imperial Conference in 1921.

THE SECRETARY OF STATE FOR THE COLONIES [the Rt. Hon, the Duke of Devonshire] stated that he wished to raise the question of the procedure for the despatch of these Memoranda to the Dominions. He understood that the Admiralty suggested that C.I.D. Paper 194-C should be circulated generally to all the Dominions, whilst the other papers should be circulated only to the Dominion actively concerned. He pointed out that this procedure would be contrary to precedent, and he was of the opinion that the whole of the papers should be placed before all the Dominion Governments.

The Committee agreed:

(a) That the Secretary of State for the Colonies should be requested to forward, under cover of an explanatory despatch, if he deemed this desirable, a complete set of the Memoranda on Empire Naval Policy and Co-operation, prepared by the Naval Staff, to the Governments of the Dominions of Canada, the Commonwealth of Australia, the Union of South Africa and the Dominions of New Zealand and Newfoundland.

(b) That the Secretary of State for India should be requested to forward a similar set of papers to the Government of India.

(c) That the papers already completed should be forwarded at once to Australia and New Zealand without waiting for the papers dealing specifically with Canada, South Africa and Newfoundland, which are not yet completed. The papers for the three latter Dominions should not be sent until the parts which specifically concern each Dominion are respectively available.

[194] *'Imperial Conference, 1923. Proposed methods for obtaining naval staff co-ordination between the British and Dominion navies', minute by Captain L.H. Haggan, Director of Training and Staff Duties*

31 July 1923

In order to ensure that the development and disposition of the British and Dominion Navies in time of peace is, so far as practicable, consistent with probable war requirements, it is considered that Naval Staff Co-operation should be firmly established between the respective Admiralty and Navy Boards.

2 Present arrangements for co-operation. Apart from the Imperial Conference the present arrangements for co-operation between the Dominion Naval Authorities and the British Admiralty are as follows:

For Australia.

| These duties are combined at the present time. | Captain R.A.N. on the Staff of the High Commissioner. Liaison Officer on Admiralty Naval Staff. Officers of the R.A.N. attend War & Staff Courses at Greenwich. Meeting of Cs-in-C. (Penang). |

For Canada.

Officers of the R.C.N. attend the War Staff Course at Greenwich.

For New Zealand.

A Commodore R.N. is in command of the New Zealand Station.

For South Africa.

C. in C. Africa Station is also C.-in-C. South Africa Naval Forces.

3   It is considered that the arrangements enumerated in para. 2 above are most valuable, but that by themselves they are not sufficient to ensure that full use is being made of the knowledge and experience available. In consequence progress towards the object in view, namely to administer and organise the British and Dominion Navies to work as a whole in time of war, is retarded.

4   Main Requirements of a System of Co-operation.

A system of co-operation should include:

(a.)   Co-operation between the Home and Dominion Governments for the framing of policy.

(b.)   Co-operation between the Navies of the Empire in order to carry out the policy decided upon.

(c.)   The co-operation aimed at in (b) above, should be such that the machinery of co-operation requires only a minimum of adjustment to enable it to function as the supreme naval authority in time of war.

Before putting forward proposals to meet the requirements as stated in para. 4 above, it must be recognised that the Dominions will continue to retain in their own hands, not only the development and disposition of their naval forces in time of peace, but also the decision whether or not to place them at the disposal of the Admiralty in time of war.

5   In regard to 4(a) Co-operation between the Home and Dominion Governments for the framing of policy. This is provided

for by means of Imperial Conferences and to a certain extent by the committee of Imperial Defence.

The assembly of the Imperial Conferences is outside the scope of this paper, but it is desired to state that the more permanent the basis on which they are established, the sounder should be the policy adopted for the Naval Defence of the Empire.

Committee of Imperial Defence

It is not considered necessary that each Dominion should be permanently represented on this Committee. When, however, matters are being considered that bear on the policy of defence, in its broader aspects, it is considered that the Dominions should be invited to send representatives. It is realised that the number of members that form this Committee is not fixed, as it rests in the hands of the Prime Minister of Great Britain to appoint, from time to time, whom he may desire to sit. This is not considered to be a defective system in dealing with matters of Empire Defence as the whole aim is to assemble an impartial body who can advise on facts and opinions placed before them.

If this proposal is adopted the Dominion Governments will be in permanent touch with the Home Government on matters of Imperial Defence Policy. This should tend to assist in the deliberations at the Imperial Conferences.

6   In regard to 4 (b) Co-operation between the Navies of the Empire in order to carry out the policy decided upon at Imperial Conferences. A system, whereby the Naval Staff Admiralty on the one hand and the Naval Staffs of the Dominions on the other can communicate direct with one another, is a necessary step towards strengthening the present regularised means.

Communication permitted on all matters other than policy.

7   The method of communication should be as shown in the diagram above.

The advantage of setting up direct communication are considered to be as follows:

(1)   That on the outbreak of war the machinery for co-operation between the Naval Staffs would be already established and automatically be available for the functioning of the Supreme Authority.

(2)   In time of peace it would act as a permanent link in advising the Dominion Naval Staffs on the development and disposition of their respective Navies.

(3)   The activities of the Naval Staffs would be enhanced by possessing corresponding data.

(4)   It would greatly assist to secure that the Navies were modeled on similar lines so far as local conditions would admit.

(5)   It would lessen the risk of the individual Navies possessing larger units of a type than would be required when the Navies worked in combination, thus producing economy of expenditure and effort.

8   The following additional means of strengthening co-operation between the Navies is also advocated. That the Dominion Navies should each attach one or more officers to the Naval Staff and/or Technical Departments at the Admiralty. The Officers selected for this duty should be additional to the officers already employed in the Divisions and Departments. They should be given whole time duty, which can always be found for them.

It is considered that they should not be appointed as 'Liaison Officers', but nevertheless they would be available for advice on those local matters which are outside the scope of policy.

9   By present arrangement, all confidential books and Fleet Orders, confidential and otherwise are issued to the Dominion Naval Boards and it is not thought that any extension of this process is required. [signature]

Minute by Captain Dudley
Pound, Director of Plans, 7
September 1923

Concur with D.T.S.D. that some system of direct communication between the Naval Staff Admiralty and the Naval Staffs in the Dominions is necessary, and that Dominion Officers should be attached to Naval Staff and/or Technical Departments at the Admiralty.

2   As the principle of the Imperial General Staff, with sections in the Dominions, was accepted by Dominion Ministers in 1907, it is thought that the easiest way to obtain the desired co-operation would be for the Admiralty to suggest the formation of an Imperial Naval Staff analogous to the Imperial General Staff.

3   Whether the proposal in (2) above is approved or not, it is suggested that concrete proposals should be prepared by D.T.S.D. for consideration by the Board, with a view to their being laid before the Imperial Conference.

4   As such a paper would probably touch on the relations between the local Commander-in-Chief and a Dominion Naval Board, it is suggested that the question of the responsibility for the control of naval matters in Canadian waters should be dealt with at the same time. Papers on this subject (M.0783/23) are attached.

[manuscript minute]

Approved as D of P 7/9/23 par[agraph] 3, ACNS 13/9/23

Minute by Captain Hugh
Tweedie, Director of Training
and Staff Duties, 28 September
1923.

In accordance with A.C.N.S. [Assistant Chief of Naval Staff] minute of 13.9.23, the attached paper dealing with Empire Naval Co-operation is submitted for the consideration of the Board with a view to it being discussed at the Imperial Conference.

2   Summarised, this paper proposes that the general principles governing Naval Co-operation which were agreed to by certain of the Dominions as a result of the 1909 Imperial Conference (vide C.I.D. paper 89c copy attached) should be reaffirmed and brought up-to-date and that two new principles dealing with Staff Co-operation should be added.

3   Only minor alterations appear necessary in order that the principles laid down may be applied to all Dominions. As regards the new principles it is proposed that the principle of a Combined British and Dominion Naval Staff and common training and inter-changeability of Staff Officers should be accepted. Such an organisation would be analogous to the principle of an Imperial General Staff accepted at the 1907 Imperial Conference and it is

for the consideration of the conference as to whether its title should not be the Imperial Naval Staff.

4 It is considered that the acceptance of these principles would go far to ensuring that close co-operation between the British and Dominion Navies rendered so necessary as a result of the acceptance of the one power standard for the whole British Empire.

5 The question which has arisen with Canada concerning the spheres of control of the Canadian Government and the C.-in-C. N[orth] A[merica] and W[est] I[ndies] should be approached on the lines of para. 4(a) of C.I.D. paper 89c, referred to, namely by laying down a Canadian Naval Station. In any case the answer to the C.-in-C. N.A. and W.I.question in para.1 of his letter 265/339 of 15th June 1923 would appear to be 'neither.' In the event of Canadian Naval Forces being placed at the disposal of the Admiralty in time of war the responsibility for Naval matters in Canadian waters would appear to rest with the Admiralty. The preparation of plans should, therefore, be carried out by the Admiralty and the Canadian Naval Staff in co-operation. [signature]

[manuscript] Concur in draft Memorandum. If the Memorandum is approved it is submitted it should be brought before the Imperial Conference. Dudley Pound D of P, 16-x-23

[manuscript] Propose to approve [indecipherable signature] 17/10/23

[manuscript] It is important under present conditions to bring before the Imperial Conference the desirability of ensuring close Empire Naval Cooperation by the setting up of an Imperial Naval Staff- concur in DTSD's memorandum as amended RM [Oswyn R Murray?] 19/10/23

[195] *'PERSONNEL', memorandum on the question of training of personnel officers and men for the New Zealand division of the Navy, by Hotham*

24 November 1923

1 At the outset I wish to state my firm conviction that the efficiency of any Navy rests in the standard maintained by the officers, and that New Zealand must not be content with anything except the best officers.

2 At present, with the exception of a chaplain and three Warrant

officers, there are no officers in the New Zealand division, and all other requirements are met by officers lent from the Imperial Navy. The Admiralty are quite alive to the necessity of sending only the very best.

3   Sooner or later, New Zealand will have to provide officers, either for a special list, or preferably to be merged into the Imperial list, but with their depot in New Zealand.

4   With the experience of the Australian and Canadian Navy before us, on no account should a list of officers be started by entering retired or active officers from the Royal Navy or officers from the mercantile Marine, but New Zealand Naval Cadets should be sent to Jervis Bay or Dartmouth, for entry into the Imperial Navy, and all other officers should be lent from the Imperial Navy, to the New Zealand ships.

5   It is only by this means that the standard will be kept up because of competition, and occasional return to the larger fleets and proximity to the central and higher technical schools. Many other reasons dealt with at length in Lord Jellicoe's other reports could be mentioned. They are well known, and need not be stated again.

6   Concerning the men who are recruited and initially trained in New Zealand, the same principles should be adopted, and though New Zealand has already started its own personnel of seamen, stokers etc., on different rates of pay etc., to the Imperial Navy, I see no great difficulty in making the question of inter-change of men, into one of fusion, using the Australian Schools for the lower non-substantive ratings, and New Zealand as a depot.

7   We already have a large number of officers and men from Dominion Navies serving in Imperial Ships and Schools and the different rates of pay, though subjects of discussion are not subjects of discontent; though this is frequently put forward as the great objection to mixing of personnel.

8   The Australian Navy has gone too far to merge its personnel, into the Imperial Navy, but New Zealand has barely started, and before any further step is taken, I would most earnestly request that the proposals be given every sympathy and close examination.

[196] *'Empire Naval Policy and Co-operation', memorandum by the Naval Staff for the Committee of Imperial Defence*

February 1924

'Secret'

The question whether the present arrangements are adequate for keeping the various Dominion Governments fully advised as to the Naval needs of the Empire with a view to obtaining continuity of Naval Policy has already been brought to the notice of the Dominion representatives in the various C.I.D. papers forwarded to them on the subject of Empire Naval Defence.

2   Any arrangements which are come to must be such as to require only a minimum of adjustment to enable a supreme naval authority to function in time of war, and it must also be fully recognised that the Dominions, unless they have decided otherwise, will continue to retain in their own hands not only the development and disposition of their own forces in peace, but also the right of decision whether or not these forces are to be placed at the disposal of the Admiralty in time of war.

3   Arrangements for co-operation between the Naval Services and Forces of the Dominions of Canada and Australia and the British Navy were agreed to at the 1911 Imperial Conference, and arrangements have, as the Services of other Dominions developed, been made with those Dominions.

4   The arrangements were severely tested during the war and they proved a fairly satisfactory basis for co-operation, but, in view of the recent reductions of the British Fleet and the possible further extension of Dominion Fleets, it is considered that they now require some elaboration and amendment; the question of Staff co-operation in particular seems to require further consideration.

5   On paper, the Empire Navy of one-power Standard, as contemplated by the Imperial Conferences of 1921 and 1923, and made up of the various Dominion Navies, together with that of Great Britain, might well appear equal to that of another Power, but there is a grave danger that its collective efficiency might fall short of that required to safeguard the Empire unless suitable arrangements exist for all parts to be treated and prepared so that they can act as one in war. The problem which has to be faced by each Government is how best to organise, train and prepare its Navy in peace so that in war all the various portions of the combined Empire Navy may be able to act together for the

defence of the Empire with an efficiency and striking power at least equal to that of any foreign fleet.

6   In all matters affecting naval fighting efficiency a Government is advised by its Naval Board or Admiralty which, in giving advice, is in turn assisted by its Naval Staff. The Naval Staff at the Admiralty is an organization apart from the administrative and technical departments, and is solely charged with questions of Naval Policy, war plans and training for war. The British Naval Staff is at present composed of seven Divisions, attached to which are a total of some 56 Naval and Marine Officers. The Divisions, and a summary of their functions, are shown in the attached table:

Intelligence Division	Collection and distribution of intelligence (World wide).
Plans Division	Preparation of plans and policy.
Operations Division	Movements of ships, &ca.
Trade Division	Plans for protection of shipping.
Training and Staff Duties	Principle of training generally. Principles of Staff organisation and co-ordination between the Divisions of the Staff.
Gunnery Division	Use of weapons.
Torpedo Division	Use of weapons.

In addition, there are two small sections, an Air section and a Tactical section, to advise on these particular subjects.

7   The setting up of a Naval Staff comparable with the above by each Dominion in the early stages of its naval development would be hardly practicable, as overhead expenses would be out of all proportion to the size of the Navy maintained. At the same time it seems clear that a Dominion Navy will require its own Naval Staff, and it is most important that the separate staffs should not work in watertight compartments.

8   These difficulties can perhaps best be overcome by the appointment of Dominion Naval Officers to the Admiralty Naval Staff and British Naval Officers to the Dominion Naval Staffs, and by arranging for a free interchange of advice being maintained between the Naval Advisers at the Admiralty on the one hand and the Naval Advisers of the Dominions on the other. The

organisation would be analogous to the Imperial General Staff, agreed to in principle at the 1907 Imperial Conference.

9   It will be recognised, however, that in order that suitable officers may be available to fill the appointments on the several Naval Staffs, Staff training will be essential, and it seems desirable that this Staff Training should be uniform. The setting up of separate Naval Staff Colleges by the Dominions would be expensive, but as a Staff College has already been established in Great Britain it is suggested that, as each Dominion Navy develops, selected Dominion Officers should be trained at this College with a view to their subsequent employment on the British and Dominion Naval Staffs. The actual strength of the several Naval Staffs would be a matter to be settled by the Governments concerned.

10   To sum up, it is proposed that the following general principles should now be adopted:

(a)   In order to obtain continuity of Naval policy and to ensure that the best advice on questions concerning the naval defence of the Empire is always available for the British and Dominions Governments, the several Naval Staffs should be built up by uniform staff training and a free interchange of officers between the British and Dominion Naval Staffs.

(b)   A free interchange of advice should be established between the British and Dominion Naval Staffs.

11   It is suggested that this development of the Staff aspect of Naval organisation should be marked by adding the title of Chief of Naval Staff to those of the first Naval Members in Australia and New Zealand and of the Director of Naval Services in Canada, and that the Chief of Naval Staff, Admiralty, should, on the analogy of the Chief of the Imperial General Staff be described as the Chief of the Imperial Naval Staff.

[197] *Committee on Replacement of Fleet Units other than Capital Ships and Singapore, note, by Hankey, on Singapore*
22 February 1924

I attach herewith a full summary of the circumstances which led up to the decision to create a naval dockyard and base at Singapore. The summary is based on the records of the Cabinet, and Committee of Imperial Defence and its Sub-Committee the Oversea Defence Committee, the Imperial Conferences 1921 and

1923, the Washington Conference, and the British Empire Delegation at Washington.

The summary is of great length, but its salient features can be presented much more shortly.

Before the War, Singapore was of secondary importance, being classified as a war anchorage, coaling station and port of refuge for merchant shipping. Hong-Kong was our principal naval base, and it was round Hong-Kong that discussions on the subject of our naval policy in the Far East took place.

After the defeat of the Russian fleet at Tsushima (1905) our own Battle Fleet was withdrawn from China Seas. Thereby Hong-Kong was uncovered. In the event of a sudden outbreak of war with Japan it could not have been relieved for a month or more. Although the Government realised the risk and decided to increase the fixed defences and local naval defences (which take a long time to complete), they deliberately declined to lock up in Hong-Kong the large garrison required to enable the fortress to hold out, in the event of war with Japan, until relieved by the Fleet. In this respect the Government relied entirely on the Anglo-Japanese Alliance, intending to reinforce the Fleet if and when that Alliance came to an end. But, as mentioned above, they maintained the defences, as well as the coaling station and dockyard (on a reduced scale) which could easily have been expanded if circumstances had made it necessary to increase the Fleet.

After the War, when our position in the Far East was re-examined, many new factors had arisen. The Japanese fleet had grown enormously and had become the third largest fleet in the world. The end of the Anglo-Japanese Alliance was in sight. Considerations of economy barred any expectation that we could maintain permanently in the Far East a fleet large enough to dominate the Japanese fleet, as contemplated before the War in the event of a termination of the Anglo-Japanese Alliance. Still less could we afford to maintain in Hong-Kong a garrison large enough to hold the fortress, in the event of war with Japan, until it could be relieved by the fleet. Moreover, our ability to send a fleet to the East and maintain it there in a mobile condition had disappeared. During the War the transition of ships from coal-burners to oil-burners had become almost complete. There was but little oil on the eastern route, and the supplies in the Far East were insufficient to maintain the Fleet there on its arrival. There were not enough oil-tankers in the world to take the Fleet to the Far East and maintain it there in a mobile state in the absence of local reserves. Moreover, the advent of the submarine had necessitated the 'bulging' of capital ships, with the result that they had outgrown the dry docks in the Far East.

In these circumstances the conclusion was reached, after exhaustive examination by the Committee of Imperial Defence and its Sub-Committee the Oversea Defence committee (the Treasury representative on the latter Committee dissenting), that Hong-Kong was too exposed to continue to be our principal naval base in the Far East. The only harbour possessing all the necessary qualifications of a great naval base – size, anchorage, secure entrance, depth, etc., was found to be at Singapore. (Not in the Defended war anchorage south of Singapore, but at a new site in the narrow strait between Singapore island and the mainland.) Moreover, Singapore was admirably situated as the base for a fleet charged with the protection of our vast Imperial interests in the Indian Ocean and in the South Pacific – a purpose for which it had been utilised during the early part of the War, before the *Emden* and *Konigsberg* and Von Spee's squadron had been disposed of.

Early in 1921, therefore, a policy was adopted in principle of creating a great naval base at Singapore with a chain of fuelling stations between that port and the Mother Country.

Shortly after this decision was taken, the Imperial Conference of 1921 assembled. The whole of the facts were communicated verbally, as well as in written Memoranda, to the representatives of the Dominions and India. Not a voice was raised in the conference against the proposal, but no formal decision was recorded, for all naval questions in the Far East were felt to depend on the results of the Washington Conference, which had come on the horizon during the Imperial conference. Nevertheless, it was made quite clear at the Imperial conference that the Singapore project was independent of the Anglo-Japanese Alliance, and this was repeated after the announcement in Parliament that the Imperial Conference was aiming at some understanding in the Pacific with the United States and Japan to replace the Anglo-Japanese Alliance.

A month or two later the British delegates at the Washington Conference were instructed, inter alia, not to enter into any Agreement 'which will in any way interfere with the development of Singapore as a British naval base, since such development is purely defensive and is already overdue.' The records of the British Empire Delegation at the Washington Conference show that representatives of the Dominions were fully cognizant of these instructions. The accompanying Memorandum disposes of the idea that any deception was practised at Washington in order to exclude Singapore from the area within which the status quo as regards fortifications and naval bases was to be retained.

After the Washington Conference the Admiralty re-examined the strategic position in the Pacific, and their Memorandum was approved by the Committee of Imperial Defence for despatch to the Dominions. In this Memorandum it was pointed out that, by the terms of the Washington Conference, we have lost the power to increase the naval facilities and coast defences at Hong-Kong; that the United States were debarred from increasing their facilities anywhere West of Hawaii; and that in effect this ruled them out so far as effective naval action in the western basin of the Pacific was concerned, and left the British Empire the sole Power to counter with naval forces any aggressive tendencies on the part of Japan. The Admiralty considered the development of Singapore essential, and urged the Dominions to assist in this task (July, 1922).

In the Autumn of 1922 Mr Bonar Law's Government came into office, and, after fresh investigations, decided to proceed with the Singapore project, although, in the financial circumstances then existing, they decided to spend very little money on it during the next two years. Their decision was discussed at great length at the Imperial Conference, 1923, which took formal note of the deep interest of Australia, New Zealand and India in the provision of a base at Singapore as essential for ensuring the mobility necessary to provide for the security of the territory and trade of the Empire in Eastern waters. Mr Massey stated that the New Zealand Parliament had already voted £100,000 towards the base, and Mr Bruce has indicated that the Australian Parliament would be asked to contribute in some form.

[198] *Governor of Newfoundland [Sir William Lamond Allardyce] to the Secretary of State for the Colonies [James Henry Thomas], telegram received*
7 March 1924, 10.10 p.m.

Careful consideration has been given by my Prime Minister [William R. Warren] to your telegram of March 6th relative to Singapore base and to that of the fifth instant from the Prime

Minister. Warren emphatic that it would be unwise not to proceed with naval base Singapore.

[199] *Governor-General Union of South Africa [Connaught] to [Thomas], received*
7 March 1924, 1.06 p.m.

With reference to your telegram of 5th and 6th March, Singapore. Following message from my Prime Minister to your Prime Minister. Begins. Your proposed statement of policy meets with my whole-hearted agreement. Purely on the grounds of naval strategy Singapore base may be sound proposal, but the authority of the British Empire as the protagonist of the great cause of appeasement and conciliation among the nations must be seriously undermined by it. I welcome the abandonment of the scheme.

[200] *Governor-General of Canada [Baron Byng of Vimy] to [Thomas], received*
10 March 1924, 9 p.m.

Following from Mr Mackenzie King [Prime Minister of Canada] for Prime Minister [Ramsay MacDonald]. Begins. Replying to your telegram of the 5th March. Realize that we are not in a position to offer any advice regarding the necessity of establishing base at Singapore, and, having thus far on that account refrained from so doing, my colleagues and I do not feel that proposed statement of policy outlined in despatch from the Secretary of State for the Colonies dated 6th March calls on our part for any comment. Ends.

[201] *Governor-General Commonwealth of Australia [Henry William, Baron Forster of Lepe] to [Thomas], received*
11 March 1924, 6 p.m.

We believe that the existence and prestige of the British Empire has been and is the greatest factor in the maintenance of the peace of the world.

To the active support backed by prestige and strength of the British Empire has been due the measure of success which has been achieved by the League of Nations since its inception.

Our strength relative to other great Powers has been the basis of the influence for peace which we have wielded in the councils of the nations and through the League of Nations.

That strength has depended mainly on the British Navy, its power and mobility. We are convinced a base in the Pacific is imperative for that mobility.

[202] *Governor-General of New Zealand [Jellicoe] to [Thomas]*
11 March 1924

'Secret. Most Urgent'

The following message is for the Prime Minister [Ramsay Mac-Donald] from my Prime Minister [W.F. Massey] – BEGINS

With reference to your cypher telegram of March 6th intimating the decision arrived at by His Majesty's Government in regard to the proposed Singapore Naval Base, I regret exceedingly that the Government of the United Kingdom do not intend to proceed with what is looked upon as one of the most important proposals connected with the defence of the Empire. The foremost Naval Authorities available have stated in no uncertain terms that a modern fleet cannot operate without a properly equipped base, and in their opinion there is no place so suitable for protecting those portions of the Empire which are situated in the Pacific and Indian Oceans as that which may be provided at Singapore. India, Australia, New Zealand and a number of Crown Colonies are intensely concerned in this matter, and are looking to the present British Government to remember that every country of the Empire and every citizen of the Empire are entitled to protection from the possibility of attack by a foreign foe. It is well to remember here that Singapore is intended for defensive and certainly not for offensive purposes, and that the establishment of a naval base at Singapore would constitute no more of a threat to Japan than Gibralter could be considered a threat to the United States of America or any other foreign power.

The New Zealand Parliament last session voted £100,000 as an earnest of its anxiety that the fortification of Singapore should be proceeded with, and it will not stop at that. America has in recent years fortified Pearl Harbour in the North Pacific, and well-qualified Naval experts say it is now impregnable and that the naval position of America has been strengthened accordingly. Fortunately the United States of America is a friendly nation and so far as it is possible to judge, will remain as such for centuries to come, and I hope for all time. We in New Zealand, separated from the heart of the Empire by 13 or 14,000 miles of sea, realise what it means to be insufficiently protected. We have not forgotten what the Royal Navy and the British Mercantile Marine suffered in the Pacific during the years of the great war, and we had hoped that the lesson taught then would not be so quickly forgotten.

You say that your 'Government stands for international co-operation through a strengthened and enlarged League of Nations'. In reply to that I must say that if the defence of the Empire is to depend upon the League of Nations only, then it may turn out to have been a pity that the League was ever brought into being. The very existence of the Empire depends upon the Imperial Navy, and if, in the event of war, the Navy is to operate successfully, it must have suitable bases from which to work and where repairs may be effected. The nearest suitable base at present is Malta which is 6,000 miles away, and therefore of no value for the purposes of capital ships in either the Pacific or Indian Oceans. An eminent authority has said that 'Unless such a base as that contemplated at Singapore is established it will be an absolute impossibility for the majority of the Empire capital ships to operate to the eastward of Suez, for the simple reason that they cannot dock either for the purpose of cleaning, and so keeping their speed, or of being repaired'. It may also be pointed out that the League of Nations although it is undoubtedly an influence for peace has not so far been able to prevent hostile action as between nations.

I may remind you that owing to the alterations in ship design since the great war, docks which could have taken certain classes of warships before 1914 will not now accommodate ships of similar tonnage, and so it is not possible to maintain the present standard of naval efficiency without the proposals regarding Singapore being given effect to.

On behalf of New Zealand I protest earnestly against the proposal to make Singapore a strong and safe naval station being abandoned, because I believe that as long as Britain holds the supremacy of the sea the Empire will stand, but if Britain loses naval supremacy the Empire may fall, to the detriment not only of its own people but of humanity as a whole, and it is surely the duty of British Ministers and the British Parliament to see that so far as it is humanly possible to prevent it, there will be no danger of such a catastrophe.

[203] *'Statement of Policy in Regard to Singapore', Appendix 'B'*
Cabinet Conclusions, 21(24)
17 March 1924

We have given careful consideration to the proposal to develop the Naval Base at Singapore. We have closely studied the reasons which led to the adoption of this project, and the arguments in support of its continuation urged upon us from the point of view of naval defence.

We had, however, to consider the matter in a wider relationship, and came to the conclusion that we could not ask Parliament to proceed with this scheme. We were convinced that if we were to do so our action would exercise a most detrimental effect on our general foreign policy. As we have repeatedly stated, we stand for a policy of international co-operation through a strengthened and enlarged League of Nation, the settlement of disputes by conciliation and judicial arbitration, and the creation of conditions which will make a comprehensive agreement on limitation of armaments possible. As stated in the Prime Minister's letter to Mr Poincaré of February 21st, – 'our task meanwhile must be to establish confidence, and this task can only be achieved by allaying the international suspicions and anxieties which exist today.'

It seemed clear, apart from any other considerations, that to continue the development of the Naval Base at Singapore would hamper the establishment of this confidence and lay our good faith open to suspicion. Whilst maintaining present standard in a state of complete efficiency, we take the view that it would be a serious mistake to be responsible for new developments that could only be justified on assumptions that would definitely admit that we had doubts in the success of our own policy. As a result we should almost inevitably drift into a condition of mistrust and competition of armaments in the Far East.

Having provisionally reached the conclusions which I have set out, we felt it our duty in view of the discussions at the recent Imperial Conference to communicate with the Governments of the self-governing Dominions. We therefore put the position before them and invited their views. I propose to give the House a summary of these views. The Canadian Government tell us that they wish to refrain from any advice on the problem. The Irish Free State has adopted a similar attitude. As to Australia, New Zealand and Newfoundland I must explain quite clearly that their Prime Ministers wish, and indeed urge, us to proceed with the Base . . .

From the above summary, I think I may claim that we have a large measure of sympathy in the Dominions with our international policy, even if all parts of the Empire do not feel able to endorse the methods by which we consider that that policy should be carried out. The criticisms are founded, in the main, on the conception that the Singapore base is essential as part of a complete defensive Pacific strategy. Should the practical necessity for putting such a strategy into operation arise by reason of the condition of world politics and a return to attempts to provide Imperial security primarily by armaments, the whole question would have to be re-considered, but in the opinion of His Majesty's Government that has not now arisen, and it is the duty of His Majesty's Government to try and prevent its arising.

We have every confidence in our policy, and we feel that a decision not to proceed with the Naval Base at Singapore will give that policy the best possible chance of success and is an earnest of our good faith.

[204] *Minute by Admiralty Naval Personnel Branch*
30 May 1924

Admiralty letter to the Colonial Office, N.958/21 of 31 July 1923 proposed some alterations in the arrangements for mobilising the R.N.R. in the Dominions and Colonies. Replies have now been received from all the Dominions except Australia. The Colonial Office have, at my request, hastened the reply from Australia and it is now proposed to consider the replies which have already been received and to clear up one or two outstanding matters so as to enable the procedure to be put on a proper footing.

Newfoundland, (N.229/24) and New Zealand (N.595/24) agree to the Admiralty proposals.

South Africa (N.268/24) agrees but postulates that R.N.R. men domiciled in the Union are not to be affected unless exemption has been granted under their Defence Act. The Union Government should be reminded that by Memorandum 'C' enclosed in Admiralty letter of the 3rd March 1922 – N.13253/21 – R.N.R. men domiciled in a Dominion are, under present regulations, discharged from the Reserve before emigration. It was stated that the question of relaxing this rule both for R.N.R. and R.F.R. was under consideration, but so far as the R.N.R. are concerned it was decided on N.886/22 not to do anything.

Canada (N.5300/23) agrees generally and proposes to distribute the Forms R.V.53C in specified ports and at the same time to issue a Press notice; but they go on to ask that Forms shall be specially printed for Canada amending the approved Form for R.N.R. men not serving in Merchant ships, and they go on to suggest that these men should be allocated to H.M. Canadian Ships and Establishments under the provisions of Admiralty Memorandum 'C', such men as could not be employed in the Canadian Naval Service to be sent to England. But Memorandum 'C' said definitely that R.N.R. men domiciled in a Dominion do not exist at all and that R.N.R. men temporarily in a Dominion are required in the Royal Navy. Proposed to remind the Canadian Government of this and to explain that it would be highly convenient as a matter of procedure to adhere to the practice which will be followed universally throughout the Empire regarding R.N.R. men in the Dominions and Colonies who are not serving in Merchant Ships, i.e., that they are to report themselves to the nearest British Naval, Dominion or Colonial Officer, that Dominion and Colonial officers concerned to whom such men report themselves will communicate the fact to the Governor of the Dominion or Colony and that the Naval Commanders-in-Chief abroad will keep the Governors of Dominions and Colonies informed from time to time as to the action which will then be taken. I should like to point out here that although this procedure is mentioned in Colonial Office Circular of 22nd October 1903 which presumably went to all Dominions as well as Colonies, although Dominions were not specially mentioned, there is no corresponding paragraph in the later despatches to Australia and New Zealand. Presumably the Commanders-in-Chief responsible for notifying the Governors General should be as follows:

Australia, New Zealand	(C-in-C, China)
Canada	(C-in-C, North America & West Indies)
South Africa	(C-in-C, Africa)

[205] *Telegram from the Governor of Hong Kong [Sir R.E. Stubbs] to [Thomas]*

16 December 1924

With the approval of the Unofficial members of both Councils I offer His Majesty's Government the sum of £250,000 representing in round figures the profits of the control of shipping during the war as a contribution in aid of the Singapore base.

[206] *Minute by Head of Naval Personnel Branch [Assistant to Second Sea Lord, Vice-Admiral Sir Michael Culme-Seymour]*
18 December 1924

As the result of the Conference held on the 8th August (vide enclosed report of D.M.D. 22.8.24) it appears that the scheme put forward by Rear Admiral Hotham [Director of Naval Intelligence ? and former Officer Commanding Australian and New Zealand Naval stations] is not so much one of 'fusion' as a wide extension of the existing practice of sending New Zealand ratings to this country for courses and other training in H.M. Ships. Subject to Naval Branch remarks of 2.9.24. and to any further observations of Departments concerned, it is proposed to submit to inform the New Zealand Government through the Colonial Office that these suggestions have been put forward by Rear Admiral Hotham and that if they commend themselves to the New Zealand Government the Admiralty will be prepared to do their part.

It is proposed also to say that there is some reason to think that the existing practice whereby New Zealand ratings are sent to this country for courses and training in time of peace and are borne on the books of H.M. Ships is open to criticism on legal grounds, and that Their Lordships are consulting their legal advisers and will be prepared, if necessary, to introduce a Bill in the Imperial Parliament to regularise the position.

Propose also to ask Treasury Solicitor to consult the Law Officers on the legal question at issue. It is pointed out that other Dominons are affected. An extract from the Australian Naval Defence Act is enclosed. There seems to be no similar provision in the Canadian Naval Defence Act, and the South African Naval Forces, apart from Officers and ratings lent from the Royal Navy, consist of the South African R.N.V.R. which is legally part of the R.N.V.R. as raised in the United Kingdom. The embarkation of their R.N.V.R. personnel on ships of the African Station for peace training is therefore covered by the Imperial Acts.

It should be pointed out that the converse case of sending Royal Navy personnel to the Dominion Services has not so far given rise to similar difficulties because it has always been possible to find volunteers. If there should ever be an insufficient number of volunteers it seems likely that similar legislation would be required for this purpose also.

[207] *Minutes of the 16th Meeting, Chiefs of Staff Sub-Committee, Committee of Imperial Defence*
24 February 1925

Singapore Naval Base. Garrison and Defences.

SIR HUGH TRENCHARD [Chief of Air Staff] stated that he was not in agreement with several points in the Report of the Oversea Defence Committee. He was aware that his views would, in all probability, not be accepted by the other members of the Committee, but he thought it would be useful to bring out those points on which there was a divergence of opinion. He considered that the suggestions which he would put forward would prove adequate for the defences and would cost considerably less than the provision of heavy guns and all that such defence involved. If it was found that his proposals were more expensive he was quite prepared to withdraw them. He agreed that if the defences were to be completed immediately the arguments of his case would not be so strong as if the defences were not to be completed for some ten or fifteen years. He considered that this question required a definite answer, in view of the fact that a new Service such as the Air Force could develop rapidly if the necessary funds were available for the provision of suitable equipment. In his opinion Singapore should be defended, not by heavy fixed defences, but by mobile forces which could proceed there in the event of necessity. He pointed out that the chain of fixed defences at British ports, such as Gibraltar, Malta, etc. were tied down to their own purely local defence, and were incapable of being pushed forward to render assistance to any other locality further afield which might be threatened. Mobile forces would be available to proceed as required to any point of danger. The rough estimate for the provision of 15″ guns amounted to £1,250,000. In his opinion this sum would only be a fraction of the actual cost involved. Men were required to man these guns. Moreover, the War Office had stated that they must have a guarantee that spotting aircraft would be provided for these guns. Such provision would necessarily increase the cost of that system of defence. Spotting aircraft required fighting aircraft to defend them. Then minesweepers would apparently be required, and also an Infantry garrison. He considered that fixed defences of a calibre of 9.2″ would be sufficient. In addition an aerodrome should be constructed and other preparations made by the Air Ministry commensurate with the Admiralty preparations, that is to say that the aerodrome should be completed pari passu with the floating dock, etc. He suggested that units of the Air Force could be despatched to Singapore if, and when, required. This latter proposal was on the same lines as the War Office proposals for the reinforcement of the Infantry garrison.

He called attention to the important statement contained in paragraph 4 of C.I.D. Paper No. 237-C, to the effect that:

> The Admiralty have further expressed the opinion that in view of the importance of the objective the Japanese would be prepared to accept the risk of damage to all their battleships in a bombardment of Singapore, with a view to its capture or destruction. The Oversea Defence Committee are in agreement with that opinion and recommend that such an assumption should be accepted by the Committee of Imperial Defence.

He pointed out that the scale of defences recommended by the Oversea Defence Committee were primarily based on that assumption. He was somewhat doubtful if such an assumption was wholly correct. He pointed out that the Japanese battleships would not come to Singapore alone; surely they would be accompanied by the whole of the Japanese Fleet with all its ancillary craft and transports carrying troops to effect a landing. If such were the case this enormous fleet would offer an easy target to attack by aeroplanes. Such attack could be carried out at a distance of 150 miles from Singapore, whereas heavy guns had a range of only 30,000 to 40,000 yards. He considered that the time factor was important. Six weeks was the period given before the main British fleet could reach Singapore, whereas the Japanese fleet could reach it in fourteen days. He pointed out that the Admiralty had already stated that in order to maintain the Fleet in Far Eastern waters it would be necessary to use the existing docking facilities even as far as Gibraltar, in addition to those which it was proposed should be constructed at Singapore. He pointed out that Japan, which was situated 2,000 miles away from Singapore, possessed none of these additional docking facilities. He expressed the opinion that reinforcements of troops would be unable to reach Singapore unless we had command of the sea. He asked Lord Beatty if he would further develop the question of the risks which the

Japanese would be prepared to accept if they carried out an attack on Singapore.

LORD BEATTY [First Sea Lord] stated that although we possessed additional repairing facilities at Malta, Colombo, etc. the Japanese repairing facilities were infinitely closer to Singapore than our own. He explained that with regard to the question of risks this was answered by a proper understanding of what a fortress and a base were intended for. The main principle from the Naval point of view was that before any Naval force could undertake operations it must have a secure base on which it could rely with absolute certainty. A base must contain all the ingredients to enable the fleet to carry out operations. There must be reserves of oil fuel; there must be the necessary repairing facilities. These items more especially the oil, were worth millions of pounds. Much of this valuable material was already on the spot, would be increasing annually and must be protected. A Commander-in-Chief must be perfectly satisfied that the base would continue to remain a base in the absence of the main fleet; that it should be absolutely safe, not only from capture, but also from attack. To afford such protection there were, at present, certain recognised methods of defence, mainly by gun-fire, the volume of which was based on the scale of probable attack. The Admiralty had already laid down the scale of attack on Singapore, and the War Office had drawn up a scale of defence to meet such an attack. It was evident that the Chief of the Air Staff did not agree with those proposals. He (Lord Beatty) pointed out that at present the Air Staff had put forward no concrete proposals as alternative methods of defence. He suggested that before the question could be further discussed the Chief of the Air Staff should prepare a paper showing his proposals. He pointed out that at the present moment Singapore was practically unprotected, and that the Japanese could to-day seize Singapore with no risk whatsoever. If they were successful they would thus obtain command of the China Sea and the Indian Ocean. In his opinion it was unquestionable that they would take the gravest risks to obtain a result of such immeasurable value. By such a course they would achieve more than they could possibly hope to achieve by a successful fleet action. In his opinion the defences of Singapore should be so strong as to act as a serious deterrent to any such attempt on the part of Japan.

[208] *Reorganisation of the Royal Indian Marine, Report of the Departmental Committee*

March 1925

President, His Excellency General Lord Rawlinson, Commander-in-Chief India.

Members:

His Excellency Rear-Admiral H.W.Richmond,
Commander-in-chief, His Majesty's Ships and Vessels,
East Indies Station;

The Hon'ble Sir B.N.Mitra, Member of the Council of the
Governor-General of India;

E.Burdon, Esq., Secretary to the Government of India, Marine
Department;

Captain E.J.Headlam, Director of the Royal Indian Marine.

The Committee met at Delhi during February 1925 and prepared their Report, which was approved in draft form by the late Lord Rawlinson before his death in March 1925.

1   Stated generally the scope of the task entrusted to us is to draw up a scheme for the purpose of putting into effect a policy defined as in the following formula:

The reconstruction of the Royal Indian Marine as a combatant force, to enable India to enter upon the first stage of her own naval development, and ultimately to undertake her own naval defence.

2   Our terms of reference, arranged for convenience in the order in which we shall deal with them, are as follows: To prepare a scheme for the reorganisation of the Royal Indian Marine so as to form the nucleus of an Indian Navy, with special reference to

(i)   The functions to be ultimately performed by the Indian Navy and methods of employment with a view to its undertaking those functions.

(ii)   The number and class of vessels that can be maintained with available budget allotment.

(iii)   Recruitment, strength, training and conditions of service of personnel.

(iv)   Relations between the higher command of Indian Navy, Government of India and Commander-in-Chief, East Indies, including the proposed appointment of Chief Naval Staff, India.

(v)   Provision for and maintenance of vessels including the continuance, or abolition, of the Royal Indian Marine Dockyard.

1 Functions of the New Navy

3   By far the most important aspect of the new force in its early stages will be its duty as a training squadron. The new personnel will need to be thoroughly trained in gunnery, minesweeping, harbour defence and seamanship. In this connexion we cannot insist too strongly on the ships of the Indian Navy becoming from the first a sea-going force. Efficiency and enthusiasm alike will melt away if the new navy remains in port, and practises nothing but harbour defence. A valuable service which we think that the Indian Navy should be able to undertake in the near future, will be the responsibility for policing the Persian Gulf in peace time, by which means the three vessels maintained in those waters by the Imperial Government will be set free.

Of the other duties at present performed by the Royal Indian Marine, we consider that the Marine Survey should be retained, as its work in peace and war is essential to a fighting sea service.

The control of the station ships at Aden, Port Blair, Rangoon and in the Persian Gulf, to attend to the conveyance of troops and officials and to supervise the work of lighting and bouying in the adjacent waters, should not be a function of the new navy. The retention of these responsibilities would not be, in our opinion, compatible with the development of a fighting force. The work of carrying troops can be contracted for commercially at rates which could hardly fail to be cheaper than the existing arrangement.

The new service should also be responsible for Marine transport, at present carried out by the Royal Indian Marine. The cost of storage and maintenance, in this connexion, will be a charge against the Indian Navy.

The functions of the new Indian Navy in peace time will, therefore, be as follows:

(a)   The training of personnel for service in war.
(b)   The services required by the Indian Government in the Indian Ocean and Persian Gulf.
(c)   The organisation of the Naval Defences at Ports which are under the control of the Indian Government.
(d)   Survey work in the Indian Ocean.

(e)   Marine transport work for the Government of India.

We recommend that in accordance with its new functions the service should be known as the 'Royal Indian Navy' and should fly the White Ensign, which is the recognised flag of the naval fighting forces of the Empire.

II Number and Class of Vessels, etc.

4   On the assumption that these will be the functions of the Indian Navy, we consider that a squadron of 4 sloops, 2 patrol craft vessels, 4 trawlers and 2 survey ships, together with one depot ship as already suggested, would suffice to begin with. We estimate that the net annual cost of maintaining such a force would amount at first approximately to Rs. 63 lakhs as follows;

Headquarters Staff	Rs 2,59,900
1 Depot Ship	3,24,851
4 Sloops	12,04,224
2 Patrol craft-vessels	3,59,616
4 Trawlers	1,23,120
Officers on leave	2,08,020
Pensions	9,10,000
Incidental expenses	3,98,740
Subsidy to Admiralty, etc.	19,00,000
Survey Ships	4,89,000
Store Department	30,820
Officers training in England and with the East Indies Squadron	52,140
TOTAL	6,260,431

This figure is exclusive of the following items:

(1)   Rupees 12,50,000 cost of Lighting and Station Ships, which should be met from lighting fees and debited to other departments. At present 2 lakhs of this expenditure is debited to Political estimates and the remaining 10 1/2 lakhs to Marine estimates.

(2)   Rupees 4,00,000 for military launches which will be included in the Military estimates.

(3)   Rupees 1,14,000 on account of transport establishment hitherto debited to His Majesty's Government.

(4)   Pension charges for ratings which will be a negligible figure for the first few years.

As soon as the Indian Navy is in a position to undertake the work of the Royal Navy in the Persian Gulf, it will be necessary to add at least 2 sloops to the force. The additional expenditure will, however, be partly set off by the saving of the existing expenditure of about 4 lakhs a year on fuel for the three Admiralty vessels, and on the refit of the *Triad*. If our proposals for leasing the dockyard (paragraph 14) are accepted, no separate provision will be required under this head, as the cost of refit is included in the figure for each vessel . . .

[209] *'Policy of the Naval Board', paper sent to the Director of Naval Intelligence [Hotham], by Acting Rear-Admiral Beal, [the New Zealand Naval Secretary], on vacating his Appointment*
21 December 1925

A. GENERAL POLICY

1   General Policy   To continue to administer the New Zealand Division of the Royal Navy on its present lines under the direction of the New Zealand Naval Board, deprecating strongly any idea of a separate Navy on the Australian or Canadian lines.

To conform to all Admiralty Regulations and Instructions and all alterations and additions thereto as far as they are applicable to the New Zealand Division.

All additions and reductions of pay which depend on the cost of living to be considered on the scale of the cost of living in the Dominion.

It is considered that there is no scope for a separate Air Force in the Dominion and that the Air Forces should continue to be administered by the Military Authorities.

2   Islands   No defence scheme appears practicable at present for the various outlying Islands included in the British Empire and situated within the boundaries of the station.

B. PERSONNEL (ACTIVE SERVICE)

3   Officers   As a general rule to obtain the services of officers by loan from the Imperial Navy. It is considered to be of great importance that the free passage of wives and families be continued, so as to widen the list of volunteers for service in the New Zealand Division.

4   Liaison Officer or Naval Representative in London   It is not considered that at present an officer could be fully employed on

these duties or is necessary in London. With the additional amount of work added to the duties of the High Commissioner's staff, it would appear that possibly an extra clerk will be required shortly, or an extra allowance for this duty made to the High Commissioner's staff.

5 *S.O.(1)* The officer occupying the position of Staff Officer for Intelligence Duties should be an officer lent to the New Zealand Government and paid by them, as in the case of all other officers. It is proposed to take over this appointment when the next relief takes place i.e. in 1928.

6 *Cadets* No scheme of entering Cadets for training in New Zealand is considered possible or advisable.

The present methods of New Zealand Cadetships at the Royal Naval College, Dartmouth, and nominations for the Public School entry, appear to fulfil all objects.

The Director of Education to deal with entries in all matters except the Medical Examination which is carried out by a Naval Medical Board. A Naval representative is attached to the Selection Committee.

7 *Entry of New Zealand Ratings and their Training* To continue entering and training Boys, Stokers and Artisans under the present scheme, the recruiting to remain in the hands of the Military. A special Naval recruiter occasionally to visit the various centres is of value.

The 'short service' entry is considered highly undesirable.

All training for the higher rates in Gunnery, Torpedo, etc., to be carried out in the Imperial schools; it is not desirable to send ratings to Royal Australian Naval Schools.

8 *Interchange of Ratings* It is considered essential that a scheme of complete interchange of ratings with the Imperial Navy be evolved, as only by this means can the New Zealand Division be kept up to the Imperial standard. It is not considered that the actual interchange of ships fully manned can take the place of this, as in a small force changes of ship, environment [sic], petty officers and messmates are essential for the maintenance of efficiency and ambition. It is however fully realised that interchange visits with the Royal Australian Navy and combined exercises with R.A.N. Ships are of the greatest value.

9 *Imperial Ratings on Loan* To continue obtaining on loan for periods of three years the required number of Imperial ratings to complete the complements of the cruisers, etc.

It is considered that these ratings should eventually be borne in the proportion of 50% Imperial to 50% New Zealand ratings in order to keep up the general experience of the ships companies.

These men should not be kept on after completing their time for pension, neither should they, as a general rule, be allowed to re-engage in the New Zealand Division for long periods or permanently.

No men should be loaned for any special positions but all (including those at the Navy Office) should be eligible for sea service.

10   *Artisan Ratings*   From the few suitable artisan ratings who offer themselves for service in New Zealand, it is not expected that a high percentage will ever be reached. Suitable artisan ratings should be entered whenever they apply (within reasonable bounds) and borne as additional until absorbed.

11   Royal Marines   The Admiralty have been asked to approve of a scheme for the local entry of recruits for the Royal Marines.

C.   PERSONNEL (RESERVES)

12   Requirements of Reserves   Officers and men are required to complete the Squadron to war complement for Convoy Staffs, Mercantile Control Service, local patrols, Minesweeping, Armed Escort Ships and D.E.M.S. [Defensively Armed Merchant Ships] Guns Crews.

13   *Reserves: Officers*   To establish a small reserve of R.N.R. Officers and to continue a yearly entry of a few young mercantile officers as Probationary Sublieutenants.

Action taken   Three Senior and four Junior Officers have already been entered.

14   *Royal Naval Volunteer Reserve*   To establish R.N.V.R. Divisions at the principal ports of the Dominion.

Future commissions only to be granted to those who have served as R.N.R. (D. class) ratings.

Action taken   The Auckland Division is already formed with eighteen officers.

15   *Emergency List (Officers)*   To keep a list giving the previous services of all officers retired or resigned from the Active List R.N.R., R.N.V.R., or other services, who are resident in the Dominion and who are willing to serve in time of war or emergency.

Action taken   A list is kept in Navy Office (Staff Officer Intelligence).

16   *Reserves: Men*   To establish a reserve of ratings corre-

sponding approximately to those of the Imperial Reserves, as follows:

Class 'A' – Active Service ratings: Pensioners etc.
Class 'B' – Persons serving in Merchant Service
Class 'C' – Persons not following the sea as a profession.
Class 'D' – Corresponding to Imperial R.N.V.R.

Entry to be continued in all these classes and R.N.R. Class 'D' Divisions to be established at the principal ports.

Action Taken   All these classes are at present in being. A class 'D' Division has been formed at Auckland.

17 *Sea Cadets* As soon as some of the R.N.R. Class 'D' Divisions are fully established, to commence a Sea-Cadet Corps scheme to act as a feed to the Divisions and generally to educate boys in sea knowledge, tradition and discipline. Such a corps will probably also assist Naval recruiting.

D. MATERIEL

18 *Cruisers* To continue the present scheme of obtaining Cruisers on loan from the Imperial Government as long as the latter are prepared to continue this very advantageous method to the Dominion.

To ask for and maintain a third Cruiser as soon as the Dominion finances will permit and, at the same time replace the Commodore commanding by a Rear Admiral.

All Cruisers so obtained to come under Imperial orders immediately on the outbreak of hostilities.

19 *Sloops* Eventually to maintain a sloop or sloops for Island duties, thus relieving the Imperial Government of a commitment which should be borne by the New Zealand Naval Defence Vote.

20 *Submarines* The expense and difficulty of maintaining a S/M Flotilla with the present resources of the dominion, do not justify the outlay that would be involved.

21 *H.M.S. Philomel* Eventually to replace 'Philomel' by a shore establishment at Devonport, which when built will be more economical and healthy.

22 *Minesweeping and Training Trawlers* Eventually to purchase three additional trawlers for minesweeping and general training purposes. These would be available for inaugurating the M.S. service in War, and in peace time for training R.N.V.R. and Class 'D' ratings at the ports where local Divisions are established.

Action taken   One trawler, the *Wakakura* has already been purchased.

23 *Oiler* To maintain for Squadron use one Fleet Oiler, preferably one on loan from the Imperial Navy.

Action taken   Oiler *Nucula* is at present attached to the New Zealand Division.

24 *Ocean-Going Tug* When considered necessary, to take over from the Wellington Harbour Board the Rescue Tug which was presented to the Dominion by the Imperial Government and is now on loan to the Board. One year's notice to the Wellington Harbour Board is required.

Action Taken   Admiralty Rescue Tug *St Boniface* from *Toia* is due at Wellington early in 1926.

25 *Battle Practice Target* To obtain and maintain at Devonport a suitable Battle Practice Target.

Action taken   Enquiries are now being made as to cost of manufacture.

26 *Armed Merchant Cruisers* (E) To arrange with the Imperial authorities to fit out complete the necessary Armed Merchant Cruisers (E) at the expense of the New Zealand Government.

To replace the ships and crews thus detailed by the Imperial authorities by ships and crews from the Dominion, these being sent to the United Kingdom or where required.

This method is considered the most efficient until sufficient 6″ guns and mountings are stored in the Dominion.

It is not proposed at present to purchase any 6″ guns, mountings or ammunition.

27 *Defensively Equipped Merchant Ships* To maintain the Imperial guns, mountings and ammunition at Devonport, and to arrange in the case of an emergency to fit out the most suitable vessels.

28 *Refits* It is considered that the most efficient method of keeping modern cruisers on loan efficient and up to date would be to send them to an Imperial Home Port for refit every 2½ years. At the same time to recommission them with the Imperial ratings required, thus economising in ocean passages. On completion of their refits they should be attached for a short period to one of the Imperial Fleets before returning to New Zealand. Only one cruiser at a time to be absent from the Station.

29 *Local Defences* To obtain and maintain at Devonport sufficient guns and ammunition to equip suitable local vessels for coast patrol duties.

It is considered that, at the present time, 15 guns are sufficient. It is proposed to purchase annually a small amount of ammunition, to form a reserve, and a requisite amount for practice purposes. No depth charges have as yet been purchased.

Action taken   The 4' guns and mountings are at present at Devonport and 5 more are being sent out.

30   *Minesweeping Stores*   To obtain and maintain at Devonport the necessary stores to inaugurate a M/S service on the outbreak of hostilities. To purchase annually a small amount of M/S gear until a sufficient reserve is formed.

Action taken   The *Wakakura* is being fitted with the latest pattern M/S gear and has ratings trained in M/S in her skeleton crew to act as instructors. Certain of the latest pattern sweeps have been ordered from the Admiralty.

31   *Mines*   No mining policy is considered necessary at present, owing to the great distance away of any possible hostile base, and to the great initial outlay and cost of upkeep of sufficient material for any efficient scheme.

32   *Booms S/M Nets* etc.   It is not considered justifiable at present to expend any money on such Harbour defence measures.

E.   SHORE BASE

33   *Barracks*   To eventually erect at Devonport sufficient accommodation to replace H.M.S. *Philomel* and, until this is done, to maintain the present recreation, sick berth and other buildings.

Action taken   A rough estimate of the cost of barracks is being prepared at the request of the Parliamentary Finance Committee.

34   *Storehouses*   To store and maintain nine months' reserve of general stores for the Squadron.

Stores should be purchased locally through the Stores Control Board if prices and quality permit, otherwise obtained from England.

Action taken   Present building is sufficient and stores are arranged for.

35   *Storage for Guns*   To maintain sufficient storing space for all spare guns and mountings for Cruisers, Sloops, D.E.M.S. and Local Patrol Vessels.

Action taken   This is at present being increased.

36   *Storage for Torpedoes*   To maintain a small testing and parting shop.

Action taken   Now being erected.

To build and maintain a suitable torpedo store to hold all spare torpedoes, with suitable workshops and sufficient staff to carry out necessary tests and alterations.

Action taken   Nothing yet done, spare torpedoes are at present stored in Australia at the expense of the New Zealand Government.

37 *Minesweeping Store*   To erect a suitable M/S Store.

No action yet taken.

38 *Magazines*   To continue the use of the present stowage at Mount Victoria and North Head, by arrangement with the Military and to maintain an Armament Supply Officer and staff to deal with the ammunition, guns and mountings in store.

As the Armament Supply Officer and staff also deal with the Imperial ammunition for the sloops and the guns and ammunition for the D.E.M.S. the Admiralty pay 40% of the whole charge.

Action taken   At present sufficient.

39 *Cordite testing Laboratory*   To equip and maintain a cordite testing laboratory at Mount Victoria, Devonport, in order to avoid having to send all ammunition to Wellington for test.

40 *Reserve of Oil Fuel*   To maintain and keep filled three oil tanks of approximately 5,000 tons each, with the necessary pumping gear etc.

Action taken   Two tanks complete have been arranged for, to contain about 4,600 tons of oil fuel in each.

To consider the erection of No. 3 tank in 1927/28.

41 *Reserve of Coal*   To maintain at Devonport a reserve supply of at least 500 tons of Admiralty mixture New Zealand coal, supplied by the Westport coal Company.

Action taken   Established.

42 *Berthing*   To increase the present wharf to accommodate one cruiser and one sloop alongside.

Action taken   This is being done.

Eventually to increase the berthage to the westward to accommodate two cruisers alongside.

43 *Reclamation*   Eventually to reclaim land in Stanley Bay as the cheapest and best method of extending the Base when the Naval Barracks etc. are built.

44 *Dock and Workshops*   To continue using the dock and workshops under the Admiralty agreement. (This agreement expires in 1940)

The idea of taking over the dry dock and workshops by the

Naval Department and managing them as semi Naval and Mercantile establishment is deprecated.

No suitable scheme of providing for a larger dry dock has, up to the present, been evolved.

F. MISCELLANEOUS

45   *Wireless Telegraphy Stations*   It is not considered that the expense of erecting a large high power station is at present justified purely as a Defence measure.

The Post Office authorities, who at present administer the W/T Stations in the Dominion, are awaiting further developments before erecting any new stations.

46   *Contracts*   To continue the scheme of making contracts, when required, through the various Government Departments concerned.

47   *Detention Barracks*   The present scheme of sending men to the Civil Prison at Auckland, appears to fulfil all requirements.

No special Naval warder appears necessary.

48   *Hospitals*   To continue to use the civilian hospitals for Naval patients.

49   *Hydrographic Work*   It is considered that the Marine Department should continue to deal with all marine survey work and any idea of the Naval Department taking it over is strongly to be deprecated. It is impossible for any extended survey to be carried out by H.M. Ships.

50   *Examination Service in the time of Emergency*   To continue to be a Military commitment.

By direction of the Naval Board.

[210] *Sir George Maxwell, Chief Secretary of the Federated Malay States, to Sir Lawrence Guillemard, British High Commissioner, Kuala Lumpur*

27 February 1926

The Malay Mail of the 23rd contained Reuter's telegram that the government of India would not contribute to the Singapore Naval Base.

Viscount Lee of Fareham [the former First Lord] was my guest at Carcosa at the time, and conversation naturally turned upon this subject. Lord Lee referred to the gift of H.M.S. Malaya, and inquired whether the [Federated Malay States] Government contemplated a contribution to the Naval Base. I replied that I understood that the construction of the Naval Base was a certainty, and that I thought that the F.M.S. Government could therefore help most by contributing to something that would be in addition to the Naval Base. I told him, in confidence, of the idea of the F.M.S. Government contributing the cost of two European regiments.

Lord Lee impressed upon me emphatically his fear that the construction of the Naval Base was very far from being a certainty, and his opinion that a substantial contribution by the F.M.S. Government might turn the scale and make it a certainty.

He referred to the £3,000,000 (more or less) which the Malaya and cost, and reminded me that earlier in the day, in conversation on the present prosperity of the country, I had expressed my confidence in a five years cycle of prosperity for the F.M.S. . . .

Mr [Arthur] Pountney [financial advisor to F.M.S. Government] and I agreed that the conversation should be reported to [Your Excellency] by me with the suggestion that if it met with Y.E.'s favourable consideration on general grounds it might be put before the Rulers by the Residents. It would not be difficult, I think, for the case to come before the Sultan of Perak in such a way that the suggestion might seem to His Highness to originate with himself.

Speaking to Mr Pountney when Lord Lee was not there, I expressed doubt as to the F.M.S. contributing as much as £3,000,000. I suggested £2,000,000 in five instalments of £400,000. I understood Mr Pountney to be inclined to agree. I did not press him for anything that was committal in any way.

I have little doubt of the Unofficials giving the proposal their support as soon as they learn Lord Lee's views.

[211] *Guillemard to Lee of Fareham*
9 March 1926

Great minds jump together. I had already in my own mind turned [down?] the British Regiment idea, which was Maxwell's idea.

The Rulers had never heard of it and it would be a grievous mistake politically. R.I.P.

What is more interesting is that I had already, after you left but before getting your letter, (? telepathy) conceived the same idea as you.

I think it will come off and that I shall perhaps bring Johore in as well as the F.M.S. and be able to offer something well over 2 millions.

To save time pending negotiations I will send a secret telegram to [Viscount] Amery [Secretary of State for the Colonies] asking whether, if our offer materialises, it would ensure the preliminary equipment.

Meantime MUM in capital letters is essential.

If any rumour leaks out some damned fool at home will suggest it as something that we ought to do.

That would take all the grace out of the offer, and might at the worst even put our fish down for good.

The Malay, like all really sporting fish, is very shy.

[212] *Admiralty minute on the South African Naval Service,*
signed Lieutenant Eastwood, NL

6 April 1926

The present position is that the South African Naval Forces are a Branch of the R.N.V.R., continuously placed at the disposal of the Admiralty and therefore when serving, in the position of other R.N.V.R., embarked. This gives their complete control to the Commander-in-Chief and doubtless the Union Government want more control of their own Forces (even in Imperial Ships), especially perhaps as their Military Service Acts allow as an alternative for military service in the case of seafaring people training in H.M. Ships (as R.N.V.R.).

Generally speaking the ordinary next development would be to apply the Dominions Naval Forces Act 1911 – as Australia & c., have done. It is not clear at present that they have done so as at present they seem to be working on the old midnineteenth century system of R.N.V.R. and Colonial Naval Forces. There are indications that they do not, probably for reasons of economy, desire to set up a Naval Board, but doubtless a Minister of Defence (with certain reservations to the Governor General) could be substituted. It would appear too at present that there is no indication as Canada and Australia have done of dispensing with the Imperial Commander-in-Chief or Imperial Squadron.

The problem has not been put to us officially through the Dominions Office as might have been expected, but semi-officially the Commander-in-Chief's Staff and the Union Authorities have overhauled the Naval Discipline Act as shown. It would seem necessary to raise the matter officially from this end and to endeavour to guide the Union Government to a system which will be as little as possible divergent from the Australian &c., model i.e., we should try and maintain the principle of applicability of the Naval Discipline Act with only such modifications as are necessary, so that Imperial Officers and ships can work with South African Officers and ships in peace (as will be necessary while the Imperial Squadron remains) and in war when we hope their ships and forces would be placed at the Empire's disposal.

It is necessary to consider how far the draft alterations of the Naval Discipline Act conform to this position.

In my opinion the drafting of this Act is wrong fundamentally. The Act describes itself (Section 84) as the Union Naval Discipline Code and in Section 85 limits its own application (except as otherwise provided) to the Union of South Africa. Throughout, the Act purports to apply only to South African personnel subject to this 'Code'.

If passed in this form it would be nothing more than a South African Statute applicable to South African personnel only and only so applicable within the territorial limits of the Union. (This is a difficult question but the solution of difficulties of Colonial jurisdiction is an important reason for the procedure of the 1911 Act as applied to other Dominion Navies). It is not therefore to our purpose to acquiesce in a local Act. It is clear from Section I of the Dominion Naval Forces Act, 1911, that the discipline of the Dominion Naval Forces is to be governed not by local and special Acts of the Dominion Legislatures but by the Naval Discipline Act, 1866, as amended by any subsequent enactment. It is true that the Dominion Legislature may by local Statute define certain modifications which are to be read into this Act in its application to the Dominion, but the Act remains the Naval Discipline Act of the Imperial Parliament and only while it so remains is it operative outside the territorial waters of the Dominion.

Consequently it is open to criticism to describe this Act as the Union Naval Discipline Code; to refer to it throughout as 'this Code:' to limit its operation to South Africa, and limit its application throughout to South African Ships and personnel. It seems also quite unnecessary to substitute 'South African Naval Forces' and 'H.M. South African Ships' for 'H.M. Navy' and 'H.M. Ships' throughout the Act as it is clear from Section 1 of the Dominion Naval Forces Act which expressly provides that the Naval Discipline Act 'shall have effect as if references therein to H.M.Navy and H.M.Ships included the forces and ships raised and provided by the Dominion.'

There are doubtless other details in the draft which will require discussion, but it is useless to proceed to details until it is made clear that this is the Imperial Naval Discipline Act modified and not a South African Statute with provisions similar to those of the Naval Discipline Act.

[213] *Memorandum 'Empire Naval Policy and Co-operation'*
7 April 1926

'Secret'

In February 1924 on C.I.D. Paper No. 221-C a memorandum by the Naval Staff dealing with Empire Naval Policy and co-operation was submitted to the Dominions. A copy is attached as Enclosure No. 2.

2.   Replies have now been received and may be summarised as follows:

Australia – Fully concurs with memorandum and proposals.

New Zealand – Fully concurs with the proposals and the principles. Considers that when the expansion of the New Zealand Division takes place the loan of a p.s.c. officer for staff duties will be necessary.

NOTE. The expansion referred to has now taken place.

Canada – The government is prepared to express general concurrence but considers that the second sentence in para. 8 of the memorandum may be taken to imply fusion of the respective staffs. An alteration of wording is suggested to make it clear that the respective staffs will be distinct entities but working in close co-operation.

Newfoundland – Concurs but as there is no naval establishment in Newfoundland is unable to offer active co-operation.

S. Africa – Ministers consider that the strength of the S. African naval service does not justify any action being taken, but the proposals will be borne in mind for future consideration.

3.   The principles and proposals with which general concurrence is expressed are contained in paragraphs 10 and 11 of the memorandum which read as follows:

10   To sum up, it is proposed that the following general principles should now be adopted:

(a) In order to obtain continuity of Naval policy and to ensure that the best advice on questions concerning the Naval defence of the Empire is always available for the British and Dominion Governments the several Naval Staffs should be built up by uniform Staff training and a free interchange of officers between the British and Dominion Naval Staffs.

(b) A free interchange of advice should be established between the British and Dominion Naval Staffs.

11 It is suggested that this development of the Staff aspect of Naval organisation should be marked by adding the title of Chief of Naval Staff to those of the first Naval Members in Australia and New Zealand and of the Director of Naval Services in Canada, and that the Chief of the Naval Staff, Admiralty, should, on the analogy of the Chief of the Imperial General Staff, be described as the Chief of the Imperial Naval Staff.

4 With regard to the principle annunciated in paragraph 10a uniform staff training already exists, but Australia and Canada should be urged to make full use of the vacancies for Dominion Officers at the Staff College offered by the Admiralty. (No Dominion Officer is attending the present course.)

As regards the War Course at Greenwich, as soon as Senior Officers are available their attendance should be encouraged.

5 As regards free interchange of officers between the British and Dominion Naval Staffs it is most desirable that, when available, Dominion officers of suitable seniority and with Staff College training should be appointed to the Naval Staff at the Admiralty. British officers are already filling appointments in Dominion Naval Staffs (Australia and Canada – 1 each). (A separate paper on this question is attached, Enclosure No. 1.)

6 To meet the wishes expressed by New Zealand, it is submitted that an R.N. p.s.c. officer now be appointed for duty with the Commodore, New Zealand Division.

7 As regards para. 10b, free interchange of advice between British and Dominion Naval Staffs now exists, where applicable, and the approval of this principle will enable its application to be extended should other Dominions develop naval services.

8 The proposals in paragraph 11 having been concurred in by the Dominions, a decision is required whether action should be initiated to bring the following into effect now, or and whether these questions should be raised at the Imperial Conference for

re-affirmation. In either case the matter will presumably be considered by the C.I.D. [in ms]

(a) The first naval member in New Zealand and Director of Naval Services in Canada to have the title 'Chief of the Naval Staff' added.

(b) The title 'Chief of the Naval Staff,' Admiralty, to be changed to 'Chief of the Imperial Naval Staff.' (C.I.N.S.)

NOTE It is presumed that the reply of South Africa may be regarded as concurring in this proposal.

(c) On page in Navy List showing composition of Australian Naval Board 'and C.N.S.' to be added after 'First Naval Member.' Australia points out that the 1st Naval Member is already officially also styled Chief of the Naval Staff.

9 With regard to the alteration to paragraph 8 proposed by Canada, the existing and proposed wording is shown below:

Existing.
The organisation would be analogous to the Imperial General Staff agreed to in principle at the 1907 Imperial Conference, when it was pointed out that this function of advice was performed by the general staff 'without in the least interfering in questions connected with command and administration.'

Canada's proposal.
The organisation would be one of frequent interchange of individual officers and continual close co-operation of the respective British and Dominion staffs, which would, however, be distinct entity under their several governments.

No objection is seen to this amendment and it is considered that this proposed alteration should [ms addition] be put forward for concurrence of the Dominions generally at the forthcoming Imperial Conference.

Enclosure No. 1   Employment of Dominion Naval Staff Officers at Admiralty
1 British Naval Staff Officers are already filling appointments on Dominion staffs (Australia and Canada) and when Dominion officers of suitable seniority and qualification are available, it is considered that the earliest opportunity should be taken to arrange through the Dominion authorities concerned for one or more to

be appointed to the Admiralty and to make this matter a routine policy, so far as possible.

2   It is important that such officers should not be merely for liaison work or 'attached.' They should be appointed to the naval staff on precisely the same lines as British Naval Officers, and should have the same access to papers as would British officers in such appointments. The restrictions imposed in M. 02594/20 as regards papers to be withheld from Dominion Officers 'attached' to the Staff should not apply.

3   Further, to obtain the best value it is considered that Dominion officers selected to serve at the Admiralty should spend their period of service in Plans, Trade, Operations or Training and Staff Duties Division or Naval Air Section or Tactical Section and not [ms additions] in a technical division or department.

4   At present, one Commander, R.A.N., one Lieutenant, R.A.N. and two Lieutenant-Commanders, R.C.N. hold the p.s.c. qualification.

It is suggested that the question of the possible appointment of Commander C.J. Pope, p.s.c. R.A.N. to the British Naval Staff should be considered and taken up with the Dominion authorities concerned.

[214] *Minute on South African Naval Service by Charles M. Pitman [Crown Law Officer]*
19 April 1926

N.L. [Naval Lord] With reference to your minute of 6th April 1926 and the papers accompanying it: The questions of Naval Discipline Legislation with regard to Dominion Forces while operating in their own and other waters, and with regard to Officers and Ratings of the Royal Navy while lent to the Dominion Naval Forces, or serving in the Dominions are questions which have in the past given rise to some little anxiety and are not easy to solve.

As I understand the matter, the objects which the Admiralty and the Legislatures of the Dominions concerned have always had before them are, in the first place to ensure uniformity in the laws relating to Naval Discipline; and in the second place to ensure that in time of war or other emergency there shall be no overlapping of disciplinary authority, i.e. that unity of disciplinary authority shall be co-extensive with unity of command. These difficulties have been practically surmounted by the Dominion Naval Forces Act

1911, which provides a scheme by which the Naval Discipline Act, with suitable modifications in each case, can be applied to those self-governing Dominions which adopt it.

The Union of South Africa is one of the self-governing Dominions specified in that Act to whom power of adoption is given. It does not appear from any of the accompanying papers that the attention of the Government of the Union of South Africa has been called to the advantage that will accrue to the Union's Naval Forces and to the Naval service of the Empire if they can see their way to solve the question they have had under consideration by adopting the Dominions Naval Forces Act of 1911. By so doing they would increase the Imperial utility of their Naval Forces, and by modifying the Naval Discipline Act in its application to South Africa, they would not only meet local requirements, but also retain control of all the purely domestic matters involved in the maintenance of discipline in their own Naval Forces in time of peace.

If they were to adopt the suggested course, the valuable work that has been done in the preparation of the proposed 'code' would be of great service in the preparation of the modifications of the Naval Discipline Act which would be necessary. I have not considered the proposed 'code' in detail, because I agree with N.L. that it is desirable that the question of principle should be settled in the first place, and that such question should, if possible, be settled by the inclusion of the Union of South Africa among the self-governing Dominions who have adopted the Dominion Naval Forces Act of 1911.

[215] *Draft paper, 'The Law of Prize in Relation to the Dominions'*

c. June 1926

His Majesty's Government are probably more interested than the Government of any other Power in maintaining to the full the right of prize, and the difficulties which would be created if any constituent parts of the Empire adopted views conflicting with or departing from the present law are plain. Even at the present time it is not always easy to convince neutrals to say nothing of belligerents, of the legality or reasonableness of British action upon the seas in time of war; and the task of the Foreign Office would not be lightened if it had to justify half a dozen codes of prize law (possibly conflicting with each other) instead of one, or was liable to have quoted against it the prize law of a Dominion whose legislation had gone whoring after strange gods. And lastly, so long as the Royal Navy continues, as it must, to be the common instrument of policy of the Empire as a whole, it is essential that in the conduct of naval operations it should be governed by a single and uniform code.

[216] *Memorandum, 'New Zealand Naval Policy', submitted by Captain W.A. Egerton, Director of Plans*
19 October 1926

It is understood that the Government of New Zealand would be willing to undertake additional commitments for Imperial Naval Defence amounting to about £250,000 p.a. The purport of this paper is to examine the alternative ways in which this sum could be best utilised to strengthen the naval defence of the Empire.

Singapore

2   It is now generally recognised that the provision of a defended naval base at Singapore is at once the best guarantee of peace in the Pacific Ocean and the essential pre-requisite of victory should the security of the Empire be actively menaced. This project has in fact been placed first in order of priority of the requirements for the Defence of the Empire: it is therefore an undertaking of Imperial importance for the benefit of the Empire as a whole.

New Zealand, being situated in the Pacific has a direct interest in seeing this undertaking brought to fruition and by assisting financially she would be directly contributing to her own security.
3   It is suggested therefore that at least some large proportion of this money should be devoted to Singapore. At the same time it is realised that, from New Zealand's point of view, assisting this scheme which is to cost some £13,000,000 is not so satisfactory as similar expenditure on some tangible asset which can be seen and appreciated by her people.

Cannot this Difficulty be Overcome Wholly or in Part?

It is suggested that it can be at least minimised if, instead of suggesting a contribution towards the cost of Singapore, the

Government of New Zealand be asked to undertake some specific item such, for instance, as the dry dock.

4   It is realised that very great difficulty would result from separate contracting for the various items, but it is felt that this is really unnecessary and that it would only be a matter of bookkeeping to differentiate between money expended on the docks or any other specific item.

Assuming that this proposal is accepted it would be easy to at once speak of the 'New Zealand' dock and to arrange for a suitable commemoration tablet recording the fact that the dock was presented to the British Empire by New Zealand to be included in the detailed design. If in addition it is understood that when the time comes the Prime Minister of New Zealand will be invited to open the dock &c., it is felt that a definite connection would have been set up which would go far towards popularising the expenditure in the Dominion.

Sloops

5   It is the accepted and settled policy of New Zealand to maintain sea-going forces, the vessels being provided by the home government but all other costs being borne by New Zealand.

At present this policy is only applied to the two cruisers and not to the Sloops attached to the New Zealand command.

This is a peculiar anachronism since the sloops, in the main, perform duties which are very specially a New Zealand commitment – viz: – the visiting of islands administered by the Government of the Dominion.

6   It is suggested therefore that, before undertaking any further commitments on sea-going forces, the New Zealand Government should take over and maintain the two sloops now on the station.

A Third 'D' Class Cruiser

7   It is understood that New Zealand desires to maintain a third 'D' class cruiser which would, in itself, use up all the available money and prevent New Zealand from either taking over the Sloops or assisting in the development of Singapore.

Apart, however, from this important objection, it is considered undesirable to encourage this desire for the following reasons:

(a)   It would not really assist in Imperial Naval Defence. Either expenditure as a whole would be increased, or we should have to reduce a cruiser elsewhere.

Given that only a certain number of cruisers can be maintained in full commission, by the Empire, the situation of New Zealand is such that 2 Cruisers really fulfil all requirements.

(b)   It is improbable that New Zealand will be able to undertake the maintenance of 3 new type cruisers. On the other hand it will be necessary to replace the 'D' Class cruisers on the New Zealand station not later than 1934 and preferably at an earlier date.

If, meanwhile, a third cruiser has been added to the N.Z. Division, she will be faced with the choice of again reducing to two cruisers or of shouldering a greater financial burden than it is fair or just to ask her to bear.

Docking Facilities

8   It is essential that New Zealand should realise that in a few years time, her cruisers will have to be replaced by modern type vessels and that, at present, she has no dock capable of accommodating them.

The provision of a dock capable of accommodating modern cruisers in New Zealand is, moreover, highly desirable on broad strategic grounds but, while this cannot be lost sight of, it is safe to say that there is no immediate urgency to consider it.

Conclusions

9   Assuming that New Zealand is prepared to devote £250,000 p.a. in addition to her present expenditure on Imperial Naval Defence it is considered that the best use to which this sum can be put is as follows in order of priority:

(a)   To provide a 'New Zealand' dock at Singapore as part of the Singapore scheme or some other selected item in this scheme.

(b)   To take over and maintain the two sloops attached to the New Zealand Division.

(c)   To provide docking accommodation for modern cruisers by 1934 at the latest.

(d)   To make all provision for the maintenance of two 'B' Class cruisers in 1934 and subsequent years instead of the 'D' class vessels now maintained.

[217] *Imperial Conference, 1926, extracts from the stenographic notes of the 12th Meeting*

15 November 1926

'Secret'

Statement of Mr Mackenzie King, Prime Minister of Canada:

Navy

Our naval activities are as yet on a small scale. It is not necessary to detail the circumstances which prevented the growth of the Canadian Navy along the lines which we anticipated in 1909. The special geographical position of Canada would have made it unnecessary to aspire to very rapid or extensive development, but had it not been for the circumstances to which I allude, we could doubtless have reasonably expected a more adequate force than as yet exists. I cannot say when that 'active and determined support of public opinion' which is so properly stated in the Committee of Imperial Defence memorandum on Empire Naval Policy and Co-operation of 1923 as being essential for the effective maintenance of naval forces will make it possible to advance to a further phase, but the question is receiving consideration.

The policy on which the naval activities of the Dominion are based at present is one of developing the local defence of the waters in the vicinity of Canadian coasts and the approaches to our ports. Also it is considered that any naval programme should, as far as possible, be one which will admit of the personnel being for the most part, and as soon as practicable, entirely Canadian. There is also in effect a system of co-operation in staff work and an arrangement of periodical service with the Royal Navy by officers and men of the Royal Canadian Navy in order that they may be trained to carry out their duties in all respects on similar lines.

In conformity with the above policy it may be stated that in the last five and a half years the personnel of the permanent Canadian Navy has been transformed from 450 officers and men borrowed or specially engaged from Great Britain, and 50 Canadians, to 40 borrowed ranks and ratings and 460 Canadians. The Royal Canadian Naval Volunteer Reserve has been organised in the last three and a half years, and is up to its full authorized strength of 1,000 officers and men. There is also a Royal Canadian

Naval Reserve of 150 officers and men. The naval training centres at Esquimalt and Halifax are efficiently equipped to train the personnel of the permanent and reserve forces. A considerable amount of work has been done by the naval service of Canada in giving the periodic training required to be carried out by the Royal Fleet Reserve men of the Royal Navy resident in Canada . . .

Statement by Mr Bruce, Prime
Minister of the Commonwealth
of Australia:

Defence Organisation in Australia

The guiding principle on which all our defence preparations are based, whether for the Sea, the Land, or the Air Force, is uniformity in every respect – organisation, methods of training, equipment, &c. – with the fighting services of Great Britain, in order that in time of emergency we may dovetail into any formation with which our forces may be needed to co-operate.

Close co-operation exists between the Home and Australian Services. Australian liaison officers are attached to the Admiralty, War Office, and Air Ministry, and a free exchange of information takes place. Every possible assistance is given to our officers so that we may be kept up to date, and valuable help is given in connection with the purchase and inspection of stores.

The organisation in our Royal Australian Navy needs no comment – it is similar to the organisation existing in the Royal Navy.

The ships of the Royal Australian Navy in commission consist of 3 Cruisers, 3 Destroyers, 3 Sloops, and 1 Repair Ship. In reserve there are 1 Cruiser, 1 Flotilla Leader, 8 Destroyers, and 1 Sloop.

The principle of interchange of cruisers has been established, and was carried out in 1925 and 1926.

I would like to say a special word of appreciation of the assistance that the Admiralty has given in connection with our exchange of cruisers. One of them came back round the world with Sir Frederick Field's Special Service Squadron, and I am certain that the opportunity it has given to our personnel of obtaining training is of incalculable benefit; it also had the effect of stimulating a much greater interest in the Navy generally, and has unquestionably assisted us in our recruiting for our Naval personnel.

Two Australian cruisers will probably visit England early in 1928 to turn over their crews to the two new 10,000-ton cruisers

Australia and *Canberra*, while H.M.S. *Renown* will be in Australian waters during 1928 . . .

Training (a) Navy

The training for our Navy is based on the traditions of the Royal Navy.

The supply of officers is provided for by the Royal Australian Naval College, where the conditions and education are similar to those at the Royal Naval College.

After completion of the College Course midshipmen are drafted to ships of the Royal Navy for a period of two years; other junior officers serve their period of training in the Royal Navy, and in all the Royal Naval training establishments, and a free interchange of Royal Naval and Australian officers exists. For the present, and for some years to come, until the Australian officers have become sufficiently senior – those, of course, are officers we are dependent on Royal Navy officers to command our ships and fill other senior appointments, although at present, and for the first time, a Commodore of the royal Australian Navy, formerly a Royal Navy officer, commands our squadron.

The present strength of the permanent sea-going Force is 5,000 men, of whom 10 per cent, are on loan from the Royal Navy. In 1913, when the Royal Australian Navy was established, over 70 per cent, were borrowed from the Royal Navy. We certainly regard that as an extraordinarily successful advance, having progressed from the point in 1913 where we were dependent on the Royal Navy for 70 per cent, of our personnel to the present time, when we are only supplied with 10 per cent.

Recruiting for the Australian Navy presents few difficulties, and it is anticipated that the call on the Royal Navy will be still further reduced . . .

Equipment

Naval equipment is on a similar basis to that of the Royal Navy, and the usual complement of reserve stores is always available.

Sydney is the main base for the Australian Fleet, and all docking and refits are carried out there at the Government dockyard.

The New South Wales Government are about to build a floating dock capable of docking a 10,000-ton cruiser. Towards the cost of this the Commonwealth Government are contributing a subsidy. The basis of the subsidy is that the dock is to be a floating dock which we can requisition at any time and tow to any part of

Australia and utilise it. That is the basis of the subsidy, and I believe it will be extremely valuable to have in Australia a floating dock capable of accommodating a 10,000-ton cruiser and capable of being taken to any part of Australia should necessity arise.

The development of Darwin as a subsidiary naval base has been under consideration; and I discussed the value of it with the Naval Officers at the Admiralty when we had an opportunity of discussing these questions with them.

Two 8,000-ton oil tanks and one 5,000-ton petrol tank are in course of erection at Darwin; two more 8,000-ton oil tanks will be commenced at an early date, while five additional 8,000-ton tanks are contemplated, the idea, of course, being to get oil supplies at Darwin in the north of Australia opposite to Singapore . . .

Singapore Base

The only other subject I want to say anything about is Singapore. We in Australia have not in any way altered the views I expressed three years ago. At that Conference we went exhaustively into the question whether Singapore was the right spot for a naval base for the Far East. We in Australia were quite satisfied that it was essential that there should be in the East a base where the British Navy could be quartered in the event of any trouble arising in the Pacific, for the purpose both of protecting the territories of the Empire and also of ensuring the keeping open of the trade routes, and our final decision was that Singapore was the place. At the 1923 Conference, I pointed out that we had not then determined upon our defence policy, but that it was essential that we should do so, and I made it quite clear that we would be quite prepared sympathetically to consider the question of contributing towards the establishment of the Singapore base. After the Conference had been held, and after the principle of the establishment of the base at Singapore had been subscribed to, the British Government changed and a different policy was adopted, and it was announced quite definitely that Britain did not propose to proceed with the building of the base at Singapore. We could delay no longer; we had to go forward in the altered circumstances and determine what our defence programme should be for the next few years. We laid down that five years' programme to which I have referred, possibly at some considerable length. I want to make quite clear the position we now find ourselves in in relation to Singapore. Under that programme of 1924, based upon the circumstances as they then were when Singapore was dropped, we went to the furthest limit that financially we could with regard to ensuring our own defence and our contribution towards the general defence of the Empire. As that programme involved the figures which I have given – a figure actually of £36,250,000 over a period of five years – it is very gravely doubtful whether we can now do anything further, particularly in view of the fact that we have got the report of the Committee of Imperial Defence with regard to the protection of our own capital city, which indicates that we ought to undertake an expenditure there of £4,000,000. I can only say, therefore, that Australia believes that the Singapore base is absolutely essential, and, while I could not for one second suggest that Australia make any contribution towards i[t]s construction in view of the commitments remaining under this five years' programme, I can promise that our position with regard to Singapore will be discussed in the Commonwealth Parliament after my return, and it will be for Parliament to come to any decision it may think fit . . .

Statement by Mr Coates, Prime
Minister of New Zealand:

Defence Policy of New Zealand (a) Navy

In so far as New Zealand is concerned, the necessity of sufficient cruisers and a base for operations in the East which can command the trade routes of the Pacific and Indian Oceans is fully appreciated in the Dominion, and the policy of provision of these can be thoroughly endorsed. That was shown at the last Conference, when Mr Massey announced New Zealand's contribution towards the Singapore base. It is not practical politics at any rate at the present time for New Zealand to build cruisers, but it is her intention to continue her policy of developing her own Division of the Royal Navy, and to maintain cruisers of modern and suitable type, thus relieving the burden of the Mother country as regards seagoing vessels kept in commission.

I have had an opportunity of discussing in detail the Admiralty proposals regarding the Singapore base, and the question of New Zealand making a definite annual contribution in this respect will be submitted for the consideration of the New Zealand Government at an early date.

As the cruisers to be taken over by New Zealand increase in size, the provision of suitable docking and other facilities for their proper and efficient maintenance in our waters becomes imperative.

On these lines, therefore, I consider that the New Zealand Government will be prepared to increase their share of the burden which Great Britain has to bear to enable our sea supremacy in any particular quarter to become more efficient, and the details regarding carrying this into effect are now under discussion with the Admiralty.

Statement by Mr Monroe, Prime
Minister of Newfoundland:

Newfoundland's Contribution of Men to the Royal Navy

Men are the only considerable direct contribution which Newfoundland can make to Empire Defence, under her existing financial conditions. Her men for serving in the land forces possess, perhaps, no virtues not common to English-speaking peoples, but experience in the Great War has amply proven, to those who should know best, that her sea-going men have qualities that especially fit them for difficult and important operations at sea. It has been admitted that in conditions that existed very frequently in the North Sea the fishermen of Newfoundland who had been trained in the Royal Naval Reserve had a fitness for rough sea work and for mine laying peculiarly their own. Their lives spent very often in a small boat fishing, and in boat work in connection with sailing vessels, had given them a fearlessness and skill such as only a comparatively small number of the Empire can attain to. It was at one time thought, and even said, that for such men the Navy had no particular need, but this belief was found to be utterly false when the test of service came.

Early in the present century a branch of the Royal Naval Reserve was established in Newfoundland, with a training ship located in the harbour of St John's, and the training and discipline which were thus afforded had a very valuable effect when the trying time came.

Training of Naval Reservists in Newfoundland

Newfoundland lost more men at sea in the Great War than the other Dominions put together, but after the war financial considerations caused the Admiralty to withdraw the training ship from St Johns's and to abandon the training of Reservists. Newfoundland contributed only a small portion of the cost, and is, unfortunately, not now in a position to offer a much larger contribution, but she is willing to continue to do her share in the work which proved so valuable, and she urges that training at St John's should be renewed. The Reserve is not needed for her particular defence, and would not be used for that purpose alone or even, in any very striking way, be valuable for her especial purposes. She feels, however, that in offering her men she is tendering a really valuable contribution to the defence of the whole Empire, and not for any selfish considerations. When the Great War began there was not in Newfoundland any regular military organisation of any kind. It was regarded at the outset as improbable that under these circumstances any large number of men would volunteer for the war but to the gratification of the Dominion her youth sprang to arms, with the enthusiasm so universal throughout the Empire; several thousands volunteered and many of them made the supreme sacrifice during service.

[218] *Imperial Conference, 1926, extracts from the stenographic notes of the 12th Meeting*

19 November 1926

'Secret'

Resolutions on Defence

7 The Conference recognises that the defence of India already throws upon the Government of India responsibilities of a specially onerous character, and takes note of their decision to create a Royal Indian Navy.

8 The Conference notes with satisfaction that considerable progress in the direction of closer co-operation in Defence matters has been effected by the reciprocal attachment of naval, military and air Officers of the Dominions to the Staff Colleges and other technical establishments maintained in various parts of the Empire, and invites the attention of the Governments represented to the facilities afforded by the new Imperial Defence College in London for the education of Officers in the broadest aspects of strategy.

9 The Conference takes note of the developments in the organisation of the Committee of Imperial Defence since the session of 1923. It invites the attention of the Governments represented at the Conference to the following resolutions adopted, with a view to consultation in questions of common defence, at a meeting of the Committee of Imperial Defence held on the 30th May, 1911, in connection with the Imperial Conference of that year:

(1)   That one or more representatives appointed by the respective Governments of the Dominions should be invited to attend meetings of the Committee of Imperial Defence when questions of naval and military defence affecting the Oversea Dominions are under consideration.

(2)   The proposal that a Defence Committee should be established in each Dominion is accepted in principle. The Constitution of these Defence Committees is a matter for each Dominion to decide.

Appendix

1   The Conference affirms that it is necessary to provide for the adequate defence of the territories and trade of the several countries comprising the British Empire.

2   In this connection the Conference expressly recognises that it is for the Parliaments of the several parts of the Empire, upon the recommendations of their respective Governments, to decide the nature and extent of any action which should be taken by them.

3   Subject to this provision, the Conference suggests the following as guiding principles:

(a)   The primary responsibility of each portion of the Empire represented at the conference for its own local defence.

(b)   Adequate provision for safeguarding the maritime communications of the several parts of the Empire and the routes and waterways along and through which their armed forces and trade pass.

(c)   The provision of Naval bases and facilities for repair and fuel so as to ensure the mobility of the fleets.

(d)   The desirability of the maintenance of a minimum standard of Naval Strength, namely, equality with the Naval Strength of any foreign Power, in accordance with the provisions of the Washington Treaty on Limitation of Armament as approved by Great Britain, all the self-governing Dominions and India.

(e)   The desirability of the development of the Air Forces in the several countries of the Empire upon such lines as will make it possible, by means of the adoption, as far as practicable, of a common system of organisation and training and the use of uniform manuals, patterns of arms, equipment, and stores (with the exception of the type of aircraft), for each part of the Empire as it may determine to co-operate with other parts with the least possible delay and the greatest efficiency.

[219] *'Dominion Naval Powers, Opinion of the Law Officers of the Crown and Mr Charles M. Pitman, K.C.', sent by Douglas McGarel Hogg, T.W.H. Inskip, and Pitman*
24 January 1927

1   We are of opinion that the Naval Discipline Act, if duly applied by the Dominion Legislature to the Naval Forces raised by the Dominion, in accordance with the Naval Discipline (Dominion Forces) Act 1911, creates a system of Naval Discipline for the Naval Forces of such Dominion which is enforceable either within or without the territorial limits of such Dominion.

2   We are of opinion that in a Dominion, such as New Zealand, which has duly applied the Naval Discipline Act, an order to a member of the Dominion Naval forces to proceed to a ship or establishment of the Royal Navy would be a lawful exercise of such Naval Discipline, and a lawful order which he would be bound to obey. The existence of a local enactment authorising the imposition of such obligation merely emphasises the validity of such an order by making it clear to a person who voluntarily enlists in the Naval Forces of the Dominion that he is liable to be sent outside the limits of the Dominion for the purposes specified. A person disobeying such an order outside the territorial limits of the Dominion could be proceeded against in England or elsewhere for an offence against the Naval Discipline Act as applied by the local Legislature.

3   We are of opinion that by virtue of the provisions of the Naval Discipline (Dominion Forces) Act 1911 and of the application in the Dominion of the Naval Discipline Act, a person so ordered, if he should in fact proceed on board one of the ships of the Royal Navy in commission, would be amenable to the Naval Discipline Act under section 87 of that Act.

4   We have nothing to add.

[220] *'Canadian Government Naval Policy', minutes by Egerton, and G.F. Cotton for the Third Sea Lord, Rear-Admiral Sir A.E.M. Chatfield*

April 1927

Minute for Head of Military
Branch of Admiralty
Secretariate [under the
Deputy-Chief of Naval Staff
Vice-Admiral Sir Frederick
Field], 4 April 1927.

Two courses are open to the Canadian Government:

(i) To build 2 destroyers.
(ii) To purchase 2 destroyers from our Scrapping List.

The Naval Staff will probably agree that (i) is the ideal arrangement. It is doubtful, however, whether the Canadian Government would face the expense. D.W.P. [Director of Weapons Procurement] can say what would be the cost of building 2 destroyers of modern type.

As regards (2), bearing in mind the importance of sending up-to-date and efficient vessels to the Dominions, it would seem reasonable to consider the sale of 2 destroyers from Reserve (or Maintenance Reserve if that scheme should be approved) which would be scrapped in the course of the next 10 years. Preferably they should be the most efficient vessels, i.e. 2 of the destroyers in Reserve with the longest lease of life. The market for warships is now limited to our own Nationals, and it is thought that it would be fair to argue that the sale value of a vessel in Maintenance Reserve would not be more than her scrapping value. If this should be accepted, 2 destroyers might be offered to Canada at scrapping value, plus the cost of refitting and retubing. D. of C. and D. of D. [Director of Contracts and Director of Design ?] can advise as to the charges that should be made under these heads.

However, even if 2 V's or W's can be brought into this category, it would not be such a satisfactory arrangement as the construction of 2 new vessels by the Canadian Government if they can be induced to take this step.

Referred for remarks [indecipherable signature]

Minute by [Egerton], Director of Plans,
8 April 1927

From the tenor of Canada's telegram, it appears clear that she does not contemplate building destroyers to replace *Patriot* and *Patrician*. If the Admiralty accept this situation now, they will be condoning, for a further indefinite period of years, the failure of Canada to develop her naval resources and acquiescing in Canada retaining a special position in this respect vis-a-vis the other Dominions.

In the Memo. on Naval Policy and Co-operation 1926 (C.I.D. 264-C), Phase II of naval development is described in the following terms:

'This phase' it says, 'involves the obtaining of one or more seagoing ships and, in the first instance, it may be necessary for financial or other reasons, to obtain both the ships themselves and a portion, at any rate, of the Personnel from Great Britain'.

Attention is particularly called to the words 'in the first instance' because Canada was in 1919/20 given a Cruiser and two Destroyers to enable her to develop her naval defence within Phase II.

She did not take steps to replace the 'Aurora' and now asks for two Destroyers, not to strengthen her position, but merely to maintain it.

2   It has further to be remembered that

(A)   Canada insists on her right to decide whether or not vessels of the R.C.N. shall or shall not be placed at our disposal in the event of war. We have, therefore, to consider the possibility of these destroyers not being available at a time when every destroyer will be invaluable. and

(B)   Canada decides, on her own volition, to maintain the Destroyer type whereas, from the Imperial point of view Cruisers and/or Submarines would be more appreciated.

3   In view of (A) above, the making over to Canada of two Destroyers with a reasonable period of remaining life implies either risking being two short of our bare minimum requirements on the outbreak of war or the building of two extra vessels to replace them, which latter expedient represents taxation of the people of this country for the benefit of Canada.

Seeing that in 1925, Great Britain spent 25s/7d per capita on

Naval defence as compared with 8d by the people of Canada, this latter appears entirely unjustifiable.

4   Finally, it is necessary to consider what the effect would be if Canada did not replace these two Destroyers at all. She would still remain responsible for her local Naval defence (for which incidentally modern destroyers are unnecessarily important units) and we should still have to build and maintain the same number of Destroyers, no more and no less.

5   In these circumstances, it is suggested that the alternatives are:

(i)   To offer the use of drawings and technical advice to enable Canada to build two Destroyers, informing her that it is anticipated that new construction destroyers of the 'A' type will have a life of 20 years.

(ii)   To offer, subject to Treasury approval, two 'S' class destroyers, informing Canada of the price. It is suggested that this price should be calculated on the estimated remaining life (based on 16 years total life) plus the cost of bringing the vessels forward for service, retubing etc.

(Note: In the circumstances referred to in paras. 1–4, no reason is seen to offer vessels at scrapping value. On the other hand, it would appear equitable that if Canada replaces the *Patriot* and *Patrician* by vessels purchased at their full value they should be allowed to retain the scrap value of the *Patriot* and *Patrician*)

It is suggested that both alternatives should be put forward with a strong recommendation in favour of alternative (i).

Minute by Cotton
for the Third Sea Lord
[Chatfield]
12 April 1927
D[irector] of P[lans] has brought out very clearly the disadvantages of presenting two destroyers to Canada which, in the event of war, might not be available for use in the Fleet.

I therefore consider we should, in our reply to the Dominion Office, suggest that every endeavour should be made to persuade Canada to develop her Naval Policy in the manner indicated in C.I.D. Memorandum 264C, emphasising that Cruisers and/or Submarines are the most suitable vessels as Canada's initial contribution towards Imperial Defence. If, however, Canada is unable to progress her Naval Policy to this extent, the Admiralty would, to enable Canadian Naval Personnel to receive some seagoing training, be prepared to loan two 'S' Class Destroyers on condition that Canada agree to pay for their refit and retubing. This loan would be on the strict understanding that these destroyers would be available for employment with the Imperial Fleet at any time it is considered that they are required.

The reason for this attitude should be made clear to the Dominion Office by embodying the sense of the appropriate portions of D. of P's submission.

[221] *'Policy Relating to Armed Merchant Cruisers on Canadian Coast 1927', Sir Edward Harding, [Permanent Under-Secretary of State for Dominion Affairs], to Flint[?]*
18 May 1927

Have under consideration the fitting out of four armed merchant cruisers on the Western Coast of Canada. In the event of a war in the Far East, in which Canada was taking an active part, the presence of such ships on the Western Coast of Canada would be very desirable, the object being largely to prevent American contraband trade with Japan. Their Lordships estimate that the very least time which would elapse between the declaration of war and the arrival at Vancouver of armed merchant cruisers equipped in England would be two and half months... Their Lordships propose that, if possible, four ships of this line [i.e. Canadian Pacific], preferably the *Empress* type, should be taken up on the outbreak of war... It is not intended to take any action as regards stiffening... Their Lordships would propose that the Canadian Naval Authorities should undertake the storage of the necessary equipment in peace, and the work involved in fitting out the ships when the occasion arose...

[222] *Minute by 'D.J.' [unidentified] for Deputy Chief of Naval Staff Field], in response to a minute by Director of Naval Engineering [Engineering Vice-Admiral Sir Robert E. Dixon, Engineer-in-Chief of the Fleet?]*
25 May 1927

I agree with D[irector of] N[aval] E[ngineering] (for Controller [Rear-Admiral Sir A.E.M. Chatfield]) that Sloop Minesweepers would be a very useful class for Canada, but looking ahead I think it would be a mistake to suggest to Canada that they should go in for a class of ships of a category lower than that already maintained by them.

Another point which must be taken into consideration is that as a result of the Coolidge Conference it may not be possible to spare 2 Destroyers for maintenance in Canada.

Propose to reply in the sense of D.C.N.S.'s minute of 12.4.27 adding that the replacement of 'Patriot' and 'Patrician' by 'S' Class Destroyers cannot definitely be agreed to until the result of the Coolidge Conference is known, but that in the meantime the question of the terms and when such vessels, if available, could be delivered will be gone into.

Propose also to take action as regards x above inside Admiralty. [manuscript addition] D[eputy] C[hief] O[f Naval] S[taff, Vice-Admiral Sir Frederick Field] would like to see draft letter.

[223] *Telegram from the Secretary of State for Dominion Affairs [L.C.M.S. Amery] to [Viscount Willingdon] Governor-General of Canada, sent*

9 June 1927, 2.30 p.m.

9th June. I am sorry that it has not been possible to reply earlier to your telegram of the 1st April. His Majesty's Government in Great Britain are much interested to know that His Majesty's Government in Canada are giving consideration to the question of future method of providing naval craft and note that in the meantime it is desired to negotiate for two destroyers of more modern type than H.M.C.S. *Patriot* and *Patrician*.

Admiralty suggest however for consideration that instead of obtaining two more modern destroyers it might be preferable for Canadian Government to order two new type minesweeping sloops of type similar to those which are about to be built for the Royal Navy; these sloops are particularly suitable for the general training of personnel and would it is thought form an essential part of any naval force which the Canadian Government might eventually decide to maintain. Two such vessels as those indicated could be constructed in this country in 18 months or in Canada in a slightly longer time. The estimated cost of construction of each ship in this country would be £100,000 and probable life estimated at not less than 16 years. Specifications and drawings follow by mail.

If Canadian Government are disposed to consider this alternative favourably and if the condition of H.C.M.S. *Patriot* and

Patrician is such that these vessels would not last until the proposed sloops were ready, Admiralty would endeavour to arrange to lend two destroyers of 'S' type on the understanding that Canadian Government would undertake to bear the cost of bringing the vessels forward for service. This is estimated at, approximately £11,000 per vessel.

[224] *Minute by T.H. Burney [Deputy Director of Plans] for Egerton*

2 August 1927

By acquiring from Great Britain two 'S' class destroyers for immediate replacement of *Patriot* and *Patrician*, Canada will be departing from the policy recommended in A.L., M.01021/27 of 13th May 1927 to Dominion Office in which it was stated (para. 9) that destroyers were unsuitable for Canadian needs. This fact was also mentioned in first para of the telegram from Dominion Office to Governor General, Ottawa.

2   In this letter it was recommended that, besides building cruisers, Canada should tackle the immediate problem of replacing *Patriot* and *Patrician* by building two new type sloops (para. 10) and that if the *Patriot* and *Patrician* would not last till the 2 sloops were completed, the Admiralty would consider lending 2 'S' class destroyers to fill the interval. (para. 11).

3   Telegram 52 from Geneva contemplates Canada not being able to start any programme of construction and anticipates purchasing two later date destroyers in lieu of building sloops.

4   In these circumstances, there seems to be no reason why we should help Canada in carrying out a policy against our advice by quoting a low price for 'S' class destroyers. Rather it is considered, the higher the price quoted the more chance is there that she will be encouraged to get on with the building of the sloops.

5   It is suggested that the price should be calculated on the following basis: Most 'S' class destroyers were completed in 1918-19. They have thus finished half of their 'official life' of 16 years. The price charged might therefore be half cost price plus the cost of reconditioning.

6   Further, it should be made clear that the question of the sale to Canada of these ships still depends on the result of the Geneva Conference.

[225] *Minute by Cotton for Head of Military Branch of Admiralty Secretary's Department [under Field]*

8 August 1927

It is suggested that the telegram from Geneva should be regarded as a simple request for information and should be replied to on the same basis without going into questions of policy.

The Director of Contracts has been consulted on the proposal made in paragraph 5 of the minute of Director of Plans and has pointed out that, as a commercial proposition, the action suggested could be challenged on the ground that the second half of the life of a destroyer could not be regarded as so valuable as the first half. It has to be remembered also that these vessels were constructed as part of the war programme and that inflated prices were paid for them, and it is felt that a figure more in the neighbourhood of one-third of the cost, plus reconditioning expenses, would be a fairer price and one less open to argument.

In the case of trawlers, it has sometimes been the practice to assume a life of 16 years and a 10% depreciation each year, i.e. 10% off each successive year's price. This would give a figure of about 44% of the cost of the destroyers. It is suggested that the price quoted to Canada for 'S' class destroyers should be assessed on this basis.

The average cost of an Admiralty 'S' class destroyer is given by D.E.A. as £213,196, and C.S.A.S. states that the value of her Vote 9 outfit (including reserves) would be £38, 800, i.e. a total cost of £251,996. 44% of this sum is £11,878. Adding £11,000 for reconditioning the total chargeable would be £121,878. It is suggested therefore that the price quoted to Canada for each 'S' class destroyer should be £122,000 including the cost of reconditioning.
Submitted.

[226] *Field to Commodore Hose, [Canadian Director of the Naval Service], c/o High Commissioner for Canada*

12 August 1927

You will recollect that while we were at Geneva the question arose of the terms upon which the Admiralty would be prepared to transfer two 'S' Class destroyers to Canada, in the event of the Canadian Government not being in a position to embark at the present time on a programme of new construction of either destroyers or sloops.

I have now been able to ascertain that the Admiralty would be prepared to sell two 'S' Class destroyers to Canada for the sum of £75,500 each, plus the cost of storing, reconditioning, and generally preparing for service with the Royal Canadian Navy, which is estimated at £11,000 per destroyer. These destroyers would be of late war time construction, and therefore about nine years old – i.e., they would have completed more than half of their 'official life' of 16 years.

P.S. The above sum of £86,500 does not of course include expenditure in connection with the passage from England to Canada.

[227] *Captain F.L. Horsey, Secretary to Deputy Chief of Naval Staff [Field], to Cotton*
17 August 1927

Commodore Hose, R.C.N. called on A[cting] C[hief of] N[aval] S[taff, probably Vice-Admiral Sir Frederick Field] this morning and discussed Mr Flint's [British High Commissioner in Ottawa] letter of the 12th August. In the course of the discussion he asked A.C.N.S. whether (1) if Canada buys outright two 'S' Class Destroyers, would payment be required in financial year in which delivery is made? He said, in this connection, that the Canadian estimates only provide enough money to pay for refit; (2) would Admiralty agree to the money received by the sale of *Patrician*, *Patriot*, *Aurora* and Submarines when scrapped to be credited against the cost of Sloops, if the latter were constructed?

A.C.N.S. asks whether you would be good enough to ascertain unofficially the probable answers to the two points raised by Commodore Hose. [signature]

[228] *Cotton, for Head of Military Branch Admiralty Secretary's Department [under Field], to Horsey*
17 August 1927

As regards (1), the normal procedure would be for the Admiralty to put in a claim for payment directly after the two vessels were transferred. If nothing were received at the end of the financial year, Canada would be asked again for payment. Of course, no pressure would be brought to bear if payment were not forthcoming, and the account might be outstanding for a long time. It took six years to get payment of a sum due from Newfoundland in respect of vessels transferred to that Dominion, and I believe India still owes us about £4,000,000 on various outstanding accounts.

If any other course should be recommended, e.g., payment by instalments, the Treasury would have to be consulted and they would almost certainly require that interest should be paid on outstanding amounts.

As regards (2), I can see no reason why one transaction should be set against the other. To credit Canada with the proceeds of the sale of the ships mentioned would be equivalent to making her a free gift of £ x. One cannot get away from the fact that it would be in the nature of a bribe. It would certainly have to receive Treasury approval, and even if the Admiralty entertained the idea I am quite certain the Chancellor would not look at it.

[229] *Telegram from the Secretary of State for External Affairs, Ottawa, [W.L. Mackenzie King] to Amery*
9 November 1927

'Secret'

Your telegram of 9th June, 1927, to Governor General relative to destroyers H.M.C.S. *Patrician* and H.M.C.S. *Patriot*. His Majesty's Government in Canada has decided to ask for an appropriation during the coming session of Parliament with a view to placing contract in England for construction of two new destroyers to be delivered one at a time but both within three years to replace H.M.C.S. *Patrician* and H.M.C.S. *Patriot* which have become no longer serviceable.

In view of unserviceable condition of H.M.C.S. *Patrician* and H.M.C.S. *Patriot* and necessity for having naval craft available for training purposes until these vessels are replaced, His Majesty's Government in Canada concur in arrangement contained in paragraph 3 of your telegram whereby two 'S' type destroyers will be temporarily placed at their disposal on the terms that His Majesty's Government in Canada will pay cost of reconditioning these destroyers amounting it is estimate to £11,000 each.

It is understood that Director of Naval Service in Canada when in London in August last was informed by the Admiralty that the two 'S' type vessels which would be allotted if arrangements were consummated would be H.M.S. *Torbay* and H.M.S. *Toreador*.

In the event of these arrangements being agreeable to His Majesty's Government in Great Britain, it would be appreciated if information could be telegraphed as soon as possible indicating estimated time necessary for reconditioning of the two 'S' type destroyers as it is desired that training and operations should proceed with as little interruption as possible.

Proposal of His Majesty's Government in Canada with regard to the intended placing of contract for new vessels should not be made public until Estimates for the coming financial year are tabled in the Canadian House of Commons.

[230] *'General Principles of Imperial Defence', memorandum by the Oversea Sub-Committee of the Committee of Imperial Defence, signed G. N. Macready, Secretary Oversea Defence Committee*

November 1927

Co-Operation in Defence by the Component Parts of the Empire.

39   The principles already stated in connection with defence against land attack apply equally to the defence of the Empire as a whole. It is essential that as large a proportion as possible of the military, naval and air resources of the various parts of the empire should be available, when need arises, for employment against the enemy's main forces. At the same time, each individual part of the Empire must have adequate local protection against such forms of attack as are likely, it being remembered that the best way to ensure local immunity is to keep the enemy main forces so fully occupied that they cannot afford detachments.

40   While it has been and remains the accepted policy that each component part of the Empire should be responsible for its own local defence, it has been also found desirable that the burden of providing and maintaining the 'mobile defence forces,' naval, military and air, should be shared throughout the various parts of the Empire according to the varying resources of the different territories. It is this aspect of Imperial Defence that for geographical and constitutional reasons has in the past presented the greatest difficulties.

For example, if the problem of Imperial Naval Defence were considered merely as a problem of naval strategy, it might be found that the greatest output of strength for a given expenditure could be obtained by the maintenance of a single Navy with a concomitant uniformity of training and unity of command. In furtherance, then, of the simple strategic ideal, the maximum of local power might be gained if all parts of the Empire contributed, according to their resources, to the maintenance of a single Navy.

41   In formulating a policy for the development and maintenance of naval strength, other considerations than those of strategy alone must, however, be taken into account. The various circumstances of Great Britain, the Dominions and the Colonies must be borne in mind.

42   Differences in their history and physical environment, and other causes, gave rise to diversity of views as to the form in which contributions towards the defence of the Empire should be made; and it was found desirable that in such matters full scope should be given for the expression of local interest and sentiment. Some Dominions, for example, preferred to assist by a simple contribution of money or material to the British Fleet, or to the formation of naval bases, while others created or are creating navies of their own. The result of the establishment of Dominion navies is to relieve the Admiralty of the responsibility for maintaining ships in certain waters in the immediate neighbourhood of the Dominions concerned in time of peace, and thus to permit of a further development of the policy of concentration.

43   Again, some of the territories, for whom local navies are out of the question, and whose security depends directly on the ability of the Navy to maintain concentrations of adequate strength in their neighbourhood, have given large grants of money or land towards the establishment of a naval base. Other Colonies are taking steps to raise locally Royal Naval Volunteer Reserve units, whose assistance in mine-sweeping, coastal patrolling, and other forms of local naval defence, will be invaluable.

The policy is for all navies, and all locally raised units throughout the various parts of the Empire, to be organised and trained on similar lines.

44   Thus it now rests with the Governments of the various component parts of the Empire to decide what forms their contributions will take, whether they will consist of naval units, financial contributions to the formation and maintenance of naval bases, military expeditionary forces (if required in an emergency), air units, or assistance in the provision of Air Bases. In order, however, that co-operation may be effective, and that the waste of effort inevitably

associated with lack of co-ordination may be avoided, certain requirements must be fulfilled.

45 Modern warfare will seldom admit of reliance solely on military forces – and still less air forces – hastily improvised subsequently to the outbreak of hostilities . . .

In a few instances, hasty improvisation may be practicable, but it is probable that, in the future, great naval, military and air events will immediately follow, if they do not precede, a declaration of war.

Any military system, therefore, to be effective, must be capable of providing, when a crisis arises, for the immediate mobilization and concentration in organised bodies of the troops required for home defence, and for at least a part of any force which the Government concerned may desire to contribute to the common cause.

Under cover of these 'first-line troops,' prearranged schemes for the mobilisation of the military effort of the various parts of the Empire can then be put into operation.

46 The second requirement is uniformity in organisation, equipment and training. The difficulties of supply and maintenance alone would be enormous if contingents from different parts of the Empire possessed different organisations, different arms and varying patterns of transport vehicles.

That the principle of uniformity has been generally accepted is shown by the following extract from the proceedings of the Imperial Conference of 1926:

'The Conference observes that steady progress has been made in the direction of organising military formations in general on similar lines; in the adoption of similar patterns of weapons; and in the interchange of officers between different parts of the Empire; it invites the Governments concerned to consider the possibility of extending these forms of co-operation and of promoting further consultation between the respective General Staffs on defence questions adjudged of common interest.'

47 Again, it would be difficult to expect uniformity in the interpretation of operation orders if the military doctrine and training varied.

For these reasons, it is generally recognised that all military forces throughout the several parts of the Empire should be organised, equipped and trained on a common basis. To facilitate this, and to further the adoption of a common military doctrine, officers from all parts of the Empire are given facilities to attend courses of instruction at various military establishments and schools in Great Britain and elsewhere, and exchanges of officers taken place between India and the Dominions, as well as between the Dominions and Great Britain.

[231] *Captain H.J. Feakes R.A.N., Australian Naval representative in London, to Brig. General Sir Samuel Wilson [Permanent Under-Secretary of State, Colonial Office]*
29 Nov 1927

The phrase 'single Navy' is a dangerous one to revive, also it is subversive of both the spirit and the letter of principles as expressed to the Dominions by the Admiralty of their wishes relative to Dominion Naval contribution. The fact that such a sentence should in this year of grace be considered suitable for inclusion in the 'General principles of Imperial Defence' suggests to me that there is great need for us to review the progress of Dominion Naval Development from its inception ... A quarter of a century ago its Naval Defence was a live question in each of the Dominions. To-day in the largest, it is sick almost unto death. It is not too much to say that this tragic fact is due to the exponents of the 'single Navy' theory. The 'single Navy' theory has much to commend it if we are considering preparation of a definite, and immediate, war problem, or, as Paragraph 40 has it, a 'simple strategic ideal'. The Dominions are not – this is recognised later in the Memorandum ... It is in realisation of this objective [of the Dominions becoming populous nations] that the 'single Navy' theory fails. It is a theory that leaves the Dominion taxpayer cold. Even should a Dominion be found prepared to subscribe to it, no population could take a possessive interest in a force that really belongs to the other fellow! And, speaking generally no political [*sic*] is concerned with a question that does not interest the voter. Certainly he that controls the Exchequer would not be.

[232] *R.V.N. Hopkins, Treasury, to Murray*
22 December 1927

Loan of two destroyers for 3 years to Canada.

On the understanding that the two destroyers which you propose taking from Reserve to lend to Canada would not other-wise be sold and that Canada pays for reconditioning, takes over in this country and bears the whole expense of the vessels from taking over to return in England, I have no objection to you going ahead in advance of your official letter. I take it that you will secure repayment of the cost of reconditioning from Canada within the current financial year if the arrangement entails payments from your Votes in the first instance.

[233] *Admiralty memorandum 'Some General Principles of Imperial Defence'*

12 March 1928

Maintenance of Sea Supremacy

9   To obtain and maintain sea supremacy is the task of the naval forces, assisted by land and air forces with whom the closest co-operation is essential. In present circumstances the major share of responsibility in regard to the maintenance of sea supremacy rests with His Majesty's Government in Great Britain. The surest method of obtaining sea supremacy is to remove that which threatens it. Hence, the primary object is the destruction of the enemy's fleets and squadrons. This destruction is not readily attained, and the most usual situation is one in which the enemy conserves his naval force by refusing decisive action. Thus, the struggle for sea supremacy is prolonged, and supremacy is seldom absolute.

10.   Our naval strategy, therefore, is based on the principle that a fleet of adequate strength, suitably disposed geographically and concentrated against the enemy's fleet, provides the 'cover' under which security is given to widely dispersed territories and trade routes. This security cannot be given by the same strength of fleet dispersed to afford local protection to particular territories or trade routes. Such dispersion would leave the enemy free initiative, and would invite the destruction of these detached forces by an enemy concentration. Dispersion of the fleet, therefore, merely defeats its own object.

11.   Whilst, however, the main fleet is the basis upon which our naval strategy rests, naval requirements are not satisfied solely by its provision. The 'cover' it can provide is rarely complete, and cases have occurred in all wars of units detached by the enemy evading the main fleet and carrying out attacks of a sporadic nature on territories and trade. To deal with this menace, cruiser squadrons are required over and above those forming part of the main fleet.

[234] *Canadian Government Naval Policy, memorandum*
M.1062/28

11 May 1928

The following are the conditions attaching to the loan of two 'S' Class Destroyers (H.M.C.S. *Vancouver* and *Champlain*) to the Canadian Government:

(a) The Canadian Government to pay the cost of reconditioning the two destroyers. These sums provide for the carrying out of somewhat similar alterations and additons to those effected in *Patriot* and *Patrician*, prior to their transfer to Canadian Service.

(b) The Canadian Government also to meet the cost of any further alterations and additions that may be carried out; in view of the service for which the vessels are to be lent, few alterations and additions have been embodied.

(c) Stores

(i) Permanent Stores (Naval and Armament):

A full equipment to be transferred with the vessels free of charge; any items required in excess of a full equipment to be supplied on repayment . . .

(e) The Canadian Government to be responsible for returning the destroyers to England on the termination of their service in reasonable condition and to be liable for all costs in connection therewith until the date on which the vessels are accepted again by the Admiralty . . .

(g) The destroyers to be available for employment with the Royal Navy, if required in an emergency.

[Noted in manuscript]
Canadian gov[ernment] obtained alteration of para[graph] 'g' to read 'The destroyers to be available for return to the Royal Navy, if required in an emergency.'

[235] *Minutes of the 8th Meeting, Committee of Imperial Defence Sub-Committee on Singapore*

10 July 1928

Mr Bridgeman [W.C. Bridgeman, First Lord] said he very much feared that there would be an outcry from New Zealand if work on the Naval Base was postponed . . .

Conclusion: While recommending no modification in the complete scheme already approved for the Naval Base at Singapore, the Sub-Committee agreed to recommend – That with a view to arriving at a decision as to whether a contract should now be entered into for the construction of a dry dock, the First Lord of the Admiralty should approach the head of Sir John Jackson's unofficially and ascertain whether and to what extent he would be prepared, in the event of his tender being accepted, to retard or slow down the earlier stages of the work so that the Government payments should fall due at later stages.

[236] *Committee of Imperial Defence, Minutes of the 199th Meeting*

2 April 1929

Draft Answer to Question (b) (Adopted at the Meeting on March 30):

'The Committee accept the view of the Secretary of State for Foreign Affairs that in existing circumstances aggressive action against the British Empire on the part of Japan within the next ten years is not a contingency seriously to be apprehended.

The Foreign Office should be responsible for warning the Cabinet and the Committee of Imperial Defence of any change in the international situation in the Far East which would necessitate a fresh review of the question.'

Draft Answer to Question (c) (As suggested by the Chancellor of the Exchequer [Winston S. Churchill] and since slightly amended.)

'Although, in accordance with recent decisions, preliminary arrangements to establish docking facilities for our largest ships at Singapore and to develop gradually the necessary oil fuel installations on our Eastern routes should proceed, there is no necessity, in view of the answer given to Question (b), during the next ten years to make preparations involving additional expenditure for placing at Singapore, for a decisive battle in the Pacific, a British battle-fleet with cruisers, flotillas and all ancillary vessels superior in strength, or at least equal, to the seagoing Navy of Japan.

This decision should be reviewed automatically by the Committee of Imperial Defence every three years, beginning in 1928, in the light of the international situation.' ...

MR CHURCHILL [Chancellor] said that there was great virtue in a ten-year period. His intention was that when the situation was reviewed every three years the ten-year period should be renewed and that it should always date from the last review of the situation. In addition to that, any new factors would, of course, be taken into consideration as they arose. They were not bound to wait the full three years before reconsidering the position . . .

MR [Austen] CHAMBERLAIN [Secretary of State for Foreign Affairs] observed that the Foreign Office, in giving their advice on this question, were undertaking a great responsibility. He had no doubt whatever as to the soundness of their advice, but he wished to make it quite clear that it was confined to the question of there being no danger of war in the next ten years, and it was not to be understood that the Foreign Office were giving advice that it was unnecessary to make any preparations for war during the next ten years. He hoped in three years' time to be able to repeat the same advice, but possibly he might have to give his opinion that it would be advisable to be ready for war at the end of ten years . . .

Conclusions

1   The Committee of Imperial Defence recommend that the policy of leaving to the Admiralty, subject to the paramount authority of the Cabinet, the responsibility for the naval defence of the Empire on the basis of the one-Power standard, should not be reversed or modified. Even before the War it was recognised that the two-Power standard did not necessarily imply that our fleets were able simultaneously to deal with the concentrated fleets of two first-class Naval Powers in different quarters of the globe as well as to provide for all other naval requirements. This is equally true of the one-Power standard to-day. The requirements of a one-Power standard are satisfied if our fleet, wherever situation, is equal to the fleet of any other nation, wherever situated, provided that arrangements are made from time to time in different parts of the world, according as the international situation requires, to enable the local forces to maintain the situation against vital and irreparable damage pending the arrival of the main fleet, and to give the main fleet on arrival sufficient mobility.

2   The Committee accept the view of the Secretary of State for Foreign Affairs that, in existing circumstances, aggressive action against the British Empire on the part of Japan within the next ten years is not a contingency seriously to be apprehended.

[237] *Sir William Clark [High Commissioner to Canada] to*
Harding

24 July 1929

I dined on Monday night alone with the Prime Minister at Kingsmere and had three hours with him in which we were able to discuss many things in that congenial atmosphere . . . His objection [to Canadian representation on the Committee of Imperial Defence] is primarily from the standpoint of Canadian politics. Representation by the Canadian High Commissioner on the Committee would, he says, be misunderstood in Canada. It would be regarded as a surrender to the supposed centralising proclivities of Whitehall . . . He admitted that the Committee was wholly advisory . . . He considers that the Committee of Imperial Defence, despite its imperial name, is really a British rather than an Empire institution, and I suppose there is some truth in this. The Canadian High Commissioner would be over-weighted by the array of British Ministers and experts, and, while he agreed that the High Commissioner need not be committed in any way . . . The Prime Minister's view is that major issues of defence policy affecting the Empire should be dealt with in the same way as other major issues of imperial policy in which Great Britain may be taking the initiative . . .

[238] *Austen Chamberlain [Secretary of State for Foreign Affairs]*
to the Earl of Balfour [Lord President]

10 May 1929

I am as much puzzled what to advise as you yourself are how to treat your unfinished correspondence with [Sir Robert] Borden [former Canadian Prime Minister]. As regards the first paper which you wrote and which is cast in actual letter form, I should have no possible criticism if it were understood that it went for Borden's own information and would not be sent on by him to the American professor. If, however, it is for communication to the professor I should be afraid that it would act as a stimulus to him to embark on a public discussion of the freedom of the seas and belligerent rights at the very moment when these questions have receded into the background and the new American Administration seems to have centred its attention upon an international limitation of armaments . . . The two things which it seems to be vital to avoid are, first, the summoning of such a conference for the re-writing or even for the codification of the laws of maritime warfare and, secondly, the public assertion by America of her intention to maintain by force in future 'the freedom of the seas' whether in its unrestricted or in one of its more restricted forms.

[239] *Memorandum respecting the conversations between Prime Minister Ramsay MacDonald and President Hoover at Washington*

4 - 10 October 1929

When the question of Halifax was mentioned, I said that this was a matter for the Canadian Government, and I could not anticipate at all what their view would be. The Caribbean and the Canadian naval stations seemed to me to be in quite separate categories, and the difficulty as regards the latter appeared to be that Canada might reasonably expect some corresponding declaration [about their future use] from the American side in regard to naval stations in the United States in proximity to the Canadian-United States frontier. On this latter point, it was intimated to me that there was little prospect of any declaration being forthcoming in regard to United States Naval stations in the Western Hemisphere.

[240] *Director of Plans [Captain R.M. Bellairs] to Deputy Chief of Naval Staff [Vice-Admiral William W. Fisher]*

3 December 1929

From the Naval Staff aspect there is much to be said for reconsideration of our recommendation to Australia to maintain submarines as well as cruisers. It is considered that it would be preferable, under present circumstances for that Dominion to concentrate on cruisers. When this matter, however, was mooted in 1928 the late First Lord expressed his strong objection to the proposal as 'unsettling to our finance and to our dockyard employment'.

Reconsideration, however, of our Empire naval policy will probably be necessary as the outcome of the Five-Power Conference and opportunity for doing this will occur at the Imperial Conference of 1930. It is suggested that this question should, in the meanwhile, be explored but no definite decision come to until the Imperial Conference. [no signature]

[241] *10th Meeting, Committee on the Fighting Services*
22 May 1930

Conclusions

1   The Committee had before them a Report (F.S.(29)28) by the Departmental Sub-Committee appointed at the previous Meeting, and a Memorandum by the Lord President of the Council (F.S.(29)27).

Referring to the suggestion at the last Meeting by the Chancellor of the Exchequer [then, Winston S. Churchill], that account should be taken of our improved relations with Japan as a result of the Naval Conference, THE SECRETARY OF STATE FOR WAR [Thomas Shaw] said that in the opinion of the War Office it would be dangerous to decide to abandon the Base because for the moment our relations with Japan were good.

Mr Orde [C.W.Orde, Counsellor, Foreign Office] (representing Mr Henderson) said that the Foreign Office had no particular point to raise in this connection. (Mr Orde then withdrew).

As regards the question of consultation with the Dominions, THE SECRETARY OF STATE FOR DOMINION AFFAIRS, [Lord Passfield] referring to a letter on the subject he had written to the Prime Minister, emphasised strongly the importance of taking the Dominions fully into our confidence before arriving at any decision, and of dealing equitably with the contributing Colonies, not only in the matter of re-paying contributions if this became necessary, but also in regard to the ultimate cost of the Singapore Base if an increase was made in the garrison. He pointed out that Malaya had been extremely liberal but they felt strongly that they should not be asked to pay for more than the garrison which was necessary at Singapore before a decision to establish an Imperial Base was arrived at.

THE FIRST LORD OF THE ADMIRALTY [A.V. Alexander] drew the Committee's attention to a recent announcement in 'The Times' by the Prime Minister of New Zealand who had said that the New Zealand Government's views on Singapore were unchanged and that they hoped that no decision reversing the present policy would be made before the Imperial Conference took place.

While there was general agreement that, if the Government's hands had been quite free, the right policy would have been to abandon the Base, nevertheless in view of the situation vis à vis the Dominions, and of the extra expense which would be involved if the Base were now abandoned and then resumed, either as a result of the Imperial Conference or of a decision by a subsequent Government, it was agreed that the present slowed down programme should continue and that arrangements should be made to obtain at the forthcoming Imperial Conference a definite settlement which would not be upset by any subsequent changes of Government.

In conclusion it was agreed: That the Prime Minister should circulate to the Cabinet on behalf of the Committee, a Report on the lines of the above agreement recommending that the question of the Singapore Base should be brought up at the Imperial Conference for final settlement.

[242] *Field-Marshal Sir George F. Milne [Chief of Imperial General Staff], Field, Air Chief Marshal Sir J.M. Salmond, [Chief of Air Staff], Chiefs-of-Staff Sub-Committee of the Committee of Imperial Defence, to MacDonald, drawn up from discussion at the 94th Meeting*

22 October 1930

6   In reviewing the previous Resolutions we have been forcibly struck by the fact that whilst it is clearly laid down that each part of the empire 'is responsible for its own local defence', no express mention is made of collective responsibility for Empire defence, and the inference, undoubtedly drawn by many of the Dominions, is that the United Kingdom would automatically shoulder the bulk of responsibility for the defence of the Empire in a major war. This has resulted in Imperial Defence arrangements being widely regarded as arrangements by the United Kingdom for the Defence of the Empire. The only plans for a major war are those prepared in the United Kingdom. The extent of co-operation by the Dominion forces, even where it is assumed that the latter would be co-operating, cannot be gauged, and the Dominion forces have to be regarded largely as an extra asset not to be taken into account.

7   The Dominions are now equal partners with ourselves in the affairs of the Empire. Considerably more than a third of the total trade of the Empire is already Dominion trade, although admittedly the bulk of it is carried in United Kingdom bottoms, and it is certain that the Dominions will continue to increase in wealth and importance.

8   For the above reasons we are strongly of opinion that the Defence Resolutions should not be regarded as immutable in principle.

If arrangements for Imperial Defence are to be a practical proposition, the Dominions will have to regard themselves as responsible for more than 'local defence', and we suggest that before long, perhaps at the next imperial Conference, serious consideration should be given to the question of allotting larger 'areas of responsibility' in defence.

An example of what we have in mind is the Defence of Singapore. It seems absurd that plans should have to be made for despatching a garrison to that fortress in time of war from the United Kingdom when it could be sent far more expeditiously from, say, Australia, which is at least as interested in its defence as ourselves. Yet this is the case.

The arrangements we envisage would not of course derogate in any way from the right of a Dominion 'to decide the nature and extent of its co-operation' when an emergency arises, but it would render possible the preparation by the various Staffs of plans for co-operation.

9   We do not suggest that any radical changes should be made in the phraseology of the Defence Resolutions this year, and we fully recognise that the Dominions did in practice co-operate with us in Empire Defence during the Great War to the fullest extent possible.

We venture to submit however for consideration that when the Resolutions are presented to the Conference a tentative suggestion to the following effect might be made:

That, as in the application of the Resolution the Dominions will no doubt desire, in the course of time, and as their circumstances permit, to extend their co-operation in the common defence of the Empire, a review of Imperial Defence arrangements should be made at the next Conference in order to ascertain the extent, if any, to which the existing principles should be modified so as to bring them more into line with the constitutional relationships between the various parts of the Empire.

10   We fully appreciate that great care and tact will have to be exercised in dealing with this question, but we venture to suggest that no opportunity should be lost to open the door, however little, to practical co-operation in Empire Defence and to relief

of the disproportionate burden at present borne by the United Kingdom.

[243] *Singapore, report of a committee to the Imperial Conference, 1930, signed by Philip Snowden [Chancellor], James H. Scullin, [Australian Prime Minister] and George W. Forbes, [New Zealand Prime Minister], Committee of Imperial Defence*
27 October 1930

In accordance with the decision of the Heads of Delegations at their First Meeting on the 1st October, representatives of the United Kingdom, the Commonwealth of Australia and New Zealand met on the 16th instant to consider the question of the Singapore Base.

2   The proposals of the United Kingdom Delegation were as follows:

(a)   That no change should be made in the present policy of ultimately establishing a defended naval base at Singapore.

(b)   That the Jackson Contract should be continued, but that, apart from this and such expenditure as is necessary for the completion of the air base on the scale at present contemplated, the remaining expenditure required for completing the equipment of the docks and for defence works should be postponed for the next five years. The precise scale of that expenditure should then be determined in the light of the results of the next Naval Conference.

The United Kingdom Delegation made it clear to the representatives of the Dominions that the continuance of the present slowing-down policy would not prevent full effect being given to the complete scheme for the Base when this is considered necessary . . .

7   The Prime Minister of the Commonwealth of Australia expressed himself as being unable to do otherwise than concur in the proposal made by the United Kingdom, in the face of the political and financial arguments which had been advanced, but stated that any suggestion for abandoning the Base would be viewed with concern by a considerable section of public opinion in Australia.

8   The Prime Minister of New Zealand, while accepting the proposals, did so with reluctance. He pointed out that the naval advice given to the Dominion, as indeed to the United Kingdom, over a series of years had been unanimous in regarding the Singapore

Base as essential to any action whatsoever by the Navy to protect the Territories, trade and shipping of the British Commonwealth in the Pacific. While welcoming the various instruments for securing the maintenance of peace, he pointed out that they had not yet been subjected to any test that would justify relying on them in an emergency of the first order of magnitude. A study of the documents which had reached the New Zealand Government from the Foreign Office during the last few years indicated a good deal of unrest in the world, and made him hesitate to share without question the confidence felt by the United Kingdom Delegates in the sure maintenance of peace during the next decade. When such vital Imperial interests were at stake he felt that a risk was being run in relying on sufficient warning being obtainable of a deterioration in the international situation. He would have much preferred, if His Majesty's Government in the United Kingdom could have found it possible, to proceed with the original programme, which provided the basis for New Zealand's contribution. He appreciated the bearing on the question of the difficult financial situation, and for this reason did not feel entitled to press his objections to the point of disagreement.

9   The recommendation of the United Kingdom Delegation, set forth in paragraph 2 above, was adopted.

[244] *War memorandum (Eastern), Admiralty*

July 1931

Part C, British Naval Strategy

50   The National Object will be the security of the interests of the British Commonwealth of Nations in the East which may be threatened by any Japanese action such as that forecasted above.

51   The primary aim of the British naval forces will be to destroy as soon as possible, and pending that to neutralise, Japan's naval forces which threaten our security.

52   A number of operations require to be considered in connection with the achievement of the primary aim. These are as follows:

(a)   To hold Singapore and Hong Kong.

(b)   To bring to Hong Kong, with the least possible delay, a Fleet superior to the Japanese.

(c)   To consolidate our position at Hong Kong by the institution of the necessary services required for the maintenance of the Fleet and for further operations.

53   Before enlarging on these, under the heading of 'British Naval Strategy', it will be necessary to review the possible courses of action open to the enemy. Whatever the strategy governing the Japanese general war plan, the opening moves are bound to be influenced by the problem of how best Japan can make use of the period before the arrival of the British Main Fleet, during which she has control of the local maritime theatre of war, to strike the heaviest possible blow against the British Empire. As our fleet represents the most dangerous menace opposed to her, it would profit Japan best if, during the period before its arrival, she were able to consolidate her position so that when the fleet arrives it will be incapable of restoring the situation in our favour. The most serious blow therefore would be the loss of our bases in the East, without which it will be impossible to maintain our fleet in the theatre of war.

54   The loss of Singapore would deprive the British Fleet of the power of acting, and would, moreover, most probably result in the loss of Hong Kong. It is to be expected that the Japanese will make raids on Singapore with a view to damaging the naval base and destroying the oil fuel reserves, but if the strength of the defences is adequate, then, on account of the distance of Singapore from Japan (2,400 miles) and the limited period in which the operation would have to be completed, as well as the risks involved, it is considered that attack with a view to capture will be undertaken against Hong Kong rather than Singapore.

55   The loss of Hong Kong will force us to undertake its recapture before any effective further operations can be undertaken against Japan. We may assume that, if captured by the Japanese, it will be heavily garrisoned and fortified, and that the ensuring campaign with the object of recapturing would be difficult and extensive. The defences of Hong Kong are limited by Treaty, and the forces which are available in peace time for its defences are, relative to the Japanese forces which can be brought to the attack, inadequate. It is possible, therefore, that under these conditions its capture before the arrival of our Main Fleet may be attempted.

69   MOVEMENT OF THE MAIN FLEET   It will be necessary for the British Fleet to operate from Hong Kong, and possibly from an advanced base further to the northward. It is therefore of vital importance that both Singapore and Hong Kong should

be in our possession and available for the use of the British Fleet on its arrival in the Far East. As the period during which Hong Kong can hold out against a determined attack on the maximum scale is limited, the British Fleet must arrive to relieve it at the earliest possible moment. The keynote, therefore, of British naval strategy during the first phase of the war will be:

TO MOVE THE BRITISH MAIN FLEET TO SINGAPORE AND FROM THENCE TO HONG KONG WITH THE LEAST POSSIBLE DELAY.

The essential requirement is mobility.

74   British Strategy   British strategy, in the event of war with Japan, must commence with the successful transfer of such forces of the Empire as are available, particularly the British Main Fleet, from the various parts of the Empire in which they are stationed to the scene of hostilities. Since our fleet cannot operate in distant waters without adequate bases, in this instance Singapore and Hong Kong, our strategy in the first instance must be directed towards the preservation of Singapore and Hong Kong.

75   [*deleted*, 'Para. 75 cancelled by M. Moor 05/33 of 26/4/1933'] At the present time (1931) the defences of Singapore are [still] inadequate, but emergency plans are being [have been] considered for the rapid provision of defences in the event of a sudden deterioration of the international situation in the Far East.[,] This investigation has not yet been completed but [and] it appears probable that sufficient defences could be installed within 12 or 18 months of the situation commencing to deteriorate to deter the Japanese from attacking Singapore rather than Hong Kong. Should the Japanese attack be directed against Hong Kong the full assistance of the naval forces will be necessary to frustrate the attack and to ensure that Hong Kong shall be available for use by the Main Fleet on its arrival. The object therefore of the British naval forces in China during the period before relief will be to contribute to their maximum ability to the security of Hong Kong.

76   From the foregoing it is reasonable to divide British strategy into three phases:

Phase I, The Period before Relief, i.e., the period that will elapse between the beginning of hostilities and the arrival of the British Main Fleet at Hong Kong.

Phase II, The Period of Consolidation After the arrival at Hong Kong the main fleet will require a period of time to consolidate the position before embarking on large scale offensive operations against the Japanese.

Phase III, The Period of Advance The final strategic Phase in which the British Fleet, in conjunction with the Army and Air Force, will commence such operations as will best serve to defeat the enemy.

[245] *Minute and Note on Indian Subsidy*
5 October 1931

Minute by C.B. Coxwell
[Principal Admiralty Clerk],
for Head of Military Branch of
Admiralty Secretary's
Department [under the Deputy
Chief of Naval Staff, Vice-
Admiral Frederic Dreyer], 5
October 1931

As stated on M. 02567/25, the Imperial Conference of 1923 approved as a guiding principle 'the primary responsibility of each portion of the Empire represented at the Conference for its own local defence.'

2 This principle involved the responsibility of the Indian Government, financially and otherwise, for minesweeping, auxiliary patrol, etc. at Indian Ports.

3 It was realised that in practice the application of this principle must be dependant on what India was able to do; this depending primarily upon the action eventually taken by the Government of India as regards the reorganisation of the Royal Indian Marine. It was hoped, however, that ultimately the Government of India would be able to accept the principle entirely in its application to Indian local defence.

4 Since that time, of course, though the reconstituting of the R.I.M. as a Royal Indian Navy has not materialised, it has been given combatant status, and the duties and functions of the Flag-Officer Commanding and Director have been the subject of agreement between the India Office and the Admiralty: under the arrangements agreed, the Flag Officer Commanding and Director is responsible for the local defence of the coasts of India in time of war.

5 The Government of India have now had under consideration

the requirements in respect of : (1) Minesweeping, and (2) Naval Control Service, and they have set out in the enclosed letter the arrangements which they propose for meeting these requirements, in so far as they expect to be able to meet them.

6   These proposals are analysed in the enclosed M[ilitary] Branch notes.

7   Briefly, the Government of India are prepared to carry out the arrangements for local defence and Naval Control Service, in so far as the necessary personnel is available, provided that their financial liability for naval defence is limited to the cost of maintaining the R.I.M. at its normal strength, plus the annual subsidy of £100,000 which they already pay.

8   The attitude of the Government of India, so far as the financial aspect of the matter is concerned, is somewhat disappointing. The Admiralty has for many years held the view that India might make a much larger contribution towards the naval expenditure of the Empire. In this connection, attention is invited to the enclosed M[ilitary] Branch notes on the £100,000 subsidy.

9   So far as local defence arrangements are concerned, it is thought that the Admiralty should adhere to the principle enunciated in 1923, and invite the India Office to press the Government of India to accept full and complete financial responsibility for its service.

10   The actual expenditure necessary in time of peace should not be great: on the general basis approved, it would presumably be sufficient, under present circumstances if one tenth of the reserves of necessary material were purchased annually.

11   As regards the Naval Control Service, the peace-time expenditure should be practically nil. In time of war, the trade and the shipping of India stand to gain from the organisation the same advantages as do those of other parts of the Empire. It would be regrettable if the Government of India definitely declined to accept any share of the cost of the system: possibly, if they realised that it involved no expense in peace, they would reconsider their position.

> Note on Indian Subsidy,
> Military Branch, ?October
> 1928, M. 4372/30

The Government of India pay an annual subsidy of £100,000 on condition that there are always three H.M. Ships operating in Indian Waters.

2   Indian Waters are defined as follows: 'The coast from Bombay to Karachi, the Makran coast, Gulf of Oman, Persian Gulf (including Shatt-el-Arab), south coast of Arabia to Bab-el-Mandeb, thence north of an imaginary line drawn to Cape Gardefui south of Sokotra to Bombay'.

3   The further conditions on which the subsidy is paid are:

(a)   No claim is to arise in India in respect of any of these subsidised ships except for fuel.

(b)   Ships are under the control of the Commander-in-Chief, but the wishes of the Government of India as to their employment are to be complied with.

(c)   They are not to be employed outside Indian Waters without the consent of the Government of India.

(d)   With the concurrence of the Government of India, a non-subsidised ship may be substituted for a subsidised one.

4   The Government of India pays the cost of all coal consumed by the subsidised vessels while employed in the Persian Gulf, or while employed elsewhere at the request of the Government of India, but not that of fuel consumed by other vessels of the East Indies Squadron unless these are substituted for subsidised vessels with the consent of the Government of India and employed in the Persian Gulf or elsewhere at their request.

5   As regards H.M.S. *Triad*, there are special arrangements, viz:

(a)   The Indian Government makes good all defects to *Triad* and bears the cost of all stores for making good defects to this vessel. The Admiralty is, however, liable for the cost involved in carrying out 'Alterations and Additions' to the vessel.

(b)   The Admiralty bears the cost of all maintenance stores for *Triad* except fuel. (Note. – The agreement in regard to repair of *Triad* applied originally to H.M.S. *Sphinx*).

6   The round figure of £100,000 per annum was finally agreed upon for ten years from 1891 and has since been continued year by year. The amount of the subsidy has been the subject of much discussion with the India Office and the Government of India.

7   A long discussion as to the justice of a larger contribution culminated in a Foreign Office conference in 1890.

8   This conference decided that India should pay the whole cost of all ships agreed by H.M. Government and the Government of India to be employed in the Persian Gulf and off the coasts of India for Indian purposes.

9   A dispute then ensued between the India Office and the Admiralty on the interpretation of the term 'Indian purposes', and the amount of the subsidy to be paid under this settlement. The India Office wished to limit 'Indian purposes' to maritime policing of the coasts, suppression of piracy and slave trade. The Admiralty held that general defence and the protection of Indian commerce on the high seas must be included in the expression. The dispute was finally submitted to the Prime Minister (Lord Rosebery), who in 1895 decided that:

'Indian purposes' must be held to include some portion at any rate of the duties devolving on the Royal Navy for the defence of India and the protection of trade in Indian Waters', and that a contribution on this basis should be settled at fixed intervals of, say, 10 years.

10   After a further dispute as to the ships* to be subsidised, a round figure of £100,000 per annum for four ships was finally agreed upon for ten years from 1891, the Admiralty waiving arrears to the extent of £183,000.
*Note. – The Admiralty proposed four ships at a cost of £117,000; the India Office, four ships at a cost of £90,000).
11   The question was considered afresh by the Royal Commission on the administration of the Expenditure of India (appointed 1895, reported 1900). Their conclusion supported Lord Rosebery's decision. They stated that the contribution of £100,000 per annum was not excessive, and that there was no case for a reduction, and that the arrangement should continue until 31st March, 1906. After protracted correspondence, it was agreed that the £100,000 subsidy should be paid after 31st March, 1906, 'without prejudice'.
12   In 1907 the situation assumed a new phase. Hitherto the Admiralty had only asked for a small contribution in respect of certain ships definitely assigned for Indian duties, but in July, 1907, they wrote to the India Office asking for a contribution of one twentieth of the Navy Estimates (one and a half millions) based on a rough calculation of the proportion of Indian trade for which no payment for naval protection was made.
13   The reply given was that India could afford no more money for Imperial services, and that, if the naval contribution was inadequate, it must be made up by a corresponding reduction in military charges.
14   The Admiralty objected to this, and proposed to refer the matter to the arbitration of the Lord Chief Justice under an

arrangement made in 1902. The India Office refused, and were supported by the Treasury.

15   In 1908, the Admiralty dropped the larger proposal for a time and, urging that the whole of the East Indies Squadron was performing Indian services at a cost of £330,000, asked for a subsidy of that amount.

16   In 1909 (24th February) the India Office declined to make any increase in the subsidy, and refused to consider arbitration.

17   During this period, Treasury sanction was obtained year by year to the acceptance from the Indian Government of the contribution of £100,000 per annum.

18   In 1913, the Admiralty again approached the India Office and suggested that India should undertake the complete protection of her own trade and the safeguarding of her interests, not only in the Persian Gulf, but also in the Arabian Sea and Bay of Bengal, and that for this purpose she should provide the cost of the construction and maintenance of a squadron in her waters and defray the cost of renewals when the vessels were worn out.

19   The Admiralty recommended a scheme under which the financial liabilities of the Government of India would be:

Capital   £1,680,000*
Annual Payment   £ 490,000

*Note.- This did not include cost of the flagship which Admiralty proposed to provide on the understanding that India would provide the capital cost of a new flagship when it became necessary.

20   In addition, the cost of special vessels which might have to be kept in the Persian Gulf would have to be met.

21   No reply was received from India before the outbreak of war in 1914, and the matter was afterwards left in abeyance, the payment of £100,000 being continued by the Indian Government year by year.

22   After the war, India was visited officially by Viscount Jellicoe who made various proposals for naval assistance by India, but did not suggest any increase in the subsidy. In commenting on Admiral Jellicoe's report, the Government of India stated that they were not in a position to increase their contribution of £100,000 per annum (8th July, 1920).

23   In reply to a suggestion from the Treasury (29th January, 1921, the Admiralty proposed, as reasonable, an increase in the contribution from India to £250,000. They pointed out that the cost

of the East Indies Squadron was about £815,000, of which £225,000 was in respect of 4 small ships for service in the Persian Gulf.

24   Time was not available for discussion of this proposal before the Imperial Conference, but the Treasury asked for a definite assurance from the Admiralty that the question of an increased contribution would be strongly pressed at the Imperial Conference.

25   The Admiralty took up this matter with the C.I.D. [Committee of Imperial Defence], but the latter did not proceed with it.

26   Nothing was definitely settled at the Imperial Council, and since then the contribution of £100,000 per annum has continued year by year.

27   The policy of H.M. Government as regards the co-operation of India in Empire Naval Policy as approved by the Committee of Imperial Defence was embodied in C.I.D. 132-D. These recommendations did not include any increase in the subsidy but included proposals for:

(1)   The re-organisation of the Royal Indian Marine.
(2)   The modernisation of Indian W/T stations.
(3)   The establishment of oil fuel reserves at Aden and Rangoon.
(4)   A contribution towards cost of Singapore.

28   The proposals for the re-organisation of the Royal Indian Marine were considered by a Committee under Lord Rawlinson early in 1925.

29   On 12th February, the Treasury were informed that no response has been received to the suggestions re Oil Fuel Reserves and Singapore, but it was understood the Indian Government was proceeding with the modernisation of the W/T Stations. Further, it was hoped that the Government of India would shortly be in a position to announce their approval of the proposals of Lord Rawlinson's Committee as to the Royal Indian Marine.

30   On the 30th December, the Treasury were informed semi-officially that the Indian Government was still proceeding with the modernisation of W/T Stations. No step had been taken, so far as the Admiralty was aware, as to Oil Fuel Reserves and Singapore.

Development of Royal Indian Marine

In 1925 a Committee presided over by Lord Rawlinson produced proposals for re-organising the Royal Indian Marine as a combatant force to be known as the Royal Indian Navy. To start with, they proposed that a few small vessels should be maintained for training purposes. They endeavoured to limit their scheme so that it should not cost much more than the Royal Indian Marine had previously cost. Their proposals, in fact, involved an annual increase of Rs.1,000,000 and an initial expenditure of Rs.900,000.

The Bill which was intended to bring these proposals into being was thrown out by the Indian Legislature, but the Indian Government found that they could carry out most of these proposals without any legislation. This is therefore being done, but the force will still be known as the Royal Indian Marine.

The original proposals for the Committee have been cut down somewhat by the Indian Government, but the fact will remain that the Government of India will now be spending a certain amount of money specifically on naval preparations. The Government of India hope, and the Admiralty have agreed, that ultimately the Royal Indian Marine should take over the patrol of the Persian Gulf. It may, therefore, prove more economical to this country to encourage the development of the Royal Indian Marine than to press for an increase in the Indian contribution.

For the present the Government of India are continuing to pay the sum of £100,000 per annum, together with the cost of coal and repairs as before and will, apparently, be prepared to continue this until they take over some portion of the patrol of the Persian Gulf.

[246] *Walker to Under-Secretary of State, Military Department, India Office*

30 November 1931

With reference to your letter M.4959/1931 of 8th September last, I am commanded by My Lords Commissioners of the Admiralty to acquaint you, for the information of the Secretary of State for India in Council, that They have carefully considered the views expressed by the Government of India in their letter No. 1047 M. of the 28th July last, with enclosures, on the subject of the Naval Control Service and the minesweeping arrangements at Indian ports in time of war.

2 My Lords concur in the opinion of the Flag Officer Commanding and Director, Royal Indian Marine, that a war with an
Eastern Power would throw a greater strain on the resources of
India than a war with a European Power, and that it is unnecessary
therefore to consider the requirements of the latter contingency.
3 The minesweeping vessels required at Indian ports (excluding
Aden) during the opening stage of a war in the Far East are 27,
i.e. 24 as stated in the Government of India's letter, plus 3 for
Nancowry. This latter addition was notified in amendments Nos.
3 and 7 to Admiralty Memorandum M.00364, copies of which
have been supplied to the Government of India. Twenty-one
further vessels are required for anti-submarine duties, as stated in
Admiralty Memorandum M.0409, a copy of which was recently
sent to Commander-in-Chief, East Indies, for issue to the Flag
Officer Commanding and Director, Royal Indian Marine, and the
total number of vessels required for local defence is thus 48.
4 The Royal Indian Marine at present consist of 11 minesweeping vessels (3 sloops and 8 trawlers) and 5 anti-submarine craft
(1 sloop, 2 patrol boats and 2 surveying vessels which will be
employed on A/S duties in war); so that it will be necessary to
take up from the mercantile marine 16 vessels for minesweeping
and 16 vessels for anti-submarine duties. The Commander-in-
Chief, East Indies, in making provisional war dispositions of Royal
Indian Marine vessels, has allocated 6 of the existing minesweepers
for service at certain ports not in India (i.e., Trincomali, Aden and
Colombo) where it will be necessary to carry out precautionary
minesweeping before the outbreak of war. It will thus be necessary
to take up 6 additional vessels for service as minesweepers at
Indian ports . . .
9 It was not Their Lordships' intention that India should be
asked to bear the cost of equipping or running the armed merchant
cruisers which are to be fitted out at Calcutta or Bombay: this will
remain an Imperial commitment. It was, however, hoped that the
Commodores of Convoy and Staffs, and the Naval Control Service
Staffs, together with water transport and office accommodation,
would be provided and paid for by India.
10 I am to point out that the object of the Naval Control Service
is the protection of the sea-borne trade of the Empire in War, and
that Indian shipping and trade will benefit from its operation in
the same way as the shipping and trade of the remainder of the
Empire. All the self-governing Dominions have undertaken liabilities in respect of this service, and My Lords had hoped that India

would adopt the same attitude. The actual cost in peace time would be confined to the cost of storing and maintaining the guns and mountings required for arming armed merchant cruisers, fast liners and defensively equipped merchant vessels, concerning which attention is directed to the correspondence resting with Admiralty letter M.02928/28 of 11th January, 1929. My Lords trust that the Government of India will on reconsideration feel disposed to accept financial responsibility for that part of the Naval Control Service which concerns them.

[247] *Committee of Imperial Defence memorandum by the Secretary of State for India [W. Wedgwood Benn], 'Control of Royal Indian Marine in War', initialed 'SH'*
December 1931

'Secret'

The point I wish to raise in this Memorandum is in itself a simple one. It is that His Majesty's Government should record their acceptance of the principle that, when in time of war the ships and forces of the Royal Indian Marine are brought, as is contemplated, under the control of the Naval Commander-in-Chief, East Indies Squadron, the Government of India should be consulted before this transfer of control is actually effected. The Admiralty have agreed to the principle, and there can be little doubt that in practice it would be automatically observed. But there is some advantage, as experience has shown, in placing His Majesty's Government's acceptance of such a principle on formal record. In the somewhat analogous case of the Army, His Majesty's Government recorded the decision (in C.I.D. Paper No. 130 D, dated 26th January 1923) that 'The Principle should be generally accepted that, except in the gravest emergency, the Indian Army should be employed outside the Indian Empire only after consultation with the Governor-General in Council.' It has proved useful – in dealing, for instance, with questions in Parliament relating to the despatch of the Indian Brigade to Shanghai in 1927 – to be able to point not only to the fact that the government of India had previously been consulted, but also to the fact that such consultation represents the standing policy of His Majesty's Government. The Government of India have asked that a similar convention should be recognised in connection with the Royal Indian Marine, and in my view it would be not only reasonable but advantageous to agree . . .

[248] *Minutes, 101st Meeting, Chiefs of Staff Sub-Committee,*
Committee of Imperial Defence

4 February 1932

1   THE SUB-COMMITTEE had before them a Note by the
Secretary (C.O.S. Paper No. 293) covering a letter from the Commandant of the Imperial Defence College, inviting attention to the
falling off in the attendance of representatives of the Dominions at
the College.

SIR FREDERICK FIELD [First Sea Lord] said that from the
naval point of view only Australia and Canada were likely to be
able to produce officers of the necessary standing for the Courses
at the Imperial Defence College, and they were reducing their
establishments. While he quite agreed that everything possible
should be done to encourage the Dominions to send officers to
the College, he thought it would be difficult for them to find naval
officers for every Course. As regards civilians he felt some doubt
as to the wisdom of inviting the Irish Free state and the Union of
South Africa to send persons whom we might be unwilling to
allow to share all our secrets.

[249] *Singapore Naval Base, visit by Sir Leopold H. Savile [Civil*
Engineer-in-Chief] February-March 1932, and Report of
interview with His Excellency the Governor [Sir Cecil Clementi]

14 April 1932

As indicated on separate submission to Board, C[ivil] E[ngineer]
in C[hief], 3070/32, before leaving Singapore on the return voyage,
I had an interview with H.E. the Governor, Sir Cecil Clementi,
who was anxious to know what H.M. Government proposed to
do about the progress of the works and completion of the Base.

He told me that there was a strong feeling locally and in Malaya
that the works should be pressed forward to completion with as
little delay as possible and that it was on this understanding that
Malaya had contributed £2,000,000 towards the cost of the scheme.
H[is] E[xcellency] impressed upon me that unless this was done,
very strong resentment would be felt even possibly to the extent
of claiming a return of this contribution, also that the Straits
Settlements who had provided the land were likely to ask for the
cost of same to be refunded.

I informed the Governor that I understood the matter was
under consideration but that so far as I was aware no definite

decision had been taken. I promised him that I would lay his representations before the Board of Admiralty on my return.

Minute by Captain J.H.D.
Cunningham [Admiralty Director
of Plans], 20 April 1932

Noted. It is not without interest that the New Zealand Delegate at the last Imperial Conference made it quite clear in the report of the Singapore Sub-Committee that it was only with the greatest reluctance that he agreed to the postponement of work on Singapore for 5 years; Hong Kong, the remaining overseas contributor to the cost of Singapore, on being consulted by telegram replied to the effect that the completion of Singapore was regarded as being essential to the security of Hong Kong. (M03437/30 in Case 2052) It would seem that the attitude of the Overseas contributors to the Singapore Scheme might have to be taken very seriously into consideration if work is much longer postponed.

It is understood that one phase of the feeling in Singapore is that the partial construction of a dockyard in the old Strait, unaccompanied, as it is, by any effective measures for its protection against attack, has exposed Singapore generally to a considerably greater risk of attack than was formerly the case when only Keppel Harbour, protected by 5-9.2' and sundry 6' guns, would have been an enemy's only objective.

[250] *Minutes by the Director of Plans, the First Lord of the Admiralty, and the Admiralty Secretary, on Australian Naval Policy and Admiral G.F. Hyde's [RAN] letter of 1 March 1932 to Chief of Naval Staff concerning replacement of Australia's destroyers*

18 July 1932

Minute by Director of Plans
[Captain J.H.D. Cunningham]

The Australian Navy is on a different basis to the New Zealand Division to which we propose to loan two new *Leanders* in due course. In the case of Australia, it would seem that any transfer must be in the form of a gift.

2 It is doubted whether the development of Dominion Navies would be best served in the long run if we were to present Australia with these five new ships.

3   The Dominions at present contribute little enough to Empire Naval Defence and a gift such as this might encourage them to lean on the British Taxpayer still further and look for more gifts when any of their ships needs replacement.
4   It is desirable that in due course Australia should herself replace her vessels as they become over-age. Admittedly the present financial situation will not permit of her doing this now, and this was recognised at the Imperial Conference, 1930. At that Conference Australian representatives stated that they wished to keep their destroyers after 1936 though they were unlikely to replace them before that date. It then became apparent that if Australia kept these vessels we should have to scrap better vessels ('V' and 'W' Class) in order to get down to the tonnage allowed to the Empire – it was out of this that the proposal to transfer the 'V' and 'W' class vessels arose, i.e. the object of the transfer was simply to avoid scrapping the better vessels and not to remove the expense of Australian replacement from Australia to this country.

Minute by the First Lord of
the Admiralty [Sir Bolton M.
Eyres-Monsell], 20 September
1932

I propose now to write to Rear Admiral Hyde asking him, in view of the official intimation that *Geranium, Mallow* and *Marguerite* have been taken in hand for scrapping, whether he can give any indication when it is proposed to commence the replacement pro-gramme as this now affects not only the sloop but the cruiser situation. The Admiralty very much hope that the replacement of *Brisbane*, which is now over age, may be commenced in 1933 and the *Geranium, Mallow* and *Marguerita* may be replaced as soon as financial considerations permit.
2   With regard to the question of replacing the present Australian destroyers, I propose to inform him that the Admiralty are nat-urally only too anxious that the destroyer strength maintained by Australia should be kept up to a requisite standard. It is of course hoped that in due course Australia will replace her destroyers as they become over age. It is, however, realised that the financial situation will not permit of this being done in the near future and this was, in fact, recognised at the Imperial Conference in 1930. At that Conference the Australian representatives stated that they wished to keep their destroyers after 1936 although it was unlikely that they could be replaced before that date. This raised the

question that, if the present Australian destroyers were retained beyond 1936 we should be scrapping better vessels (V and W class) in order to keep within the tonnage allowed to the Empire as a whole.

3   This could be avoided if Australia were to replace her present S. Class destroyers and *Anzac* by V. or W. Class destroyers now in reserve in England and there is no doubt that, if such an arrangement were reached, a very suitable Flotilla Leader and four destroyers could be selected which would be a decided improvement in the strength of the Royal Australian Navy. This would have an additional advantage in that the shell part of the outfits of ammunition is already in Australia as the shell for S. Class is the same as for V. and W. Class.

4   Such an arrangement would, however, raise the question of both policy and finance. The development of Dominion Naval strength has all tended in the direction of the Dominions providing their own ships and, although I fully realise how the present financial stringency is felt in Australia, the position in this Country is precisely similar. A free transfer of 5 vessels from this Country to Australia would therefore not only be a retrograde movement in naval development but Treasury sanction would be very difficult, if not impossible, to obtain.

5   It is possible, however, these difficulties might be overcome by Australia paying scrap value for the vessels transferred, provided of course that Treasury sanction were forthcoming. I understand that there is practically no market for scrapped war vessels in Australia so that, if *Anzac* and the four S. Class destroyers could be prepared for sea at reasonable cost, it might be the most economical arrangement if they were steamed to England with the officers and ships companies of their relief vessels and sold for scrap on arrival – the amount so obtained could be set off against the scrap value paid for their reliefs and this would save cost of transport of ships companies to England (approximately £28,730). The cost of the passage of the relief vessels from England to Sydney would be approximately:

Fuel and lubricants	£13,300
Sea Stores	£ 2,200
Suez Canal dues	£ 970
	£16,470

6   I propose to conclude by asking Admiral Hyde to think over this question and suggest to him that the next step must be through the ordinary official channels in the form of proposals from the Navy Office.

Will Secretary please remark and say whether he concurs in so far as finance is concerned.

Minute by Sir Oswyn Murray
[Permanent Admiralty
Secretary], 24 September 1932

The arrangements proposed in paragraphs 4 and 5 of your minute seem to me to be rather complicated financially, and may have the appearance to Australia of being less favourable than those which we made with Canada in somewhat similar circumstances in 1927.

It will be remembered that we then, with Treasury sanction, agreed to lend two Destroyers (the *Torbay* and *Toreador*) to Canada until two new Destroyers had been built, the terms of the loan being as set forth in the attached copy of our letter to the Treasury dated 24th January 1926.

The advantages, financial and otherwise, of lending these ships to Australia until she is in a position to replace her Destroyers seem to be, if anything, greater than in the Canadian precedent case, and I do not see any reason why the Treasury should not be willing to agree to generally similar arrangements, except that in lieu of condition (e), providing for the return of the vessels to this country, we should stipulate that, as it will be necessary for the ships to be scrapped in Australia at the conclusion of the loan, the Commonwealth Government should undertake to pay the difference between the sum actually realised by us there and the sum we should have realised had they been scrapped in Great Britain at the prices then ruling. The provisions as to stores would also have to be modified to conform to this alteration in condition (e).

If the arrangements were on these lines, it would be entirely for Australia to decide how and where the Anzac and the 'S' Destroyers should be disposed of, and how the crews to take over the loaned Destroyers should proceed to England.

[251] *Minutes, 103rd Meeting, Chiefs of Staff Sub-Committee, Committee of Imperial Defence*

2 June 1932

[SIR JOHN LATHAM, Attorney-General of the Commonwealth of Australia] They had also worked out the possibility of invasion from the shipping point of view, and he believed they had reached the conclusion that a convoy of something like 100 vessels or more would be required. They thought it would be very difficult indeed for Japan to obtain this number of vessels on the outbreak of war. The conclusion was that for many years it would be very difficult for Japan to provide the physical basis for an invasion of Australia. Further, such a fleet would be exposed to very great risk of attack on the long journey to Australia. It therefore appears to the Australian Government that it would be better to provide efficient protection against raids rather than inefficient measures against invasion.

With regard to raids on the coast or on commerce, the first consideration was that any attack must be seaborne, and therefore Australia ought to be provided with naval forces to repel this form of attack as well as means to repel a seaborne air attack. Here a difference of opinion arose as the Air Force claimed a larger role for an independent air force.

It appeared to the Australian Government that such air fighting as would have to be done would only be against machines flown from ships, and that there would, therefore, be distinct limitations on the numbers and on the performance of aircraft which could attack. The number of carriers was strictly limited, their cost was great, they were very important to the fleets which possessed them. It therefore seemed unlikely, at any rate unless the British Main Naval Force was defeated, that carriers would be sent far away from a base and from main bodies of naval forces to be used against Australia.

It therefore had to be considered whether more value would not be obtained by some modification of the present system of contributing to naval defence. For example, greater efficiency might perhaps be obtained with better results for the expenditure if the ships were provided by the Royal Navy and maintenance and provision of personnel were undertaken by the Commonwealth . . .

SIR FREDERICK FIELD [First Sea Lord] said he thought that no hard and fast line could be drawn between the needs of local and Imperial defence. He agreed, from the naval point of view, that attack would be by seaborne forces, but he declared emphatically that the one thing which could save Australia from such attack on any scale was the main Fleet. Whether it would be possible for the Fleet to give the necessary degree of protection to Australia depended largely on whether Singapore was adequately defended. If this was so, he entirely agreed that the risk of invasion could be ignored. Whether any enemy carriers or raiders would have a reasonable chance of attacking Australia depended on the forces available to intercept them. Therefore, it was of paramount importance that Australia should keep sufficient naval forces for local protection and for contributing to Imperial defence . . .

SIR FREDERICK FIELD said he would like to add that the Admiralty viewed with great concern the reduction in the Australian Navy in recent years. Personnel had been reduced from 5,000 to 3,500 and expenditure from about £5 millions to less than £1 1/2 millions. In this connection he reminded Mr Latham that the naval forces of the whole Empire were treated as a single unit by treaty and we were bound to maintain a Navy of a certain size. If Australia was unable to replace certain vessels, this threw an additional burden on Great Britain. He understood that there were a large number of ships in reserve and only three ships in full commission at the present time. He had also heard that the sloops would soon reach their age limit and have to be scrapped. The Brisbane and six destroyers would also reach the age limit next year and he understood that at the present time Australia had no intention of replacing them. He emphasised strongly that this would mean that Great Britain would have to do so. The Admiralty strongly hoped that Australia would be able to maintain two cruisers in full commission, two cruisers in reserve, and to replace the *Brisbane* beginning next year, as well as the three sloops when due for scrapping. They also hoped that the Commonwealth would consider taking some of our destroyers and maintaining them if necessary in reserve when the present Australian destroyers reached the age limit. In this way there would, at any rate, be destroyers on the spot to be manned by reserve personnel, etc. in an emergency. It appeared to the Admiralty that with the money at present voted, the maintenance of the present ships could be continued, and, with a slight increase annually, the *Brisbane* could be replaced.

[252] *Memorandum by the Marine Department of the Government of India*

Simla, 8 September 1932

'Secret'

In paragraph 19 of our Marine Despatch No. 3 of 1923 we said that we thought that a Naval Volunteer Reserve force would have to be formed in this country before India could do much towards playing the part assigned to it in the Admiralty Naval Control Scheme. I do not think that there was any more said about this Reserve until Walwyn [Vice Admiral Sir H.T. Walwyn, Flag Officer Commanding and Directing the Royal Indian Navy, Bombay] put up a scheme in April 1929. He then suggested that a Reserve should be formed with a limited strength of sixty officers, with its headquarters at Bombay, which he considered to be the best place for its general administration and likely to provide more candidates than other ports. He recognised, however, that officers required for vessels at other ports should, as far as possible, be residents of those places in view of the importance of local knowledge.

2   Having regard to the procedure followed in forming the Indian Army Reserve of Officers, we came to the conclusion that no legislation would be required for the constitution, with the approval of the Secretary of State, of the contemplated Royal Indian Marine Volunteer Reserve or the preparation of detailed regulations to govern the composition and conditions of service. In order to provide for the discipline of the Reserve, we were advised that it would be necessary to amend clauses (a) and (b) of sub-section (1) of section 2 of the Indian Marine Act, 1887, by the insertion of a provision corresponding to those in clauses (9) and (10) of Section 175 of the Army Act relating to the Indian Army Reserve of Officers and the Army in India Reserve of Officers.

3   In March 1930 the proposal to form the Royal Indian Marine Volunteer Reserve was accepted in principle and we asked Walwyn to submit detailed regulations to govern the composition and conditions of service in the Reserve on the lines of the 'Regulations for the Army in India Reserve of Officers' as contained in Appendix XXX to the Regulations for the Army in India. A draft of the regulations was received in September 1930 and was scrutinised in consultation with the Military Finance Branch. I enclose a copy of the draft rules in the form in which they have

been accepted by all concerned. In our Marine Department letter No. 1047-M of the 28th July 1931 we told you that we had the constitution of a Reserve under consideration, and a reference to it was also made in our letter No. 1006-M of the 11th July last.

4   It is estimated that the scheme will cost Rs. 47,000/- in the first year and Rs. 20,000/- per annum thereafter; and Rs. 47,000 was entered for the purpose in this year's Marine Budget. In June, however, Council decided that in view of the necessity for economy the further consideration of the question of forming the Reserve should be postponed. Walwyn was informed of this decision and was told that the proposal would be considered further as soon as financial condition[s] improved. He then protested and urged that the decision should be re-considered. We cannot of course meet his wishes entirely, but we think that the necessary legislation might precede, and not wait for, an improvement in financial conditions, so that the Reserve can be formed as soon as funds are available. The Army Department, therefore propose to ask Council to agree to legislation being undertaken at an early date, and, when their approval has been obtained, we shall address you officially in the matter. While waiting for our official letter would you mind making a preliminary examination of our proposals, in consultation, if you think necessary, with the Admiralty?

[253] *Imperial Defence Policy, Annual Review for 1932 by the Chiefs of Staff Sub-Committee*

12   The above events prompt inquiry as to our own readiness to face sudden aggression by Japan. The position is about as bad as it could be. Our naval forces in the Far East include nothing larger than 10,000-ton Cruisers armed with 8-inch guns. Most of these are moored in the Whangpoo River, together with the bulk of the international forces in the Far East, all of which are incapable either of effective resistance to a Japanese fleet of capital ships or of escape from this cul-de-sac, unless they receive sufficient warning to enable them to reach the open sea. The Hong Kong* gun defences are out of date, and such essential elements as mines, anti-submarine defences, boom defences and aircraft are lacking. There are insufficient anti-aircraft guns. One of the three Battalions of the peace time garrison of Hong Kong and part of the movable armament are at Shanghai. Singapore is not in much better case. The defences of Keppel Harbour are out of date and

not sited to defend the naval base. Apart from one or two old 9.2-inch guns, which are unsuitably sited, and two 6-inch guns, a single squadron of 12 torpedo-bombers and a squadron of four flying-boats constitute the sole defence for the floating dock and large oil reserves. There are not enough anti-aircraft guns and no boom defences, anti-submarine defences, or mines. The garrison consists only of two battalions (one of which is a battalion of Burma rifles located at Taiping) and volunteer units. It is by no means certain that India could supply the force of one Division required as a war garrison, but even were it made available it seems certain that we could not bring it to Singapore in time. At Trincomali also the naval oil supplies required for the movement of the Main Fleet to the Far East are totally unprotected; there is not even a garrison. [* Owing to Article XIX of the Washington Treaty for Limitation of Armament, 1922, no increase can be made in the coast defences of Hong Kong, and action is limited to bringing the existing armament up to date.]

13   In a word, we possess only light naval forces in the Far East; the fuel supplies required for the passage of the Main Fleet to the East and for its mobility after arrival are in jeopardy; and the bases at Singapore and Hong Kong, essential to the maintenance of a fleet of capital ships on arrival in the Far East, are not in a defensible condition. The whole of our territory in the Far East, as well as the coastline of India and the Dominions and our vast trade and shipping, lies open to attack.

14   There is nothing new in this situation. The records of the Committee of Imperial Defence show that every year since 1919 the weakness of our position in the Far East has been brought to the attention of successive Governments in Reports by the Committee of Imperial Defence, its Sub-Committees and the Staffs of the Defence Departments, including, since 1926, our own Annual Reviews. The vital importance of Singapore, in particular, has been emphasised again and again both to the Government and to the Imperial Conference of 1921, 1923, 1926 and 1930, with the result that New Zealand has made an important contribution to the development of the naval base . . .

15   One of the most disquieting features of recent events is the suddenness with which Japan took action and the success with which her intentions were concealed, notwithstanding the glare of world-wide publicity to which she was exposed, first at the Assembly of the League of Nations, when the Manchurian episode began in September 1931, and later in connection with the

Shanghai incident during the meetings of the Council and of the Disarmament Conference. If Japan were ever to prepare for operations of wider scope it must be assumed that these preparations would be concealed with equal care and her blows delivered with equal suddenness, in order to gain the maximum advantage at the outset. This would be consistent with the Japanese attack on the Russian fleet at Port Arthur, 48 hours before the outbreak of war in February 1904. While we have no reason to impute aggressive intentions to Japan, unless she is goaded into precipitate action, we must point out that the unavoidable concentration of international naval and military forces in a cul-de-sac at Shanghai and the weak state of the defences of Hong Kong and Singapore, to say nothing of the complete absence of defences for the naval oil fuel reserves at Trincomali, would provide a tempting opportunity to a Power in the mood to resort to an aggression.

16   Normally, our Battle Fleet would require 38 days from the zero hour in which to reach Singapore. That is the basis of which the theoretical scale of defence of Singapore and the other ports in the Far East has been worked out. But in their present weak state these ports would be liable to capture, or at least to the destruction of their facilities, before the arrival of the Fleet. Unless these ports have been strengthened adequately and reinforcements arrive in time, on a calculation of reasonable probabilities we should have to assume either that they would be captured or that their facilities would be destroyed in the first month of war. The position would then become one of the utmost gravity. Improvisations would be required on a vast scale before the Fleet could move to the East, and we cannot calculate how long would elapse before an attempt could be made – and then under very adverse conditions – to re-establish our naval supremacy in Eastern seas. In the interval our vast territorial and trade interests in the Far East, and our communications with the Dominions and India, would be open to attack. What the political reaction in India and in the various Colonies would be we leave to experts to determine. We have no hesitation in ascribing this highly-dangerous weakness mainly to the assumption that at any given date there will be no major war for ten years.

[254] *Colonel G.L. Pepys [Staff Officer attached to the Military Department] India Office, to the Admiralty Secretary*
13 January 1933

'Secret'

I am directed by the Secretary of State for India to inform you that he has consulted the government of India on the question raised in your letter of the 30th November 1931, No.M.02165/31. The Lords Commissioners of the Admiralty are strongly of opinion that the guiding principle approved by the Imperial Conference of 1923 in relation to local defence should be accepted by each portion of the Empire. The Government of India on the other hand, represent that India is not in the same position as the Dominions, but pays an annual contribution for the general defence of its shores and the protection of trade in its waters. The Government of India state that they will endeavour to maintain the Royal Indian Marine at its peace strength in the state of efficiency required to fit it for service in war, and when financial conditions improve they hope to consider the expansion of the service, but that they must emphasise the fact that at present they are unable even to afford the cost of normal replacements.

2   The Secretary of State for India agrees with and endorses the views of the Government of India.

3   With regard to the details of the scheme for minesweeping and naval control, the Government of India report as follows.

4   Owing to the decision to dispose of 7 out of 9 trawlers hitherto maintained on the strength of the Royal Indian Marine and to scrap H.M.S. Ship 'Palinurus', only the following eight vessels will be available in India:

For minesweeping duties. H.M.I. Ship *Clive*, H.M.I. Ship *Hindustan*, H.M.I. Ship *Lawrence*, Trawler *Madras*, Trawler *Rangoon*.

For anti submarine duties. H.M.I. Ship *Cornwallis*, Patrol boat *Baluchi*, Patrol boat *Pathan*.

5   It is understood that His Excellency the Naval Commander-in-Chief, East Indies Station, has allocated these eight vessels for duty outside India in time of war. To meet the requirements of Indian ports during the opening stage of a war in the Far East it will, therefore, be necessary to take up from the mercantile marine 27 vessels for minesweeping duties and 21 vessels for anti-submarine duties, as against 16 and 16, respectively, previously

computed. It will also be necessary to take up one additional vessel to complete the number (six) required for service outside India.

6   The minesweeping winches, gallows and other gear in the seven trawlers to be disposed of will be retained and maintained in the Royal Indian Marine Dockyard at Bombay; they will be available for use in the vessels that may be taken up in case of war or emergency.

7   Twelve 12 pounder guns and twelve mountings are in reserve in the Royal Indian Marine Dockyard at Bombay. Consequently, there is a shortage of 26 guns and 26 mountings, not 26 guns and 38 mountings.

8   It will not be possible to provide the requisite personnel from the existing Royal Indian Marine cadre, which is barely sufficient to meet the normal requirements of the service. As previously reported, the question of the formation of a Royal Indian Marine Volunteer Reserve was under consideration, but it has recently been decided that the proposal must be postponed for financial reasons. Until a reserve is formed, the Government of India do not see how the officers required for vessels to be employed on minesweeping and anti-submarine duties can be found in India. In order to provide the ratings required for these vessels and for the five armed merchant cruisers, apart from their peace-time crews who will be re-engaged as far as possible for combatant service, the Flag Officer Commanding and Director, Royal Indian Marine, is examining the possibility of forming a Drafting Reserve and a Fleet Reserve. If these reserves eventually come into existence it might be possible to meet the requirements in respect of ratings, but until then, the majority of the ratings required will have to be provided from sources outside India.

[255] *Admiralty minute addressed to First Lord of the Admiralty*
[Sir Bolton M. Eyres-Monsell] for Field, author unknown
18 January 1933

The First Sea Lord and Admiral Sir Ernle Chatfield interviewed Mr Bruce, Australian Minister in London, today 18th January, on the subject of the replacement cruiser for *Brisbane*.

Mr Bruce was quite unable to give a definite guarantee that Australia would be able to commence a replacement ship for *Brisbane* (Leander Class) in the 1933 Estimates, even though he is aware that money would not have to be spent until the Spring of 1934.

He considers that there would be strong representations to build her in Australia but realises this would entail much increased cost. To avoid this increase of cost, which will always recur if Australia replaces existing units, the Australian Government are considering a scheme by which Great Britain would build ships to replace Australian units and Australia would increase [added in ms 'and {illegible}'] her annual expenditure by 'maintaining' the number of vessels that will balance the annual cost of replacement.

As no money will be spent on the cruiser replacing *Brisbane* until the Spring of 1934, he suggests she should be left in the programme while the above matter is under consideration. He also feels fairly certain that the Australian Government will be prepared to make some contribution towards her construction; exactly what she will be able to do can be definitely settled before the British Government have to spend any money on the ship.

Admiral Chatfield and I thought it much better that you should hear what Mr Bruce has to say before the Cabinet meeting and we have arranged, therefore, for him to see you tomorrow fore-noon, 19th January.

We both think also that Mr Bruce could perhaps be pressed to obtain the consent* of the Australian Government to share the responsibility for the ship being put into our Naval estimates. [signed illegibly, for First Sea Lord]

* Manuscript addition by W.B.
Coxwell [Private Secretary to
the First Lord] [18 January]

The object of this would be to make it clear that the ship was being built by us on behalf of Australia, & so facilitate a request to Australia to make a contribution later, or something of the sort. I have ascertained this from Admiral Chatfield.

[256] *Minute by Captain H.R. Moore, Admiralty Director of Plans*
24 January 1933

The High Commissioner for Australia (Mr Bruce) recently dis-cussed with the 1st Lord and the Chief of the Naval Staff the question of the replacement of the *Brisbane* and other Australian Cruisers. Further to this conversation a letter, copy attached, has been received setting out tentative proposals from Australia House.

2   The Chief of the Naval Staff has directed that a reply be sent if possible by Friday, 27th January. Mr Bruce has been informed that it will not be possible for the Admiralty to give him a reply for a few days.

3   Advance copies of the enclosures have been distributed to Members of the Board by [Deputy Chief of Naval Staff, Vice-Admiral Charles C. Little].

4   The following remarks are offered: The proposal is worked out on the assumption that two cruisers will be maintained by Australia at a yearly cost of £200,000 each. For some years to come only 'D' class cruisers will be available, as it has been decided that the first Leanders should do their first commission in home waters, being of a new type. Furthermore we have already committed ourselves to replacing the two 'D' class cruisers in New Zealand by Leanders about 1936–37. The maintenance cost of the 'D' class is understood to be about £135,000 and therefore in order for naval votes to get the full quid pro quo, which presumably the Treasury will insist on, it appears necessary to allot three 'D' class. The alternative of sending more 10,000 ton cruisers to Australia is not considered desirable.

5   The proposal would therefore cause us to have 6 or 7 cruisers, out of our total of 50 cruisers in 1936, whose availability in time of war would be dependent on the decision of the Commonwealth Government. This would increase the difficulties of distributing the Fleet in peace time to meet possible war requirements.

6   In order to transfer to Australia the full cost of maintenance of the extra cruisers it will be necessary that they shall be 2 (or 3) of those now employed in full commission. From the experience of the *Brisbane* in China in 1925 (M.01735/25) it is apparent that the restrictions which would be imposed by the Commonwealth Government would seriously restrict the stations on which it would be wise to employ them.

7   The acceptance of a scheme of this type by the Admiralty would appear to be giving hostages to fortune as it is at least doubtful whether the present Australian Government could commit their successors for a period of say 20 years.

8   From the point of view of the disarmament negotiations, presumably the additional two ships would have to be included in the Australian quota in any general Disarmament Convention (as are the two New Zealand Division cruisers), and a corresponding

reduction would have to be made in the U.K. figures. The acceptance, therefore, of the Australian proposal must be dependent on the inclusion of a satisfactory transfer arrangement in the General Convention. It is by no means certain that this can be obtained and the difficulties of the question are about to be referred to the D.C.(M) Committee.

9 It is satisfactory that by putting forward this tentative plan Australia appears to be ready to accept the responsibility for the replacement of her four cruisers, but the method she proposes to adopt suffers from a serious objection.

Broadly speaking the plan would have the effect of permitting Australia to control and maintain 6 cruisers at the cost of maintenance and period replacement of her existing force of 4 cruisers; whereas the U.K. would lose control of two cruisers without any financial relief.

It has always been our policy to encourage the Dominions to take a greater share of the burden of Imperial Naval Defence. This scheme, however, does not relieve our burden and at the same time increases the risk that the full strength of the Empire Fleet would not be available in every emergency.

Appendix (P.D.04238/331)
'Extract from the Standing
Instructions for the guidance
of the Commander-in-Chief on
the Mediterranean Station'

Employment of War Vessels of Commonwealth Government of Australia

16 In the event of circumstances arising necessitating the employment of War Vessels of the Commonwealth Government of Australia or crews of such vessels in operations, during the conduct of which force may have to be used, the following instructions are to be observed:

(i) If the urgency of the case requires it you are permitted to use any Australian Warship, attached to your command, for operations designed solely for the protection of British lives or property without reference to any high authority. If possible, however, the sanction of the Commonwealth Government should be obtained beforehand, and that government should invariably be informed of any action taken. A full statement of the circumstances should be sent to the Commonwealth Government by telegraph in either case.

(ii) You should not, however, use H.M. Australian Ships for operations beyond those required expressly for the protection of British lives or property until the whole circumstances of the case have been laid before the Commonwealth Government and their assent has been received.

(iii) Procedure for communicating with the Commonwealth Government is as follows:

(a) All communications that may be necessary with the Commonwealth Government under par. (i) and par. (ii) of these orders should be addressed direct to the Prime Minister, Canberra, Australia.

(b) When seeking authority to take action under par. (ii) of these orders, you should forward to the Prime Minister direct a full report of the situation together with your proposals as to the action to be taken.

(c) All communications under (a) and (b) should be repeated to the Admiralty.

(iv) If ever you foresee the possib[ilit]y of H.M. Australian ships being required to take action of the nature described in par. (i) and par. (ii) you should inform the Commonwealth Government and the Admiralty of the situation and keep them informed of the movements of the warships.

(v) All communications with the Commonwealth Government under the preceding paragraph so far as they relate to the employment of Australian Warships, should be made in the manner laid down in par. (iii). Communications relating to the movements of such vessels and any administrative details should be addressed to the Naval Board.

Note. The above instructions were issued as a result of correspondence with the Commonwealth Government in connection with H.M.A.S. *Brisbane* in China in 1925 and H.M.A.S. *Melbourne* in the Mediterranean in 1926.

[257] *Minute by Rear-Admiral C.J.C. Little [Deputy Chief of Naval Staff]*

26 January 1933

The views expressed by Director of Plans [Captain H.R. Moore] are generally concurred in. There are almost insuperable difficulties in the suggestion that Australia should maintain two of our Cruisers. This is mainly due to our being quite down to the minimum of Cruisers in service and that two could not be spared to serve in Australian waters.

It was decided at the Board this morning that these difficulties should be pointed out to Mr Bruce and that the following alternative suggestion should be made.

The replacement ship for *Brisbane* to be built in this Country to our cost for about £1,600,000.

That Australia should reimburse us by five yearly payments of £320,000.

Although this is not a sound financial arrangement we should have the advantage of

(a)   The construction of the ship in this Country.
(b)   Ensure the replacement of *Brisbane*.

The Australian Government would have the advantage of spreading the cost over a term of years.

[258] *'The Far Eastern Situation', memorial by Chatfield 25 February 1933, prepared for the 107th Meeting, Chiefs of Staff Sub-committee, Committee of Imperial Defence*
28 February 1933

The safety of the Naval Bases at Hong Kong and Singapore is continually under consideration by the Naval Staff and at a time such as the present calls for special anxiety. The availability of both these bases in time of war in the Far East is essential.

2   Up till the year 1926 the China Station War Orders envisaged falling back on Singapore on the outbreak of war with Japan, with the whole strength of the naval forces, and abandoning Hong Kong. In 1928 the Commander-in-Chief China (Admiral Tyrwhitt) recommended 'that the whole strength of our naval forces on this Station should be directed to delay and frustrate attacks on Hong Kong, for I am convinced that no enemy (and most assuredly not the cautious and slow-moving Nation against whose aggression our plans are presumably made) would ever attempt to capture Singapore without first reducing Hong Kong'.

3   After considering this recommendation, the Admiralty decided, and informed C-in-C China (December 1928) 'that in view of the satisfactory state of the political outlook, and the strengthening of the Singapore defences, you may assume that

Singapore is less liable to attack by the Japanese than Hong Kong. Their Lordships therefore concur with you that the forces under your command should be directed to delay and frustrate enemy attacks on Hong Kong.'

4   In consequence of this decision the China Station War Orders now state that 'the object of the British Naval forces in the Far East Will be the security of Hong Kong' and these forces would be concentrated for operations in this neighbourhood.

5   At the time of this decision (December 1928) it was intended to complete the defences of Singapore under a regular programme. Subsequently in 1930 work on the defences was stopped and they have since remained in much the same condition as they then were. The consequence is that the defences have not been materially strengthened since the war.

6   In March 1932, as a result of disturbances in the Far East, the Far Eastern situation was appreciated by the Chiefs of Staff Sub-Committee. As the result of this appreciation and the '10 year rule' being withdrawn, and also in consideration of the decision of the Coast Defence Sub-Committee of the Committee of Imperial Defence, Singapore was given again Priority I on the list of ports to be defended and measures are being recommenced for the improvement of the defences. Nevertheless the measures in question cannot take effect for at least 2 years and Singapore remains totally inadequately defended.

7   The Admiralty has recently received secret information from the Far East that the Japanese have a definite scheme for an attack on Singapore by a force consisting of one division, probably with some light artillery, to be carried in 18 transports and to be routed via Pelew Islands (a Japanese base) – South of Philippine Islands – North of Borneo – Singapore. The estimated time from Japan to Singapore is 8 – 10 days. The information stated that this force was actually ready to embark at the time of the Shanghai crisis of 1932 and is possibly ready now.

8   It now appears that many purchases by Japan of British ships, ostensibly for scrap, have been completed during the last 12 months. Information from Japan shows that some of these ships are being retained in Japan for use as military transports, thus increasing the supply of Japanese shipping already laid up and available for transports.

9   China Station War Orders make no arrangements to meet such an attack on Singapore. The cruisers, submarines and destroyers concentrated in the neighbourhood of Hong Kong, and operating

in the neighbourhood of the Formosa Channel and the Bashi Channel to the South of Formosa, would be quite unable even to locate the expedition referred to sent by an evasive route North or South of Borneo.

10   The capture of Singapore would be a disaster of the first magnitude and prevent the main fleet from proceeding East of the Indian Ocean until it was recaptured. During this time Hong Kong also would be captured with its garrison. It is just possible that the Japanese expedition to Singapore might confine itself to destruction of the fuel and base facilities only, and then withdraw. In this case the Fleet would still be able to use the anchorage, although the loss of fuel and docking facilities would very seriously hamper operations in the China Seas. The high strategical importance to Japan of the capture of Singapore and the ease with which it can be effected is therefore very evident, and lends weight to the 'intelligence' referred to in paragraph 7.

11   The naval forces on the China Station, quite insufficient in themselves, certainly could not be divided for the protection of both Singapore and Hong Kong. If the present orders were modified by withdrawing all naval forces from the Hong Kong area to protect Singapore, it would mean complete abandonment of Hong Kong by the Navy, leaving Hong Kong exposed to almost immediate capture. (NOTE. Half of the garrison of Hong Kong is at present in Shanghai.)

12   The defences of Hong Kong are far from complete and the time it could hold out, even aided by the full force of the China Squadron, is most uncertain. A full scale Japanese attack would probably eventually succeed if the Fleet was unduly delayed, but to abandon the fortress at the outset is unthinkable.

13   It has therefore been considered necessary to point out the existing strategic situation in order that the Government may decide whether immediate strengthening of the Singapore garrison by military and air forces is possible during the present time of uncertainty, and if not, that a time table should be made out of the time before hostilities necessary to strengthen the garrison. The present policy of attempting to defend Hong Kong, and leaving Singapore open, threatens the loss of both bases, and is a great temptation presented to the Japanese of striking, almost undeterred, a blow of the first magnitude.

14   Failing reinforcement of the Singapore garrison, there appears no alternative but to alter the role of the China Squadron to that of defending Singapore.

[259] *'The Situation in the Far East', Report by the Chiefs of Staff Sub-Committee, CID 1103-B. Signed by Chatfield, General Sir D.A. Montgomery-Massingberd, [Chief of Imperial General Staff] and Salmond*

31 March 1933

1   In accordance with the terms of our Warrant, under which we are instructed to submit for consideration any matter relating to Imperial Defence on which, in our opinion, further enquiry or investigation is necessary, we have been re-examining the strategical situation in the far east in the light of recent events.

2   We are well aware that His Majesty's Government are seized with the importance of avoiding any step liable to precipitate a crisis in which this country whether as a Member of the League of Nations or alone, would be involved as a belligerent in the Far East. The Secretary of State for Foreign Affairs made our position in the matter perfectly clear in the House of Commons as recently as the 27th February, 1933.

3   In every year since 1919 the weakness of our military position in the Far East has been brought to the notice of the Committee of Imperial Defence. The last occasion was a year ago when we submitted two Reports on the subject (C.I.D. Papers Nos 1082-B and 1084-B). Since then nothing has happened to alleviate the situation, and it is therefore unnecessary to describe it in much detail. It is sufficient to mention that the Naval Base at Singapore is the pivot of our whole naval strategical position in the Far East: that, until provided with adequate defences, the Naval Base and its facilities are liable to capture or destruction by a coup de main before our main fleet can arrive on the scene; that its re-capture would be a major operation of the greatest difficulty; that, in the meantime, the important Naval Base at Hong Kong would be liable to share the same fate as Singapore; that British territory in the Far East, including the coasts of India, Australia, New Zealand, and of the various Colonies and Protectorates, would be exposed to depredation; and that our trade and communications in the Eastern Hemisphere would be, to a large extent, at the mercy of the enemy. With the aid of the Dominion of New Zealand, the Colonies of Singapore and Hong Kong and the rulers of the Federated Malay States, some progress has been made for the provision of a Naval Base at Singapore. Up to now, however, the question of the defences has not been tackled in earnest, with the result that the resources of the Naval Base offer

an inviting objective for an enemy desiring to dislocate our entire system of Imperial Defence in the Pacific before the arrival of the Main Fleet.

[260] *Cabinet Minutes*

12 April 1933

5   The Cabinet had before them a Most Secret Note by the Secretary . . . circulating by direction of the Prime Minister a Report by the Chiefs of Staff Sub-Committee of the Committee of Imperial Defence, dealing with the situation in the Far East, and an Extract from the draft Minutes of the 258th Meeting of the Committee of Imperial Defence, held on April 6, 1933, at which the Report was considered. Particular attention was called to the fact that the Committee of Imperial Defence, while approving generally the Report, subject to the amendments set forth in the Conclusions, reserved for the decision of the Cabinet the proposal in paragraph 11 for the provision of a second aerodrome at Singapore. The Conclusions reached by the Committee of Imperial Defence were as follows:

1   To approve generally the Report of the Chiefs of Staff Sub-Committee (C.I.D. Paper No. 1103-B), subject to modifications . . .

The Cabinet agreed

(a)   To approve the recommendations of the Committee of Imperial Defence as set forth in the Reports before them and summarised above:
(b)   To note, in particular, that the recommendation for the withdrawal of troops from Shanghai to Hong-Kong was limited to one Battalion, and to invite the Secretary of State for War to ask the concurrence of the Secretary of State for Foreign Affairs when the moment arrived for giving effect to the recommendation:
(c)   Subject to agreement between the Chancellor of the Exchequer and the Secretary of State for Air as to the necessary financial adjustments, to approve the proposal for the construction of a new aerodrome at Singapore.

[261] *Note of a Conference held at the Admiralty*
19 September 1933

'Secret'

Present: The First Sea Lord, Sir M. Hankey, Sir H. Batterbee, Mr Barnes, Mr Coxwell, Mr S.M. Bruce, Mr Shedden.

THE FIRST SEA LORD [Admiral Sir A. Ernle M. Chatfield] recalled a proposal which Mr Bruce had made to him in January, 1933, viz. that Great Britain should build and pay for the replacement cruisers for the Royal Australian Navy, and that Australia should compensate Great Britain by taking over the manning and maintenance of two cruisers now manned and maintained by Great Britain.

The First Sea Lord remarked that he had (on the 31st January) sent to Mr Bruce [Australian High Commissioner] an Admiralty memorandum setting out certain objections to that proposal, and making an alternative proposal. [note in ms lost in gutter] To this latter proposal, no reply had as yet been received.

MR BRUCE [Australian High Commissioner] concurred, and proceeded to comment on some of the objections raised in the Admiralty's memorandum.

1 The Admiralty had assumed that if the Australian proposal were adopted, Australia would wish the two additional cruisers to be employed in Australian waters: to this the Admiralty had raised practical objections.

Mr Bruce expressed the view that this difficulty was not a very fundamental one, and could be resolved.

2 The Admiralty had foreseen serious difficulties if the two additional cruisers were to be not entirely at the disposal of the Commander-in-Chief to whom they were allocated.

Mr Bruce admitted that this objection, if no mutually satisfactory solution could be found, was serious enough to end all naval co-operation between Great Britain and the Dominion. But he did not himself think the problem insoluble.

3 The Admiralty had pointed out that at present the Commonwealth Government reserved the right to decide at the time whether the R.A.N. should be placed at their disposal in the event of Great Britain being at war, and that it would be a very serious matter for Great Britain to increase the number of cruisers which would be subject to such decision from four to six.

Mr Bruce expressed the view that to raise such a question as this, in a document actually forwarded to Australia, would be very dangerous, for the following reasons:

(a) If Australia were to be definitely limited to a maximum of four cruisers, she would probably feel compelled to overhaul the whole question of her present contribution (the R.A.N.) to the Naval defence of the Empire, and would perhaps decide to give up the R.A.N. and concentrate on her own defence by military and air forces.

(b) If the Admiralty expressed uncertainty as to what Australia might do in the next war, Australia might reply by expressing uncertainty as to what Great Britain herself might do – e.g. as to whether she would actually send a Naval force into the Pacific.

Mr Bruce thought it would be very undesirable to raise such questions, more especially (b), in a communication to Australia, and he had therefore not forwarded the Admiralty memorandum to the Commonwealth.

He suggested that if the Australian proposal had to be rejected, it could be rejected on technical grounds, and not on grounds of higher policy, and that the reasons for rejection should be stated at greater length.

THE FIRST LORD asked whether it were not the fact that Australia had declined to be a party to the Locarno Treaty, because of her wish to retain complete freedom of action?

To this MR BRUCE replied that Australia had not really thought very much about that aspect of the matter.

SIR H. BATTERBEE [Secretary of State for Dominion Affairs] remarked that of course Australia would not necessarily keep out of a future war.

THE FIRST LORD observed that the Admiralty memorandum had of course been sent to Mr Bruce for his personal consideration, and not for transmission to Australia. He now gathered, however, that Mr Bruce rejected the Admiralty proposal, and that we were back at the original proposal of the Commonwealth.

MR BRUCE agreed that this was the position, until that proposal had been finally rejected. He added that he did not think there would be any difficulty in the two additional ships working outside Australian waters.

As regards the counter-proposal made by the Admiralty, the Commonwealth would not like the idea of being limited to four cruisers.

THE FIRST LORD replied that the Admiralty had no desire

to limit Australia in the number of cruisers which she might be willing to maintain. Questions of limitation arose only through Treaties.

The First Lord proceeded to explain that the replacement of the BRISBANE was now possibly to be complicated by a new factor. Owing to the building of large cruisers by Japan and the U.S.A., we might have to alter our 1933 Programme, building 3 cruisers instead of 4. The 3 to be built would be either improved (and larger) LEANDERS, or 2 of these & 1 ARETHUSA: the Treaty limitations would in any case prevent our building more than 3 such vessels. This would give us at the end of the London Naval Treaty 49 cruisers, instead of our bare minimum (which the Admiralty had only accepted temporarily and under special conditions) of 50.

A Note had been addressed to the U.S.A. which might have the result of making this alteration of programme unnecessary, but if it were necessary it would undoubtedly have a bearing on the questions now under discussion: to be quite frank, it would increase the Admiralty's objections to the original proposal of the Commonwealth, and it might also make it necessary to propose a postponement of the replacement of the BRISBANE for a year.

MR BRUCE said that Mr Shedden would shortly be going to Australia, and he suggested that the best thing would be for Mr Shedden to have the whole position fully explained to him by someone at the Admiralty, so that he could inform the Commonwealth Prime Minister as required.

THE FIRST LORD expressed his personal concurrence in this course, and asked Mr Barnes to initiate arrangements accordingly.

The First Lord emphasised the fact that the Admiralty had no desire to tie Australia, as a matter of principle, to 4 cruisers.

MR BRUCE fully accepted this, but reiterated that the Admiralty memorandum, had it been forwarded to Australia, would have created the impression that such a limitation was intended.

THE FIRST LORD enquired whether many people in Australia wished to give up the maintenance of a Navy. To this MR BRUCE replied in the negative, but said that the partisans of the Australian Army and Air Force would be glad to see such a step taken.

Mr Bruce went on to explain that a great difficulty under which the Commonwealth Government were labouring was that of deciding how best to allocate the money available for defence as between naval, military and air forces. Some day this question would have to be faced and settled, and it was of the very greatest importance that right decisions should be taken.

SIR. M. HANKEY referred to the Report of the Chiefs of Staff in 1932 as giving their views on the subject.

MR BRUCE having referred to the extreme importance, in the view of Australia, to the completion of Singapore, the First Lord expressed a hope that Australia did not think the Admiralty were under any delusions about that matter. Actually, we were doing a lot to improve the position there, and accelerate progress.

MR BRUCE said that he fully appreciated this, but pointed out that at present no British naval force could be sent to the Pacific. He also repeated that Australia might well have doubts whether even with Singapore completed such a force would certainly be sent in an emergency threatening the Commonwealth: there might be a popular cry in England against denuding our own coasts of naval defence, or a new Government might be unwilling to take such a step.

Reverting to the main question under discussion, SIR H. BAT-TERBEE said that the Secretary of State for Dominion Affairs was in agreement with Mr Bruce that the existing Admiralty memorandum should not be transmitted to Australia.

THE FIRST LORD intimated that he accepted this view, and added that the best course was that already agreed, viz. that Mr Shedden should obtain from the Admiralty a new statement which he could take to Australia.

In conclusion, MR BRUCE intimated that he would be very grateful if he could be given, confidentially and informally, some further advice as to the best allocation of the funds available in Australia for the defence of the Commonwealth. He suggested that possibly the Chiefs of Staff, in conjunction with Sir M. Hankey, might be willing to give such advice.

SIR M. HANKEY expressed his personal willingness, after speaking to the Prime Minister, to assist in the way suggested, and thought that the Chiefs of Staff would also be willing to help.

THE FIRST LORD concurred.

The conference then terminated.

[262] *S.M. Bruce [Australian High Commissioner] to Chatfield,*
First Sea Lord,

Australia House, London
21 February 1934

Many thanks for sending me the memorandum with regard to our conversations last Saturday. It served my purpose and gave me the necessary information on which to base a cablegram to Australia.

I have indicated to them the impossibility for strategic reasons of increasing the number of 8″ [15″ sic] Cruisers on the Australian station and have put to them the alternatives of a 'Leander' or an 'M' giving the dates when these classes would be respectively available.

I have also dealt with the point of the possibility of the limitation of 8″ [18″ sic] cruisers disappearing after the Naval conference next year, and pointed out to them the necessity of our having some arrangement with the British Government in the event of the decision being to take over one of the 6″ class . . .

[263] *Keith Officer, Department of External Affairs, Commonwealth of Australia, to A.S. le Maitre [Admiralty Principal Clerk]*

6 March 1934

'Most Secret'

With reference to the discussions you had with Mr Bruce [Australian High Commissioner] before he left for Australia, I am now asked to advise you that the Commonwealth Government propose to announce that they intend to make provision for the replacement of the *Brisbane*, and the Minister for Defence will accordingly submit to the Cabinet in the near future a proposal for the provision of a cruiser of the 'Leander' type. This is being done on the assumption that one of this type will be available for the Royal Australian Navy from the four vessels completing during 1935, and I am asked to confirm that this assumption is correct.

[264] *Memorandum prepared jointly by the Admiralty and the Foreign Office for the 1935 Naval Conference*

23 March 1934

6   In the last 12 years our naval security has been seriously jeopardised, so that the despatch to the East of a fleet sufficient to meet that of Japan, combined with a distribution of cruisers to ensure the security of our sea communications against Japanese attack, would leave us with a strength in Europe and Home Waters definitely inferior to that of the strongest European naval power. In capital ships alone we should have a bare equality, but in other classes of ships the position is serious, as we should have a mere

handful of cruisers and destroyers left to meet the powerful cruiser and submarine forces which could be brought against our Atlantic and Mediterranean trade. It is therefore vital that the following question should be most carefully considered:

If we have to send out to the Far East an adequate fleet, how is our security in Home Waters to be obtained, and what is the minimum strength of our naval forces in those waters that can be accepted?

7   It is fully realised that it is our policy to remain on good terms with France, but the Naval Staff emphasise that the French naval strength provides, and will provide, the measure for the strength of the strongest European naval Power and that the present position of France may be assumed by a Power whose interests are opposed to ours. There are reasons why a long naval agreement is highly desirable and so our foreign policy must envisage, from the point of view of naval security, our relationships for a long time ahead. The standards of relative strengths agreed to in 1935 will bind us and determine our relative strengths in, say, 10 years' time.

8   If we are to accept definitely that it is an impossible financial task to build up a sufficient naval strength to face the strongest European Power when we are already engaged with Japan, that is a 'Two-Power Standard,' we must also accept the fact that the Admiralty cannot guarantee the security of our vital sea communications in Home Waters against attack by sea. It seems that we must either trust to a naval combination with some other Power to give us security at sea against such aggression, or we must keep the balance of our forces remaining in Europe sufficiently strong to prove an effective deterrent to any interference, namely, a 'One-Power Standard.' . . .

11   It has been assumed in this paper, therefore, that our minimum strategical requirement for security can be stated as follows:

We should be able to send to the Far East a fleet sufficient to provide 'cover' against the Japanese fleet; we should have sufficient additional forces behind this shield for the protection of our territories and mercantile marine against Japanese attack; at the same time we should be able to retain in European waters a force sufficient to act as a deterrent and to prevent the strongest European naval Power from obtaining control of our vital home terminal areas while we can make the necessary redispositions.

It is on this strategical requirement that these proposals are based.

Foreign Office Comment on Part
I (General Strategical
Requirements for Security)

It may be assumed that, in the years covered by any future naval treaty, it would be the policy of His Majesty's Government, with their eye on dangers nearer home, to spare no pains to improve relations with Japan and to avoid possible causes of friction. Nevertheless, in the absence of any Cabinet ruling that the Admiralty will not be called upon to send a fleet to the Far East capable of engaging the Japanese fleet, the Naval Staff are clearly justified in basing upon the above premiss their proposals for the 1935 Conference ...

Part IX – Dominion Quotas

88   The Inter-Departmental Disarmament Committee gave exhaustive consideration, at the instance of the Admiralty and the Dominions Office, to the question of the procedure to be followed in the future in regard to the application of naval limitation to the Members of the British Commonwealth of Nations. Their report is contained in paper D.C. (M) (32) 32 of the 10th February, 1933, which has not, as yet, been considered by the D.C. (M) Committee. The conclusions in that report may be summarised as follows:

(a)   There is no option for the time being but to adhere to the present system i.e., a Commonwealth quota, or some modification of it which has the same effect, e.g., separate Dominion quotas with transfer within a Commonwealth total.

(b)   That if the latter plan be adopted, the following would be a suitable formula for inclusion in the Convention: 'It shall be permissible for the tonnage limitation figure of any member of the British Commonwealth of Nations to be exceeded, provided that that excess is offset by a corresponding deficiency in the permitted tonnage of other Members.'

(c)   The ideal solution of this problem, from our point of view, would be that the naval strength of other Powers should be calculated in relation to the naval strength of the United Kingdom only, the naval strength of the Dominions being left out of account in such calculations. The level of the United Kingdom naval strength, which is based on practical needs, would require to be fixed sufficiently high to safeguard our naval position in the event of one or other of the Dominions not building up to its full quota, or not being available for naval assistance in certain contingencies. (d)   While such a solution of this problem is not regarded as practicable at the present time, it is, nevertheless, an objective towards which we might work as and when opportunity offers.

The Naval Staff is in general agreement, with these conclusions, and regard the scheme in (c) above as being the true objective at which we should aim now.

[265] *George Stanley, Governor of Madras; Field Marshal Sir P.W. Chetwode, Commander-in-Chief India; Khan B.M. Fazl-i-Husain, Vice President Governor-General's Executive Council, and others, to Sir Samuel Hoare, Secretary of State for India*
14 June 1934

All that we do contend is:

(1)   that India's defence expenditure as a whole is very large, even taking into account the contribution of £1 1/2 millions that His Majesty's Government has recently decided to make towards it;
(2)   that we have of recent years very considerably increased our own expenditure on naval defence; and
(3)   that, the amounts that we now pay towards the cost of His Majesty's Navy, however small they may appear, are yet sufficient to enable us to carry out the major portion of our responsibilities for local naval defence and would therefore be more usefully employed if placed at our disposal, instead of forming a comparatively negligible contribution towards the revenues of the United Kingdom.

The proposals that follow are all based on the understanding that it is a transfer, and not a reduction, of expenditure that we have in view and that the funds that may be set free as a result of our recommendations will be utilised for naval purposes and not to lessen the burden of our defence expenditure.
8   The first of the three contentions advanced in the previous paragraph is not one that requires development . . . The most important points as they appear to us, are:

(a)   The contribution itself has always been expressly stated to be open to periodic revision.
(b)   At the time when it was fixed at £100,000 in 1895, the Royal

Indian Marine was a non-combatant force costing about Rs. 40 lakhs, exclusive of the contribution. Apart from the fact, therefore, that the ships and men of the Royal Indian Marine constituted a potential reserve, which might be made available for combatant duties after the necessary equipment and training in time of war, the contribution represented the sole payment made from the revenues of India towards the cost of naval defence.

(c)   In 1913 when His Majesty's Government raised the question of increasing the contribution, the arguments set forth at length in our Finance Department Despatch No. 260, dated the 23rd July 1914, were accepted without question as sufficient to rebut the claim.

(d)   In 1923, in connection with the question of retaining the transports of the Royal Indian Marine, the Admiralty stated in their latter No. M.O. 1073, dated the 24th December 1923 – 'The only condition under which the Admiralty would consider it desirable to scrap the transports is that the Indian Government should proceed forthwith with constructive steps for the formation of a sea-going combatant Naval Service, starting for example with one or two sloops, the idea being that these may be maintained definitely as men-of-war. If these vessels can be maintained as efficient men-of-war, they might replace one or more of the vessels of the Imperial Navy in the Persian gulf, thereby reducing the amount of the subsidy paid to the Imperial Government in connection with such vessels'.

(e)   The Royal Indian Marine, so long as it remained a non-combatant force (and its cost as such had increased by 1914 to nearly Rs. 67 lakhs including the contributions) carried out services that were of direct value to the Government of India in the way of transport, station duties, lighting and bouying, etc. The conversion of the Royal Indian Marine into a combatant force (the cost of which including the contributions, originally came to some Rs. 68 1/4 lakhs), not only deprived the government of India of the services referred to, which now have to be performed by other agencies, but also meant an increase in their expenditure on naval defence by the whole amount of the Marine budget, less the contribution previously paid, that is to say an increase of about Rs. 50 lakhs or £375,000.

10   These facts by themselves would appear to present an unanswerable case, on its merits, for a revision of the present arrangements and the actual abolition of the contribution; for if it was accepted, as it was, that £100,000 was a sufficient contri-

bution in 1914, it can hardly be contended, even when due allowance has been made for the alteration in the value of money since that date, that a contribution of nearly five times that amount is not more than sufficient in 1934.

As already explained, however, it is not our object to reduce our expenditure, but merely to change its destination; and quite apart from the arguments stated above, there is another reason of substance why in our opinion such a course would be justified.

11   In addition to the actual contribution of £100,000, there are various other items that we now pay on Admiralty account, the average annual cost of which comes to about £32,600. These are shown in Appendix III and it will be seen that the majority of them are connected with the maintenance of the subsidised vessels of the East Indies Squadron in the Persian Gulf. With regard to one particular item, namely oil fuel, the arrangement is obviously uneconomical, because we are compelled to enter into a small separate contract for the supply of this oil and therefore have to pay for it at higher rates than those secured by the Admiralty under their own larger contract for the same article. Apart from this, it seems to us to be wrong in principle, and it is a matter of particular difficulty under the stabilised Defence budget system, to pay for supplies over the amount of which we can exercise no control. For example, in 1929 all the prospective savings in the Marine budget were swallowed up to meet an entirely unforeseen demand for Rs. 46,000 for excess coal. Above all, it seems to us that the arrangements regarding the subsidised vessels in the Persian gulf, which date from 1869, were based on a balance of interests in that region which has been entirely upset by the course of events during the present century. It may have been a fact that in 1869 purely Indian interests predominated in the Persian Gulf, but the position of Persia herself, the independence of 'Iraq and the importance of the oil fields and the air routes have rendered a new orientation necessary; and it would be difficult to deny that the presence of a naval force in that area is today far more a matter of Imperial concern. It would indeed, as we stated in our Despatch No. 17, dated the 4th September 1930, be a matter of considerable difficulty to reconcile Indian opinion to the employment of the Royal Indian Marine in the Persian Gulf as one of its principal functions; the recent debate on the Indian Navy (Discipline) Bill disclosed the suspicion with which the proposal to confer full naval status on the force is still regarded; and it is at least open to question whether the payments under discussion

would not invite serious criticism if their existence were more widely known than happens to be the case. Here again, therefore, we see strong grounds for a revision of the existing arrangements. 12   We must now turn to the task that lies before us if we are to discharge what we take to be our legitimate responsibilities in the matter of naval defence. The first requisite in our opinion is to provide the equipment needed for the execution of local naval defence schemes and the personnel to use that equipment. It has already been estimated in consultation with the Admiralty that one Mark II set and twenty-three Mark III sets of minesweeping gear, including winches, will be required to equip the trawlers and other vessels which will be taken up from civil sources on mobilisation for minesweeping duties at the Defended Ports, while a total of 38 guns will be required for these vessels.

The next question is the provision of the requisite personnel. As you are aware, the proposal to constitute a Royal Indian Marine Volunteer Reserve to provide a reserve of officers, has been engaging our attention for some considerable time, but the seriousness of the financial situation has so far precluded us from putting the scheme into force, although legislation was undertaken in 1933 to provide for the discipline of such a Reserve (see Act No. I of 1933). The scheme, it is estimated, will eventually lead to a total extra expenditure of about Rs. 30,000 per annum, but with a view to reducing its initial and recurring cost for the present it would be possible to make a start with a smaller number of officers and to work up gradually to the full establishment as circumstances permit. In addition, the Flag Officer Commanding and Director, Royal Indian Marine, has submitted to us a scheme for the formation of a Fleet Reserve of ratings intended primarily to man in part the auxiliary craft to be taken up on the outbreak of war in the East and to provide the trained key ratings for the operation of the minesweeping gear and guns who, according to our naval advisers, will be indispensable if we are to obtain good results from the commencement of any emergency. This scheme is estimated to cost Rs. 83,000 per annum if annual training is to be provided. The cost of providing accommodation for the guns and mountings that we should require to maintain for the various classes of auxiliary vessels to be taken up when the Naval Control service is put into operation in an emergency or for the arming of merchant ships is now estimated at about Rs. 2,70,000 initial and Rs. 27,000 recurring per annum, including maintenance charges.

We propose to make further enquiries to see whether these estimates can be reduced.

13 Hitherto we have confined ourselves entirely to the question of seaward defences. Shore and sea defences are necessarily complementary, and in discussing the naval aspect of defence, it is relevant to point out that in connection with the former the Defended Ports technical committee in their report of 1929 made a series of recommendations, which, if carried into effect in their entirety, would involve us in expenditure amounting to some Rs. 90 lakhs. Expenditure on any such scale, as we pointed out in our Army Department letter No. 32801/1 (G.S.M.O.-1-B.) dated the 26th November 1931, to your Military Secretary, is beyond the sphere of practical politics, so far as we are concerned, but we shall have to spend at least Rs. 30 lakhs to give effect only to such recommendations as regards shore defences as it is absolutely essential for us to adopt if our seaward defences are not to be much reduced in value.

14 Finally there is the question of the expansion of the Royal Indian Marine itself. If the Force is to progress and grow on the lines which have been recommended to us by Vice-Admiral Sir Humphrey Walwyn and which we ourselves favour, we should like to look forward to a fleet of six sloops and four minesweepers within a reasonable number of years. It must be remembered that at present the shore appointments which are available under the Commerce Department constitutes one of the main attractions of the Service and provide convenient openings for the more senior officers. In course of time it seems to be inevitable that these appointments will be held by Indian members of the Mercantile Marine Service and it will become increasingly difficult to provide suitable appointments for senior officers within the combatant cadre of the service if it remains at its present size. This, however, is an aspect of the case of which we must necessarily, and can afford to, defer consideration for the present.

15 It now remains to sum up our conclusions and put our practical proposals before you.

We feel that the time has come when, if India is to make any progress in undertaking her own naval defence, we should be given more direct control over our naval expenditure than we possess at present. We agree that we should undertake direct responsibility for the practical preparation and execution of local naval defence schemes. We cannot afford to do so, so long as we continue to pay as much as Rs. 18 lakhs out of our Marine budget of Rs. 65 lakhs, towards the cost of His Majesty's Navy. We consider that there is a case for the total abolition of these payments, on the understanding that we should ourselves devote an equivalent sum towards naval defence measures. We are, however, of opinion that the best course would be to draw up as a start a definite programme designed to liquidate the most urgent and important portion of the responsibilities referred to above; and we request that we should be placed in a position to carry through such a programme by being given a remission to the required amount out of the rs. 18 lakhs that we now pay annually to His Majesty's Government. On the completion of this programme we propose that the whole subject should again be comprehensively reviewed.

16   We attach as Appendix IV an outline of the measures noted in paragraphs 12 and 14 with their estimated cost and also a draft five-year programme as a basis for discussion . . .

The execution of this programme will involve a capital outlay and recurring expenditure amounting to Rs. 52½ lakhs spread over the five-year period and thereafter annually recurring expenditure amounting to Rs. 3½ lakhs a year . . .

If it would help towards the acceptance of our proposals, we should also be prepared to consider the question of replacing our annual cash payments to some extent by actual service in the Persian Gulf, e.g., by employing one of our sloops there in conjunction with His Majesty's ships . . .

17   Finally, we must make it clear that unless some such course as we suggest is accepted, we shall be unable, on financial grounds alone, to alter the attitude that we have hitherto adopted in the correspondence referred to at the beginning of this Despatch, in which case, in the event of a sudden emergency, the Admiralty might find themselves faced with the necessity for incurring the expenditure involved at short notice.

[266] *Memorandum by J.G. Laithwaite, Principal Secretary Political Department, India Office*

7 August 1934

The Government of India's suggestion that conditions in regard to the Persian Gulf have so changed since 1869 that it is no longer appropriate that they should make the specific contribution to naval expenditure in that area represented by their provision of fuel for H.M. sloops is plausible. But while there is something in it, I doubt very much if their case is really a good one.

2   The arguments on which the Government of India base their suggestion are the position of Persia, the independence of Iraq, the importance of the oilfields and the importance of the air routes, and the fact that 'purely Indian interests' no longer predominate as they are alleged to have done in 1869.

3   While Indian interests may no longer occupy quite so dominating a position as they did 60 years ago, the fact remains that they are still very considerable. Indian dhows still carry a very large proportion of the gulf trade, both on the Arab and on the Persian littoral. This argument is, at best, I think, of very doubtful value in the present connection.

4   The development of the oilfields has undoubtedly given the Home Government a closer interest than in the past in the position of Persia. But the position of Persia is and must continue to be one of considerable interest to the Government of India. The unfriendly Persia under Soviet control or in alliance with the Soviet might constitute a very serious military menace to India. The primary importance of the Gulf is that it commands the western flank of approach to India, and developments in the Gulf or in Persia which would make it possible for a foreign power to establish a naval or air base there or to acquire a dominating position on the South Persian shore are as important to India to-day as they ever have been.

5   The oilfields, as stated above, are undoubtedly a new factor and one of interest primarily to the Imperial Government.

6   The independence of Iraq is not, I suggest, of real relevance in the present connection.

7   The air route argument is, in my view, one which weights the scales against India rather than the reverse. The Gulf with the development of air communications has become an essential link in the air route to India. It has been generally accepted in recent discussions among the interested Departments here that it is of first class strategic importance. But while this, of course, [is] important as a link in the air route to Singapore, Australia and New Zealand, it is primarily of importance as being on the flank of India. India has not so far been pressed for any contribution to air expenditure in the Persian Gulf, either civil or military. But if the question of the Indian contribution to defence expenditure is reviewed as a whole, whatever chance there may be of our securing some reduction in our naval contribution, there is little

question that we may sooner or later be faced with a demand for an air contribution in respect of the Gulf area.

8   Other points of some general importance in assessing the value of the argument put forward by the Government of India are the following:

(i)   Up to, I think, 1923 or 1924 the Government of India provided a despatch vessel for the Political Resident in the Persian Gulf. They have now been relieved of this obligation, the Resident being carried by H.M. ships. There has also been a considerable saving – say Rs. 98,000 (PZ. 184/33) – on the contract grant of Rs. 120,000 previously set aside for the Resident's touring thanks to the assistance given by H.M. Ships, under arrangements reached in 1933. The saving no doubt enures to the Political Department grant, but it is relevant in any general appreciation of the present question.

(ii)   It has been accepted as between the Government of India and the Secretary of State in discussions that have recently taken place that it is very desirable to keep the political control of the Persian Gulf in the hands of the Secretary of State and of the Government of India. Partly for this reason we have hitherto refrained from pressing for any contribution from the Imperial Departments to minor items of Gulf expenditure. It would be consistent with our attitude to let existing arrangements in regard to the contribution to the cost of the sloops stand. The more the Imperial Government pays the more its interests (in so far as they may conflict with those of the Government of India) must predominate in Gulf affairs.

9   I suggest that on a review of the situation as a whole we should be ill-advised to make use of the argument now under consideration in discussions with the Admiralty. I see no reason why we should not endeavour to secure certain minor adjustments – e.g. that naval contract rates should apply in the case of the oil fuel supplied by the Government of India to sloops. But I doubt whether we should have much hope of persuading the Admiralty to abandon this contribution, and I think there is very little question that if we did the victory might be rather an expensive one; that it might react on the degree of control at present exercised by the Government of India in the gulf, and that it might give rise to consideration of a request for a contribution to air expenditure in that area from Indian revenues.

[267] *Debates of the Indian Legislative Assembly on the Indian*
Navy (Discipline) Bill

29 August 1934

Statement by Lieut.-Colonel.
A.F.R. Lumby, Deputy Secretary
Army and Marine Department,
Government of India

Sir, I should like to thank the House very much indeed for the sympathetic way in which Honourable Members have spoken of this Bill, particularly as only a short six years ago the House decided that they would not touch this very Bill with a pair of tongs. As regards the question of future expansion, what I said was merely that this Bill does not commit this House or anybody else to spending any more money upon the expansion of the Indian Navy. But I also said that I had little doubt that Government would be only too glad to fall into line with any suggestion backed by Indian public opinion to increase India's navy, provided always that the money was available. I meant merely to give the impression that Government do not intend to increase the cost of the navy unless they have public opinion behind them.

My Honourable friend, Colonel Gidney, referred to the excessive size of the officer staff of the service. Some 25 per cent of this staff is employed in Mercantile Marine appointments under Local Governments, as Nautical Surveyors and Principal Officers in the various Ports. These officers are borne on the cadre of the Royal Indian Marine, but they are not paid from the Royal Indian Marine budget. They perform a very necessary service to India, and there is nobody else in any other service as yet who can perform the various duties they perform. Ultimately, when the Indian mercantile marine has been placed on a more definite footing, thanks to the training of suitable officers in the Training Ship *Dufferin*, these duties will undoubtedly be performed by officers of the mercantile marine. For the present, however, the Royal Indian Marine officers are filling a gap which could otherwise only be filled by importing officers from outside India.

As regards the question of Indianisation, it was definitely stated during the debate in 1928 that the ratio of one in three which had been accepted as the basis of Indianisation would mean only one Indian officer a year added to the staff of the Royal Indian Marine. In actual fact the first officer was taken on in 1928 and in six years we have got 14 officers and officers designate instead of the six that might have been expected (Mr S.C.Mitra: 'Very Good'), and we are looking round for four more this year. So I really think that we have not done so badly. As I said earlier in the day we have to judge of the correctness of the steps that we are taking as regards recruitment and training, and we have not got very much on which to base our judgment at present. It must be remembered that the earlier training of these officers takes time, it takes five years to train an engineer officer. But I can assure the House that, when we have more out of the 14 officers actually serving with the fleet, we will consider very carefully whether we can increase the proportion above one in three. (Cheers.)

The impression was given by one speaker that we were not making use of the Training Ship *Dufferin* as a source from which to obtain officers for the Marine. Our trouble has been that we once gave a guarantee in this House that we would obtain our Indian officers for the service by open competitive examination. The open competitive examination failed us, and, comparatively recently, we had to start taking cadets from the Training Ship *Dufferin*. I may say that the boys we have taken from that source have done extraordinarily well. (Hear, hear.) In October next, we are having another examination; and at least one vacancy will be open to the cadets of the *Dufferin*.

My Honourable friend, Mr Gaya Prasad Singh, referred to the question of racial discrimination. I do not think I need say very much on that subject. There is no racial discrimination in the service. (Hear, hear.)

Mr Gaya Prasad Singh: I myself said so.

Lieut.-colonel A.F.R. Lumby: I am just giving the guarantee he asked for. As regards the question of the Admiralty commandeering the ships of the Royal Indian Marine for purposes unconnected with the defence of India, I should like, if I may, to read an extract from the speech that I made in an earlier stage of this Bill on this very subject. What I said was: 'In addition, as was announced in the House during the last debate on this Bill, it is the intention of the government of India to consult the Legislature in future, so far as may be possible, whenever any question arises of lending the Indian Navy to His Majesty's Government for operations other than in the defence of India. I assure the House that this is a definite pledge and not merely a formula of words.' (Cheers.)

[268] *Press release, Information Officer, India Office, 'The Indian*
Navy (Discipline) Bill'

5 September 1934

This bill was passed by the Legislative Assembly in India on
the 29th August, 1934, without a division and is expected to come
up before the Indian Council of State this week. If and when
information comes to hand through the usual press sources that
the bill has passed the Council of State, editors may find the
following unofficial note to be of interest.

The Bill gives the full status of a Navy to the Royal Indian
Marine, and provides that the discipline of the Force shall be
regulated, like that of Dominion Navies, by the British Naval
Discipline Act, suitably amended to meet local conditions. This
will replace the Indian Marine Act of 1887 which is out of date
and limited in its scope to Indian Waters.

The Royal Indian Marine can claim to be the oldest service in
British India. It traces its history back to the formation of a
squadron in 1612 by the East India Company for protection
against European rivals and Indian pirates. During the 17th
Century it was called the Honourable East India Company's
Marine, and in the 18th the Bombay Marine. In 1830 it was given
the title of Indian Navy which it held until 1863, when it became
again the Bombay Marine. The title of Royal Indian Marine was
finally adopted in 1892.

During the first sixty years of the 19th Century the force was
employed on almost continuous War Service in places as far apart
as Burma, Mauritius, Java, Zanzibar, the Red Sea, Egypt, Arabia,
and the Persian Gulf. Yet at the same time it carried on, round
the coasts of India, Burma, and the Persian Gulf, the arduous
work of Marine Survey for which it has been distinguished since
the latter part of the 18th Century.

In 1863, after the transfer of the Government of India to the
Crown, the Service was reconstructed on a non-combatant basis,
its duties being generally confined to surveying, transport work,
harbour inspection, etc. But it continued to see active service when
Indian forces were engaged, and during the Great War its ships
and personnel were absorbed into the Royal Navy.

In 1926 it was decided, on the advice of a Departmental Committee under Lord Rawlinson, to reorganise the Service as a
Combatant force. It consists at present of 5 armed sloops (of which
one is under construction), 2 patrol vessels, a surveying ship and

a depot ship. The administrative task of reconstruction is already finished, and the necessary Imperial Act was passed by Parliament in 1927. When the Bill becomes law it will make the Royal Indian Marine a Navy in the full sense of the term.

[269] *S.K.Brown [Assistant Secretary, India Office] to Lumby*
7 September 1934

We are sending you an official telegram to map out the course of events after the passage of the Indian Navy Act. We shall want a breathing-space between the G[overner] G[eneral]'s assent and the issue of the operative notification in order to obtain H.M.'s consent to the use of the title 'Royal Indian Navy' and also to settle various administrative questions such as that of the method of switching over the personnel from the R.I.M. to the R.I.N. I do not know at all what procedure, if any, is necessary for this purpose with regard to ratings, and we shall probably leave this to you to decide. As regards officers, our strong preference is to issue a general notification declaring that they are henceforth officers of the R.I.N., in just the same way as a general notification was issued in the London Gazette in the War declaring that all officers holding Dominion Commissions in the Dominion Forces were to be regarded as holding commissions in the regular land forces. There would then remain the question how we are to commission new entrants after the Indian Navy has come into being. At present, of course, the commission is issued in the name of the S[ecretary] of S[tate] for I[ndia], and is signed by the S[ecretary] of S[tate] and two members of Council and countersigned by the King. As regards the immediate future, I see three possibilities:

(1)   To bring the R.I.N. commission forthwith into line with the new commission in the Indian land forces for all new entrants, both British and Indian;
(2)   To do this forthwith for Indian entrants only;
(3)   To make no change for the present.

My own view is that we should have to change the present form of commission in any case under the new Constitution, because it would hardly be appropriate that the S[ecretary] of S[tate] for I[ndia] as such should continue to exercise the prerogative in this particular case when in all other cases it will either continue to be exercised direct by the King (British officers of the Indian Army) or will be exercised by the Governor-General (Indian Commissioned Officers). Since the R.I.M. is already, and the R.I.N. will continue to be, an exclusively local force unlike the Indian Army as a whole, I myself think that under the new Constitution the commissions ought to be issued by the Governor-General in the name of the King-Emperor for all new officers, both British and Indian. With that in mind, I should be strongly inclined to wait for the new Constitution and then make a comprehensive change – possibility No. (3) above. It is true that this would cut across the principle of the G[overnment] of I[ndia]'s proposal to give the new Dominion type of commission to all future Indian entrants to the Army after the 1st January 1935; and also it might be argued that since the R.I.M. is already an exclusively local force, the change could more readily be made. On the other hand, we have no question here, as in the case of the Army, of breaking away from the grant of full commissions to Indians in the regular forces as such; and since the G[overnment] of I[ndia] did not themselves mention the R.I.M. in connection with their Army proposals, I presume that they do not feel strongly about it. More-over, it would I think be even more invidious to give the Dominion type of commission now to new Indian entrants alone in the case of the R.I.N. than in that of the I.M.S., on which we commented fully in dealing with the G[overnment] of I[ndia]'s Army pro-posals. There is, of course, possibility No. (1), viz. to give the Dominion type of commission to all new entrants, both British and Indian; but I should be very reluctant to do this in advance of the new Constitution. If we do it now, we shall be accused of marking the occasion of the elevation of the R.I.M. into the R.I.N. by the issue of an inferior brand of commission. If we wait for the new Constitution, we shall have a much better reason to give, viz. that the change is itself a direct consequence of the major constitutional changes.

[270] *Colonel Sir Clive Wigram [Private Secretary to the King]*
to Hoare

15 September 1934

I return herewith the submission for the 'Royal Indian Navy' which the King has signed with pleasure and wishes every success to the Royal Indian Marine in its new role.

[Enclosed] Samuel Hoare to
King George V, 12 September
1934

Sir Samuel Hoare presents his humble duty to Your Majesty and has the honour to submit for Your Majesty's approval that the Royal Indian Marine shall in future be styled the 'Royal Indian Navy', and that this change of designation shall take effect from the date on which the Indian Navy (Discipline) Act is brought into operation.

[Manuscript minute:]

App[roved] G.R.I.

[271] *Decipher of telegram from Government of India, Army Department, to Secretary of State for India*
20 September 1934

Important, Second Part His Majesty the King Emperor has been graciously pleased to approve that the Royal Indian Marine Service shall henceforth be designated the 'Royal Indian Navy'.

The Governor-General in Council is pleased to direct that, with effect on date on which the Indian Navy (Discipline) Act, 1934, comes into force, the designation Flag Officer Commanding and Director, Royal Indian Marine, shall be 'The Flag Officer Commanding, Royal Indian Navy'.

The Governor-General in Council is further pleased to direct that, with effect on the said date, Warrant Officers and rating serving in the Royal Indian Marine shall become Warrant Officers and ratings of the Royal Indian Navy with the rank and seniority that they held in the Royal Indian Marine.

[272] *Minute, Brown to the Under-Secretary of State for India*
7 October 1934

The Indian Navy Bill has now passed both Chambers of the Legislature. I submit a draft telegram for the reasons given in Mr Clauson's note below, in order to make sure that a number of administrative questions arising from the conversion of the R.I.M. into the Royal Indian Navy will be properly considered before the new Act is brought into operation by a Gazette notification.

The point on which this draft requires submission is the proposal in paragraph 1 that H.M. should be asked to approve the title 'Royal Indian Navy'. This title was informally approved by H.M. in 1926 – please see Lord Stamfordham's letter of the 22nd January 1926 – , but as the Assembly rejected the Bill on that occasion, it could not of course be proceeded with. Now that the Indian Legislature has passed the Bill, it is presumed that the original intention will be pursued.

It may be recalled that up till now the Royal Indian Marine has been governed by the Indian Marine Act of 1887, which although it is based substantially on the provisions of the Imperial Naval Discipline Act, is a purely local Indian Act and does nothing in itself to link the R.I.M. with the Royal Navy or the other Navies of the British Commonwealth. The standing tradition is that a local sea-going force can only be styled and treated as a 'Navy' if its discipline code is brought into direct relation with the standard naval code embodied in the Imperial Naval Discipline Act. Australia, for instance, was empowered by the Imperial Parliament in the Australian Constitution Act to adapt the Naval Discipline Act for this purpose. In 1927 Parliament passed the Indian Navy Act which similarly empowered the Indian Legislature to adapt the Naval Discipline Act; but the Indian Bill founded on this Act was rejected on that occasion. This is the Bill which has now been revived and passed. Consequently the Indian Naval Forces hitherto known as the Royal Indian Marine can now assume the style of the Indian Navy. H.M.'s specific approval is, however, required for the use of the title 'Royal Indian Navy'.

It is desirable to issue this telegram as early as possible, in case the Government of India have overlooked any of the questions it raises.

[273] *Minute, Brown to Sir Findlater Stewart, Permanent Under-Secretary of State for India*

c. 7 October 1934

The procedure proposed in this case may have to be considered in relation to the new Constitution.

The recent Indian Navy (Discipline) Act, like the British Naval Discipline Act but unlike the British and Indian Army Acts, contains no provision for the enrollment of ratings (corresponding to the attestation of enlistment of soldiers). In the Army, enlistment is regulated by legislation; in the Navy, by exercises of the prerogative. In the one case, Parliament has established the doctrine that a standing Army cannot be maintained without the consent of Parliament, and has accordingly legislated for the recruitment of such an Army. But it has not extended this doctrine to the

Navy, and consequently provision in this respect is still made in the Navy by exercise of the prerogative.

The recent Indian Navy Act is not a self-contained piece of Indian legislation but a deliberate adaptation of the naval Discipline Act. It follows that on this occasion at any rate we must provide for enrollment by exercise of the prerogative and not by legislation. At the same time we do not want to do anything that might preclude the Indian Legislature indefinitely from legislating on such a matter. There is no reason why we should perpetuate in India a form of procedure derived from historical circumstances that are peculiar to this country. There is also no reason on the merits why the Indian Legislature should not legislate for enrollment in the R.I.N. just as it legislates for attestation in the Indian Army. At the moment we have to proceed on the basis that, for accepted reasons of policy, it has omitted to do so in the recent Act.

The first question is whether, by obtaining the proposed Order in Council, we should in any way prejudice the rights of the Indian Legislature hereafter. I doubt myself whether we should. As I understand the position, the effect of the new Constitution Act will be to vest all executive power and authority in the Federation save in regard to such subjects as may be expressly reserved in the Act itself. Unless, therefore, the Act reserved or excluded enrollment in the R.I.N., power to deal with this matter would become vested in the Federation, with the consequential power to legislate thereon, notwithstanding the fact that it will previously have been exercised through the prerogative. If this view is correct, and if occasion should arise hereafter to legislate in India on this subject, it might then be necessary to regularise the position by obtaining H.M.'s consent to legislate on a matter touching the prerogative; but that should present no difficulty.

There is, however, a separate point in the proposed Order in Council which may have a bearing on the above. The draft asks H.M. in Council to authorise the G[overnor] G[eneral] of I[ndia] in C[ouncil] to issue regulations 'for the good order and discipline of the Royal Indian Navy'. This is intended to cover the same ground as is covered in the Army by King's Regulations as distinct from legislation, viz. discipline as between the various ranks. But unfortunately 'discipline' has two meanings; it also means discipline in the matter of punishments, which is a proper subject of legislation. It is used of course in the latter sense in the title of the recent Act – 'The Indian Navy (Discipline) Act'. We must take care that the terms of the Order in Council do not provoke the criticism in India that we are resuming under the prerogative a power that ought to be exercised by the Indian Legislature . . .

[274] *India Office minute, Brown to 'Legal Adviser'*
14 October 1934

You may be interested in these papers relating to the form and contents of the Government of India's Gazette notification bringing the Indian Navy (Discipline) Act into operation. We thought it advisable to get off our telegram of the 10th September as quickly as possible in order to warn the Government of India at once that they would not be able to publish on the 15th September and also that we ought to consider the points raised in our telegram. I do not think there is any point of legal importance except that relating to the method of transferring ratings of the R.I.M. to the R.I.N. The reason why we raised the question whether re-enlistment might be necessary is that serving ratings of the R.I.M. have all been enlisted under the Indian Marine Act of 1887 as subsequently amended. Presumably enlistment under an Act of this kind is of the nature of a contract; and although minor amendments of the Act from time to time would not impair the contract, any more sweeping changes might make it necessary to enter into a new contract. The Indian Navy (Discipline) Act repeals the Indian Marine Act of 1887 in toto, and enacts a new set of provisions in its place. Thus there has been a complete change in form at any rate of the original code of discipline. On the other hand, there has not been much change in substance, since the Indian Marine Act of 1887 itself substantially reproduced the provisions of the Naval Discipline Act, and thus the new Act does not differ materially from the old. We felt that, since the G[overnment] of I[ndia] are primarily responsible for the enlistment of R.I.M. ratings, it would be largely for them to decide this question, although we thought it as well to bring it to their notice.

[275] *Minute by Captain F.A. Buckley, Admiralty Director of Training and Staff Duties*
23 October 1934

3   D.T.S.D. [Director of Training and Staff Duties] considers it essential in the interest of training that both European and Indian

Officers of the Indian Navy should serve from time to time in ships of the Royal Navy. On the other hand, it is most undesirable that Indian Officers should be given R.N. commissions which will place them in 'unsupervised command and disciplinary control' of white ratings. It is therefore recommended to take the same action as proposed by the War Office, where the problem is exactly similar, i.e. not to issue R.N. commissions to Indian Officers of the Indian Navy, but to allow the C.-in-C., East Indies, to accept such officers in his squadron for training, under the same conditions as are now in force for junior R.I.N. officers at Home . . . , but only on occasions specially authorised by the C.-in-C. to allow them unsupervised command of white ratings.

[276] *J.A. Simpson [Deputy Clerk of the Council of India and Principal Officer, Military Department], India Office, to the Secretary of the Admiralty*
14 December 1934

I am directed by the Secretary of State for India to refer to Admiralty letter No.C.W.5200/27, dated 29th August, 1927, in which it was stated that officers of the Royal Indian Marine should rank and command with officers of the Royal Navy according to their rank and seniority in the Royal Indian Navy, provided that the courses of instruction prescribed for them are such as, in the opinion of the Admiralty, to justify such a procedure. This letter was written on the assumption that the Indian Navy (Discipline) Bill was about to pass the Indian Legislature. When, however, the Indian Legislature rejected the Bill, the Lords Commissioners of the Admiralty felt compelled to reconsider their position, and their revised views were stated in Admiralty letter C.W. 4207/28, dated 6th July, 1928.

Now that the Indian Navy (Discipline) Act, 1934, has been brought into effect, and a Royal Indian Navy, subject to the Navy Discipline Act, suitably modified to meet Indian conditions, is in being, the Secretary of State trusts that their Lordships will see their way to re-affirm the undertaking which they gave in their letter of the 29th August, 1927. If so he would be glad to learn their views regarding the courses of training necessary for the Royal Indian Navy.

[277] *Hoare to His Majesty King George V*
1 January 1935

Sir Samuel Hoare presents his humble duty to Your Majesty and has the honour to submit for Your Majesty's approval a proposed notification for issue in the London Gazette providing that, with effect from the 2nd October 1934, officers serving in the Royal Indian Marine shall be officers of the Royal Indian Navy with the rank and seniority which they held in the Royal Indian Marine.

[278] *Decipher of telegram from the Government of India, Army Department, to Hoare*
9 January 1935

Your telegram of November 9th 2669.
2 Transfer of ratings. Flag Officer Commanding is strongly opposed to individual re-enrollment, which he considers ratings would regard with suspicion as introducing into their contracts a change which they had been led to believe did not exist. We agree and consider such a course would have unsettling effect more particularly as 50 per cent of personnel were enrolled on April 1st last year. As regards difficulty that a transfer by general notification would give no opportunity to individuals to object to variation of their contract, we invite attention to our Army Department notification No. 524 of October 2nd, which announced that officers, warrant officers and ratings serving in Royal Indian Marine, would with effect from aforesaid date become officers, warrant officers and ratings with corresponding rank and seniority in Royal Indian Navy, and that steps would be taken to provide accordingly.

During the three months that have elapsed no one has objected. We suggest that (?continuance) [sic] in the service without objection for this period amounts to acceptance of the new conditions and that no further opportunity for objecting is necessary. If you agree, we propose that, as soon as general order effecting the transfer of commissioned officers is published in the London Gazette, we should publish a similar order in India regarding warrant officers and ratings. For this purpose we should be glad to learn in advance the text of the home notification.

[279] *Commander Edward L. Hastings RN, Canadian Department of National Defence (Naval Intelligence and Plans Division), for the Canadian Chief of Naval Staff, Commodore P.W. Nelles, to Little*

Ottawa
23 April 1935

'Personal'

Commodore Nelles has been away on leave and on his return has been put on the sick list. He asks me to reply to your letter to him.

The replacement of *Vancouver* and *Champlain* is a matter of great concern to this Department, and every effort has been made to obtain the approval of the Canadian Government to lay down two Torpedo Boat Destroyers. The difficulties to be overcome have so far proved insuperable, and though our efforts will be renewed I am by no means sanguine that we will obtain the necessary approval. The lives of *Vancouver* and *Champlain* could be prolonged for five years if they received a D 2 refit this year but we have, so far, failed to obtain approval for the necessary money to be spent on them. Failing this refit, I consider that the useful lives of these ships will expire in from one to two years.

In view of these facts we welcome your suggestion that an application to transfer two 'C' Class Destroyers to the Royal Canadian Navy in lieu of *Vancouver* and *Champlain* would receive favourable consideration by the Admiralty. The increased complement necessary for these ships would increase the Naval Vote by about $100,000 and as financial stringency at this time causes opposition to any increases in the Estimates, it will militate against the whole hearted acceptance of the generous help which you suggest may be possible.

Mr Bennett has left for England and our Minister has seized the opportunity of writing to him on this matter, which he hopes may result in approaches being made to the Admiralty.

I hope you will treat this letter as quite unofficial as no approval has yet been obtained.

[280] *Admiralty minute, author unknown*

April or May 1935?

As lately as November 1934, we told the Canadian Authorities that *Vancouver* and *Champlain* would probably not have to be scrapped until the last day of 1936, but ought to be kept seaworthy until then. We also invited them – this was the Dominions Office contribution – to consult with us 'when the probable results of the 1935 Conference could be more clearly estimated'.

We are therefore barred from making definite proposals at the present, but it seems important to have our ideas clear against any approaches from Canada, which may be expected if Sir M. Hankey is right.

According to the terms of the loan the Canadians are bound to return *Vancouver* and *Champlain* with full equipment. Whether they sail the ships over with crews for their reliefs, or scrap them locally and credit us with the average U.K. price for such vessels, is all one. The replacement of the ships will start from scratch.

With an Imperial quota of destroyers as opposed to U.K. and Dominion quotas it seems difficult to sustain D[irector] of P[lans]'s argument that transfer of two under age destroyers to Canada will make it necessary for us to build two more seven years earlier than we should otherwise. It assumes that two 'C's in full commission in the Canadian Navy are of less value than 2 'C's in Reserve as Flotilla reliefs at home. This may be true but the Canadians might not appreciate it. In any case it might be argued that our building is limited by treaty and not by the disposition of our ships on the various Naval Stations within the Imperial (as opposed to Royal) Navy . . .

[281] *Admiralty minute, author unknown*

9 May 1935

[It should be possible to trace the author, as he called on Prime Minister Bennett. It may have been the First Sea Lord.]

I called upon Mr Bennett, the Prime Minister for Canada, at Claridge's Hotel, at 4.15 today, Wednesday, at his invitation to discuss the question of the replacement of *Vancouver* and *Champlain*.

Mr Bennett commenced by repeating to me the feeling which a large section of the Canadian population hold in regard to avoiding additional expenditure on the Canadian Navy, but on the other hand if he 'lived' long enough, in the political sense, he would like to see their destroyer force not only maintained but augmented. He asked me whether I was in a position to say what the financial liability would be if the proposal to replace these two destroyers by two of our 'C' Class was implemented. I said that this proposal so far had not gone beyond the Naval Staff and the Treasury had not been consulted, but that if the transaction were carried through it would undoubtedly show considerable financial saving to Canada on the alternative, namely to build two new destroyers for the Canadian Navy. I told Mr Bennett in reply to his enquiry that I thought the exchange could be carried through if the Canadian Government really wished it.

I found that Mr Bennett was personally acquainted with all four of the Canadian Destroyers having inspected them at sea and he described *Vancouver* and *Champlain* as only being fit for scrapping. I said that we had formed a high opinion of the efficiency of the Canadian Destroyer Division as a result of the reports we had received from the Commander-in-Chief, Home Fleet, when they had joined up with the Home Fleet in the West Indies in the Spring of 1934 and of 1935.

I said to Mr Bennett that I understood Ministers had already acquainted him with their anxiety in regard to the possibility of a war with Germany and I explained that the Naval Staff was apprehensive of our Naval position should the Japanese take such an opportunity for interfering with some British interests in the Far East. Although such an event might not lead to war, it might well necessitate taking up the defensive position in the East which would entail three-quarters of the British Navy leaving European waters. Under these circumstances only slender forces would remain in Europe and we should to some extent be dependent on the French and the Naval Staff therefore attach the utmost value to the Canadian Destroyer Division for assisting to maintain the sea communication with Canada unbroken. Mr Bennett said that he readily understood the importance of this strategic conception to Canada and that on his return he would ask the Minister for Defence to communicate with the Canadian High Commissioner in London with a view to the proposed replacement of *Vancouver* and *Champlain* being placed on an official basis with the Admiralty.

Propose that the Dominions Office should be informed semi-officially of what has taken place.

[282] *S.H. Phillips, Assistant Secretary in the Military Branch, Admiralty, to E.E. Bridges, Assistant Secretary, Treasury, copy*
10 May 1935

'Confidential, Immediate'

The Admiralty have had under consideration a proposal put to us by the India Office as a result of representations made by the Government of India regarding the development of local naval defence at Indian ports. This is a matter of first class importance from our point of view, since not only is Indian naval defence a rather embryonic stage of evolution, but also it has hitherto proved impossible to induce the Government of India categorically to accept as a proper charge on Indian resources those naval defensive measures which the Admiralty regard as being clearly an Indian responsibility under the generally accepted rule that Dominion and colonial Governments are responsible for their own local defence requirements.

2   As you are aware, the Government of India make an annual contribution to Navy Votes of £100,000 on account of the services of certain selected vessels of the East Indies Squadron in 'Indian waters'. In addition, they contribute services (fuel, repairs, etc.) for those vessels when employed in 'Indian waters', amounting to approximately £30,000 a year.

3   Briefly, the Indian Office proposal which is now before the Admiralty is that, in consideration of an abatement, in each of the next five years, from the contribution of £100,000 annually due from the government of India, the latter will undertake to devote an equivalent amount to the defensive equipment of Indian ports, etc., comprising the following items of local defence:

Minesweeping gear; Guns for local defence vessels; Storage accommodation for guns, etc.; Searchlights for auxiliary vessels; W/T Sets; Depth charges; Reserves of ammunition; R.I.N. Volunteer Reserve (personnel); Construction of two special minesweepers or one sloop;

and will also provide for the storage at Indian ports of guns intended for armed merchant cruisers and defensively equipped merchant ships which are now stored, so far as space permits, at Colombo. These ships are to fit out at Indian ports and moreover, space at Colombo is required for other purposes. The programme which is designed to extend over five years, has been examined in the Admiralty and is generally in accord with the approved naval policy for Indian local defence.

4   The approval of the full programme would involve a reduction of the annual contribution to Navy Votes from India of approximately £78,000 a year. More than half this amount, however, is on account of the construction and maintenance of two special minesweepers, or alternatively, a sloop. We feel that this question requires further consideration, and so we are not at this stage raising it with the Treasury. Moreover, the guns required for the local defence vessels of a total value of £64,000 are available from Admiralty reserve stocks, and we consider that it would be reasonable to supply them free of cost as a separate transaction, though still part of the scheme, since it would not be proposed in the circumstance to replace them in the Admiralty reserve. This, however, is really a matter for Treasury consideration, but it would simplify matters in some ways not to regard them as paid for out of the subsidy; the net effect on Navy Votes would, of course, be the same whichever course were adopted.

5   If the minesweepers or sloop and the guns are excluded from the programme submitted to us, the annual sum required to carry out the necessary development would on the average, according to the approximate estimate furnished by the Government of India, be about £21,000, and the Admiralty are inclined to accept the proposal, as modified, subject to a maximum annual abatement from the contribution of £25,000, or a correspondingly higher amount if payment for the guns is insisted upon. This would be on the understanding that payment of the full contribution would be resumed after five years and that the Government of India would accept full financial responsibility for maintenance and other recurring charges due to the programme of development.

6   At first sight, this does not seem much of a bargain from the Admiralty point of view, but there are several considerations, both of policy and finance, which have led us to regard the proposal with more favour than its face value would appear to warrant.

7   In the first place, the essential work, long overdue, of putting the Indian ports in a proper state of naval defence will be tackled. This will not only be of the greatest value to the Admiralty, but should also give considerable encouragement to the development of Indian Naval defence. Moreover, the Government of India have made it very clear that in the present state of their finances, they are unwilling to find the required funds themselves.

8   Secondly, the acceptance of their proposal, even on the modi-

fied scale which we have in mind, will deprive the Government of India of any excuse in the future for evading their responsibility for local naval defence. Indeed, they have given us to understand that they will not attempt to do so.

9   Thirdly, with a great constitutional change impending, it seems above all things desirable to avoid any action, or lack of it, which might result in embarrassment for the Government of India. This is a consideration to which the Treasury will doubtless attach due weight.

10   Finally, while, as already stated, it is agreed that, at the end of the five years, payment of the full contribution will be resumed by India, the India Office do not exclude the possibility of India regarding the equipment thus provided as being subject to repayment of cost at some future time.

11   So far as the actual provision of money for the programme is concerned, we have not allowed in the current estimates for any abatement of the Indian contribution and it is impossible at this stage to say whether the deficiency could satisfactorily be met, if the development were to be begun in this financial year. In the meantime, however, we should be glad to know whether the Treasury would be prepared to give favourable consideration to our negotiating an arrangement with the India Office on the lines suggested, whatever is agreed to being then formally submitted for Treasury sanction.

12   As the India Office have been pressing us for an answer, we should be glad if you could treat the matter as urgent. If there are any points or details which you would like to discuss, I should be very pleased to come over and see you.

[283] *Minute by J.S. Barnes, Principal Assistant Secretary,*
Admiralty
19 June 1935

The Public Accounts Committee have been very critical of the recent transfers to Australia of the destroyers and the Cruiser *Sydney.* They suspect that we were not getting an adequate quid pro quo, and that the changes were being made at the expense of the British taxpayer. It was possible to satisfy the Committee about the arrangements made in the case of the *Sydney,* but they were more doubtful about the destroyers. It is very desirable, therefore, that the transfer now contemplated should be on a good financial basis.

Having regard to the Admiralty proposals for international agreement upon destroyer tonnage and to the fact that we contemplate the retention of a certain amount of overage tonnage, it is thought reasonable to take a life of 20 years for each destroyer (16 years under age and 4 years overage) and to fix the price at first cost less 5% depreciation per annum. This would give a figure of £229,222 plus £51,215 (armament and stores) for *Crescent* and £232,447 plus £51,215 for *Cygnet*, a total of £564,200 or, say, £564,000.

The change would have the effect of reducing the United Kingdom underage tonnage by two destroyers and increasing the Canadian underage tonnage accordingly. The arguments for the transfer may be summarised as follows:

As regards the Admiralty. We are concerned that the Canadian force should be maintained and the old destroyers replaced. This is a favourable occasion for effecting the replacement, and it may not recur. We cannot employ these destroyers as usefully as the Canadians. Canada will, we hope, keep them as part of a Division in commission; we should use them occasionally as flotilla reliefs, otherwise keeping them in reserve.

As regards the Treasury. We shall get full value for the destroyers and may not have to find the money to replace them for many years, if at all. (Owing to the international situation the future building programmes are entirely vague, but it is unlikely that we shall build two destroyers only, our practice is to build not less than a Flotilla at a time). The transaction is thus in a nature of a windfall for the Exchequer.

As regards Canada. They save a difference between the cost of two new destroyers and the price now to be charged, namely about £80,000 per vessel, or £160,000 in all.

P.A.S.(M.E.) concurs.

[manuscript addition]

If these proposals are approved it is submitted to approval [by] the Treasury at once. The Canadian High Commission is very anxious for a reply this week.

[minuted]

Concur. Oswyn Murray 19/6

[284] *Nelles to Little*

Ottawa
27 June 1935

'Secret and Personal'

Please understand that this letter is entirely unofficial, as being from one sailor to another – a personal matter.

Your letter of 26th March led me to hope, (being an incurable optimist), that in the event of the Canadian Government being unable under existing financial conditions, to replace *Vancouver* and *Champlain* with new construction, the Admiralty might be willing to transfer two 'C' class Torpedo Boat Destroyers to Canada under similar conditions, as finally agreed upon, for *Vancouver* and *Champlain*.

The telegram just received through External Affairs does not propose a 'transfer' but a sale of *Cygnet* and *Crescent* to Canada at a cost of £564,000 and has therefore come as a severe disappointment.

In the letter which Hastings wrote you he pointed out the difficulty of even increasing the estimates by $100,000 a year for the upkeep of these ships, and I would regard it as a miracle if the Government voted £564,000. to acquire them.

Under these circumstances I greatly fear *Vancouver* and *Champlain* will not be replaced unless the Admiralty can see their way to making a more generous offer.

[285] *F.L.C. Floud, High Commissioner to Canada, to Harding*

Ottawa
23 July 1935

With reference to your letter of the 27th May, No. 6254/120, and to the subsequent semi-official letter from the Admiralty to MacLeod of Canada House dated the 21st June, I had an opportunity last light to discuss the question of the replacement of *Champlain* and *Vancouver* with Grote Stirling, the Minister of Defence, and Commodore Nelles, Chief of the Naval Staff. I find that they are rather staggered by the proposal that they should purchase the two destroyers which the Admiralty are prepared to hand over in place of *Champlain* and *Vancouver*, as they had assumed that the transfer would be made on loan, as was the case with their present ships. Grote Stirling said that in any case it would be quite impossible for any government on the eve of an election to propose expenditure of more than half a million pounds on two new destroyers and that accordingly he had suggested to the Prime Minister that he should ascertain whether the Admiralty would not make the transfer on loan. Commodore Nelles also said that he was not prepared to agree to the suggestion that the life of a destroyer should be taken at twenty years. In the meantime, a supplementary estimate of $100,000 has been passed which will enable certain work to be carried out on 'Champlain' and 'Vancouver' to keep them in good seaworthy condition.

As you are aware, there is always a good deal of opposition here to any expenditure on naval or military defence, and I do not think, therefore, that there is the least prospect of this or any other government being prepared to agree to the proposals of the Admiralty. The people here are under the impression that the Admiralty would be prepared to make a loan of the new destroyers but that the Treasury had insisted that the transaction should be on a purchase basis.

[286] *J.A. Simpson to the Secretary of the Admiralty*
5 August 1935

I am directed to refer to this Department's letter of 14th December, 1934, No. M.8171/34 and subsequent enquiries on the question of the relative rank and powers of command of Royal Naval officers and officers of the Royal Indian Navy and also to Mr Brown's [S.K. Brown, Assistant Secretary, India Office] demi-official letter to Mr Barnes [Principal Assistant Secretary, Admiralty] of the 22nd February, 1936, No. M.7043/35. The latter communication referred also to the subsidiary question of the form of commission to be granted to R.I.N. officers after the coming into operation of the new Government of India Act.

As the Secretary of State for India in Council ceases to exist from the 1st April next when the New Government of India Act comes into operation a new form of commission for R.I.N. officers will be necessary. It is felt that the grant of commissions by the Secretary of State for India would not be appropriate under the Government of India Act, 1935 and would also not be in accord with the new status of the Royal Indian Navy which is now a combatant force under Naval discipline. It is therefore proposed that in future all Royal Indian Navy officers should be granted commissions by the Governor General 'in His Majesty's Fleet,' in a form similar to that granted in the Australian and Canadian

Navies. I am to enquire whether the Lords Commissioners concur in this proposal.

The Secretary of State understands that in the view of the Lords Commissioners such an alteration in the form of commission would carry with it no change in the powers of command of the officers of the Royal Indian Navy vis-à-vis Royal Naval personnel since this matter is determined by King's Regulations and Admiralty instructions. The above proposal in regard to the form of Commission should, however, in the Secretary of State's opinion, be dependent on the Lords Commissioners' agreement to the previous proposal that Royal Indian Navy officers should rank and command with, but immediately below, Royal Naval officers of the same rank and seniority.

Both Vice Admiral Sir Humphrey Walwyn [Flag Officer Commanding and Directing, Royal Indian Navy, Bombay] and Vice Admiral Bedford [Flag Officer Commanding, and C-in-C Royal Indian Navy] have attached great importance to the grant of equal powers of command to officers of the Royal Indian Navy. Such powers are granted temporarily to officers seconded for training with the East Indies Squadron and would presumably have to be accorded in time of war if Royal Indian Navy vessels serve under the command of the Commander-in-Chief, East Indies.

The question of relative powers of command is liable to arise as a practical issue whenever the Royal Indian Navy takes part in joint operations or manoeuvres with the East Indies Squadron and may therefore arise with increasing frequency. It would be valuable enhancement of the prestige of the service and conducive to its general efficiency if such powers were granted to Royal Indian Navy officers under King's regulations, as is the case with officers of the Dominion Navies. The proposal is strongly advanced by the Government of India and the Secretary of State trusts that the Lords Commissioners will, now that the Indian Naval Discipline Act has been passed, be able to reaffirm the decision given in 1927 and accord equal powers of command to officers of the Royal Indian Navy.

Provision with regard to the grant of Commissions to officers in the Royal Indian Navy should be included in the Letters Patent of the Governor-General of India under the new constitution. This document is now being drafted and I am, therefore, to request that the Lords Commissioners will be good enough to furnish their views on the question of relative rank and powers of

Command and the form of commission of Royal Indian Navy officers at a very early date.

[287] *Vice-Admiral Arthur Bedford, Navy Office, Bombay, to Little*

September 1935

'Personal and Confidential'

Thank you very much for your letter of 11th May and also for all you did for 'Indus' while she was in Home waters.

With regard to the local Naval Defence of India, I am sending you a copy of an Appreciation which has been prepared after discussions which have been taking place during the last two months. This appreciation was made out for the Army – Commander-in-Chief and I understand a copy is being sent to Rose.

Paragraph 7(a) on page 3 was put that way to strengthen the case, but we realise that when patrols could be instituted in the event of an Eastern War they would be on the line Malacca Straits – Australia.

Unless the R.I.N. is to be abolished or die of stagnation it must go forward at least sufficiently far to carry out local Naval Defence properly. The proposed expansion is necessary to raise the reserves required and to make the service sufficiently attractive to exist.

The R.I.N. as are all other Indian Services, is committed to the partial Indianisation of Officers.

The efficiency of Indian Officers as Commanders at Sea has yet to be proved and it will be many years before this can be done. There are at present no executive officers above the rank of Sub-Lieutenant.

Given full initial training the necessity for sending Indian officers to R.N. Ships would be overcome and the prejudice of European officers and ratings against taking orders from these officers would be obviated.

I imagine that it could always be arranged that when R.N. and R.I.N. ships were in company the Senior Officer should be European, even if it were necessary to give him acting rank.

With regard to the Financial aspect it should be realised that the taxation of half of India is not controlled by the Government of India, also that the proportion of her budget allotted to Defence is greater than that of England or any Dominion.

The Indian politician is already alive to this and has, with some justification I think, a grouse.

The attitude of the Government of India is that they already pay enough for Imperial Defence by maintaining a large part of the British Army as much on Imperial account as on their own and they do not see why they should spend more on Imperial defence by a subsidy towards the R.N.

[manuscript] Many congratulations on the 'K' [i.e. knighthood] and your new appointment. I hope you will let me know when you come through Bombay.

[288] *Neville Chamberlain, Chancellor of the Exchequer, to the Marquis of Zetland, Secretary of State for India*
30 September 1935

You are no doubt aware that for a long time past the Committee of Imperial Defence have been much concerned at the backward state of the local naval defences at Indian ports, and that representations have been made from time to time to the Government of India urging them to accelerate the completion of the agreed defence schemes, but hitherto without result.

Recently the Treasury have had before them a proposal by the Government of India that part of the cost of improving these defences should be found by a reduction over a period of 5 years of some £20,000 a year in the contribution of £100,000 a year which India makes to Navy Votes.

These proposals have been the subject of discussion between representatives of your Department, the Admiralty and the Treasury, and I understand that, notwithstanding the strong objections thereto from the point of view of the Exchequer, the proposal is still pressed.

I am therefore going to ask you to consider very carefully the following observations. In the first place the inadequacy of the existing contribution does not, I think, admit of question. You will no doubt be aware of the persistent attempts made by the Admiralty over the last 30 years or so to induce the Government of India to increase their contribution to something more in accordance with the true value of the Navy's protection and of that Government's equally persistent refusal to meet them: but throughout the negotiations the inadequacy of a contribution first fixed in the previous century does not appear to have been contested; indeed, at the 1921 Imperial Conference Mr Montague clearly contemplated the possibility of an increase when conditions improved.

Again, it will not be disputed that India, in common with the Dominions and Colonies generally is responsible for the cost of her own local defences; the proposal to provide such local defences at the expense of her (already inadequate) contribution to Navy Votes amounts in fact to an evasion of that responsibility.

The attitude of India in this matter is in sharp contrast to that of other members of the British Commonwealth. While the Dominions generally have increased substantially their effective contribution to the naval defence of the Empire by maintaining naval units of their own and in some cases also by cash subventions to particular defence services, e.g. Singapore, and are at the same time spending large sums in bringing their coastal defences up to date, India's annual contribution has stayed at the same level for some 40 years, while the deficiencies in her coastal defences remain virtually untouched.

Your answer to this will, I imagine, be that the Government of India have very great difficulties to face in securing the necessary funds for defence expenditure. I appreciate that this is so, but I am sure you will not forget the other side of the picture. In recent years we have assumed a large additional liability by way of grant in aid of the defence of India; we have also agreed to take over sole financial responsibility for the defence of Aden. Notwithstanding these factors, and the growing burden of our commitments at Singapore (to which India has not contributed) I can fairly claim that we have taken a not ungenerous line in recent adjustments between the two Governments. I do not wish to multiply examples, but I might mention, for instance, the grant in respect of the Quetta earthquake and the contingent undertaking in regard to Commonwealth Trust properties.

In the light of the above, I feel bound to make my protest against the apparent unwillingness of the Government of India to accept responsibilities which are unquestionably theirs. You may not know that the Public Accounts Committee have recently shown an increasingly critical attitude towards financial adjustments of expenditure on defence between the Imperial Government and the Dominions, India and the Colonies, on the ground that the mother country does not always insist on a good enough bargain from her point of view, and I am extremely unwilling to put before them what I think they might well consider an unusually flagrant case of the inequity of which they complain.

I hope very much therefore that you will use your best endeavours to induce the Government of India to take a broader view of their general defence obligations and to embark without further delay on the agreed port defence schemes . . .

[289] *Admiralty Secretary [Murray] to JA Barlow, Treasury, draft*
October 1935

You will remember that when we were discussing the Government of India's five year plan for Naval defence (Phillips' letter to Bridges of the 10th May last, M.03347/34), you referred to the proposal to include a sloop in the scheme, and you asked whether, provided an assurance could be given that the sloop, if built, would be placed at the disposal of the Admiralty in time of emergency, we should be able to reduce our own shipbuilding programme by one sloop, and save the cost of its maintenance in time of peace.

2   The answer is in the affirmative: should the Indian Government construct a sloop which, in the event of any war, would be placed at the disposal of His Majesty's Government in the United Kingdom for service not necessarily in Indian waters, the number of R.N. Sloops to be built and maintained could be reduced by one vessel.

3   At the same time, we feel bound to tell you that we are very doubtful whether the Government of India would ever give us an assurance that we may employ the vessel in emergency outside Indian Waters, and, further, we are not at all sure that it would be proper to make such a request.

4   Admiral Bedford, the present Flag Officer Commanding the Royal Indian Navy, has written recently pleading for an early decision, and pointing out that unless the Royal Indian Navy is to be abolished or die of stagnation it must go forward at least sufficiently far to carry out local defence properly. With his letter he enclosed an appreciation drawn up for the Army Commander-in-Chief in which he refers to the accepted policy for the development of Dominion Navies as laid down in C.I.D. Paper O.D.C.537M, dated 19th March 1928. He points out that the four stages of development are:

Phase I – Local Naval Defence in its narrowest sense.
Phases 2 & 3 – Local Naval Defence in its widest sense.
Phase 4 – Units able to join the Royal Navy Main Fleet.

He makes it clear that the sloops are intended for local naval defence in its widest sense – Phases 2 and 3 – and not for Phase 4, Units able to join the Royal Navy Main Fleet.

5   It seems to us that this is a fair statement of the position. It is of first importance to the United Kingdom that provision should be made for Indian local naval defence. If it is not made by the Government of India the liability will in the long run have to fall on the United Kingdom. In the view of the importance of India to the Empire, there is everything to be said from the point of view of the Admiralty, and we think also from the point of view of the Government as a whole, for accepting this scheme for providing for this necessary service, in spite of its financial drawbacks.

6   As you know, we limited the scheme which we put before you to the earliest phase, and did not include the sloop, but we should undoubtedly prefer, if it were possible, to adopt the full scheme, covering phases 1, 2 and 3.

7   May we ask you, in the light of this further information, to give the question very early consideration?

[290] *Murray to J.A. Barlow*

8 November 1935

With reference to your letter of the 24th October concerning the Government of India's five year plan for Naval defence, I write to say that the intention of the India Office, as we understand it, was that the cost of one sloop should be recovered by India by means of a deduction from the Indian annual contribution towards the cost of Naval Defence. You will remember that Stewart pressed the point that the Government of India, during the five years, were building another sloop at the cost of Indian funds to replace the *Pathan*.

We estimate that the financial effect of including the cost of construction and subsequent maintenance of a sloop in the five year scheme would be to make the annual abatement from the Indian contribution £66,000 out of £100,000 a year instead of £21,000 out of £100,000 a year.

The only condition on which we could accept the reduction of our own shipbuilding programme by one sloop would be that the Government of India should agree to place the sloop, which they include in the 5 year scheme, unreservedly at the disposal of H.M. Government in the United Kingdom in the event of any war. It would not be enough that the Government of India should place the vessel at our disposal for use only in Indian waters.

We note that the Chancellor has written a personal letter to the Secretary of State for India concerning the proposal, and hope that this may lead to a speedy settlement of the question, as while the discussion goes on the unsatisfactory state of the local defences of India continues.

[291] *Zetland to Neville Chamberlain*
14 November 1935

I must apologise for the delay in replying to your letter of the 30th September regarding the proposal that the Government of India should be allowed to divert some £78,750 annually for a period of five years (I assume that the £20,000 mentioned in your letter is a typist's error) from the £100,000 contribution which they at present pay towards the cost of the Royal Navy to the completion of their scheme of naval port defence. I have given my most careful attention to the considerations set forth in your letter and I do not deny that, as marshalled by you, they make a considerable showing. But they leave a good deal to be said on the other side; and when this too is taken into account I feel justified in maintaining the proposal . . .

As regards the distinction which you draw between India and the Dominions to the disadvantage of the former, I would say that the inadequacy of the contribution of certain Dominions to naval defence is notorious, while even those who maintain units of their own would be virtually powerless without the backing of the Royal Navy to the cost of which they, unlike India, make no direct contribution. Further, when it is said that India's contribution has remained at the same figure for some 40 years, we have to remember the great increase in the cost of the Royal Indian Marine (now the Royal Indian Navy) and its conversion into a combatant force of direct value for naval defence.

[292] *Neville Chamberlain to Zetland*
3 January 1936

In effect therefore I am prepared to relax the attitude which we have previously adopted to the extent of agreeing that one half of the cost over the next two years of making good India's local naval defences should be met by a diversion from India's naval contribution, on the understanding that the other half is borne on Indian funds. We should of course require to be consulted as to the total sum to be diverted from the naval contribution in each of these two years. I sincerely trust that you will find no difficulty in accepting this proposal which seems to me to go a very long way to meet your difficulties. For my part I should be ready to look at the position again after two years. But you will of course understand that we do not in any way recede from our view that the contribution of £100,000 paid by India towards the cost of the Navy is inadequate. I must reserve for further consideration the question whether a later improvement in India's financial position may not justify some reimbursement of the sums which it is proposed should be temporarily diverted under this agreement.

[293] *Minute by J.A. Simpson to Lieutenant-General Sir J.F.S.D. Coleridge, [Secretary, Military Department, India Office] and Under-Secretary of State for India*

6 January 1936

I venture to submit that the reply to the Chancellor should take the form of acceptance in principle of his half-and-half offer combined with an appeal to him to extend the Imperial contribution to the whole cost of putting India's local naval defences in order instead of confining it to the first two years. I read the Chancellor's offer as relating to neither the £20,000 nor the £64,000 mentioned in the earlier paragraphs, but to the cost during the first two years of a comprehensive plan on the general lines of the Government of India's five year plan, the Treasury being consulted regarding the manner in which the amount of expenditure during the two years in question is arrived at. It would, for instance, obviously not be open on the Chancellor's offer to the Government of India to concentrate all the expenditure in two years or even three years instead of five and then ask the Treasury to pay half.

Simultaneously with the reply to the Treasury I suggest that India should be requested to obtain from the local naval authorities a fresh plan, which in the light of the recent discussions would show what is now considered to be the minimum necessary to put the local naval defences in working order. That a fresh plan is necessary is clear. Mr Tottenham [Secretary of the Government of India, Army Department] has confessed to me privately that the 5 year-plan attached to the Despatch was mistaken in that it arranged for guns and mine-sweeping gear before it had arranged for trained personnel to use them: The intensive examination of the question which has taken place during the present international crisis will almost certainly result in a plan that is better worked out. This must be done in Bombay. It cannot be done in Whitehall.

There seems to be a good deal of confusion on the subject as a whole, and I think that the local naval authorities should be invited to clear their minds on the following points.

Does the decision that the vessels of the R.I.N. are to be handed over in war to the Naval Commander-in-Chief, East Indies, necessarily mean that none of them are to be available for the purposes of local naval defence? I do not see that it necessarily does, and, although the decision that the vessels should be handed over is clear enough, the question of what the Commander-in-Chief should do with them on being handed over is very obscure. He was supposed to consult the Government of India, but the first sentence of paragraph 5 of the Government of India's letter of 11th July 1932 (flagged), a very important letter, seems to show that the Government of India were themselves in the dark. It will be seen from the Government of India's letter that the allocation by the Commander-in-Chief of all the R.I.N. vessels to duty outside India, involves the taking up of more mercantile marine vessels for the purposes of local naval defence, and that the R.I.N. has not the trained personnel to man them. This paragraph should be read in the light of General Coleridge's submission on M.40 attached.

[294] *Minute by F.W. Mottershead, [Admiralty Assistant Principal Clerk] for Head of Military Branch of the Admiralty Secretary's Department [under the Deputy Chief of Naval Staff, Vice-Admiral William M. James]*

9 January 1936

Propose to submit to seek the concurrence of the Treasury to the loan of the *Crescent* and *Cygnet* to the Canadian Government on terms similar to those laid down for the loan of *Toreador* and *Torbay* in Admiralty letter M.03237/27 of 24th January 1928, the cost of refitting and any necessary alterations to the vessels being borne by Canada. It would be pointed out to the Treasury that the refit will be necessary on account of past service with the R.N., but that nevertheless it is understood that Canada would accept these terms. Also to point out the advantages of the transfer to imperial defence as a whole, and the saving to the U.K. as stated by D.O.C. in his minute of 15.10.35 on M.03355/35.

If the above is approved, a draft letter will be prepared and referred to Departments and Branches concerned, and submitted. It would be proposed to send a copy of this letter to the Dominions Office.

The *Crescent* and *Cygnet* should be relieved by H Class Destroyers about September 1936.

If the loan is approved a decision will be necessary later as to the disposal of the *Champlain* and *Vancouver*. Possibly the Canadians may desire that these vessels should be navigated to the United Kingdom by the crews who will take over the *Crescent* and *Cygnet*. The two old destroyers would then presumably be scrapped.

[295] *Memorandum by J.A. Simpson for the Under-Secretary of State for India*

26 January 1936

To get the present position clear in regard to the handing over of the R.I.N. to the Admiralty or the Commander-in-Chief East Indies it is necessary, I think, to make three distinctions:

(1)   We must distinguish between a statutory obligation or permission to hand over the R.I.N. and an administrative agreement to hand it over.

(2)   We must distinguish between agreement that the R.I.N. should be handed over and agreement as to how it is to be used after being handed over.

(3)   We must distinguish between (a) handing over to the Admiralty in the sense of a transfer of the R.I.N. to the uses of the Home Government and (b) handing over to the Commander-in-Chief East Indies in the interests of unity of command of the naval forces in Indian waters.

The difficulties which now trouble us appear to begin with the decision to reorganise the R.I.M. as a combatant force. In the Report of the Departmental Committee appointed to prepare a scheme occurs the following sentence 'In war time unity of command is essential, and we therefore recommend that in war the ships and personnel of the Indian Navy should automatically come under the direct control of the Commander-in-Chief, East Indies'. This recommendation was embodied in the Government of India's proposals for the conditions of appointment of the Flag Officer Commanding, Royal Indian Navy, and accepted by the Admiralty and the Secretary of State for Indian apparently without discussion. (vide papers at back of collection 80/153 1926–27).

Next comes the Indian Navy Act of 1927 as a result of which inter alia the following two sections were added to the Government of India Act.

Section 44 A. 'Any naval forces and vessels which may from time to time be raised and provided by the Governor-General in Council shall be employed for the purposes of the Government of India alone, except that if the Governor-General declares that a state of emergency exists which justifies such action, the Governor-General in Council may place at the disposal of the Admiralty all or any of such forces and vessels, and thereupon it shall be lawful for the Admiralty to accept such offer.'

Section 22 (2). 'Where any naval forces and vessels raised and provided by the Governor-General in Council are in accordance with the provisions of this Act placed at the disposal of the Admiralty, the revenues of India shall not, without the consent of both Houses of Parliament, be applicable to defraying the expenses of any such vessels or forces if and so long as they are not employed on Indian naval defence.'

Meanwhile discussions had been going on in India as to the interpretation of the decision regarding the position of the Royal Indian Marine (N.) on the outbreak of war. The original orders of the Government of Indian were that in time of war the Flag Officer Commanding, Royal Indian Marine should place the vessels of the Royal Indian Marine with their personnel at the disposal of the Naval Commander-in-Chief for employment on the duty for which the reorganised force was primarily intended and which the Commander-in-Chief would superintend i.e. the local defence of Indian coasts and harbours. While it was so employed, Indian revenues would continue to bear all expenditure and the vessels and forces would still be at the disposal of the Government of India. This apparently was found unsatisfactory. The amendment at first proposed reads as follows 'In time of War the Flag Officer Commanding will place the vessels of the R.I.M. with their personnel at the disposal of the Naval Commander-in-Chief to be employed under the latter's command for the purposes of the defence of India. While the forces are so employed Indian Revenues will continue to bear all expenditure.' A most important paragraph follows 'The intention is that the Naval Commander-in-Chief should be able to employ the ships of the R.I.M. in whatever way would be best during War, so long as the main object, namely the defence of India was subserved; and it would only be when the ships were diverted from this purpose that the question of placing them formally at the disposal of the Admiralty, and thereby relieving Indian Revenues of their cost, would arise.'

The Naval Commander-in-Chief in commenting on this observed 'It is essential that in wartime the Admiral Commanding, East Indies should have the unfettered control of all ocean going vessels of the R.I.M. Their disposal, on the outbreak of war, is, however, a matter for previous arrangement.' The Naval Commander in Chief then went on to state that he intended to dispose of the vessels of the R.I.M. at Henjam, Trincomali, and Aden, and requested the Government of India's concurrence. The Government of India's final draft was then produced which runs as follows:

'The Naval Commander-in-Chief will assume command of all naval forces in Indian waters including the vessels and personnel of the R.I.M. . . . .

The Flag Officer Commanding . . . will place the personnel and vessels of the R.I.M. at the disposal of the Naval Commander-in-Chief in accordance with a pre-arranged plan approved by the Government of India.' [The underlining is mine.] At the same time the Government of India accepted the plan of disposal of the vessels of the R.I.M. mentioned in the preceding paragraph.

The Naval Commander-in-Chief while accepting the above stated in his reply that the 'ultimate decision as to the disposition of all ocean going warships must rest with the Naval Commander-in-Chief'.

These proposals were then formally sent home by the Government of India and accepted by the Secretary of State in Council and the Admiralty . . .

[296] *Minute by J.H. James [Admiralty Principal Clerk] for Head of Naval Branch [under Second Sea Lord, Vice-Admiral Sir Martin Dunbar-Nasmith]*

12 February 1936

It is proposed to send the following sketch of general Admiralty views to the Dominions Office who have agreed to take the attached Parliamentary Question next Wednesday, and are anxious to have a statement from the Admiralty to help them.

It is requested that this paper may be passed very quickly.

Entry from the Dominions into the R.N.

(1)   South Africa: Here there is a permanent recruiting organisation at Simonstown under the Captain-in-Charge who arranges the final entry of recruits and sends them on to England in one of H.M. Ships for training. The maximum number of recruits for all branches whose entry is permitted in this way is 20 per annum.

(2) Other Dominions: The Admiralty have no wish to encourage entries from the Dominions for the two main reasons that:

(a)   The uncertainty and difficulty of entry which necessarily follows from the fact that there are no Admiralty Recruiting Stations in the Dominions and that therefore it is impossible to say whether any man is fit and suitable for retention in the Navy until he has been brought to England.

(b)   There is no present shortage of candidates for entry into the Navy from sources at home and therefore no reason is seen for going to additional trouble and expense in order to obtain Dominions entries. In addition drafting difficulties would be created by such a form of entry.

The Admiralty therefore do not wish to increase entries from the Dominions. They are, however, naturally unwilling to discourage entries and for this reason, if a candidate shows great keeness to enter or (more often) if representations are made through any official channel, arrangements are if possible made for the provisional examination of any candidate's fitness for entry when one of H.M. Ships happens to be in the appropriate locality. It still remains necessary for the candidate concerned to bear the cost of his own passage to England if he desires to enter unless a passage can be provided for him in one of H.M. Ships returning to England, and in the case of candidates who are provisionally examined and passed every effort is made to arrange such a passage.

The Dominions Office might be informed that the Admiralty concur with their general view of what the answer to this question should be, and suggest that the terms of it might be (roughly) that the Admiralty are themselves satisfied with the present position, and that though they would willingly consider any suggestion which might be made by any of the Dominions there are nevertheless considerable difficulties in any extension of the present arrangements which it would be difficult to overcome.

Referred for remarks prior to submission.

[297] *Coleridge to Murray*

15 February 1936

In reply to your letter No. M.05504/35 of the 1st January 1936, I am directed by the Secretary of State for India to inform you that in view of Their Lordships assurances, as conveyed in paragraph 6, that the danger of submarine attack at Karachi and Bombay is negligible, the Government of India agree that two sloops of the Royal Indian Navy would be more usefully employed at Aden should the present emergency develop.

2   With reference to the general question of policy in regard to the allocation of sloops of the Royal Indian Navy in the event of war, I am to say that this will form the subject of a separate communication in due course.

3   I am to convey the thanks of the Secretary of State for Their Lordships' approval of the loan of mine-sweeping equipment, guns, and ammunition, as reported in paragraph 5 of your letter.

[298] *Stewart to General Sir Robert A. Cassels, Commander-in-Chief India*

23 March 1936

Many thanks for your private letter of 14th February on the subject of the despatch of two sloops of the Royal Indian Navy to Aden. I quite agree that it is better to clear up the misunderstanding demi-officially in the first instance, but in the light of the explanation which I am able to give you I do not think that there is any reason why you should not reply officially in due course with reference to the Admiralty letter of January 1st when you have received and considered the official letter which should issue shortly from us on paragraphs 8 and 9 of that letter. The Admiralty are quite friendly and understanding, as exemplified by their offer of loan of equipment, and I do not think that a reasonably worded rejoinder on your part is likely to cause any resentment of bad feeling.

The main point is to make clear exactly what the Admiralty case is. You complain that you were ordered to employ the Royal Indian Navy sloops on local naval defence and then attacked by the Admiralty because in obedience to these orders you found yourself unable to hand over any of the sloops to the Naval Commander-in-Chief. The explanation is that our telegram of November 7th, owing to the necessity of urgent action, received only oral concurrence from an officer in the Admiralty, and that the final and considered view of the Admiralty is that expressed in their letter of 1st January which attaches greater importance to the despatch of two sloops to Aden than to the carrying-out of the anti-submarine portion of local naval defence. These facts are, of course, known to the Admiralty and I think if you read paragraph 2 and paragraph 6 of their letter in the light of the above explanation you will acquit them of any intention of being unfair to you on this point.

I am inclined to think also, that while the government of India telegram 2951 stated that all H.M.I. ships were to be employed on anti-submarine patrols, in the absence of any emphasis of the fact that as a result no ships could be placed at the disposal of the Naval Commander-in-Chief, the full significance of this statement did not impress itself on the Admiralty at first sight.

Nevertheless the Admiralty certainly hold that they have a case to be met, and as I interpret it, the case is as follows. The Government of India when they established the Royal Indian Navy on a combatant basis agreed to place the sloops at the disposal of the Naval Commander-in-Chief on the outbreak of war. (On the exact meaning and conditions of this undertaking our official letter will have a great deal to say, and I pass over it at the moment). The Admiralty have framed their plans on that undertaking, and consider themselves entitled to rely on it, but find on the present occasion that the Government of India are unable to honour it in full and at the same time to take what are, or might be, the necessary measures for local naval defence. That, as I understand it, is the case; I do not commit myself to saying whether it is substantiated or not. The same consideration applies to paragraph 6 of your letter. His Majesty's Government's attitude, as I understand it, is not a demand that India must maintain a sea-going fleet which can be put at their disposal when they want it, but an expectation that when India has created such a fleet at her own free will, and has, again at her own free will, agreed in certain circumstances to place it at the disposal of the Naval Commander-in-Chief, she should not default on this undertaking because of her neglect of the primary duty of putting her local naval defences in order.

I hope that in all this you will not think that I am backing the Admiralty and His Majesty's Government against India; I am merely trying to make clear to you what appears to me to be the real nature of the case that they put forward.

I have at the same time written to Murray, (the Secretary of the Admiralty), to make clear to him, though I think the Admiralty quite understand this, that you are not trying to be obstructive and that the decision that the sloops would not 'be allowed to leave the ports of India' was not your decision.

P.S. The point in paragraph 5 about reserves might well be put in your official reply. I have, indeed, reason to believe that the Admiralty are not ignorant of the position and fully expect a reply from you on the point.

[299] *Minute by J.H. James for Head of Naval Branch [Under Dunbar-Nasmith]*

4 June 1936

The question of the restoration of the Newfoundland Royal Naval Reserve was raised by Mr Winston Churchill in the Debate on the Supplementary Estimates on Monday May 4th, and the extract from Hansard is attached; while the question of the enlistment of the Newfoundland recruits has been the subject of Parliamentary questions – the latest of which was on Wednesday February 12th – and correspondence between the Parliamentary Secretary and Lord Winterton [Unionist MP]. The present occasion of a letter from Sir R. Walwyn [Governor of Newfoundland and formerly Flag Officer Commanding and Directing, Royal Indian Navy] to the First Lord on the means of using Newfoundland personnel for the Navy is therefore probably an opportune moment to investigate the whole question . . .

[300] *Notes for First Lord [Hoare] on points that Naval Staff wish to raise with Mr [the Hon. Oswald] Pirow [Union of South African Minister of Defence]*
Lunch, 17 June [1936]

Admiralty Requirements

1 The Admiralty wish to see the gun defences of Capetown, Simonstown and Durban modernised and improved as recommended by the Joint Oversea and Home Defence Committee.
2 The Admiralty wish to install six 12,000 ton oil tanks at Cape Town.

Union Government's Requirements and Admiralty views

3 At Cape Town Mr Pirow wishes to go further than the recommendation of the Defence Committee by installing 15″ guns. As justification for this he argues that when his projected harbour development scheme (the Craig scheme) is complete the port will assume added importance. His real reason is, however, understood to be the wish to be the father of a 'big' project.

In view of the Imperial importance of Capetown Mr Pirow wishes for help from H.M.G. to meet the cost of the defences – such help might be in the form of the loan of a 15″ gun and mounting.

He says he has no money to modernise Simonstown (to the proper defence of which South Africa is committed in a written agreement).

4 The Admiralty consider 15″ guns to be over-insurance, but nevertheless, if the Union is so keen on it we should not obstruct them. Improvement of Simonstown defences, however, must not suffer thereby. A case can be made out for departing from the principle of Dominions being responsible for their own local defences, and providing some assistance from the Imperial Government.

5 As regards the oil tanks, the Union Government were not inclined to undertake the work themselves or to allow the Admiralty to do so. They were prepared to consider a proposal that a holding Company should erect the tanks financed by the Admiralty, and the Union Government would lease the necessary land to the Company, but said the arrangement must be kept secret for political reasons.

6 The Admiralty dislike the idea of secrecy over the oil tank

question, since they do not consider secrecy can be maintained in practice, and the unrestricted use of the tanks in all circumstances could not be guaranteed owing to legal difficulties. The Admiralty dislike any agreement which is not entirely open and want South Africa to come out in the open about it so that we can get ahead installing the tanks as soon as possible.

7   Imperial agreement with the 15″ gun defence project of Mr Pirow might go a long way towards his acquiescence regarding the oil project. Consequently the Admiralty would be prepared to support a proposal that the Imperial Government should provide some assistance towards the Union's scheme of Cape Town's defences, though this is held to be over insurance.

Other Naval Proposals

8   It is desired to put forward the following suggestions for naval contributions by the Union Government.

(a)   To increase the number of candidates for Dominion cadet-ships in the Royal Navy, as is being done in the case of New Zealand;

(b)   To complete, as soon as possible, provision of minesweeping outfits already contemplated; establish a small local reserve of anti-submarine material; and provide and maintain anti-torpedo nets for Cape Town harbour.

[301] *Nelles to the Secretary of the Admiralty [Sir R.H. Archibald Carter]*

30 July 1936

Be pleased to inform Their Lordships that this Department has in the past received numerous applications from residents of Canada with regard to the procedure which they should adopt to join the Royal Navy.

All such applicants have been advised that they would have to proceed to England at their own expense and take the chance of rejection upon arrival there. Some of the applicants have for-warded enquiries to the Director of Recruiting, London, S.W., and have been referred back to this Department.

If Their Lordships so desire, arrangements could be made for applicants for entry in the Royal Navy resident in Canada to fill in the attached Questionnaire amended as indicated, together with the certificates and credentials required therein, and also for such candidates to be provisionally medically examined in Canada at their own expense. Application and Report of Medical Examination could then be forwarded to the Admiralty for provisional acceptance, after which the candidates would follow such instructions as may be issued to them with regard to joining the Royal Navy.

It is requested that the wishes of Their Lordships in this matter may be communicated.

It is also requested that you please supply this Department with the latest edition of Admiralty Recruiting Regulations. [signature]

[302] *Secretary of State for Dominion Affairs [Malcolm MacDonald] to the Governor-General of New Zealand [Viscount Galway]*

4 August 1936

'SECRET. No. 93'

Your Secret Telegram No. 56 of the 1st June[.] His Majesty's Government in the United Kingdom are very glad to hear that His Majesty's Government in New Zealand have under active investigation the problem of the best measures for securing the effective defence of the Dominion and for co-operation in imperial defence plan(s) and that they contemplate making increased provision in the coming Budget for the necessary services. Without wishing to be in any way unnecessarily alarmist His Majesty's Government in the United Kingdom feel the international outlook to be such that considerations of common prudence dictate the maximum possible strengthening of defence position(s).

His Majesty's Government in the United Kingdom note from your telegram quoted above that His Majesty's Government in New Zealand contemplate a re-organisation of the whole of the Defence Estimates of the Dominion apparently entailing a reduction of naval commitments in order to ensure maximum expansion on air, and that in particular His Majesty's Government in the United Kingdom are asked to reconsider the advisability of and necessity for keeping to the existing intention that His Majesty's Government in New Zealand should maintain two cruisers *Achilles* and *Leander*.

His Majesty's Government in the United Kingdom feel most strongly for reasons stated below that any decrease in the efforts hitherto made by New Zealand on the naval side would have most serious effects on the security of the Dominion especially at the present time.

The strategic dangers to which New Zealand could be exposed arise from sea-borne attack. One type of danger would be incurred if an enemy could count upon security of its communications between its home country and New Zealand for an indefinite period. Another type of danger would be incurred if an enemy enjoyed this security for a short period only. The strategic problems of New Zealand are therefore how to meet these two types of danger from sea-borne attack and how to be able also to co-operate more widely in the defence of the British Commonwealth particularly in the Pacific area.

If any enemy could secure his communications to New Zealand indefinitely no local defence measures which New Zealand could take could give permanent protection against aggression. New Zealand's security against the greatest danger which could threaten her depends therefore upon the ability of the British naval forces to command the communications in the Pacific.

The fundamental basis of the defence of the Empire lies in the maintenance of communications so that the forces of the Empire may be moved to any threatened point. It is only the fact of our ability to do this which permits the restriction of local forces and defences in many territories of the British Commonwealth to a scale which it is possible for the taxpayers to maintain.

If the Navy were unable to fulfil its responsibilities the whole basis of Imperial Defence would crumble and no local forces on the lines suggested could save New Zealand in the long run. The relatively few aircraft which New Zealand could maintain for the same cost as the *Achilles* and *Leander* could not effect the ultimate issue of a conflict with Japan. New Zealand's main security in a conflict against Japan must be the ability of the British naval forces to limit the period during which New Zealand could be seriously attacked.

The very heavy increase on naval expenditure now being made by the United Kingdom Government should provide sufficient evidence that the ability of the British Navy to fulfil in the future the role it has played in Imperial Defence in the past will in all circumstances be maintained. This is a time at which any serious reduction in her contribution to naval defence by any Dominion relying for her security primarily upon the supremacy of the British naval forces would appear as a failure in solidarity of co-operation between all members of the British Commonwealth in their defence plans. His Majesty's Government in the United Kingdom would view with dismay any reduction of a Dominion's naval efforts which would add still further to the very heavy burden being borne by the United Kingdom.

Command of the sea by an enemy for a limited period would expose New Zealand to the second form of danger. New Zealand's problem in this case is to ensure by her local forces the security of her coasts during the period before relief. The distance between Japan and New Zealand is so great as to make the likelihood of a sea-borne expedition against her in these circumstances extremely remote. Even if an expedition were attempted the proved fighting qualities of the New Zealand forces are such as to make it at least doubtful whether the Japanese would succeed in securing and maintaining a footing on the Islands. The maintenance of efficient local forces in New Zealand even on a small scale should afford security from this form of danger. In the organisation of these local forces up to the limit(s) of the financial possibility(ies) a correct balance between the nucleus of navy[,] army and air forces should be kept not only for efficiency in local defence but also in order that New Zealand should possess power to expand her forces for co-operation in the defence of the British Commonwealth.

Should your Ministers so desire we should of course be most happy to make more detailed suggestions as to the composition and precise role of such forces as could in our view be most usefully maintained.

His Majesty's Government in the United Kingdom note the desire of His Majesty's Government in New Zealand for a maximum expansion on air defence and if the financial situation will admit of this would welcome such a expansion which if beyond the needs of local defence would be a contribution to the general defence of the Commonwealth. My recent telegram(s) regarding the application of Pan American Airways will have indicated the importance attached here to the Pacific from the air point of view and any extension of the air defence of the Dominion which would mean a strengthening of the British air position in the Pacific would in the view of His Majesty's Government in the United Kingdom be a most valuable contribution to Imperial Defence in that area.

A further portion of this message will follow tomorrow.

[303] *Report of Interview between Mr Mackenzie King, Canadian Minister of Defence, and First Sea Lord [Chatfield]*

6 August 1936

'Secret'

Mr Mackenzie called on me today to discuss Naval Defence questions. He informed me that he thought Naval matters in the Dominion were on the mend. Increased money had been voted for the Estimates this year, and, curiously enough to him, without any criticism whatever.

Canada was aiming at a force of 6 Destroyers and some Sloops. He said they were considering building 2 new Destroyers and recorded that no decision had yet been arrived at as to also purchasing the 2 'C' Class Destroyers from this country. Their idea was to complete the 6 Destroyers during the next two or three years. He asked how long it would take to build 2 Destroyers, and I said about 2 years from order to delivery. He said, 'then we ought to start pretty soon.' I pointed out to him that the 2 'C' Class Destroyers, about whose condition he asked, were quite good vessels but, of course, they were getting older each year and if they wanted them we wanted to know as otherwise we might find other uses for them; so the sooner they decided if they were going to buy them the better. I understood there had been certain questions as to the price, but that was a matter which had to be settled by the Dominions Office and Treasury and was beyond Admiralty control. He said, of course, their other 2 Destroyers, *Vancouver* and *Champlain*, only had 2 years more life and therefore it would be necessary that they should be replaced also now. I said it seemed to me that what they wanted was to buy the 2 'C' Class from us, and also to build the 2 new Destroyers, and he generally agreed. He said he would take the matter up as soon as he got home and hoped to let us know something soon.

I pointed out to him that, of course, it was not our advice to Canada to have Destroyers: in our opinion Cruisers and Sloops were the most important vessels. If they could not provide Cruisers then, no doubt, our Cruisers would have to do the work and their Destroyers might then join up with our own forces in exchange, which he said he fully understood.

I then raised the question of their West Coast defences and pointed out to him, as I had already done at the Chiefs-of-Staff interview with him and Mr Dunnin [Charles A. Dunning?, Liberal MP] that if we were in trouble in the East we should want to cut the communications of the Japanese with America, and one of our best ways of doing that was to base a force of cruisers on their West Coast; that was why we were anxious about the defence of Esquimalt and Vancouver. He said that they had had this in their mind for some time and they realised the importance of improving their gun defences, which they were going to do in conjunction with the War Office. He also thought that they were taking steps as regards their net defences.

I told him that the practice in other Dominions was to ask the Admiralty for technical advice in such circumstances, they then would be certain of getting up-to-date material and laying it out in an up-to-date manner. It was also important, if our ships were to use their ports, that the Admiralty should know what the efficiency of the defences would be before we risked our ships in those ports. Did he think, therefore, that it would be possible for the Canadian Naval authorities to get into touch with us on this matter? He said he fully understood the importance of that and entirely agreed in principle, and he saw no objection to it at all so long as it was purely technical information exchanged between the two Staffs. He would take the matter up as soon as he got back, with that intention.

With regard to Sloops he asked whether it would be possible for Canada to build them over there. They had good shipbuilding facilities and would, of course, prefer to do it that way. I said I thought there would be no difficulty in that, though, of course, technical material would have to be supplied by this country. He asked what the Sloops would be like and how they would be armed, and I told him their chief value was for local defence and local convoy, hunting submarines or minesweeping. He said they were anxious to have minesweepers as well. I informed him of the general equipment of our own Sloops.

My impression was that he was anxious to be up and doing and is an active minded man who will not let the grass grow under his feet. Apart from the unsatisfactory fact that the situation as regards purchasing the 2 'C' Class Destroyers is still unsettled. I think the interview on the whole was very hopeful.

[304] *Rear-Admiral G.C.C. Royle, Naval Secretary, to Hoare*
24 August 1936

The main issue is whether steps are now to be taken to create with the Royal Indian Navy a state of disciplinary unity on the Dominion model.

The Board of Admiralty take the view that the grant of disciplinary unity is not called for at present, and is not necessary for the efficiency of the two Services either in peace or war.

The proposed letter for your approval is drafted on these lines, i.e., the Board do not wish that Indian Naval Officers should have command over, or sit on Courts-Martial on, white Naval personnel.

[305] *'Indian Naval Defence', note by India Office as a provisional basis of discussion, prepared by Captain A.G. Maundrell RIN, Chief of Staff to the Flag Officer Commanding, Royal Indian Navy*

not dated, [*c.* mid-1936]

1   The average budget for Indian Naval Defence at present, in round figures, is:

(1) Standing Charges for upkeep of Dockyard Depot, 5 Sloops, 1 Patrol craft, 1 Survey ship, Pensions, etc.	Rs. 50 Lakhs	(£375,000)
(2) Payment to His Majesty's Government. (Contribution Rs. 13 lakhs, subsidy Rs. 4 lakhs)	Rs. 17 Lakhs	(£130,000)
(3) Contribution to Sinking fund for replacement of ships.	Rs. 14 Lakhs	(£107,500)
Total	Rs. 81 Lakhs	(£612,500)

2   The payment of item No. 2 above dates from many years ago when India maintained no combatant naval force. The old Royal Indian Marine, which performed certain non-combatant duties for the Government of India, was converted into a combatant force about 12 years ago, by arming the existing vessels of the Royal Indian Marine and training their crews in combatant duties. Since that time two modern sloops have been built at India's expense, one to replace an old ship converted into a sloop and the other to replace a small Patrol vessel.

3   Meanwhile, the accepted first stage in the development of 'Dominion Navies' – namely the provision of Local Naval Defence – assumed greater importance.

4   Realising that the provision of Local Naval Defence is a domestic responsibility, India is now prepared to finance an organisation for this purpose from her own resources. She is anxious to make a start as soon as possible by providing a number

of small vessels for minesweeping, purchasing the necessary equipment and training the necessary reserve personnel. The broad lines of the organisation proposed have been approved by the Admiralty and it has been roughly estimated that it will cost something in the neighbourhood of Rs. 14 lakhs a year for the next five years, i.e. something approximately equal to the Sinking Fund payments referred to as item (3) in paragraph 1.

5   Meanwhile one of the existing sloops (Cornwallis) and the remaining Patrol Craft (Pathan) require very early replacement. India would like to replace these two vessels during the next 3 or 4 years by two new sloops of the most up to date type, at an estimated cost of about Rs. 66 Lakhs (£500,000).

6   Whilst fully admitting their liability to continue the payments to the Admiralty referred to in paragraph 1, the Government of India's position is that they are under no similar obligation to replace the two old vessels by two new sloops, and, in fact, owing to the political and financial situation in India, they are quite unable to replace them at all while at the same time continuing their cash contributions to His Majesty's Government. On the other hand, they are most anxious to maintain and, if possible, expand their seagoing fleet on the definite understanding that the ships would be placed at the disposal of the Admiralty in war. The two additional sloops would to some extent ease their own problem in the provision of Reserves for Local Naval Defence: but the main point is that, in their judgment (and they have reason to believe that the Admiralty agree with them) such a course would represent the best possible contribution they could make towards Imperial Defence.

7   What it comes to, therefore, is that India offers His Majesty's Government the choice between receiving during the next four years a sum of approximately £500,000 in cash (as represented by India's annual contributions of £130,000) and seeing the Royal Indian Navy reduced in size by two seagoing ships or allowing this sum of money to be devoted towards the construction of two new sloops for the Royal Indian Navy, on the understanding that all the sloops would be placed at the disposal of the Admiralty in an emergency. A reduction in the number of seagoing ships would be a most retrograde step and would seriously affect the efficiency and morale of the Royal Indian Navy as a whole. The Government of India feel themselves – and here again they believe that they have the support of the Admiralty –

(a)   that the cash contribution is in itself a somewhat antiquated arrangement which is hardly consistent with India's own development and her position vis-a-vis the Dominions, and

(b)   that its conversion into a contribution in kind – for that is what the proposal really amounts to – would in practice prove of greater value both to India and to His Majesty's Government.

It seems possible that, if the two new sloops were not added to the Indian Navy, the Admiralty might find it necessary to provide and maintain them as part of the Royal Navy. On all these grounds, therefore, the Government of India press for the remission of the contribution of £130,000 for the next four years.

8   The ultimate outcome of this policy might be the permanent remission of the contributions on the bases of a declaration made and accepted by India that her long term policy would be to provide her own Local Naval Defence and maintain a seagoing force of not less than six up to date sloops which would be placed unreservedly at the disposal of the Admiralty in war. India herself would be fully prepared to make such a declaration of policy and would welcome an arrangement on these lines as affording the most satisfactory possible settlement of the whole problem. But she admits that the question of permanent remission would not arise until she had proved her ability to maintain a fleet of that kind. Meanwhile, however, she asks to be given a chance of proving it by means of a temporary diversion of the contribution.

9   As an alternative, India would agree, if it were preferred, that His Majesty's Government should provide India with the two new sloops free of cost. The most important point, in India's opinion, is that they should be enabled to go straight ahead with their Local Naval Defence proposals and that an order should be placed at any rate for the first of the two new sloops at the earliest possible moment.

[306] *Minute by J.A. Simpson, Acting Joint-Secretary of the Military Department, India Office*

26 August 1936

Simultaneously with the receipt by me on 11th August of M.4112 with the notes by the Under Secretary of State I received the Government of India's letter of 27th July enclosing the report and Plan on Local Naval Defence (India) prepared under the direction of the Flag Officer Commanding, Royal Indian Navy, and the Report thereon of the Sub-Committee of the Indian Defence Committee.

These documents revolutionise the situation so completely, that I think it advisable to make a fresh submission which will cover the whole ground in the light of Captain Maundrell's [Chief of Staff to the Flag Officer Commanding, Royal Indian Navy] Report. This Report which, in the words of the Sub-Committee, 'represents the first real effort to examine in practical detail the actual requirements in men, money and material, needed to place on a satisfactory footing the local naval defence of India's seven major ports', is an invaluable document to the preparation of which Captain Maundrell and his staff must have devoted a remarkable amount of labour and talent. I suggest that in replying to the government of India the Secretary of State should express his warm appreciation of the work done by the Royal Indian Navy Staff. For practical purposes the summary of the Staff Report given in the Report of the Sub-Committee is adequate, and the latter Report should be read in full.

[307] *'Nine Years' Plan' for the Development of the Royal Indian Navy*

? 1936

1 Policy. The provision of 6 modern Escort Vessels, the provision for Local Naval Defence including the formation of Naval Reserve Forces, together with the requisite instructional establishments and personnel for training.

2 Plan. Based on the foregoing policy a plan of development spread over a period of nine years has been evolved, the salient features of which are contained in four parts:

I. Personnel.
II. New Construction.
III. Depots and Material.
IV. L.N.D. Equipment.

3 Description. In order to provide a Squadron of six modern Escort Vessels and the Local Naval Defence Forces necessary to meet the scale of attack laid down by the Committee of Imperial Defence the following are considered the minimum requirements:

(a)   Under Part I the requisite personnel have been provided as follows:

(i)   The strength of the Active Service is increased as shown below:
Commissioned Officers from   101 to   169.
Warrant Officers from   52 to   104.
Ratings from   1140 to 2562.
(ii)   Reserves are necessary to man auxiliary Vessels and provide personnel for harbour defence in war. After thorough examination of the whole problem the following numbers are recommended as representing the minimum requirements for this purpose:
R.I.N.R. Officers   252
R.I.N.R. Men   912
R.I.N.V.R. Officers   71
R.I.F.R. Men   593
Communications Reserve   286
(iii)   Adequate training cadres for boys and apprentices in the skilled trades are required to be provided for completely separately from those allowed for the Active Service.

In the case of boys this principle has already been accepted.

Boys take 22 months to train from the date of recruitment and Artificer and Artisan ratings 4 1/2 years.

(b)   Under Part II provision has been made for the replacement of Cornwallis, Lawrence and Clive, the construction of a sixth Escort Vessel and the rearmament of Hindustan and Indus. Further in order to implement the requirements of Part I (ii) six small Minesweepers have been provided for. Without the provision of adequate ships and material it is not a practical proposition to undertake the training of Naval ratings in any shape or form.

(c)   Part III makes provision for the requisite training establishments, instructional equipment and stores without which training cannot be carried out.

It is considered that the main manning and training depots should be situated at: Bombay, Cochin, Vizagapatam, with sub Depots for subsidiary training of Reservist officers at Calcutta, Madras and Karachi.

(d)   Part IV deals entirely with Local Naval Defence Equipment and is based on the requirements laid down by the Committee of Imperial Defence. It makes provision for the equipping of 48 auxiliary vessels to be taken up on the event of hostilities. Of these vessels 25 are required to be fitted as Minesweepers and 23 as anti-Submarine vessels.

4 It is in addition proposed to acquire a flotilla of 8 Motor Torpedo Boats as an offensive measure of Local Naval Defence. Provision for these together with their stores etc. and the additional personnel required has been made separately in the financial summary attached.

5 It is also proposed to make provision for the replacement of ships by means of a sinking fund, as shown in Appendix II.

[308] *Message to First Sea Lord [Chatfield] from Naval Service Headquarters*

Ottawa
4 September 1936

Important   (?Urgent) Following secret and personal message from Honourable Ian Mackenzie Ministry for National Defence begins. Recalling our pleasant conversations would appreciate your opinion by W/T (? as to whether) Admiralty (?would) transfer by purchase now two C. Class destroyers at one million dollars each (corrupt group) and transfer to us on loan 2 remaining C. Class (? destroyers) as soon as manning situation will permit us to take them over. Ends. 1515/3

[309] *Minute by Le Maitre for Head, Military Branch of Admiralty Secretariate [under William M. James]*

4 September 1936

The Canadian Minister for Defence unofficially offers 1,000,000 dollars each for the *Crescent* and *Cygnet*; this offer totalling £400,000 for the two ships is the same as the figure which we have just put to the Treasury. Unofficially M[ilitary Branch] learns that this proposal has been agreed by the Treasury although we have not yet had their reply. [manuscript addition: 'It is on its way over now']

So far as *Crescent* and *Cygnet* are concerned the answer to Mr Mackenzie seems to be 'Yes'. But he has joined to his suggestion a proposal that the other two 'C's' should be transferred on loan as soon as the manning situation permits of their being taken over.

There are certain difficulties about this. The Treasury have agreed to the sale of the *Crescent* and *Cygnet* aged four years, but it is by no means certain that they would agree to the loan of *Comet* and *Crusader* aged seven years, nor is it certain that the Dominions Office would view the proposal favourably. The Dominions Office dislike the idea of lending *Crescent* and *Cygnet* to Canada, and although *Crusader* and *Comet* would be three years older the possible difficulties with other Dominions might still be regarded as a bar.

Further it is our intention to use the other two 'C's' to attend on aircraft carriers. This means that we should have to find some other modern destroyers not part of any running flotilla to do this job in the absence of the two 'C's'. Assuming a three years' interval it is possible that modern destroyers would be falling out of the running flotillas and perhaps at least one flotilla would have to be broken up for various employments, and this might be an opportunity for relieving the two 'C's'. It might very well be that after four years in a running flotilla and three years attending on aircraft carriers the two 'C's' would not be in a position to undertake serious service in the Canadian Navy without a retubing refit, which presumably the Canadian Government would have to undertake.

In view of the difficulty of committing ourselves to being able to spare *Comet* and *Crusader* in three years' time, and the difficulty of committing the Treasury to let them go on loan, it seems that the only possible reply to Canada would be in the following sense:

That the offer for *Crescent* and *Cygnet* is accepted with great pleasure, the ships to be taken in the condition in which they are at present; the Admiralty thoroughly sympathises with Mr Mackenzie's desire to take over the other two 'C's' when they can be manned but (a) at this stage we cannot commit the Treasury to a transfer on loan and (b) we have already earmarked these ships for aircraft carrier attendance and the question of their relief in say three years must be carefully gone into, since we have no other modern destroyers available without breaking up a flotilla and thus necessitating a re-organisation of our plans.

Suggest that the purchase of *Crescent* and *Cygnet* be put in hand immediately and that the second proposal be treated separately when it has been fully considered.

[310] *J.A. Simpson to the Admiralty Secretary [Carter]*
23 September 1936

I am directed by the Secretary of State for India to refer to Admiralty letter of the 26th August, 1936, No. C.W.7447/36 on the subject of the relative rank and powers of command of Royal Naval and Royal Indian Naval Officers.

2   The Secretary of State is glad to note that the Lords Commissioners of the Admiralty are prepared to agree that Officers of the Royal Indian Navy should take rank with Royal Naval Officers according to seniority.

3   As regards command, their Lordships observe that it is unnecessary to lay down any general rule so far as peace arrangements are concerned, since for practical purposes the position can be regulated as proposed in paragraphs 3 and 4 of their letter. The Secretary of State does not wish to dispute that as regards the strict practical necessity of the case, it is possible for the two Forces to co-operate upon this footing. He feels bound to observe, however, that such an arrangement would fail to meet the strongly expressed wishes of the Government of India, which he desires to press once again on Their Lordships' consideration, that the Royal Indian Navy which has now been raised to combat status and placed under naval discipline, should assume a relation to the Royal Navy more nearly approximating to that held by the Dominion Navies.

4   The assumption has throughout been entertained by this Office and the Government of India that the Lords Commissioners of the Admiralty would grant equality of powers of command as a result of the passage of the Indian Naval Discipline Act. The Secretary of State considers that the previous history of the subject has given ample justification for this assumption, which indeed was inherent in the whole policy underlying the Indian Navy Act of 1927 and the consequential legislation in India. Thus in their letter of the 29th August 1927, No. C.W.5200/27, the Admiralty agreed that such powers should be granted. In paragraphs 2 and 3 of their letter of the 6th July, 1928, the Lords Commissioners stated that the failure of the Indian Legislative Assembly to pass the Indian Naval Discipline Act 'will preclude Royal Indian Marine officers from exercising command in the Royal Navy' and that therefore the proposals in the Admiralty letter of August 1927 should be canceled and the question be 'raised again in the event of the Royal Indian Marine [sic Marines] being converted

into a Navy with Naval status and discipline'. The requirements thus stated by the Admiralty have now been fulfilled and it is not unnatural that the Government of India should expect that the decision given in 1927 would be reaffirmed. Moreover the provisions of Section 90 C of the Indian Naval Discipline Act, the text of which received the approval of the Admiralty, provides natural ground for the assumption that the Admiralty would be prepared to take steps necessary to enable Royal Indian Navy officers to exercise command in relation to Royal Navy officers and men.

5   Moreover, the Secretary of State considers it important that the reactions on the general morale and prestige of the Indian Service should not be lost sight of. There is no doubt that officers of the Royal Indian Navy feel a genuine grievance as to their present position. Under article 176(2) of King's Regulations and Admiralty Instructions, it is definitely laid down that officers of the three Dominion Naval Forces shall rank and command with Royal Naval Officers of the same seniority and rank. Officers of the Royal Indian Navy feel, not unnaturally, that the exclusion of their Service from this provision reflects upon their status and efficiency. It is difficult to combat this belief when the other Naval Forces of the Empire are thus placed upon an equal footing, and the Royal Indian Navy alone is not.

6   Apart from these considerations, the Secretary of State would point out that the recognition of the Royal Indian Navy as being on an approximately equal footing with other Dominion Navies is of particular importance in securing the support of Indian political opinion for the development of the Royal Indian Navy; unless this support can be retained there will be increasing difficulty in developing and expanding the Royal Indian Navy at the expense of Indian Revenues.

7   On the other hand the Secretary of State appreciates that there are practical objections of considerable weight to any proposal to put the Royal Indian Navy in exactly the same relation to the Royal Navy as the Australian and Canadian Navies. He realises, for example, that officers of those Services are automatically entitled to sit on Courts Martial and exercises summary jurisdiction on British Naval personnel. The Secretary of State considers that the Government of India could have no reasonable objection to the withholding of such powers from Royal Indian Navy officers.

8   In the circumstances, the Secretary of State would be glad to know whether the Lords Commissioners would agree to embody

the arrangements proposed in paragraphs 3 and 4 of your letter under reply in some general provision, which although falling short of the degree of reciprocity prevailing as between Royal Naval and Dominion Naval Officers, would none the less satisfy the considerations above referred to. He would suggest that provision might be made in a general Admiralty Order that, whenever ships of the Royal Indian Navy are in company with ships of the Royal Navy or when officers of the Royal Indian Navy are serving for training or otherwise in Royal Navy ships, Royal Indian Navy officers shall have powers of command customary to Royal Naval officers of the same rank and seniority, except on such occasions and for such purposes as the Commander-in-Chief of the Station concerned may direct. Orders of this kind would correspond with the position in the Army in India where Indian Commissioned Officers have powers of command in relation to British Army personnel with such restrictions as the officers commanding the station, brigade, command, etc. concerned may direct.
9   In the event of the Lords Commissioners being able to accept this proposal, the Secretary of State would be glad to learn whether any objection would be seen to the grant of commissions in His Majesty's Fleet to Royal Indian Navy Officers.
10   In view of the necessity of including provision in regard to the form of commissions in the Governor-General's letter patent, I am to request the favour of an early reply.

[311] *Minute by Head of Military Branch of Admiralty Secretariate [under the Deputy Chief of Naval Staff, William M. James]*

30 September 1936

The letter from the High Commissioner shows that the Canadian Government have mixed up our two proposals for the sale of *Crescent* and *Cygnet*.

The original proposals were that Canada should pay £564,000 for the two destroyers, which included approximately £98,000 worth of stores etc. The present proposal is that Canada should take the two ships as they lie for £400,000.

Canada has now officially accepted this offer, but (possibly by a clerical error) they have added £98,000 to the price.

It is suggested that a semi-official signal be made now to the Naval Headquarters Ottawa, explaining that we intended the £400,000 to include the stores and suggesting that if they have made an error a telegram should be sent to the High Commissioner in London asking him to delete the words 'and approximately £98,000 for their necessary armament and sea stores'.

Although this error may seem to present us with a windfall, it is suggested that having got a ruling out of the Treasury based on a certain scale of depreciation we may well want to use it again. For instance the Treasury are unlikely to sanction the free loan of the second two destroyers that Canada want to have later, but might allow them to be sold on the depreciation scale which would give them a value of about £80,000 each in three years. If we now save the Canadian Government the £98,000 which they have offered in error, there may well be a greater likelihood of arriving at a satisfactory arrangement for the second two ships in three years time.

[312] *A.E. Gloyn Cox, Principal Clerk, Admiralty Military Branch, to D.J. Wardley, Treasury, draft*
October 1936

'Secret and Immediate'

With further reference to your letter of September 4th (S.40048) to Le Maitre [A.S. – , Principal Clerk], I write to say that the Canadian Government have accepted our offer of the *Crescent* and *Cygnet* in their present condition, for £400,000. Arrangements for the transfer are now being made.

In accepting these two ships the Canadian Government also asked whether they might have our remaining two 'C' Class destroyers, the *Crusader* and *Comet*, on loan as soon as they were able to man them. This, we gather, is likely to be in about 2 or 3 years' time.

We realise the objections to lending warships built by the U.K. to the Dominions, and we feel that as a first step we should ask Canada to buy the *Crusader* and *Comet*, the price being calculated on the basis set out in Le Maitre's letter of September 1st for the *Crescent* and *Cygnet*. If Canada took the vessels over about the middle of 1939, when they will be 7 years old, the price for the two vessels would be about £217,000 on the assumption that no material change takes place in the meantime in shipbuilding costs. It would, of course, be more or less than this according as Canada took them earlier or later than this date.

We therefore propose in the first instance to offer to sell the ships to Canada on this basis, and we should be very glad to have your very early concurrence in this proposal.

Should the Canadian Government reply, however, that they are unable to buy the ships the Board of Admiralty strongly recommend that we should transfer the vessels on loan. The Board was very gratified at Canada's acceptance of the *Crescent* and *Cygnet* especially as it is not long since the Canadian Government was inclined to do away with their Navy altogether. We are therefore very anxious to encourage this fresh sign of Canada's desire to strengthen and develop her naval forces.

The effect on Navy Votes of transferring the vessels on loan would be the saving of the cost of two destroyers in maintenance reserve, plus the cost of a major refit to the two vessels (£15,000 each) which would be about due, in any case, at the time of the transfer and would, if the vessels were transferred, be borne by the Canadian Government instead of us. Moreover, what is perhaps more important than the financial aspect, Canada's interest in her own defence would be strengthened and encouraged and Imperial Defence as a whole would be the stronger by having two destroyers in running commission instead of two in reserve.

May we therefore have further sanction to offer the two vessels to Canada on loan, should the Canadian Government be unwilling to buy them.

I am sending a copy of this letter to Wiseman at the Dominions Office, in case he has any remarks to offer.

[313] *Wardley to Gloyn Cox*

29 October 1936

'Secret'

Thank you for your letter of October 23rd with the information that the Canadian Government have accepted your offer of the *Crescent* and *Cygnet* in their present condition, for £400,000.

As regards the remaining two 'C' Class destroyers which Canada wish also to acquire, we should have no objection to your agreeing to sell them outright two or three years hence at a price calculated on the basis set out in Le Maitre's letter of September 1st. But we do not feel disposed to agree to the alternative of transferring these vessels on loan. Is it necessary at this stage to ride at this fence?

[314] *S.H. Phillips [under William James] to Brown*
9 November 1936

'Secret and Immediate'

The Naval Commander in Chief, East Indies in a letter dated the 7th June last states that 'it is understood that it has been officially stated that, if Congress is sitting, no Royal Indian Sloop will be sent out of Indian waters until Congress has been informed of the proposed movement'.

We have not yet replied to this letter, and we had had the intention to leave it over until we received from you officially the Government of India's reply to your despatch 912/36 of the 23rd March last, a copy of which was forwarded to Murray by Stewart for information.

The Commander in Chief is however pressing for a reply before he visits Bombay early in December and I think we ought to endeavour to give him an assurance that nothing will be done which will cause any delay in giving effect to the agreed arrangements for the employment of the Royal Indian Navy in war.

You will remember that the convention of 1932 ... defined the functions and duties of the Commander in Chief and the Flag Officer Commanding in peace and war, and included a provision to the effect that the personnel and vessels of the Royal Indian Navy would be placed at the disposal of the Commander in Chief in accordance with an agreed plan.

As I understand it, this convention still remains in force and the current discussions with the Government of India relate only to ways and means of carrying out its provisions (or alternatively a method of formal loan to the Admiralty) under present day conditions. I should be glad to have your assurance that this accords with India Office views, so that we may let the Commander-in-Chief know.

[315] *Brown to S.H. Phillips*
received 16 November 1936

Your secret letter dated 9th November, No. M.03558/36.

2   The actual text of the statement made in the Indian Legislature to which the Naval Commander-in-Chief evidently refers is as follows 'With the approval of the Secretary of State I am therefore authorised to announce that it is the intention to consult the Indian Legislature, so far as may be possible, whenever any ques-

tion arises of lending the Indian Navy to the Admiralty for operations other than in the defence of India'. This statement was made by the Secretary to the Government of India in the Defence Department on the 27th February, 1934, and has subsequently been confirmed. The two important points to note are: (1)that the words 'so far as may be possible' do not necessarily mean that the Legislature will invariably be consulted whenever it happens to be in session, as the Commander-in-Chief's version implies, but that consultation will depend on other factors such as the urgency or the secrecy of the occasion; and (2)that the statement deliberately does not refer to occasions when the Royal Indian Navy is placed under the command of the Naval Commander-in-Chief for purposes connected with the defence of India. You will remember that the distinction between placing the Royal Indian Navy under his command for the latter purpose, and a formal loan of the Royal Indian Navy to the Admiralty for general purposes, whether directly Indian or not, was discussed in our letter of the 23rd March last, No.M.912/36 to which you refer.

3   As regards your last two paragraphs, the Government of India, as you know, have put forward a plan whereby the Royal Indian Navy vessels would be earmarked for local naval defence, which we discussed at our conference on the 20th October. You undertook to let us have the considered criticism of the Admiralty, and it is quite possible that as a result the Government of India's plan may be drastically revised on technical grounds (incidentally, it would help us to get on with this problem if you could let us have a statement of your views fairly soon). But in any case we must, I think, assume that the present arrangements between the Government of India and the Naval Commander-in-Chief hold good unless and until some other plan is definitely approved which would supersede them; though, as you will have seen from Maundrell's [Captain, Chief of Staff to the Flag Officer Commanding, Royal Indian Navy] report, difficulties may arise from the existing shortage of trained personnel.

We hope that when the Naval Commander-in-Chief visits Bombay he will be able to get into touch with representatives of the Government of India and discuss the whole situation with them. We are sending a copy of the present correspondence to the Defence Department with that object in view.

Perhaps I might add that 'Congress' in India stands for a political party; not for the Legislature. [signature]

[316] *S.H. Phillips to Brown, draft*
11 December 1936

As agreed at the meeting of 20th October, the scheme for the local naval defence of Indian ports has again been examined in the Admiralty and an alternative scheme, which would have the approval of the Admiralty, is now enclosed.

Admiralty comments on Captain Maundrell's [Chief of Staff to the Flag Officer Commanding, Royal Indian Navy] scheme are as follows:

Captain Maundrell's report was, on the instructions of the Sub-Committee of the Indian Defence Committee on Local Naval Defence, drawn up on certain assumptions, the second of which was:

'The Flag Officer Commanding, Royal Indian Navy,
was to assume that the active service vessels of
the Royal Indian Navy would be at his disposal
(i.e. for local naval defence) and not, as
hitherto, be earmarked to come under the orders of
the Naval Commander-in-Chief in the event of war'.

This assumption has naturally coloured the whole report, and in the view of the Admiralty it has led to what, in their opinion, are the fundamental weaknesses of the scheme propounded.

The Royal Indian Navy has, in recent years, achieved combatant status and made tremendous strides. During this period its aim has been to be a seagoing force, capable of taking its place in war at the side of the Royal Navy and the Dominion Navies, and this aim has, in the view of the Admiralty, been the main incentive which has led to the remarkable progress achieved.

In the opinion of the Admiralty a decision to confine the force to Local Naval Defence would be a retrograde step and a fundamental mistake and could not fail to act most unfavourably on the spirit, outlook and morale of the whole Service.

Apart from these most important considerations, the Admiralty do not consider that such a decision would be practicable, nor could it be maintained in war. The vessels required for the local Naval defence of ports are minor vessels, and their duties are restricted. In the opinion of the Admiralty, the fine new Escort vessels of the R.I.N. (and particularly destroyers should these be provided in future, as suggested in the Report) would be wasted, unless they were diverted to more important duties, nor indeed would their officers and men be content to carry out a minor role.

For local defence at the most important Naval operational bases, the Admiralty use specially constructed vessels, i.e. minesweepers (until recently known as 'Sloop Minesweepers'), but for all other ports they rely on vessels taken up on emergency and fitted with minesweeping equipment (the equipment being provided in peace). At the commencement of a war, only vessels already in commission are available, and it is for this reason that escort vessels are fitted for minesweeping and those serving in peace throughout the world are earmarked for particular minesweeping duties (in the way that the R.I.N. sloops have been earmarked for minesweeping duties at certain ports in Indian Waters). It is the intention, however, that, with a few exceptions, such vessels shall be relieved as soon as possible by the vessels taken up after the outbreak of war.

In Home Waters, the type of vessel available for taking up in numbers is the trawler, and for most of our requirements in these waters, we rely on these vessels. In ports abroad, the most suitable vessels available are earmarked.

The Admiralty see no reason why Indian Local Naval Defence should require more expensive vessels than those which will be provided for the ports of the United Kingdom and other ports of the Empire. If local Naval Defence for all such ports were suggested on the scale put forward in the report for India, it would be an impossibly high commitment . . .

Outline Scheme for the Local
Naval Defence of India,
M.04635/36

Object of Local Naval Defence. The object of Local Naval Defence is to provide protection for shipping in, and in the approaches to, ports.

This is achieved by M/S and A/S vessels and by seaward defences.

M/S and A/S Vessels In the opinion of the Admiralty, escort vessels and minesweepers of the types included in the Royal Navy are, at ports other than Naval Bases actually in the area of operations, unnecessarily large and costly for local defence purposes and involve the employment of larger numbers of trained personnel than can be justified by the services involved. A less expensive type of vessel is desirable for M/S and A/S duties at these ports and, without constructing special vessels, the trawler type (which exists in considerable numbers for commercial purposes) is the best readily available.

India's requirements in local defence vessels have been assessed as 48. These have been apportioned as follows:

Bombay	10
Calcutta	14
Karachi	5
Madras	6
Cochin	5
Vizagapatam	4
Chittagong	4
	48

Of these, 19 suitable vessels and 11 unsuitable vessels have been earmarked locally. The deficiency is, therefore, 18, or 29 if the unsuitable vessels are excluded.

It is recommended that the Indian Government of India should aim at acquiring by purchase a sufficient number of trawlers to make good the deficiency in local defence vessels: the majority of them should be placed in reserve at the ports which they will serve in war time and the remainder maintained for training purposes . . .

[317] *Minute by Captain T.S.V. Phillips, Admiralty Director of Plans*

10 February 1937

The Commander-in-Chief wants to get the arrangements for placing the Indian Navy at the disposal of the Admiralty more cut and dried than they are at present, and he is prepared, according to paragraph 3 of his letter, to accept that consultation with the Indian Legislature might be necessary for this purpose.

The Admiralty, on the other hand, in their letter of the 18th November, 1936 (M.03558/36) were prepared to accept the present arrangements which do not necessarily mean consulting the Indian Legislature, but are subject to the qualification that the R.I.N. is used 'for the purposes of the Government of India alone'. The Admiralty view was that these words were not likely in practice to be unduly restrictive of the employment of the R.I.N. vessels in war. It is considered that, on the whole, the Admiralty view is a wiser one to take. In seeking for theoretical perfection, as the

Commander-in-Chief suggests, it seems quite possible that difficulties might be encountered with the Indian Legislature, and it is considered far better to accept the present situation where we can in all probability do all we want without consulting the Indian Legislature rather than to raise the matter afresh.

In this connection it must be remembered that a similar problem arises in connection with all the Dominion Navies, and in their case we have no theoretical watertight assurance that the Navies will be placed at the Admiralty disposal, nor in fact that the Dominions will join in a war at all.

The position with regard to the Indian Navy is, therefore, even in the worst way of looking at it, better than that of the Dominion Navies. This is another reason for leaving well alone.

[318] *Memorandum by Royle to First Lord*
17 February 1937

The Government of India have for some time past been pressing the Admiralty that the Royal Indian Navy (which has recently been raised to combatant status and placed under Naval discipline,) should assume a relation to the Royal Navy more nearly approximating that held by the Dominion Navies.

The King's Regulations lays down that Officers of the three Dominion Naval Forces (Australia, New Zealand and Canada) shall rank and command with R.N. Officers of same seniority and rank.

Officers of the Royal Indian Navy feel that their exclusion from the above regulation reflects on their status and efficiency.

The line taken by the Admiralty up to the present has been to avoid a decision on this rather delicate subject – at any rate during peacetime. In wartime it was recognised that unified command would be necessary. (At present the senior Indian Officer in the R.I.N. is a Sub-Lieutenant.)

The Secretary of State for India has now raised the whole question again in his letter marked 'z'. He realises the practical objections to Indian Officers sitting on Courts Martial and exercising summary jurisdiction on British Naval personnel, and considers that the Government of India would have no reasonable objection to withholding such powers from Indian Naval Officers, and he suggests that the Admiralty should issue the following order:

Whenever ships of the Royal Indian Navy are in
company of ships of the Royal Navy, or when
Officers of the Royal Indian Navy are serving for
training or otherwise in Royal Navy ships, the
Royal Indian Naval Officers shall have powers of
command customary to Royal Naval Officers of the
same rank and seniority, except on such occasions
and for such purposes as the Commander-in-Chief of
the Station concerned may direct.

The above order corresponds with the position in the Army in
India where Indian Commissioned Officers have power of
command in relation to British Army personnel, with such restrictions as the Officer Commanding the Station may direct.

Board opinion (15h & 25h) on the above suggestion is as
follows:

1   An Officer of the Royal Indian Navy when serving for training
on board a R.N. ship is not put in a position where it is necessary
for him to exercise any command, so that question is not of
importance as it is unlikely to arise. There is, therefore, no objection to it being left in.

2   As regards the other point, i.e. when ships of the Royal Indian
Navy are in company with ships of the Royal Navy, the Board
agree with the proposal of the Secretary of State, but suggest that
before final approval the Commander-in-Chief, East Indies, should
be given the opportunity to offer his remarks.

Your approval to the above is requested at A.

[319] *'Review of Imperial Defence by the Chiefs of Staff Sub-
Committee', Chatfield, Marshal of the Royal Air Force Sir
Edward L. Ellington, [Chief of Air Staff], and Field Marshal
Sir Cyril J. Deverell [Chief of Imperial General Staff]*
22–25 February 1937

88   The rearmament of the forces of the United Kingdom has
been initiated on a large scale primarily as an insurance against
the dangers inherent in the present policy of Germany, and to
enable the British Navy to provide an adequate fleet to meet
liabilities in the Far East, while leaving in home waters a fleet
materially superior to that of Germany. Possible directions in
which the Dominions might co-operate with the United Kingdom,

in various contingencies, are summarised in the Annex to this report.

In contrast to what is possible under the autocratic regimentation of industry in dictator States, rearmament under conditions which do not dislocate our peace-time economic system must be a comparative slow process, showing little output at first. Once effective output starts, it will gather volume rapidly, and the provision of equipment should cease to limit the efficiency and expansion of our forces.

In the meantime, our forces inevitably remain inadequate to meet the full weight of the liabilities which they may have to face. 89   The task of rearmament is an immense one and the risks during the period of rearmament are great, not only because we have not achieved the strength at which we aim, but because the period of rearmament is a period of some disorganisation and dislocation. The danger that British security and interests may be challenged is already a real one and is becoming increasingly apparent as Germany's military strength increases.

In these circumstances, the danger of war is in inverse proportion to the strength which we are believed to represent. There are signs that the world has been deeply impressed by the steps already taken to strengthen the defence forces and defences of the Commonwealth, and we believe that the greatest service in the cause of peace which the Members of the British Commonwealth of Nations can give at the present time is to demonstrate to the world a determination to rearm and a readiness to defend their vital interests.

[320] *101st meeting Chiefs of Staff Sub-Committee, Committee of Imperial Defence*

16 March 1937

SIR ERNE CHATFIELD [First Sea Lord] said he was not quite clear to what extent the new strategical idea implied in the increase of the New Zealand air forces, to the exclusion of the two cruisers which they had up to now maintained, was related to the advice which had been sent out in the telegram of the 5th August.

MR NASH [W. Nash, Minister of Finance, &c, New Zealand] explained that the New Zealand Government were naturally limited in the first instance by the total amount of money which was available. They had come to the conclusion that the maintenance of naval forces was probably not the best way of defending their shores. The Leander and Achilles, they thought, could do little to defend New Zealand territory, especially if raiding forces of any size came down into their waters. Moreover, the two ships might be elsewhere at the time.

SIR ERNE CHATFIELD said that the two cruisers would not leave New Zealand waters until the main danger of the Japanese fleet in the Pacific had been cleared away by the arrival of the British Main Fleet. Not till then would Leander and Achilles join the cruiser forces of the Main Fleet.

During the period before relief of Singapore, which would probably be about two months, Australia and New Zealand would be liable to raids, but raids by light forces only and not by capital ships. The Australian and New Zealand cruisers, amounting to 2 – 8″ cruisers and 4 or 5 6″ cruisers would form a fairly strong force which would be in all probability quite sufficient to deal with anything which the Japanese could spare, bearing in mind the ultimate necessity for them of having to meet our Main Fleet. As long as these cruisers were in New Zealand waters the Japanese could do little more than cause inconvenience to New Zealand. If, however, there were no naval forces in New Zealand waters, very small and weak Japanese forces could do a large amount of damage.

MR NASH said that the New Zealand Government thought that the air would be a better protection against such raiders than would cruisers.

SIR ERNE CHATFIELD said that the Chiefs of Staff did not concur with that view. He quoted Paragraph 24 of the telegram of the 5th August:

'If any enemy could secure indefinitely its communications to New Zealand no local defence measures which New Zealand could take could give permanent protection against aggression. The security of New Zealand against the greatest danger which would threaten her depends therefore upon the ability of the British Naval Forces to command communications in the Pacific.'

It was of vital importance that we should have command of the Pacific communications, and New Zealand's contribution in the form of two cruisers are most valuable in this respect. The twenty-four aircraft would provide purely local protection and would not help in obtaining this general superiority in the Pacific. The two cruisers, besides providing local defence, would be of Imperial value in achieving this superiority.

MR NASH pointed out that the twenty-four aircraft could be used to reinforce Singapore, which was only 36 hours away by air. Surely this would be a contribution to the larger Commonwealth interest. New Zealand was a small country at a great distance from the United Kingdom but she could not isolate herself from the general interests of the Commonwealth. Nevertheless she wished first to concentrate on her own local defence.

SIR ERNE CHATFIELD enquired as to the local defence interests New Zealand had in mind – whether defence against raiders or invasion or the severance of her communications.

MR NASH said that his government felt that raiders were the most pressing danger. Communications were of less importance. New Zealand was more or less self-contained so far as foodstuffs were concerned and could hold out for a long period even if her communications were cut.

[321] *Minute by J.A. Simpson to General Wilson, and Rear-Admiral Herbert FitzHerbert RN [Commander-in-Chief, Royal Indian Navy]*

16 April 1937

The common ground between the Government of India and this Office, and, to a large extent, the Admiralty, is that there is a reasonable case for requiring the Government of India to fulfill only two of the three obligations specified above, (2) being expressed in broader terms as the maintenance of a sea-going fleet to be placed at the disposal of the Admiralty on the outbreak of war. The fundamental weakness of the Government of India's case, which it seems impossible to make clear to them, is that they persist in offering to fulfill only one of the obligations. The reason for this is that owing to the total absence of reserve personnel the sloops of the Royal Indian Navy must be used for Local Naval Defence during the many years that must elapse before the alternative scheme can be brought fully into operation, and that therefore no guarantee can meanwhile be given to the Admiralty that the sloops will be available to them on the outbreak of hostilities.

[322] *Admiralty Board Memorandum on the New Standard of*
Naval Strength

26 April 1937

Summary of Conclusions and Recommendations

1   No approval has yet been given for the adoption of the new standard of naval strength recommended by the Defence Requirements Sub-Committee in 1935. But the building programmes, necessary to reach such a standard, approximate to the actual programmes for 1936 and 1937.

2   With the rise of modern navies in Europe our present standard of naval strength would not only be insufficient for the defence of our position in the Far East if we were already at war in Europe, but it would preclude at any time the despatch of a sufficient fleet to the Far East to deter Japan from aggression. The Admiralty, therefore, attach the greatest importance to the adoption by His Majesty's Government of the proposed new standard of strength.

3   It has been assumed that increased forces will be maintained by the Dominions, but this of course is not a matter for the Admiralty to decide. It is recommended that this aspect of the question should be explored at the forthcoming Imperial Conference.

4   An early decision on our future standard of naval strength is also necessary for reasons of administration and planning.

5   To implement the proposed new standard of strength, the Naval Staff recommend on purely strategical grounds, that the fleet should include:

20   Capital Ships
15   Aircraft Carriers
100 cruisers (of which 10 might be maintained by the Dominions)
22   Destroyer Flotillas (of which 3 might be maintained by the Dominions)
82   submarines.

6   Subject to certain reservations, the 'stabilised' cost of such a Navy (if run on lines similar to the present day Navy) is estimated at about £104 millions per annum.

7   By effecting certain economies and reductions in Fleet strength the Admiralty hope to reduce the cost to about £97,000,000 per annum.

8   The basic assumption on which the strength of the Navy has been calculated amounts in practice to readiness for simultaneous

war with Germany and Japan. It takes no account of the possibility of Italy being also hostile.

9 If an agreement could be reached with Japan, leading to improved relations, further reductions could be accepted, bringing down the cost to perhaps £94,000,000 per annum.

[323] *MacDonald to Canada, Commonwealth of Australia, and New Zealand, draft*

26 May 1937

I have the honour to state that the Lords Commissioners of the Admiralty are desirous of extending to other Naval Stations the following system of recruiting for the Royal Navy, which is at present in force on the Africa Station.

2 The Admiralty propose that an appropriate Officer under the Commander-in-Chief of the Naval Station concerned should be charged with the duties of Recruiting Officer for the Station and that the Commander-in-Chief for the Station should be authorised to enter recruits at his discretion up to a maximum annual number. The recruiting of men under the scheme would be subject to the normal recruiting regulations, and candidates could be given no undertaking that they would be drafted to any particular station. They would, on final entry, normally be accommodated on board, and brought to England in His Majesty's Ships.

3 If the scheme is favourably received the Admiralty propose that the several Commanders-in-Chief, as responsible for recruiting within the compass of their Stations, should establish a direct contact with the appropriate local authorities with a view to making general and local arrangements; and subsequently communicate directly with them from time to time, as necessary. The authorities concerned would for their part be able to assist materially by disseminating recruiting literature, advice and information of all sorts, with which they would be provided.

4 Artificers are excluded from the categories of recruits owing to difficulties attending their entry outside the United Kingdom. There is, however, at present a certain shortage of Artificers in the Royal Navy, and the Lords Commissioners do not in the least wish to discourage Dominion candidates from applying for entry. Special arrangements therefore would, when possible, be made for the examination and entry of particular applicants for entry into the Artificer branches.

(Canada only)   5   I should be glad to learn whether His Majesty's Government in Canada would be prepared to concur and co-operate in such a scheme. In that event it is proposed that in the case of the America and West Indies Station the maximum annual number of recruits under the scheme should for the time being be 60 Special Service Seamen, Stokers and Royal Marines and 40 Continuous Service Boys.

(Australia only)   His Majesty's Government in the Commonwealth of Australia would be willing to recruit men on similar lines for the Royal Navy from the Australian Station. In that event a further communication would be sent as to the details of the proposed scheme.

(New Zealand only)   Your Excellency's Ministers would be prepared to concur and co-operate in such a scheme. In that event it is proposed that in the case of the New Zealand Station the maximum annual number of recruits under the scheme should for the time being be 10 Special Service Seamen, Stokers and Royal Marines, and 5 Continuous Service Boys.

(all)   6   As the Admiralty are anxious to lose no time in putting this scheme into operation as early reply would be appreciated.

[324] *Committee of Imperial Defence, Chiefs of Staff Sub-Committee: Questions Relating to General Defence Problems in the Far East, submitted by the Australian Delegation, signed by Chatfield, Ellington, and Deverell*

4 June 1937

Part I

Question

9   The Commonwealth Government asks for a clear definition of the strategical object of the Empire forces in a war with Japan or with Japan and another first-class Power.

Answer

10   In the following paragraphs we outline the policy which should govern the conduct of a war with Japan:

(a)   When we are at peace in Europe, but must retain in Home waters a Fleet sufficient to neutralise the German Fleet.

(b)   When we are already engaged in war with Germany before Japanese aggression occurs.

When we are at peace in Europe, but must retain in Home waters a fleet sufficient to neutralise the German Fleet.

The basis of our strategy lies in the establishment of our Fleet at Singapore at the earliest moment after the outbreak of hostilities. Our own security and our ability to bring pressure to bear upon Japan are equally dependent on this action. It is, therefore, of paramount importance that the Singapore base should be available on the arrival of our Fleet. To ensure this, reinforcements must be moved to Singapore and the security of their routes safeguarded upon any state of tension developing.

In addition, all measures necessary for the security of our defended ports and our air and other routes to the Far East must be put into operation.

Before the arrival of our fleet at Singapore we should, where necessary, rely on evasion for the security of our commerce, and upon local defences for the security of our ports and territories, coupled with operations against the raiders by the forces immediately available. Once our main fleet is in the Far East we should rely on its presence to give adequate security to our communications.

Although our situation at Hong Kong is inherently weak, we should, as long as our policy is to hold the fortress, take any opportunity to reinforce the garrison during a period of strained relations, but not at the expense of reinforcements considered necessary for the security of Singapore.

Although we cannot foretell what the situation will be when the British Fleet arrives at Singapore, it is clear that, so long as Hong Kong is still holding out, the Fleet will be required to go forward either to reinforce or to evacuate the garrison. The decision as to which course of action is to be adopted must be made at the time in the light of all information then available.

The attitude of China will be an important factor in this situation, and we must from the outset take all possible steps to encourage the Chinese in active hostility to Japan. At the same time, we must so restrict our action in China as to avoid any liability of becoming involved in major land and air operations.

We must primarily rely upon the exercise of economic pressure to enable us to defeat Japan. With our Fleet based at Singapore or at any base further north the restrictions which we could impose on Japanese trade give good prospects of breaking Japanese powers of resistance in the course of two years provided that at the same time we can sever her trans-Pacific trade. To this end, in addition to our naval operations, we should do all that is possible to enlist the sympathy and assistance of America.

A successful fleet action is our only apparent means of forcing an early decision. Not only would this give us complete control of sea communications, but the psychological effect on Japan might well be such as to force her to make terms. The decision to seek or decline a fleet action must rest with Japan. Our policy, therefore, should be to seek a fleet action under adequately favourable conditions and, as a corollary, to avoid exposing the fleet to any serious risk of attrition.

While this war will make very full demands on all our naval resources, and may continue for two or three years, the policy which we recommend does not visualise the employment of army or air forces on a national scale.

When we are already engaged in war with Germany before Japanese aggression occurs

Our policy for a war against Japan would remain unchanged except in the following respects:

During the period immediately following the outbreak of war with Germany and before the outbreak of war with Japan, we should immediately avail ourselves of the opportunity of carrying out preparatory measures in the Far East without interference by Japan.

At Singapore we should, during this period, not only put into effect the existing reinforcement plans, but should also increase the reserves of supplies to the maximum extent possible.

We should concentrate our outlying battalions from North China at Hong Kong, but should send there no further reinforcements except those necessary to enable the existing fixed defences to be manned adequately. We should also increase the reserves of supplies at Hong Kong. Although we cannot, in this case, rely on our fleet being able either to reinforce or to evacuate our garrison at Hong Kong once war has broken out with Japan, our policy would still be to hold Hong Kong as an outpost for as long as possible.

Once war with Japan has broken out, our policy must be governed by the consideration that, until the issue with Germany has been settled, we cannot count on being able to support anything more than a defensive policy in the Far East. Economic pressure would remain the essential feature of this policy, but owing to the heavier demands on our naval forces its action is bound to be slower.

Effect upon our Policy of Intervention of the U.S.S.R. or Italy

It seems unlikely that Soviet intervention on our side would introduce any serious change in our policy for war in the Far East.

The intervention of Italy against us would at once impose conflicting demands on our fleet.

In this situation our policy must be governed by the principle that no anxieties or risks connected with our interests in the Mediterranean can be allowed to interfere with the despatch of a fleet to the Far East . . .

14 In a Far Eastern war even the Commonwealth of Australia, the Dominion nearest to Japan, is still so far distant that no Japanese Government could face the responsibility of committing a large expeditionary force for service in Australia unless the command of the sea line of communication was assured for a sufficient period to enable the object to be achieved.

In the opinion of the General and Air Staffs an invader must, in addition, be confident of being able on arrival, to operate air forces adequate to ensure air superiority during the landing and subsequently to protect the expedition and its reinforcements and supplies on arrival at their destination, against action by the air forces of Australia. It is not considered possible to establish such air superiority with ship-borne aircraft against adequate land-based aircraft in so large a country as Australia.

15 A Japanese overseas expedition aimed at Australia may consequently be said to be a highly improbable undertaking so long as our position at Singapore is secure, and the fleets of the British Empire are maintained at such a strength as to enable a force capable of containing the Japanese fleet to be despatched to the Far East, should the occasion arise. We propose, therefore, to examine the question of the strength of the fleet that we can send to the Far East, and the time required for its arrival in that area, in the event of war with Japan.

16 In order to relate our examination to the practical political

conditions of the world as they exist at present, we propose to consider war against Japan in the following cases:

(a)   When we are at peace with Germany, but must retain in home waters a fleet sufficient to neutralise the German fleet.
(b)   If war with Japan breaks out when we, allied with France, are already at war with Germany.

We will then briefly outline the possible effects of intervention by the U.S.S.R. and Italy.

When we are at peace with Germany, but must retain in home waters a fleet sufficient to neutralise the German Fleet

Strength of Fleet for Far East

17   The strength of the fleet that we could send to the Far East must be governed by consideration of our home requirements. In the new standard of naval strength we would aim at:

(i)   Placing a fleet in the Far East fully adequate to act on the defensive, and to serve as a strong deterrent against any threat to our interests in that part of the globe.
(ii)   Maintaining in all circumstances in home waters a force able to meet the requirements of a war with Germany at the same time.
NOTE – Included in (i) and (ii) would be the forces necessary in all parts of the world behind the cover of the main fleets, to protect our territories and merchant ships against sporadic attacks.

18   When considering what naval strength we must retain in home waters as a counter to Germany, we might count on France as an ally in the event of war actually breaking out. In their Annual Review for 1935 the Chiefs of Staff stated: 'Although His Majesty's Government in the United Kingdom would never, we presume, confide the entire protection of this country (i.e., the United Kingdom) and its vital sea communications to a foreign navy in the absence of our main fleet, yet if France were our ally, her naval forces could undertake part of this responsibility. A British capital ship, cruiser and destroyer strength in home waters equal to that of Germany is probably the least we could accept.' On this basis we review below the situation as regards our heavy ship strength vis-à-vis Germany and Japan.
19   The following table shows a provisional allocation of heavy

ships between the Home and Far Eastern Fleets during the next three years to meet the increasing Germany Navy and the nine modernised heavy ships of Japan. The withdrawal for modernising of three of our heavy ships introduces an element of weakness in the situation, particularly from the spring of 1938 to the summer of 1939, but even during this period we can regard our Far Eastern Fleet as at least equivalent in fighting value to the Japanese Fleet.

Heavy ship strength available to meet the German Fleet and nine modernised Japanese heavy ships

Summer 1937 to Spring 1938	Spring 1938 to Summer 1939	Summer 1939 to Spring 1940
German Fleet/		
3 Deutschlands	3 Deutschlands	3 Deutschlands
	2 Scharnhorsts	2 Scharnhorsts
		2–35,000 ton battleships*
Our Fleet at Home/		
Hood	*Hood, Repulse,*	*Hood, Repulse,*
Repulse	*Malay, Barham*	*Renown,*
		2 Nelsons
Our Far Eastern Fleet/		
2 Nelsons	2 Nelsons,	3 Warspites,
Warspite (fully	*Warspite, Royal*	Malaya, Royal,
modernised)	*Oak,*	Oak, Barham,
Malaya, Royal	4 Revenges	4 Revenges
Oak (partly		
modernised)		
Barham, 4	Total:	Total:
Revenges		
(unmodernised)		
Total: 10 ships	8 ships	10 ships
Out of action (modernising)/		
Renown, Valiant,	*Renown, Valiant,*	
Queen Elizabeth	*Queen Elizabeth*	

* 1 at the end of 1939; 1 in the spring of 1940.

Although it will be seen from the above table that our battleship strength will be at a low ebb in 1939 compared to Germany and Japan, at no time up to 1940 will we be unable to send a fleet to the Far East; and with the completion in 1940 of five new battle-ships (already laid down) for this country the period of weakness will be past, provided we retain all our existing ships and the crews to man them.

20   As regards classes of ships other than heavy ships, the numbers available will vary according to the year in which war breaks out. Taking conditions in 1939, we might retain at home two Aircraft Carriers (excluding 1 for training and 1 refitting), six 8-inch and twelve 6-inch cruisers, and four destroyer flotillas. This would leave available for the Far East and trade protection four aircraft carriers, seven 8-inch and forty-eight 6-inch cruisers and five destroyer flotillas. When the demands for cruisers to work with the fleet have been met, there will be a shortage of cruisers for the protection of our own trade and action against that of Japan. In these circumstances, we shall have to rely largely on the seventy-four armed merchant cruisers to be taken up on the outbreak of hostilities. It appears most improbable that we shall be able to spare any aircraft carriers for operations on the trade routes.

Passage of Fleet to Far East

21   As regards the time of arrival of the fleet in the Far East, for the actual passage from home waters and the Mediterranean, the combined Home and Mediterranean Fleets would require 28 days to reach Singapore via the Suez Canal and Malacca Straits. If the Sunda Straits are used, an extra 2 days must be added. If the fleet has to use the Cape of Good Hope route, the passage time would be 45 days. Apart from passage time, however, an allowance of 10 days must be made for preliminary preparations, and 15 days for possible fueling delays and weather conditions en route. Plans for the necessary fueling arrangements on passage have been made by the Admiralty. Hence, the maximum time that must be allowed for the arrival of the fleet at Singapore is 70 days.

If war with Japan breaks out when we, allied with France, are already at war with Germany

Our Naval Situation

22   Our naval situation will be largely governed by the time at which war with Japan breaks out, and the action which the German

naval forces have taken up to that time. In the best case, war with Japan might not occur until our naval resources had been fully developed, while German naval forces had acted entirely on the defensive, being held back in the Baltic and North Sea. In the worst case, war with Japan might occur before our naval resources had been fully developed, while German naval forces had assumed a vigorous offensive against our trade in the Atlantic, and possibly elsewhere. The time at which we can send a fleet to the Far East, and its strength, will depend upon the conditions which have actually arisen. Again, as shown in paragraphs 19 and 20 above, the year in which war breaks out will also determine the allocation of naval forces between the European and Far Eastern theatres. We are thus faced with many varying factors, and can only endeavor to make some general deductions.

23   To take a probable situation in the European theatre, our naval forces may be operating at strength in the Atlantic, considerably dispersed. French battle cruisers may be assisting in these operations, while a proportion of our heavy ships may be assisting with French African convoys. If, in these circumstances, we have to deal with Japan, a very considerable period may elapse before the progress of our operations against Germany and the redistribution of our forces permit of a fleet arriving in the Far East.

24   Thus, the strength of the fleet for the Far East, and the time within which it would reach Singapore, must be variable factors, dependent both upon naval and political considerations. Nevertheless, the basis of our strategy will lie in establishing at Singapore, at the earliest possible moment after the outbreak of hostilities with Japan, a fleet whose strength, as a minimum, will enable it to act on the defensive and to serve as a strong deterrent against any threats to our interests in the Far East.

The problem of Invasion from the Japanese point of view

25   It may be of value now to examine the problem of invasion as it might appear to Japan. Although in the absence of the British main fleet from the Far East Japan can obtain control of sea communications, she must take into account the possible presence and operations of our China, East Indies, African, Australian and New Zealand naval forces, amounting to seven 8-inch and nine 6-inch cruisers, one aircraft carrier, fourteen destroyers and fifteen submarines (June 1937). So long as these forces, or a proportion thereof, are in being, Japan will have to take adequate naval

measures for the security of the passage to Australia of her expeditionary force, and for its maintenance and reinforcement. This landing to be successful, would necessitate the defeat not only of the Australian army, having in mind the great power of defence in modern warfare, but also the air forces of Australia, which by 1939–40 will amount to a total of twelve squadrons R.A.A.F.

26   To estimate the probable size of the initial expeditionary force sent to Australia from Japan, its point or points of landing, the scale of its reinforcements, and the subsequent military operations in Australia, would demand an extensive appreciation, which might best be prepared by the Australian General Staff and Air Staff. It is noted that the Australian General Staff consider that the Japanese political object might be achieved most readily by the capture of Sydney and/or Melbourne, possibly preceded or accompanied by hostile air attack on those cities. The capture of cities, however, presents a problem of considerable military difficulty, and would not, in our opinion, be of itself necessarily decisive.

27   Without making a detailed appreciation it appears reasonable to assume that Japan would not undertake the initial act of invasion with a land force of less than two divisions as an absolute minimum.

This would necessitate about 400,000 tons of shipping or from 40 to 70 ships. From Japan to Sydney is a distance of 4,400 miles. From the Pelew Islands, where the force might be assembled, to Sydney is a distance of about 3,000 miles.

In view of our remarks in paragraph 25 above, Japan, despite her initial superiority in the Far East, would be faced with a considerable naval commitment in escorting, maintaining and reinforcing such an expedition.

In view of our remarks in paragraph 14 above, it is also clear that Japan would be faced with a considerable air problem; and if she relied on ship-borne aircraft she could not hope for air superiority.

28   Assuming the successful landing of the expedition in the vicinity of Sydney or Melbourne, Japan would have to establish herself in sufficient depth to build up her forces for her subsequent operations. The length of the communications between Japan and Australia, and the necessity for their protection, militates against the rapid reinforcement of her initial expedition. It is difficult to see how Japan could build up in Australia an adequate force

to embark on operations on the necessary scale in any period under six months. In effect, Japan would require to be certain of ample time if she hoped, by means of invasion, to force Australia to sue for peace.

29   Japan, however, can never be certain of being allowed time. She cannot prevent the passage of a British fleet to the Far East, nor rely on circumstance being or remaining such as to prevent its despatch. With a British fleet established at Singapore, and operating therefrom against the Japanese lines of communication with Australia, the whole position of the Japanese invading forces in Australia would become precarious, unless Japan could decisively defeat the British fleet. It is in the highest degree unlikely that Japan would engage upon operations for whose success she would have ultimately to rely on a decisive victory in a fleet action fought under conditions not of her own choosing and possibly disadvantageous to her.

30   In effect, if Japan wishes to ensure time for the invasion of Australia, she must endeavor to prevent the British fleet form operating in the Far East; one means to this end would be for Japan to capture Singapore. The defence of Singapore has been fully examined, and the main conclusion drawn was that deliberate operations held out the best hope of success for Japan. These, however, require time, and though the possibility of such operations could not be definitely excluded, it appeared unlikely that the Japanese would undertake them even if a 60 to 70 days' delay in the arrival of the British main fleet could be anticipated.

This conclusion was based on the fact that Japan can never exclude the possibility of having to fight a fleet action in the Singapore area in support of their deliberate operations against Singapore; and, as in the case of the invasion of Australia, Japan would be in the highest degree unlikely to commit her forces to an operation for whose ultimate success she might have to rely on a decisive victory in a fleet action fought under conditions, in this instance, certainly disadvantageous to her.

Even if Japan captured Singapore, she could not absolutely rely on preventing operations in the Far East by the British Fleet, despite the great difficulties with which it would be faced in such circumstances.

31   To sum up, apart from the magnitude of the task involved in any effort to overcome the military and air opposition to her landing and subsequent operations in Australia, Japan, from the strategical point of view, would be unjustified in committing large

military and air forces to such an operation unless the British Fleet had first been defeated and largely destroyed.

Intervention of the U.S.S.R.

32   The intervention of the U.S.S.R. against Japan would involve Japanese land and air forces in major operations in Manchukuo and the countries round the Sea of Japan. In these circumstances the possibility of Japanese action against Singapore, or to invade Australia, would become even more remote than under the conditions already examined.

Intervention of Italy

33   The intervention of Italy against us would at once impose conflicting demands on our fleet. In this situation our policy must be governed by the principle that no anxieties or risks connected with our interests in the Mediterranean can be allowed to interfere with the despatch of a fleet to the Far East.

Conclusion

34   With the naval forces of the British Empire at the present strength, and maintained in the future at the standard laid down in paragraph 17, His Majesty's Government of Australia need not regard the danger of invasion as a real one.

Part III

SCALE OF RAIDS AND REVIEW OF PROGRESS IN FIXED DEFENCES

Sea-borne land attack

Question

35 The Commonwealth Government has asked for an interpretation of the term 'raid' as applied to defence against sea-borne land attack in Australia.

Answer

36   The essential characteristic of a sea-borne land raid is that it is carried out by forces that are self-contained and do not depend on a line of communication for maintenance during their operation.

It is characteristic of a raid that the troops executing it would be landed with a specific task; it would be hoped that the troops would be able to re-embark after completion of their task, but in many cases appreciable risk of the loss of the raiding party might be accepted.

The following are examples of specific tasks which might be undertaken by raiding parties: putting isolated coast defence guns or searchlights out of action; damaging or destroying stores of oil, cable landings or W/T stations; causing a diversion to draw defending troops away from another area which is about to be attacked; creating alarm and causing the retention of unnecessarily strong forces for home defence.

The strength of raids may vary from large forces to a boat load of armed men. Parties landed from a single cruiser or armed merchant ship would seldom exceed 200 men.

Question

37   The Commonwealth Government has also asked for an estimate of the maximum and minimum scales of sea-borne land attack to be expected against the defended ports of Sydney, Melbourne, Fremantle, Newcastle, Darwin, Brisbane, Hobart and Adelaide respectively.

Answer

38   Numerous factors may govern the scale of land attack to be expected at a given port; the chances of attack and its possible strength will be influenced by the naval situation at the moment, the strategic importance attached to the success of the particular raid, the enemy's information about the defences and many other factors.

It is therefore of little value to attempt an accurate forecast of the maximum and minimum scales of sea-borne land attack to be expected at a particular port. The Committee of Imperial Defence has found it necessary to adopt a uniform formula describing possible scales of attack and, after considering the known factors affecting the liability of a particular port to attack, to define the scale of attack upon which its defence measures should be based in accordance with this formula.

In the case of sea-borne attack, only two scales of attack are considered:

(a)   A landing in force, aiming at the capture of the defended area.

(b)   Attack by raiding parties landed from war and/or merchant vessels.

In Part II of this paper [not reproduced] we have shown that a landing in force in Australia is not to be expected. The Committee of Imperial Defence have recommended that the scale of attack referred to in (b) above should be adopted as the basis for defence measures against sea-borne land raids at all the eight ports referred to by the Commonwealth Government.

Although the defence measures against sea-borne land attack to be taken at all defended ports in Australia would thus be determined in relation to a common scale of attack, there would be other factors to take into consideration. The importance to Japan of the objective offered at a port, any special local conditions which might affect the probability of attack at a certain port and the distance of a port from Japan might all affect the importance of competing defence measures at particular ports. The Commonwealth Government would naturally allow for these factors when deciding upon the priority of work as between the various ports and details of the measures to be taken . . .

[325] *Committee of Imperial Defence, defence questions raised by the New Zealand Delegation, signed by Chatfield, Ellington, and Deverell*

4 June 1937

17   A Japanese overseas expedition aimed at capturing New Zealand may consequently be said to be a highly improbable undertaking so long as our position at Singapore is secure, and the fleets of the British Commonwealth of Nations are maintained at the standard which is outlined in paragraph 10 above . . .

The Role of the New Zealand Cruisers

Question

21   The New Zealand Government enquire what part is to be played by the New Zealand cruisers in the new war plan on the outbreak of hostilities in the Far East. (Item 7 of questionnaire.)

Answer

22   In view of the above considerations and the fact that the danger from this form of attack from raids is probably greatest in the period before the arrival of the British Fleet at Singapore, His Majesty's Government consider that the primary role of the New Zealand cruisers in the initial stage of a war against Japan will be to assist in providing local protection in New Zealand waters. Once the Fleet is established at Singapore the situation can be reviewed and any necessary change in dispositions can be effected. This does not necessarily mean that New Zealand cruisers would even then be required to leave home waters, as their reinforcement in that area for purposes of trade protection might conceivably prove to be necessary.

[326] *Minute by T.S.V. Phillips*

8 June 1937

As will be seen from the attached note by the India Office, the Government of India will probably be prepared to reconsider their attitude towards naval defence on the following lines:

(i)   Recognise that the provision of local naval defence is a domestic responsibility, and take early action to provide it.

(ii)   Replace two old vessels of the R.I.N. with modern escort vessels.

(iii)   Come to a definite understanding that the R.I.N. ships would be placed at the disposal of the Admiralty in war.

2   These proposals are a considerable advance on India's attitude hitherto, but (ii) and (iii) are made conditional on the annulment of the annual cash contribution totaling about £130,000 which the Government of India make to His Majesty's Government towards naval defence. It is further suggested that the eventual outcome of this policy might be the permanent remission of the cash contribution in exchange for an undertaking by India to provide her own local naval defence and to maintain a seagoing force of

not less than 6 modern escort vessels, which would be placed unreservedly at the disposal of the Admiralty in war.

3   It is submitted that a policy on these lines would be much to the Admiralty's interest, but the prospects of its adoption will turn largely on the Treasury's attitude towards the proposed remission of the cash contribution. A meeting on the subject is to be held at the Treasury on Thursday, 10th June . . .

[327] *Minute by J.H. James for Head of Naval Branch [under Dunbar-Nasmith]*

11 June 1937

There would of course be no objection at all to Newfoundland being included in the general Imperial recruiting scheme, and as it is framed it will enable her to be included without any difficulty.

However, the Dominions Office have already been informed in our official letter which explained the scheme, the each C-in-C will be allowed to recruit up to certain maximum number on the station under his command, and the Dominions Office I think might be put right over their assumption that Newfoundland could be given a special quota. Any recruits derived from that Dominion would form part of the quota allotted to the C-in-C, America & West Indies, and I suggest that there is no reason to raise the quota for this command, which is already generous, because of the inclusion of Newfoundland.

There is at present little Naval need for Dominion recruits; Newfoundlanders particularly are reported to be dull, whereas our only present requirement is for men intelligent enough to be specialists in one line or another. The scheme now under consideration, apart from its obvious imperial merit, is a machinery for recruiting, whose production can be expanded if and when the need arises. But owing to many difficulties canvassed in attached papers, and to its expense owing to the lack of naval facilities for bringing recruits to England, the numbers produced by it must at present be restricted. Consequently, I doubt whether many men would be recruited from Newfoundland, since ships seldom call there.

This is after all, a Dominions Office baby and if they are not prepared to find any money for its upkeep I don't see why the Admiralty should.

[328] *Zetland to Sir John Simon, Chancellor of the Exchequer*
28 June 1937

In view of the findings of past Imperial Conferences the Government of India realise that the provision of adequate Local Naval Defence is their primary responsibility, and they are now prepared to fulfill it at their own expense. But they cannot do this and at the same time continue the cash contribution and provide an up-to-date fleet of sloops. My contention is, and I believe the Admiralty are generally prepared to agree with me, that India has now reached the stage, which certain of the Dominions reached in the past, at which she can make a more useful contribution to the Naval Defence of the Empire by the provision of a sea-going fleet than by a payment in cash. It should be remembered that if India did not maintain sea-going sloops the Admiralty would have to build and maintain a corresponding number at the expense of the British tax-payer; and the cost of maintenance would be appreciably higher because the British seaman cost a good deal more than the Indian.

My proposal is therefore that His Majesty's Government should remit both the direct cash contribution of £100,000 and the various payments amounting to between £20,000 per annum and £30,000 per annum made by India on account of the Royal Navy sloops in the Persian Gulf. In return the Government of India will keep in being a sea-going fleet of at least six modern sloops which will be at the disposal of the Admiralty in an emergency. They will also put India's Local Naval Defence in an efficient state in consultation with the Admiralty; and in return for the remission of the Persian Gulf payments they will release the Admiralty from the obligation to keep three Royal Navy sloops in the Persian Gulf. The actual arrangements for the policing of the Persian Gulf, whether by Royal Navy or Royal Indian Navy sloops, will presumably be a matter for arrangement between the Naval Commander-in-Chief, East Indies, and the Flag Officer Commanding, Royal Indian Navy . . .

I have only one caveat to make regarding the placing of the Royal Indian Navy at the unreserved disposal of the Admiralty in an emergency. It is understood that the use made by the Admiralty of the Royal Indian Navy will in all probability be such as will enable the Government of India to state that it is being used in the defence of India in the sense, for instance, of the protection of India's sea-borne trade in the Indian Ocean. If, however, the

Admiralty wished by any chance to use the Royal Indian Navy for a purpose other than the Defence of India in this sense, it should be recalled that the Government of India are under an obligation in such a case to consult the Legislature if circumstances, i.e. urgency and secrecy, permit. Even so, the obligation to consult the Legislature is not unrestricted but only 'if circumstances permit'.

[329] *A. Duff Cooper [First Lord] to Simon*

12 July 1937

'Secret'

Looking at the matter firstly from the political point of view, we feel at the Admiralty that the system of paying a contribution is very much out of date. At one time, Australia, New Zealand and South Africa all paid annual contributions towards the cost of the Royal Navy, but these contributions have all disappeared one by one, the last to go being that of South Africa in 1922, and the principle is now accepted that the Dominions shall contribute towards Empire Naval Defence by providing for their own Local Naval Defences, and, so far as their means permit, by providing units which would form part of one empire Navy in time of war. It is true that India has not advanced so far along the road of self-government, but I think we must have regard to the fact that our declared policy takes her to the same goal, even if the way may be different.

If, therefore, we are offered now what seems to be a reasonable bargain in existing circumstances, we think that we should be well advised to consent to it, since the conditions for continuing the contribution are almost bound to become less and less favourable as time goes on . . .

There remains the question of the Escort Vessels. If it is a choice between our giving up our cash contribution or India giving up the Escort Vessels, we are in favour of the former course, for the following reasons:

Firstly, to give up the Escort Vessels would be little short of catastrophe to the Royal Indian Navy. It would most adversely affect morale, and efficiency would be bound to suffer very greatly, since the Royal Indian Navy would be reduced from a sea-going to little more than a Harbour Force. All chances of expansion would disappear, and the valuable training now given by the Escort Vessels would be lost.

Secondly, the reactions on public opinion in India would be very unsatisfactory, and the result would almost certainly be that India would take less interest in Naval affairs, while feeling against the contribution would, as suggested above, be likely to increase as time went on. We should, therefore, find it more and more difficult to persuade India to continue the contribution, and less and less chance of getting an adequate quid pro quo, if we agreed to give it up.

Thirdly, by being relieved of the necessity of always keeping three vessels in the Persian Gulf, we should hope to be able to send a vessel occasionally from there to the Red Sea, and thus strengthen Naval representation in the latter, which is very desirable under present conditions, without the necessity of having to have an additional Escort Vessel in full commission for that purpose.

Finally, the keeping up by India of these six Escort Vessels in time of peace, which will be placed at the disposal of the Admiralty in emergency, would, when our own building programme is complete, enable us to reduce the total number of vessels maintained by us (including vessels in reserve) by the same number. The cost of building these vessels is such that replacement alone would cost £91,500, and other costs, even for vessels in reserve, would amount to about £40,000 a year, a total of over £130,000 a year, or £10,000 more than the £120,000 a year which India is now paying. If, as explained in the previous paragraph, we are able to cut down by one the number of Escort Vessels we maintain in full commission, a saving on this one vessel alone, apart from the annual charge for replacement, would be about £45,000 a year.

If, therefore, we can get a watertight agreement under which we can rely upon India putting her Local Naval Defences in order, and maintaining them in an efficient condition, and also supplying and maintaining at least six modern Escort Vessels, and placing them at our disposal in time of war, we think that we should do well to accept it.

[330] *Simon to Zetland*

26 July 1937

The First Lord of the Admiralty has also written to me, stating that he is in favour of acceptance of these proposals.

After considering the matter very carefully I have reached the conclusion that your proposals provide a satisfactory basis for a settlement. But before I finally commit myself I would ask you to consider the following two points.

I do not think that it would be right for H.M. Government to agree to the permanent waiver of the fixed contribution from India unless steps were taken to satisfy ourselves that no difficulties are likely to arise later which will make it impossible for the Government of India to carry out the proposed assurances about her long-term policy in this matter. I should be glad therefore if you could arrange in such way as seems best to you for Indian opinion to be sounded regarding these assurances so that we may ascertain whether they are generally acceptable. [Marginal note: Does he mean public opinion, e.g., the Legislative Assembly? Z.]

In the second place I consider that there should be a written agreement between the Government of India and His Majesty's Government to the effect that if India failed to implement her assurances His Majesty's Government would be at liberty to reopen the question of a cash contribution. I do not think that it is necessary that this point should be brought out when you are sounding Indian opinion as suggested above but this is a matter which I leave to your discretion.

There is one further minor point which I should mention, namely the definition of the term 'defence of India'. I understand that the Admiralty anticipate no difficulty, even in an European war, in confining the use of the Indian escort vessels to the Indian Ocean, assuming that this expression includes the Persian Gulf and the Red Sea and extends as far as Singapore on the east. I hope this definition will be acceptable to you.

Provision is made for the payment of the contribution in this year's estimates, and I assume that it is contemplated that the contribution should cease as from the end of the current financial year.

[331] *Zetland to Simon*

3 August 1937

I am sure you will appreciate that the Indian Legislature has no constitutional power, either now or after Federation, to veto action proposed to be taken in the sphere of Defence; and even with reference to the pledges that have been given to the Legislature as to consultation, much depends, as I explained in my letter of the 28th June, on the connotation given to the term 'the defence of India.' Speaking for myself, I should say that your description of the probable range of employment of Indian escort vessels by the Admiralty adequately meets the point raised in the last paragraph of my letter. But I imagine that you have in mind the possibility that, apart from any question of constitutional powers or of the interpretation of pledges already given, the Government of India might conceivably be faced with political obstruction of such a kind as to embarrass them in carrying out the policy. I hope that we shall find ourselves able to give you a satisfactory assurance on the point, and I am consulting the Government of India as to how best it can be done.

[332] *Eastwood to Under-Secretary of State for India, Military Department, India Office*

18 August 1937

With reference to your letter of the 23rd September, 1936, M.6492/36, I am commanded by My Lords Commissioners of the Admiralty to acquaint you, for the information of the Secretary of State for India, that they have had under further consideration the question of the powers of command of officers of the Royal Navy and Royal Indian Navy in relation to the other force.

2   My Lords desire me to explain that the present arrangements concerning command between the Royal Navy and the Dominion Navies, which were agreed to in 1911, make a distinction between those occasions when the vessels are ordered to co-operate by mutual arrangement between the Governments concerned, and those occasions when they are otherwise in company, e.g., when visiting a port at which a vessel of the other force is lying. On the former occasions full command is exercised by the Senior Officer of the ships concerned, irrespective of which service, but on the latter occasions such command is exercised only in respect of ceremony or international intercourse, or where united action is agreed upon. These principles are embodied in Article 938(a), King's Regulations and Admiralty Instructions, a copy of which is enclosed.

3   My Lords are prepared to extend a similar arrangement to the Royal Indian Navy except in so far as the position when ships of the two forces are lying together in foreign ports is concerned, and to some extent, in relation to international intercourse. They accordingly propose that the Article referred to should be ampli-

fied by the addition of the following to clause 4: 'The above instructions also apply to the Royal Indian Navy subject to any special orders that may be issued'.

4   As ships of the Royal Navy and Royal Indian Navy are unlikely to meet in foreign ports, the points on which special instructions are likely to be needed will not be numerous and no difficulty is anticipated in covering them.

5   My Lords still consider it unnecessary at present to specify the power of command of individual officers, as officers of the Royal Indian Navy will, in general, only serve in ships of the Royal Navy under appointments for training only.

6   My Lords agree that the form of commission for officers of the Royal Indian Navy should be for service in 'His Majesty's Fleet'.

[333] *Telegram, Viceroy of India [the Marquis of Linlithgow] to Zetland, received 6 September*

3 September 1937

'Private and Personal'

Your secret letter dated August 3rd. Indian naval contribution. I am satisfied after consultation with those best qualified to advise me, that there is nothing to be gained by sounding present Assembly. I discussed situation, as a whole, in Council yesterday, and no difficulty was seen in giving an assurance such as that suggested in paragraph No. 2 of your letter that, so far as could be reasonably foreseen, the Government of India do not think that there is any real risk of being prevented by political obstruction from adhering to long-term policy. Nor is there any difficulty about a written agreement such as suggested in Simon's letter to you of July 26th.

[334] *Hankey to Colonel H.L. Ismay, [Deputy Secretary to the Committee of Imperial Defence]*

7 January 1938

My own very strong view is that we should continue to assume that the Dominions will come into a war in which we are engaged. If once we begin to assume the contrary, and especially if we start talking to them about it, we shall give the whole show away point by point.

[335] *Rear-Admiral Percy Nelles to Chatfield*

15 January 1938

'Secret and Personal'

Though somewhat early to predict definitely, there are signs which lead me to believe the Canadian Government will shortly accept the kind offer to sell us the *Crusader* and *Comet* complete with Asdics, armament and sea stores, for £327,000.

As the various telegrams exchanged on the subject of price quote the month of June for taking over, we in this Naval staff presume that month is as early as the transfer can be conveniently arranged, and that we may anticipate the refit will commence early in that month. From our point of view only, and subject to the passing of our Estimates, the earlier the transfer takes place the better. Provided it be convenient, we would prefer the refits of both ships take place at one port and, due to its closeness to the Admiralty and the High Commissioner, that the refitting port be Chatham.

3   Due to the complements of *St Laurent* and *Fraser* being each one third larger than that required by their predecessors, the *Champlain* and *Vancouver*, we shall have difficulty in adequately manning these ships at first, but anticipate being able to provide some two-thirds complement, which will at least permit us to steam them to Canada. This will necessitate utilising the some 100 R.C.N. ratings now serving in the Royal Navy afloat and those doing courses.

4   Another matter in which I wish to solicit your assistance is with reference to Commander H.A.C. Lane, R.N., presently acting as Director of Naval Intelligence, Ottawa. This officer completes his loan of two years in July 1938. He has done, and is doing good work, and I will shortly submit an official request that his time with us be extended by one year. I hope you will be able to grant this request.

5   We are grateful for all the assistance you continue to give us.

[336] *Hankey to Harding*

22 April 1938

I myself feel that there is a moral danger in admitting, even to ourselves, the possibility of any of the Dominions standing aside if we were fighting for our lives, and I feel even more strongly that it would have a quite deplorable effect if any of the

Dominions should come to know that we thought this possibility so real that we were actually modifying the provisions of our War Book to meet it.

[337] *S.H. Phillips to Under-Secretary of State for Dominions Affairs*

16 May 1938

'Most Secret'

I am commanded by My Lords Commissioners of the Admiralty to refer to the semi-official discussions which took place at interdepartmental meetings in the Admiralty in January and February of this year concerning the difficulties which might arise, in the event of an emergency, if Canada did not institute the precautionary stage, or the war stage, simultaneously with the United Kingdom, and I am to state that the Admiralty planning arrangements have now been reviewed in the light of this eventuality, with the result that My Lords have reluctantly reached the conclusion that it is impracticable to devise further alternative plans.

> Note on a discussion at the War Book Sub-Committee Meeting on 18th May, 1938, on the desirability of amending the War Book to meet the case where one or more of the Dominions may delay their entry into the war, or remain neutral.

1   MR DODWELL [C.M. Dodwell, Principal Clerk] stated that the Admiralty had only considered the case of Canada but they had come to the conclusion that alternative plans could not be prepared. Their reasons were that the hypotheses were too nebulous and that it would be impossible to prepare alternative plans without its coming to the knowledge of the Dominion Government. The Admiralty were satisfied that the necessary ad hoc plans to meet such a situation could be made at the time.

He gave as examples of the difficulties – (a) Victoria B.C. was essential for contraband control in a war against Japan. If Canada were neutral no other alternative port could be found. (b) Esquimalt B.C. was essential as a supply base in a Far East war. If Canada were neutral there was no alternative port. (c)The stores for equipping armed merchant cruisers for action in Canadian waters were held in Canada in peace. (d) Halifax was required as a fuelling base for cruisers, and as a convoy centre. As a less satisfactory alternative port in Newfoundland could be used. (e) The West Indies Intelligence Service, which was essential for the routing of merchant ships, relied for its reports on Consuls in Canada. If Canada were neutral these reports would not be available. It was impossible to establish an alternative Intelligence Service in Canada in peace without its coming to the knowledge of the Canadian Government.

[338] Hankey to Harding

20 May 1938

We are counting on Halifax as a fuelling base for cruisers and as a convoy centre in time of war. As a less satisfactory alternative, ports in Newfoundland could be used, but it would obviously be most undesirable that we should take any overt action now which could in any way give the Canadians cause to suspect that we have in mind that Halifax would not be available.

[339] [Admiralty Secretary] to the Secretary, HM Treasury, revised draft

May 1938

I am commanded by My Lords Commissioners of the Admiralty to acknowledge your letter of 28th February, S.40048, communicating the covering sanction of the Lords Commissioners of the Treasury for the arrangements already made in respect of Naval stores, guns and gun mountings of the Destroyers *Vancouver* and *Champlain*, and their authority for the arrangements proposed in respect of the Naval armament stores.

2   My Lords note that Their Lordships of the Treasury consider that their prior sanction should have been sought before the loan of the two vessels was extended for a period longer than that mentioned in Admiralty letter of 24th January 1928, M.03237/27. On this point I am to state that in mentioning a period of about three years, which was the time expected to elapse until the new destroyers contemplated by the Canadian Government could be completed, My Lords merely intended to give some idea of the policy involved rather than to imply a definite restriction of the period of loan, and they assumed that Their Lordships of the

Treasury would interpret the correspondence in this sense. Apart from this consideration I am to add that in 1930, towards the completion date of the two destroyers then being built for Canada, the question of the return of the *Vancouver* and *Champlain* to the Royal Navy was raised. In reply the Canadian Government stated that the matter would be considered after the results of the Imperial Conference of 1930 were known.

3   At the Conference His Majesty's Government in the United Kingdom suggested that, in the event of His Majesty's Government in Canada finding it possible to take a further share in the task of naval defence, this might more advantageously take the form of maintaining one or more cruisers (either new cruisers built to order of the Canadian Government or cruisers on loan from the Navy of the United Kingdom). Failing the provision of cruisers, it was suggested that the construction of sloops might be proceeded with. It was also suggested that the destroyers *Champlain* and *Vancouver* if not scrapped on completion of *Saguenay* and *Skeena*, should in any case be scrapped before the 31st December 1936, (see C.I.D. paper 997B of 19th June 1930). The views of the Canadian representative were summarised as follows in C.I.D. paper 1028B of 29th October 1930:

> Having regard to local defence requirements on the widely separated seaboards, and to the needs of training, it was considered more important at the present stage of development of the Royal Canadian Navy to maintain a number of small vessels rather than a few large ones, and for this reason the provision of cruisers was not under consideration at present.
>
> Two new destroyers specially designed for Canadian conditions were under construction and the replacing of the existing trawlers by sloops was being considered.
>
> There was no decision yet regarding the scrapping of the two old destroyers when the new ones were completed, but as the need for such scrapping would not arise before 1936, this matter could well stand over.

This C.I.D. paper came to the Committee of Imperial Defence at their 241st meeting on the 28th November 1930, at which also a general discussion on imperial defence took place, and was duly taken note of. The prime minister presided at this meeting, at which the Prime Minister of Canada was present, and the matter having thus been settled for the time being on a question of high policy, it would have been impossible for the Admiralty to have re-opened it.*

[340] *Memorandum 'The Co-operation of the Colonial Empire in Imperial Defence', signed by V. Sykes, Secretary, Oversea Defence Committee*

8 June 1938

Co-operation of the Colonial Dependencies

4   The Colonial Dependencies in general do not dispose of resources in man-power, munitions or wealth, comparable with those of the United Kingdom or the self-governing parts of the Empire. Nevertheless, many Colonies have recently made notable efforts and incurred very considerable expenditure, not only in increasing their own security but also in some cases in organising their local forces so as to be ready to assist in the defence of other parts of the Empire which are more open to attack by potential enemies. The Committee desire to take this opportunity of recording their high appreciation of these contributions towards Imperial Defence.

5   Every single Dependency, however, from the smallest and least well-endowed to the largest and most wealthy, has its own part to play in Imperial Defence, and the Committee in this Memorandum propose to indicate in general terms some of the ways in which the active co-operation of Colonial Dependencies is particularly desirable at the present time.

6   In a general Memorandum of this nature it is clearly impossible to give specific advice applicable to each individual territory. The Committee cannot hope to do more than give a few indications which may be helpful to Governors in reviewing the defence preparations of the territories committed to their charge. The relative importance of one form or another of defence must be determined with regard to local conditions in every case, and the Committee are ready at any time to give their advice on particular questions referred to them for their opinion.

7   From the defence point of view, the Colonial Dependencies may be broadly divided into two categories:

---

* This paragraph was amended in draft, and is much less apologetic than is the original draft, let alone a draft of September 1934 belatedly requesting Treasury sanction for the continued loan of *Vancouver* and *Champlain* to Canada. [Ed.]

(a)   Those which contain Imperial garrisons for the defence of fortified naval bases.

(b)   Those which depend for their local defence upon local forces only, whether regular or auxiliary.

8   Unless our naval bases are secure against enemy attack, the strategic mobility of the Fleet will be seriously prejudiced in war, and it might be unable to give proper protection to the sea communications of the Empire. In such circumstances, not only would our trade be liable to interruption, but we should also lose the freedom, which we have always enjoyed in the past, to redistribute our land forces to meet any contingencies which might arise. Even the mobility of air forces depends on the freedom of sea communications, since air forces cannot be kept supplied for operations on a large scale unless their fuel and maintenance stores can be brought to their bases of operation by sea. Large sums of money are now being expended by His Majesty's Government on the completion and modernisation of the defences of the various ports overseas. These efforts on the part of the United Kingdom require to be supplemented by the active co-operation of the Colonies in which the ports are situated, in such matters as the maintenance of local forces to supplement regular garrisons, and the civil measures necessary to ensure the security of the fortresses.

9   In the Dependencies in which no Imperial garrisons are stationed, two considerations arise. In the first place, as has been frequently stated in previous Memoranda of the committee, the defence forces which are under the immediate control of His Majesty's Government in the United Kingdom must be employed in time of emergency in the defence of those areas which are of vital importance to the whole Empire from the strategical point of view. The local defence of other territories of lesser importance must, therefore, in the first instance at least, be the concern of the local Government. At the present time the forces of which the United Kingdom Government disposes are barely sufficient to meet the numerous threats which confront the Empire in all quarters of the world. In these circumstances it is more than ever necessary that Colonial Dependencies should put forth their greatest efforts to ensure their own local security. In this way they will reduce their potential demands upon the Imperial Forces, and thus make an indirect but very definite contribution towards the security of the Empire as a whole.

10   In the case of those Dependencies which are liable to attack by an enemy on a large scale, with a view to capture and use as advanced bases, it might be argued that any attempt at resistance by small local military forces unsupported by artillery and modern weapons would be of no avail and would merely result in a useless loss of life; and that in these circumstances their best course would be to 'go to ground' and trust to the ultimate victory of British arms to free their territory from the invader. As a general proposition, the Committee are quite unable to subscribe to this point of view, although there may be special cases in which His Majesty's Government might decide that resistance to attack would not be justified. Moreover, the value of small forces resolutely handled, particularly when operating in undeveloped territory, should not be underestimated. There have been many examples in history of such forces causing a diversion of effort on the part of the enemy altogether disproportionate to the size of the defending forces. It may suffice to recall the drain on our own resources in the last War caused by the gallant resistance of the small and entirely unsupported German forces in East Africa. It may well be, therefore, that the resistance of small local forces would have a most important effect upon the course of the whole war. The enemy would probably have to employ much larger forces to establish and protect an advanced base than would be necessary if no resistance at all were offered. Moreover, it need not be assumed that these small local forces would necessarily be left entirely unsupported. In certain circumstances the air arm provides a means of rapidly and powerfully reinforcing small centres of resistance . . .

Naval Forces

15   At every defended port local naval forces have to be maintained for minesweeping and other local defence duties. For these purposes Naval Volunteer Reserve units are particularly valuable, since they enable a corresponding economy to be made in regular naval personnel, who can be more profitably employed in other duties requiring a higher standard of technical training.

16   The uses of local Naval Volunteer Reserve units are by no means confined to defended ports. Their presence is equally desirable in ports of purely commercial significance which may attract the attention of enemy commerce raiders. Mining in the approaches to a port is a simple operation for an enemy raider, particularly when the port is undefended, and if effectively carried

out it would paralyse the trade of the port until the channels had been swept clear. It is unlikely, particularly in the early stages of a war, that the Royal Navy would be able to spare minesweepers for commercial ports which, though important from a trade aspect, have not particular naval significance; and, even if such vessels could be spared, they are slow and might take a long time to arrive. In these circumstances, the value of local minesweeping units, available on the spot, needs no emphasis.

17   Craft suitable for adaptation to minesweeping purposes are usually available in commercial ports, and the minesweeping gear itself is comparatively inexpensive. It is essential, however, that the equipment should be available on the outbreak of war and that the crews should have been trained beforehand in its use ...

Duties of the Royal Navy in War

28   The Royal Navy in war has a dual role to perform – that of safeguarding the freedom of our own sea communications, and of strangling the seaborne commerce of our enemies by the exercise of economic pressure. Colonial Dependencies have an important part to play in assisting the Royal Navy to carry out these tasks. Every maritime Dependency has to provide officials to act as Detaining Officers and Reporting Officers, and some would also contain branches of the Contraband Control Staff and the Naval Control Service.

Treatment of Enemy and Neutral Shipping: Duties of Detaining Officers

29   In a Colonial Dependency the Detaining Officer is generally responsible for preventing the departure from his port of enemy merchant ships or neutral shipping suspected of carrying contraband or prohibited exports. His services are therefore of considerable moment in the attack on enemy trade. Unless the appointed official is fully conversant with his duties, important information regarding such vessels may not be passed on to the right quarters during the precautionary period; and if careful arrangements have not been made beforehand to co-ordinate his activities with those of the police and local forces, such vessels may succeed in slipping through the net after the outbreak of hostilities.

Naval Reporting Officers

30   The work of Reporting Officers in war consists mainly in feeding into the Naval Intelligence system all information which they can obtain concerning the movements of enemy warships and merchantmen and of suspicious neutral shipping. By this means our own naval forces are enabled to seek out and intercept such vessels. Reporting Officers have also the duty of issuing route instructions to British merchant ships to ensure the safe continuance of their voyages and of informing Naval Commanders-in-Chief of their movements, in order that our patrolling vessels may afford them such protection as may be possible. This duty will continue throughout a war at the smaller ports, but at the larger and more important ports it will be taken over by the Naval Control Service which will be established as soon as possible after the outbreak of war.

31   The instructions for these officers are contained in O.D.C. Memorandum No. 662-M and Admiralty Confidential Book C.B. 3000 A (38) respectively. Opportunities are taken to exercise Reporting Officers in their duties during the visits of H.M. Ships to their ports, but in the more isolated Colonies these visits must necessarily be somewhat infrequent. The Committee would suggest that enthusiasm will be engendered and efficiency increased if Colonial Governors show a personal interest in the activities of the officers responsible for these duties in their territories.

Contraband Control

32   For the closer examination of merchant shipping which might be trading with the enemy and which has been intercepted on the high seas, Contraband Control Bases are established at certain focal points on the trade routes. Personnel for these bases will normally be provided under Admiralty arrangements from the United Kingdom, but before their arrival a nucleus staff of local personnel will have to be employed under arrangements to be made by the local naval authorities. The majority of the small craft required at these bases will have to be obtained locally, and in the Colonies in which Contraband Control Bases would be located, the active co-operation of the civil authorities would be of great assistance.

Naval Control Service

33   No less important would be the co-operation of the civil authorities in the work of the Naval Control Service which, as mentioned in paragraph 30, will be set up as soon as possible after the outbreak of war at all the more important ports, and which has the duty of routing our own shipping so as to avoid enemy commerce destroyers, and of organising and forming up convoys, if convoy sailing is instituted. The personnel of this service is drawn from retired naval officers, selected where possible so that their war stations are near to their places of residence in peace.

Censorship

34   Another defence measure in Colonial Dependencies which is of high importance from the naval point of view is that of censorship. Wireless messages relating to the movement of shipping would be of particular value to enemy commerce raiders if intercepted by them. Censorship, however, is of equal interest to all the Defence Services, for, unless it is effective, information regarding the movements of our armed forces may leak through to the enemy either direct or through neutral countries. Such a leakage might well prove disastrous in the case, for example, of an expedition intended to carry out a surprise on the enemy.

35   The development of communications, and particularly of wireless, will, as a general rule, necessitate the imposition of complete censorship throughout the Empire on the outbreak of a major war. This must be applied simultaneously, immediately on receipt of the appropriate telegram. For this reason it is of the utmost importance that in every Colony all details should have been planned in advance in time of peace. This involves the detailing of personnel and the instruction of at least a proportion of the staff in the duties they have to perform in war. The Committee are by no means satisfied that censorship arrangements have been worked out in every Colony in sufficient detail to ensure that censorship would be immediately effective in an emergency, and they would commend this matter to the attention of Colonial Governors.

Defence Security

36   Finally, attention must be drawn to the need for good intelligence security measures. In certain territories which contain a considerable alien element in their population it is most important

to ensure that adequate arrangements are made beforehand to deal with the possibility of hostile action by 'the enemy within the gates.' Defence against external aggression is primarily the function of the three Fighting Services, whereas defence against espionage, sabotage and internal disorder fomented by evilly-disposed inhabitants of the Colony, is primarily the responsibility of the Civil Government working through the police authorities. Security against these forms of internal attack demands a sound system of police intelligence, the maintenance of a 'black list' of suspected persons, and detailed plans for the immediate internment of such persons when an emergency arises.

37   In this connection the Committee desire also to emphasise the need for good arrangements for the custody of secret documents. These are of importance not only to the security of each Colony itself, but also to that of the whole of the Empire. Secret documents containing details of our defence arrangements must of necessity have a world-wide distribution, and if they fall into improper hands, the unfortunate consequences which might result might be equally widespread. No less dangerous is indiscreet conversation on the part of those connected with defence preparations. Such indiscretions often obtain wide publicity through the press and though they may seem to refer to matters of trifling importance, they may often provide valuable information to the well-organised Intelligence Service of a potential enemy. The Committee recommend that all concerned should be warned of the need for restricting to the minimum any discussion of the contents of secret defence documents or information received by secret means.

[341] *Harding to Floud*

14 June 1938

As you are aware, it is regarded as most important that no indication should be given to the Canadian Government that the possibility of their standing aside in an emergency is being seriously considered here.

[342] *Nelles to Chatfield*

16 June 1938

'Secret and Personal'
Now that we are about to accomplish the acquisition of H.M.

Ships *Crusader* and *Comet* we are considering next year's estimates and I propose to recommend to my Minister that we continue our expansion to the fullest extent within our means.

Due to the comparatively large expansion of R.C.N. personnel in the last two years, – which still continues – our destroyers are in effect doing the duties of training ships. This state of affairs must continue for at least the next twelve months.

After that it is, in my opinion, essential that the training of our officers afloat should be brought to a higher plane and that we should tackle the problem of tactical training. To do this efficiently we must obtain craft capable of accommodating the minimum staff of specialists at least. I therefore propose to recommend to our Minister when the time arrives, that one of our further forms of expansion be to buy from England a flotilla leader.

Before making such a recommendation it is essential for me to know that, if this proposal is made by the Canadian Government, the Board of Admiralty would be prepared to recommend such a sale.

To be specific, I should like to recommend that Canada buy H.M. Ship *Kempenfelt*. I particularly say *Kempenfelt* because it is well known here that she was the leader built for the 'C' class and it will enable us to advance this proposal as a logical sequence of our acquisition of the four 'C' class destroyers. It is our intention to base the four 'C's' on Esquimalt.

This will arrive almost at the completion of your appointment and I should like to add that I heartily appreciate the help you have given us during your time in office.

[343] *Floud to Harding*

21 June 1938

'Secret'

The Prime Minister himself [W.L. Mackenzie-King] in his desire, as he has more than once mentioned to me, to keep his eye on what he considers to be the main objective, viz. the preservation of Canada's unity, is content again and again to insist on the supremacy of Parliament as the interpreter of the people's wishes if and when the time comes. He refuses resolutely to take any other line and it is clear that however unsatisfactory this may be for those who are charged either here or at home with working out anything in the nature of Imperial defence plans, we cannot hope, under the present regime, to get any further . . .

I believe myself that there has been in recent months a marked change of view. As you know the present policy of the United Kingdom Government in regard to Europe is one with which the Prime Minister finds himself strongly in sympathy. He has gone out of his way, or almost out of his way, to make some very complimentary remarks on the way in which the present difficult situation is being handled . . . Mackenzie King associated himself with Lapointe's [Ernest Lapointe, Minister of Justice and Attorney-General] affirmation that the United Kingdom had saved and was saving the world from war . . . All I myself really fear is a period of hesitancy, and I am afraid that we cannot necessarily count on Canada being in with us from the very beginning.

[344] *Hankey to Harding*

11 July 1938

'SECRET'

I am very glad to find that both you and he [Floud] agree that there is a slight improvement in the situation. That was my own impression. The agreement to recruit and train airmen in Canada is very significant. I have also read Ashton Gwatkin's [F.T.A. Ashton-Gwatkin, Counsellor, Foreign Office] memorandum and noted the significance of the last few paragraphs.

As you know, the impression I formed at the end of 1934 when I visited Canada was that, while the situation was an anxious one, they would eventually come down on the right side. I had in mind the sort of reasons that Floud expresses so well.

If we can get some more orders placed in Canada and develop the recruiting scheme on sound lines, the effect in the long run should be in the right direction.

[345] *J.A. Simpson to the Deputy Secretary of the Admiralty*

24 September 1938

'Secret'

4 As a general guiding principle the Secretary of State has, however, provisionally agreed with the Chancellor of the Exchequer that it would be reasonable to regard vessels as being employed for purposes connected with 'the defence of India in the broad sense' whenever they are employed within the confines of the Indian Ocean, assuming that area to cover the Persian Gulf

and the Red Sea, and to extend to Singapore in the East. This point was raised by the Chancellor of the Exchequer with particular reference to the Government of India's obligation to consult the Indian Legislature, where circumstances permit, in cases in which it is proposed to employ vessels of the Royal Indian Navy for the purposes other than the defence of India in the broad sense. It appears to the Secretary of State, however, that this principle is of general application and would therefore be relevant, e.g., in deciding the incidence of cost under Section 150 of the Government of India Act, 1935. He would be glad to know as soon as possible whether the Government of India are prepared to accept it.

[346] *Captain V.H. Danckwerts, Director of Plans, to the Deputy Chief of Naval Staff [William James]*

27 September 1938

'Secret'

General Muspratt of the India Office telephoned to say a private and personal warning telegram had been sent to the Viceroy of India referring to the procedure for placing Sloops R.I.N. under the control of the C-in-C, East Indies. He had received a reply that when the request is made, the Government of India will be prepared to place the 3 Sloops at the disposal of the C-in-C, East Indies.

2   Propose to inform C-in-C, East Indies, accordingly.

[347] *Cipher telegram from Zetland to the Government of India, Defence Department*

28 September 1938

Immediate 1674 Reference War Book India Chapter 2 Section I.K. Admiralty have requested that the ships of the Royal Indian Navy should be placed under the command of the Naval Commander in Chief, East Indies. (This request supersedes that made in Secretary of state's personal and private telegram 404 of 26th September.) Do you agree? If so please take the necessary steps in direction communication with Naval Commander in Chief, East Indies, for distribution of vessels and personnel of Royal Indian Navy between needs of Local Naval Defence and specific Admiralty needs, informing this department of distribution of vessels by

name, and date from which vessels are placed under command of
Naval Commander in Chief.

[348] *Paraphrase telegram from the United Kingdom High
Commissioner in the Union of South Africa [Clark] to [the
Secretary of the Admiralty] received*
29 September 1938, 5.7. p.m.

No. 141 'Secret'
I was asked by Naval Commander in Chief to help in getting the
Defence Department to implement their commitment in regard to
the issue of notices on page 8(a) and (b) of African Station War
Organisation Book. Mr Pirow [Union of South African Minister
of Defence] objects strongly on purely political grounds to issuing
notices having regard especially to bye-election in five weeks. He
was also inclined to dispute genuine obligation but realised that
this could not carry conviction.
I thought it unnecessary in view of easing crisis, to insist and
have consulted Admiral again by telephone. The latter says he
has communicated with all reservists etc. by telegram and thinks
that they will probably come in. He is therefore prepared not to
demand issue of notices at present. Mr Pirow insists that authority
must be obtained from Hertzog if we have to demand issue.

[349] *Cipher telegram from the United Kingdom Acting High
Commissioner in Canada [-----], to [the Secretary of the
Admiralty], immediate, dated and received*
29 September 1938, 9.45 p.m.

No. 218 'Secret'
Your telegram 28th September No. 232. Royal Naval Reserve
personnel. I have taken action accordingly.
2   You should however know that I understand most privately
that the authorities here have been put in some difficulty by the
receipt by the Naval Service Headquarters at Ottawa of a signal
containing instructions that the Naval Control Service is to be
established, which would involve action on their part in respect
of Naval reservists. I believe the view taken here is that in the
absence of any specific or general request from the United
Kingdom Government to the Canadian Government concerning
action of this nature it would in any case not be proper or possible
for effect to be given to the Admiralty instructions on the basis

of those instructions. (In fact, having regard to the new developments in the general position, it is not to be expected that action by the Canadian Government for the establishment of the Naval Control Service would be taken at any rate in the immediate future).

3   Admiralty signal referred to above is, I understand, thus thought to be premature in the continued absence of a message from Government to Government, e.g. the third message contained in Appendix I to O.D.C. 664-M. I assume myself that no message in this or similar terms has been sent to the Canadian Government if only in order that the Prime Minister of Canada may be able to assure Parliament if and when it is summoned in connection with the present crisis that he has received from the Government of the United Kingdom no request for any commitment of Canada in that connection. This would be in accordance with the statement mentioned in paragraph 5 of my telegram of the 27th September No. 214 and seems to accord too with the 6th paragraph of your telegram of the 17th October 1935 No. 128. Please see also last paragraph of my telegram of 27th September No. 210.

4   I understand that Canadian authorities would, as I believe to have been the case in 1935 in connection with the Abyssinian crisis, find less difficulty in the immediate application of the Naval Control Service Scheme, so far as action on their part is required, to places outside Canadian territories only which would necessitate merely the distribution of co-ordination documents now in their charge.

5   In view of the latter part of paragraph 2 of above, you may well think it unnecessary to take any action but I felt that I should bring the foregoing points to your notice. It is of course most important that it should not be known to the Canadian Government that I have any information as to the Admiralty signal or as to the way in which their minds have been working in this respect.

[350]   *Decipher of telegram, Government of India, Defence Department, to Zetland*

30 September 1938

Immediate, 1567. Defence Department. Your telegram No. 1674 dated September 28th.

We agree to place four ships Indus, Hindustan, Clive, and Lawrence at the disposal of Naval Commander-in-Chief, East Indies with effect from September 30th. Orders have been issued accordingly to Flag Officer Commanding, Royal Indian Navy.

[351] *Cipher telegram from the United Kingdom Acting High Commissioner in Canada [-----], to [the Secretary of the Admiralty] received*
30 September 1938 1 October 1.23 a.m.

No. 222 'Secret'

Your telegram No. 234 of the 30th September I have seen Skelton [O.D. Skelton, Deputy Minister of External Affairs] and explained that Admiralty's signal in question was addressed to Canadian Naval Service Ottawa in error. I did not of course disclose that I had any previous knowledge of the point at issue. Skelton informed me that although the authorities here had expected appropriate message addressed to the Canadian Government would have preceded any such Admiralty signal addressed to Canadian Naval Service Headquarters Ottawa it had nevertheless been decided by the Prime Minister that necessary arrangements should be made since they represented only precautionary measures. He very readily accepted my apology for any inconvenience which this incident might have caused to Canadian authorities and did not ask for a written communication on the matter.

2   My most private information is that, as foreshadowed as a possibility in my telegram No. 221 of the 29th September, effective action on the above decision was this morning suspended in view of the improvement in the International situation.

[352] *Copy of telegram, Admiralty Secretary to Commander-in-Chief, East Indies [Vice-Admiral James F. Somerville], repeated to the Flag Officer Commanding, Royal Indian Navy [FitzHerbert]*
3 October 1938

Request you will arrange for H.M.I.S. *Indus, Hindustan, Clive* and *Lawrence* to be returned to control of Government of India. Report date.

[353] *Vice-Admiral Sir Sidney Meyrick [Commander-in-Chief, America and West Indies] to Admiral Sir Roger Backhouse, First Sea Lord*

Admiralty House, Bermuda
14 October 1938

I sent a personal cypher message to you after meeting the Prime Minister of Canada [W.L. Mackenzie-King] here at Government House. I had quite a long conversation with him in the course of which he said there was no doubt in his mind that the mobilisation and readiness shown by the British Navy was a very big factor in the avoidance of war.

He asked what ships I had out here and how they were situated during the crisis, which gave me the opportunity of telling him the extent to which I was relying on the assistance of the Canadian destroyers, and how important it was that all of us should be on our war stations during a critical period, and before hostilities actually broke out, in order to be in the best position for the immediate protection of trade, etc.

He fully agreed and told me that Nelles [Rear-Admiral Percy, Canadian Chief of Naval Staff] had been to see him as to action to be taken by the R.C.N. as soon as Ottawa heard that the Navy had mobilised. I gathered they were expecting proposals from London, but he did not actually say so. He ended up by saying that anyway, with regard to the future, we must see that the way is paved to proper collaboration and understanding.

As a result of this conversation I also feel that there must have been some misunderstanding about supposed difficulties in the way of the distribution of N.C.S. [Naval Control Service] books, which is hardly a hostile act; anyway I was assured by Lane [Commander H.A.C. Lane, R.N., Director of Naval Intelligence, Ottawa] last spring that, 'as far as he could say' there would be no difficulties over the distribution of these books.

I do not know if you have ever met Mackenzie King but both the Governor and I liked him very much, also his outlook. Amongst other things he told me that he hoped Canada would soon be able to put a stronger naval force into the field and agreed that the right place for Canadian ships was on the ocean and not in their home ports – which is the accepted belief of a great many Canadians that I have met, who comforted themselves with the idea that when they can see their destroyers they are being properly protected.

I feel you will agree with me it will clear the air a lot if, as a result of this last spasm, we can get all doubts removed of Canada coming into line to make her contribution to Empire defence felt. I am confident myself the spirit of the majority is there.

P.S. Since writing this letter I have spent several hours with the Prime Minister at Admiralty House when he again told me how necessary he considered it was that preparatory plans should include R.C.N., which I think the last paragraph of the enclosed letter confirms.

[354] *Admiralty memorandum, 'Brief history of South African Co-operation in Imperial Defence'*
14 October 1938

'Secret'

South Africa

The following is a brief history of South African co-operation in Imperial Defence. In 1898 the Cape of Good Hope commenced an annual contribution of £30,000. In 1902 an agreement was reached whereby South Africa contributed as follows:

Cape Colony   £50,000
Natal   £30,000

These grants continued until 1921/22.

2   After the war a policy was initiated whereby South Africa was to commence to raise and train her own naval forces, beginning on a small scale. Accordingly the trawlers *Foyle* and *Eden* were lent to South Africa and renamed the *Sonnenblom* and *Immortelle* respectively. These were returned to H.M. Government in the United Kingdom in 1935 and the South Africa naval forces abolished.

3   South Africa maintains an R.N.V.R. which is trained at Simonstown and in H.M. Ships. Ratings are accommodated in the Naval Sanatorium which has been renamed Klaver Kamp. This is a very successful force.

(2)   Papers relating to South Africa, General Naval Policy.

C.I.D. paper 154-C. The Smuts-Churchill Agreement. By this, in 1921, South Africa agreed to keep the defence of Simonstown in an efficient state for use by the fleet in war.

C.O.S. paper 468. This was prepared for the visit of Mr Pirow

[Union of South African Minister of Defence] in 1936 by the Chiefs of staff on the general policy which should guide South Africa in Naval Defence. It indicated the alteration in the strategical position created by the Italian conquest of Abyssinia, and the enhanced importance of the Cape route to the East owing to the increased threat to our communications in the Mediterranean and Red Sea.

The specific Naval proposals were:

(i) South Africa might increase the number of candidates for Dominion cadetships for the Royal Navy.

(ii) South Africa should proceed with the following local defence measures:

(a) To complete 30 sets of minesweeping outfits.
(b) To establish a local reserve of A/S material.
(c) To provide and maintain an anti-torpedo net for Capetown.

Recommendation (i) has not been acted on, although Mr Pirow is sympathetic. The position regarding (ii) is given in 'Local Naval Defences' below.

E(37)I (in E(37) series). Naval Appendix to Chiefs of Staff review for the Imperial Conference 1937.

In addition to general naval advice to Dominion Governments the following specific recommendations were made to South Africa:

(1) A Single anti-torpedo net should be provided for Capetown.
(2) A reserve of anti-submarine gear should be built up.

Recommendations were also made regarding oil fuel stowage.

For (1) and (2) above, the position is as stated in 'Local Naval Defences' below and for the last the position is as stated in 'Oil Fuel Storage' below.

(3) Paper D.C.O.S. 67.

The S.O. (1), Capetown, occupies a special position vis a vis the South African Government by virtue of the right of the Royal Navy to use Simonstown. It was proposed that he should be made a 'Defence Liaison Officer' to represent the other two Services but this met with opposition from the Army and R.A.F. Accordingly the status is to remain officially as before but, to facilitate co-operation which would involve frequent periods of absence from Capetown, the S.O.(1) will be provided with an assistant.

The Union Government contribute £1,200 a year towards the

cost of the S.O.(1) and provide offices in the Defence Headquarters, Capetown.

This matter is settled unless Mr Pirow wishes to make more use of the S.O.(1) as a Defence Liaison Officer or make any increased financial provision for two officers.

(4)   R.N.V.R.

The establishment of this force is 911 ratings and 65 officers. In paper M.04956/38, C-in-C, Africa's report on the S.A.N.S. for the year ended 30th June 1938, it was stated that the strength was 60 officers and 835 ratings. The ratings entered from Capetown, Durban, Port Elizabeth and East London in 1937-38 attended an annual camp at Klaver Camp, Simonstown, in September-October 1937. During 1937/38, 40 officers and 715 ratings were embarked for training in H.M. Ships.

In a report on the Union of South Africa in relation to Naval requirements (C-in-C, Africa, No.213/037 of 15th April 1937) on which notes were prepared in Plans Division, the commander-in-chief gave the commitments and resources as under:

	Officers	Men
Union Government commitments	249	1769
Admiralty commitments	23	114
Total commitments	272	1883
Total resources	190	1245
Deficit	82	638

Although it has not been recommended specifically that the R.N.V.R. should be increased it is for consideration whether this should now be done and a greater S.A.N.S. war reserve [be] built up.

On paper M.04956/38 it is recommended that the number of communications ratings should be increased. This would involve an increase of permanent staff instructors.

(Paper M.04956 contains the most recent recommendations of C-in-C, Africa, on personnel and manning.)

(5)   The Hydrographic Survey Section, S.A.N.S., has carried out surveys of Port Nolloth and other places and the Department of Defence has agreed that the whole area in Admiralty Chart 642 should be re-surveyed on a scale of not less than 1/40,000. This is satisfactory and no further action is required. (Paper M.04956/38).

[355] *'South Africa, Local Naval Defence', Admiralty memorandum*

[c. 1938]

The Union Government have undertaken to provide by local requisitioning 30 vessels as auxiliary minesweepers.

(M.02795/37) 2 The major part of the stores and equipment required have been provided under a four-year plan and with certain exceptions should be complete this year.

(M.04781/38) 3 Certain stores, depth charges, etc., have not yet been supplied and C-in-C, Africa, has urged their supply in the coming year.

(M.04781/38) 4 Hitherto no provision has been made for auxiliary A/S vessels. A reduction in the requirements for 22 to 12 has been proposed by C-in-C, Africa, and concurred in by Admiralty (M.02124/38). The Union Government have been asked by C-in-C. to consider provision of stores and equipment for 12 auxiliary A/S vessels.

5   Guns are available at Simonstown for 30 M/S vessels (also 2 for A/S) but these have not been taken over by the Union Government.

(M.02795/37) 6 A summary of naval commitments which would be undertaken up to the limit of available resources is given by C-in-C, Africa, in M.02795/37.

Boom Defences, Capetown

(M. 01180/37) 7 Additions to the A/T defence of Capetown Harbour, consequent on the extensions thereto, have been proposed, and recommendations regarding this and the provision of a double line of A/T, with estimates of cost, were forwarded to C-in-C, Africa, for information of the Union Government, in M.06761/37.

Details asked for by the High Commissioners for South Africa were forwarded in T.01180/37.

[356] *Notes for conversations with the Hon. O. Pirow. Oil Fuel at the Cape*

14 October 1938

(Extract) At the 1921 Imperial Conference, the Admiralty stated that they considered a reserve of 24,000 tons was necessary in South Africa, and recommended that the Union should provide and maintain this stock.

This reserve of 24,000 tons has been provided by the Union of South Africa in two 12,000 ton tanks at Simonstown.

It is the Admiralty policy now to arrange that a further 72,000 tons of oil fuel from commercial stocks at Cape Town would be available, to meet the requirements of the Fleet, in the event of it being required in an emergency to proceed to the Far East via the Cape.

With this additional amount making a total of 96,000 tons available in an emergency, the Admiralty consider that the situation in South Africa as regards naval oil reserves would be satisfactory . . .

In point of fact the storage at Cape Town had, in 1937, already been commenced but it was desired by Mr Pirow that the general impression in South Africa should be that this was for commercial purposes.

The outcome of negotiations with the government of South Africa and Mr Pirow was, that the Admiralty in the absence of any practical alternative and on the advice of the Dominions Office, decided to adopt the Union Government's suggestion that the tankage at Capetown should be provided through the medium of an oil company, the site for the installation being leased by the company concerned from the Union Government at a nominal rent of £1 per acre per annum. Accordingly arrangements were made with the Consolidated Petroleum Company (a holding of the Anglo-Iranian and Shell Company's) for the construction of the tanks and necessary pipelines at Yserplasts, about 5 miles from Cape Town.

This tankage is now almost completed but there is some difficulty over the method of filling them. If the impression that these tanks are for commercial use is to be preserved it is necessary to use commercial oilers to fill them and this method is actually being employed. This involves payment of extra port duties. During the emergency it was decided to complete the filling of these tanks at once. Commercial tankers were obtained on an exchange basis, and by November, 1938, this installation will be nearly half filled.

Remarks

It is suggested that Mr Pirow be asked to agree that the necessity for secrecy is no longer imperative and that it should be openly admitted to be an Admiralty installation.

Should he still consider secrecy is advisable, it is not desired to press the point.

[357] *South Africa, Coast Defences. Brief History of South African Co-Operation in Imperial Defence*
14 October 1938

C.I.D. Papers

309-C – Recommendations for Coast Defences.
332-C – J.D.C. comments on dispatch from South Africa giving decision to postpone work on defences.
419-C – Considerations of Mr Pirow's proposals for defence of Capetown.
455-C – C.O.S. Memorandum on the Monitor question.

The present coast defences of South Africa consist of:

Capetown	$2 \times 9.2''$ (15°)
	$2 \times 6''$ (15°)
	$2 \times 4.7°$
Simonstown	4×9.2 (15°)
	$6 \times 6''$ (15°)
Durban	$2 \times 6''$ (15°)

In January, 1928, the C.I.D. recommended to the Union Government the following defences (C.I.D.309-C):

Capetown	$2 \times 9.2''$ (35°)
	$2 \times 6''$ (15°)
Simonstown	4×9.2 (35°)
	$6 \times 6''$ (15°)
Durban	2×9.2 (35°)
	$2 \times 6''$ (15°)
Saldanha Bay	$2 \times 6''$ after the outbreak of war.

2   The latest proposals produced in South Africa as an alternative to the 1928 C.I.D. scheme are as follows:

Capetown	$2 \times 15''$ (35°)
	$2 \times 9.2''$
Duiker Point	$2 \times 6''$ (15°)
Simonstown	4×9.2 (35°)
	$6 \times 6''$ (15°)

                    2 × 6 pdr. twin A/M.T.B.
                    2 Concrete platforms for 2 × 6″ Mk. XIX
Durban   MONITOR 2 × 6″
Saldanha Bay   2 × 6″
East London   2 concrete platforms for 2 × 6″ Mk.XIX
Port Elizabeth   2 concrete platforms for 2 × 6″ Mk.XIX
Achilles   Alternative position for MONITOR
                    1 or 2 × 9.2″ guns ex Capetown
                    2 × 6″ on Arrol-Withers platforms.

3   As a result of the 1928 recommendations South Africa ordered two 35° mountings for the 9.2″ guns at Simonstown. These were delivered early in 1937, but after a great deal of talk, culminating with a definite assurance that work would commence on their installation in June, 1936, nothing has been done on the plea that the estimated cost of work is now given as £158,000 instead of the original estimate of £78,000 and Pirow could not justify this in Parliament unless he could show some quid pro quo.

4   In April, 1935, the Union Government proposed that instead of carrying out the 1928 recommendations, Capetown should be defended on a higher scale. They proposed that 15″ guns should be installed on Robben Island and that part of the expense should be borne by the U.K. Government. If these proposals were not acceptable to the U.K. Government, then the Union Government were prepared to proceed with the modernisation of the defences as suggested by the C.I.D.

5   The Union Government had reported that the 9.2″ battery at Capetown could not be fired owing to the great damage it would do to the surrounding buildings.

6   The C.I.D. framed alternative proposals, accepting the inability to fire the Capetown 9.2″ guns from their position at Lions Head.

7   Proposal (i). To provide complete protection against bombardment by capital ships by the installation of two 15″ batteries, one on Robben Island and the other at a site South of Capetown. This would, of course, involve very large expenditure.

    Proposal (ii). To provide partial protection against capital ships by siting a 15″ battery on Robben Island and a 9.2″ battery about Duiker Point, covering the area out of reach of the Robben Island battery. This would be a somewhat unsatisfactory solution, since a bombarding capital ship could take up its position in the area not covered by the 15″ battery, where the 9.2″ battery would be comparatively ineffective against it.

Proposal (iii). To provide complete protection against 8″ gun cruisers, by installing 9.2″ guns at the same sites as proposed in (i) for the 15″ guns.

Proposal (iv). To provide reasonably good protection against 8' gun cruisers by siting a 9.2″ (35°) battery on Robben Island, another 9.2″ (35°) battery with an all round arc of fire in the Simonstown area, and a 6″ (45°) battery at Duiker Point. This would have the advantage of providing protection for Simonstown in addition to Capetown.

Proposal (v). To provide partial protection against 8″ cruisers by siting a 9.2″ (35°) battery on Robben Island, and a 6″ (45°) battery near Duiker Point. This solution would again be illogical like that in (ii), since an enemy cruiser would be unlikely to move into the area covered by the 9.2″ guns, and would remain where it had only to face 6″ artillery.

8   The C.I.D. stated, through the C-in-C, that the 15″ guns were an over-insurance and not recommended.

9   These proposals were considered by Mr Pirow but he was adamant regarding the installation of 15″ guns for the defence of Capetown. He said that the Union Government were now prepared to bear the whole cost both of providing and of installing one 15″ gun and also the cost of installing the second, if the gun itself with the mounting could be lent by the Imperial Government.

10   The J.D.C.'s [Joint Defence Committee] reactions to this were that, while adhering to their technical opinion that the 15″ gun project was an over-insurance and therefore unnecessarily expensive, but having regard to the high importance from the Imperial point of view of ensuring that the opportunity for co-operation in a scheme of Imperial defence was not lost, it was considered desirable on broad political grounds to reach some comprehensive agreement on the matter, notwithstanding that it involved departure from the general principles of responsibility for Imperial Defence.

11   The visit of Mr Pirow to London in May, 1936, resulted, when the whole matter was discussed at length.

12   At this time the question of the loan of a monitor, in default of the provision of 15″ shore guns, was broached. The result of Mr Pirow's visit and the subsequent history is clearly set out in the Chiefs of Staffs' memorandum in C.I.D. paper 455-C.

13   It will be seen that Mr Pirow was given the choice of the *Erebus* or the *Marshal Soult*. He chose the *Erebus*. The conditions

of loan were all set out in a memorandum sent to Mr Pirow by the First Lord on 30th June, 1936 (copy attached).

14   Owing to an incorrect range being quoted in the memorandum (33,400 yards instead of 29,000 yards) certain difficulties arose, but these were overcome by the Admiralty obtaining approval from the Treasury for the replacement of the existing guns in the *Erebus* by new guns from stock and the provision of the latest type of ammunition without any additional charge to the Union.

15   A number of details concerning the *Erebus* were given by the Admiralty to Colonel de Waal, a Staff Officer who had been sent over from South Africa by Mr Pirow. Copies of some of these letters are attached.

16   Many technical difficulties arose over the monitor scheme, e.g. manning, mooring, fighting and docking, and no decision was arrived at by the Union Government.

17   Mr Pirow made a suggestion that a breakwater should be built off the East coast of Robben Island for the permanent berthing of the monitor and that in addition the *Terror* should be loaned on the threat of emergency arising, to be moored in Hout Bay.

18   The Admiralty remarked, however, that they were not in a position to release *Terror* for this purpose as she was required elsewhere.

19   In a telegram from C-in-C, Africa, dated 7th October, 1936, the C-in-C refers to a memorandum outlining the new artillery policy . . .

[358] *Minute by C.B. Coxwell, Admiralty Assistant Secretary*
                                                              17 October 1938

Any difficulties which arose in regard to the Dominions in the recent crisis arose, I think, wholly from the fact that the Fleet and the Naval Reserves were mobilised without the prior issue of a warning telegram initiating a precautionary period.

2   Had the warning telegram been sent, the Dominions Office would have sent to the Dominions the prepared telegram about the calling out of the Reserves. Actually, the latter telegram was not sent, because the Dominions Office took the view that it could not be sent until the beginning of a precautionary period. The telegrams which the Dominions Office did send to the Dominions about mobilisation were not sent until 10 p.m. on that day. The W. List messages 'Mobilise' and 'Mobilise Naval

Reserves', on the other hand, had been sent soon after 11 a.m. and had been repeated in accordance with the approved W. List procedure to Melbourne, Wellington and Ottawa. The order 'Establish Naval Control Service Staffs' had been sent in the afternoon and this also had been repeated to the Dominion Naval authorities above mentioned.

3   I entirely agree with the view expressed by the Dominions Office that the standing arrangements with regard to such matters as the calling up of the Naval Reserves and the establishment of the Naval Control Service should be reviewed in the light of the experience of the recent crisis.

4   It may be necessary, as a result of such review, either (1) to delete the Dominion Naval Authorities from the list of those who receive W. List messages, or (2) so to amend the W. List as to make it clear that W. List messages sent to the Dominions are sent purely for information . . .

7   The only really satisfactory course, however, is to ascertain at the earliest possible stage of an emergency – and whether a precautionary stage has been formally instituted or not – whether the Dominion Governments are in fact prepared to co-operate in the precautionary measures laid down in the War Book for execution during that period. How precisely this is to be ascertained in any future emergency is a matter for consideration. The warning telegram, in the form in which it goes to the Dominions, automatically produces, if it is sent, the information required: it contains a direct request for co-operation, and if this resulted in a refusal to carry out pre-arranged plans, we should be able to devise alternative arrangements. The practical problem now before us is the discovery of a means of ascertaining the intentions of the Dominions in a case such as that which occurred recently – one in which the Warning telegram is not in fact issued . . .

[359] *S. L. Holmes [Principal Officer, Dominions Office] to Commander (E) H. Dixon [attached to Second Sea Lord's Office, Little]*

18 October 1938

'Secret'

I think that we thus should be clear in our own mind that full credit is due to the Canadian Government for the fact that they were prepared to take all the action under the [Naval Control] scheme for which they had accepted responsibility, and this despite the fact that there was some doubt in their own mind as to whether they had been approached in the proper way. I have tried to find out for your information in what other direction the Canadian government is entitled to credit. I am afraid the list of points is extremely meager but our information is no doubt not exhaustive.

Some Royal Canadian Air Force activities were apparently noticeable to the newspapers, and clearly this amounted at least to the reinforcement of Halifax from Ottawa. I believe that some other activity took place at Halifax by way of seeing that the coast defences were in order and possibly this extended also to arrangements for the reception of cruisers. In this latter connection I should record that I understood most privately that a message was received by the Canadian naval authorities from the Commander-in-chief indicating that in an emergency he would base two cruisers at Halifax. This gave me and my informant some anxiety, since it seemed to presuppose the automatic alignment of Canada with the United Kingdom and, whatever the precise theoretical obligations of Canada to us as regards the availability of Halifax for Royal Navy purposes, the sudden arrival of two cruisers, at a time when Canada had not arrived at a formal decision to co-operate with us, might have proved very embarrassing to the Government.

What is perhaps more important than anything else is the information which reached me from two sources that in a great hurry the Prime Minister had authorised the expenditure of rather more than $6,000,000 for defence purposes on the basis of a Governor-General's warrant. How this will come out in the wash I do not know, but it was clearly a very significant action and it fits in well with the impression which I gathered from several quarters that Mr Mackenzie King himself had clearly been perturbed at the very incomplete character of the Canadian precautionary arrangements designed to meet an emergency of this sort. The practical effect so far as we were concerned was illustrated by the sudden interest which the Department of External Affairs took in certain defence matters on which progress had been negligible for some time; a case in point is the Prize Court procedure and the provision of the necessary documents (in this connection see my telegram No. 215 of the 28th September). Both Chiefs-of-Staff and the Senior Air Officer have assured me privately that the emergency has shaken up the Department of National Defence itself and still more the Department of External Affairs, and the Chief of the General Staff told me that he had been able, though very much at the last moment, to assure his Minister that every nut and bolt was in place, so far as they existed at all.

[360] *Unattributed Cabinet Paper*

November 1938

'Secret'

At the beginning of this year, just before Sir Francis Floud came on leave, he sent us a letter in which he expressed serious doubt whether, in time of emergency, the Canadian Government would, in fact, take any action to put into force the 'precautionary stage' (i.e. the arrangements necessary in the period of strained relations immediately preceding an outbreak of war), or even the 'war stage', simultaneously with the United Kingdom Government. He subsequently explained in conversation here that he did not think that there was a serious risk of Canada formally declaring neutrality in a war in which the United Kingdom was engaged (which would, of course, raise very serious questions going to the foundation of the British Commonwealth relationship), but he did think that there was a definite possibility that, while Canada would eventually be prepared to participate in War Book measures, the Canadian Government would take no action for a period which might, perhaps, amount to a month or six weeks . . .

In fact, experience during the crisis of September 1938 indicated that, if there had in fact been an outbreak of war, Canada at any rate would not have adopted an attitude of extreme 'neutrality'. From information received from the United Kingdom High Commissioner in Canada it appears that the Canadian Government were preparing for certain measures of co-operation. (1) The Canadian Government made some preliminary arrangements not likely to attract any publicity in connection with the establishment of the Naval Control Service. These consisted of summoning to Ottawa one officer who would have been appointed to the key position under the scheme at Quebec and of preparations for the distribution of the necessary documents to ports both in Canada and in other parts of North America. (2) There was some reinforcement from Ottawa of the Canadian Air Force at Halifax, and steps were taken to see that the coast defences at Halifax were in order. (It is possible, though not certain, that steps were also taken to prepare arrangements at Halifax for reception of cruisers of the Royal Navy). (3) The Prime Minister of Canada is reported to have authorised the expenditure of rather more than $6,000,000 for defence purposes on the basis of a Governor-General's warrant. (4) The Department of External Affairs displayed a sudden interest in the question of Prize Court procedure and asked urgently for the provision of a number of documents bearing on the matter.

[361] *Vice-Admiral G.H. D'O. Lyon, Commander-in-Chief, Africa Station, to the Secretary of the Admiralty, 'Report of Proceedings on Africa Station during Emergency Period', number 819/0128*

3 November 1938

'Secret'

Be pleased to lay before Their Lordships the following brief narrative of events on the Africa Station during the recent emergency . . .

7 Tuesday 27th September   The High Commissioner for the United Kingdom, who was at Pretoria, together with Union Government Officials, was informed that mobilisation of reserves would probably be ordered next day and he was asked to confirm that calling up notices would be issued by Union Government as laid down in Africa Station War Orders. This question was the subject of further conversations, and is being dealt with separately . . .

8 Wednesday 28 September   Mobilisation warning orders received and communicated to the High Commissioner, Pretoria, and R.N.O. Freetown. The latter was given information also of retired officers in West Coast Colonies who would be ordered to report to him.

9   His Excellency the Governor General left Capetown for Pretoria P.M. on 28th September and I had an interview with him before leaving to discuss the question of the issue of proclamations for mobilisation. This is dealt with in a separate submission. I discussed with His Excellency the question of South Africa participating should war be declared and pointed out to him that the four Escort vessels of the Africa Squadron were already on their way to their war stations, which were out of the Africa Station, and that I was expecting to move off at any moment with the cruisers. During my absence, British shipping, which carried 90% of the commerce of South Africa, would be at the mercy of enemy raiders and A.M.C.'s whose work of destruction would be

facilitated by Union neutrality. His Excellency replied that he expected his government knew this, but possibly did not realise that ships of the Africa Squadron would not be there to protect shipping in South Africa.

I also informed His Excellency that I was concerned at the absence of anyone at Defence Headquarters with any knowledge of naval affairs and that I considered, in the event of the Union coming into the war, that there should be a Naval Officer in close touch with the Minister of Defence who could represent to him the naval needs of the Union. I mentioned that Rear-Admiral Guy Halifax (retired) was in South Africa and that I thought he would be a suitable officer for such an appointment. Should there be any likelihood of the Union Government requiring Admiral Halifax's services, I would undertake not to call on him for any other employment but that in the meantime I would arrange for him to make himself conversant with South African Naval affairs. His Excellency agreed that such an appointment would be a necessity and that Admiral Halifax, who, as secretary to the last Governor General, was 'persona grata' with most of the Government, would be a most suitable choice. He undertook to discuss the question with the Minister of Defence.

10   At 1600 on 23th September, orders to mobilise were received, together with instructions re requisitioning of shipping and retention of time expired men. I informed the High Commissioner of the Mobilisation Orders. Calling up notices were sent to pensioners and R.F.R. class B individually by Captain-in-Charge, Simonstown . . .

Thursday 29th September   *Rochester*'s arrival at Walvis Bay was welcomed with a sigh of relief by the South African population who were considerably exercised by the possibility of trouble with Germans in South West Africa.

Fortunately or not, the impression has undoubtedly been given that, on the outbreak of war, Escort vessels will proceed to Union Ports for protection of South African interests. It is for consideration that this impression should be removed . . .

13   Two hours before I sailed in *Amphion* I received a telephone call from the Chief of the General Staff, Pretoria, Major General Sir Pierre van Ryneveldt [Peter van Rynfeld], who stated that it had been decided to carry out an exercise (he repeated 'an exercise') to mobilise the Coast Defences. All four batteries at Simonstown, the batteries at Capetown and the six-inch guns on the Bluff at Durban would be manned. Two mobile six-inch guns

were being sent to Durban and two to Port Elizabeth. The batteries and guns would be manned, 1/3 by Permanent Forces and the remainder by the Coastal Artillery Brigade.

Further, it had been decided to carry out an exercise of the local Seaward Defences of Capetown, Simonstown, Port Elizabeth and Durban. Owners of whalers and trawlers had been requested to place all their available boats at the disposal of the Government.

It was proposed to call on R.N.V.R. personnel to volunteer for this exercise and he asked me if I could call for these volunteers. I informed him that, although I was responsible for the administration and training of the R.N.V.R. I thought that it would be necessary for the Union Government to call for volunteers for an exercise of this nature. He then told me that it was intended to do the exercise thoroughly and to fit the whalers and trawlers with guns and minesweeping gear, and he asked me if the Admiralty would lend the guns for the exercise. I answered him that they would be lent.

I told him that, as I was leaving in *Amphion* in two hours' time, he should communicate with the Captain-in-Charge, Simonstown, for further details. He impressed on me several times (presumably with his tongue in his cheek) that it was purely an exercise.

The above conversation gave me the impression that, despite the refusal of the Minister of Defence to issue proclamations calling up reserves ... the Union Government had suddenly woken up to the fact that, should war come, they would be totally unprepared to defend their own coastline and that they could not rely on any support from the British Navy.

Since the emergency, however, the somewhat grandiose exercises planned have dwindled and the old 'political expediency' bogy has arisen once more. When I sailed from Simonstown for the West Coast on 19th October, the Coast Defence exercise, which consisted of manning the batteries with full complements of untrained recruits, was being carried out, but the requisitioning of whalers and trawlers had hung fire and no effective progress had been made ...

15   At 1200 the N.C.S. Staff at Capetown reported they had established themselves in accordance with instructions in Admiralty message 1538/28th September (in offices loaned by the Union Castle Shipping Company) and on an entirely unofficial basis – officers wearing plain clothes. N.C.S. Base at Durban under similar conditions was being organised.

[362] *Memorandum 'Co-operation of Dominions in Defence Action Required during Precautionary Period', forwarded by Coxwell to Admiralty Secretary*

4 November 1938

'Most Secret'

This matter was considered in some detail, as regards Canada, on the attached file M.00660/38 etc.

It is not very practicable to dissociate measures of co-operation required in the precautionary period from the steps necessarily taken in peace to ensure that such co-operation will be possible, nor from those measures which do not become effective until war has broken out, though preparatory steps are required at an earlier stage.

Attention is invited to a memorandum prepared by the Dominions Office in January, 1938, copy tabled in M.00658/38, which gives a very fair picture of the war measures which depended, under present arrangements, upon Dominion co-operation in order to be fully effective. These measures can conveniently be considered under five headings, viz:

(i)   Intelligence Arrangements.
(ii) Protection of Trade.
(iii) Attack on enemy trade.
(iv) Mobilisation of Personnel.
(v)   Miscellaneous.

(i) Intelligence Arrangements

The world-wide naval intelligence organisation includes sections in Canada, Australia, New Zealand and South Africa, each with its own centre in the Dominion, controlled and paid for by the Dominion concerned. (Note: – This is only partially true of South Africa where the Intelligence Centre at Capetown is run by the Royal Navy, though the Union Government meet a considerable part of the cost. In South Africa, however, as in the other Dominions, the Reporting Officers are local Government officials).

In the precautionary period, we look to the Dominions' sections of the Intelligence Organisation to co-operate in all intelligence matters e.g. in tracing the movements of the potential enemy's warships and merchant vessels including merchant vessels which may fit out as raiders. In addition, this organisation is responsible for trade protection duties pending the provision of special staffs (See (ii) below).

Censorship and control of aliens are other matters which can be considered as embraced by intelligence in its widest sense. Steps to institute the former, in particular, are to be taken during the precautionary period and Dominion co-operation is necessary for full efficiency.

(ii) Protection of Trade

The organisation which is to be established for trade protection purposes is known as the Naval Control Service. It includes –

(a) Routing of merchant vessels.
(b) Convoys
(c) Arming of merchant vessels.
(d) Issue of confidential books to merchant vessels.
(e) Issue of war warnings to merchant vessels at sea, and therefore necessitates the provision of
(f) N.C.S. Staffs, Commodores of Convoys (and staffs), gun's crews for D.E.M.S., provision and arming of Armed Merchant Cruisers including the provision of crews, guns and other stores, storage of books for merchant vessels and for N.C.S. and other staffs, and use of W/T Stations.

In the precautionary period we look to the Dominions to provide and appoint the staffs mentioned, to issue books as necessary, to commence the equipment of Armed Merchant Cruisers, to start routing (using intelligence staffs in the first place – see (i) above), to issue war warnings through W/T Stations in the Dominions, and generally to take all the steps necessary to ensure the safety of shipping once war has broken out.

Note – Canada, Australia and South Africa have all accepted responsibility for equipping and manning a number of Armed Merchant Cruisers on Imperial account.

While Dominions generally are responsible only for the organisation within their own boundaries, Canada is responsible for a considerable portion of the U.S.A. for N.C.S. and intelligence purposes.

(iii) Attack on Enemy Trade

The medium for this duty, so far as the Royal Navy is concerned, is the Contraband Control Service. This Service necessitates the provision of bases and patrolling vessels, including the necessary staffs.

In the precautionary period, we are looking to the Dominions concerned (Australia, South Africa and Canada) to initiate the steps for establishing Contraband Control Bases, including the provision of staffs and the provision and equipment of patrolling vessels. In addition, we expect all Dominions to prepare for the seizure of enemy merchant vessels in their ports on the outbreak of war (including the provision of prize courts), and if Days of Grace are granted, for the routing of vessels concerned to their home ports.

(iv) Mobilisation of Personnel

All the action required will (we hope) take place in the precautionary period. We expect the Dominions to assist in the following fashion:

(a) Promulgation of notices about mobilisation i.e. issue of proclamations.

(b) Disposal of officers and men who report themselves (and are not reserved for Dominion use), provision of travelling facilities, advances of money, etc.

(c) Issue of war appointments to officers in accordance with wishes of Commanders-in-Chief concerned.

(v) Miscellaneous

It is not possible to give a complete list of other measures involving Dominions co-operation, but the following are some of the more important. Most of them necessarily involve action during the precautionary period.

(a) Dispositions of Dominion warships in accordance with the general strategical plan (such as placing of certain vessels at Admiralty disposal and their despatch to war stations, e.g. Australian vessels to Singapore).

(b) Provision of protected bases available for use of Royal Navy, e.g. Port Darwin and Esquimalt, where in certain circumstances Flag Officers of the Royal Navy may be appointed in control of squadrons of H.M. Ships based on the ports.

Note: Simonstown is in rather a different category.

(c) Provision of stores for naval purposes. A special example is the supply of victualling stores to Singapore from Australia.

(d) Provision of shipping for Royal Navy purposes, e.g. it is hoped that Canadian vessels will be taken up for service at Jamaica as Armed Boarding Vessels in the Royal Navy.

No reference has been made to Eire in these notes, since the position regarding co-operation with the Government of Eire is not yet sufficiently advanced.

[363] *Record of a conversation held in the Rt. Hon. Sir Thomas Inskip's room [Deputy Chairman of the Committee of Imperial Defence and Minister for Co-Ordination of Defence] at the House of Commons*

10 November 1938, 5 p.m.

Monitor

Mr PIROW [Union of South Africa Minister of Defence] said that he would like to have the monitor for the defence of Cape Town as had been proposed. Sir Roger Backhouse had suggested that the personnel to man the vessel should be trained by the Royal Navy. It might be necessary at first for the Navy to supply the whole complement, but he wished to have as few men from the Royal Navy as possible. Reconditioning the vessel would take some time, and during this period personnel could be collected in South Africa and some perhaps could be sent to the United Kingdom for training. In effect, therefore, the Union Government might have to call upon the Admiralty to supply the full complement at first, but the numbers obtained from them would be reduced and replaced by South African personnel at the earliest possible moment.

[364] *J.A. Simpson to P.E. Marrack, [Admiralty Assistant Secretary]*

28 November 1938

Will you please refer to your letter dated 13th August, 1938, and connected correspondence regarding the training to be given to Royal Indian Navy Cadets? . . .

Under the existing system Indian cadets are relegated to the position of apprentices pure and simple. They live on their own in lodgings in the town, and usually go home for their meals in solitude. They have hardly a chance of getting to know anybody worth knowing outside their own small circle and are, in consequence, thrown too much on their own resources. If they are to play any games they have to make their opportunities by joining local clubs: there is at present no opportunity open to them of joining in with the Royal Navy cadets in any of these activities. All this is very discouraging to the boys themselves. They are immensely keen on joining the Royal Indian Navy; they have travelled over 6,000 miles to come here for their training, and when they arrive here they find that they are to be treated like artificer apprentices for five years.

At the end of this period the cadets are examined orally by an interviewing committee and if they are considered to have completed their course satisfactorily they are approved for promotion to the rank of Engineer Sub-Lieutenant.

The last occasion on which Indian Engineer Cadets were examined here was last May when Engineer Captain D.J. Hoare, Royal Navy, Assistant Engineer-in-Chief, represented the Admiralty. On that occasion we interviewed three cadets, and though the Committee were dissatisfied with the standard reached by them, they felt it was not the fault of the Cadets but of the system ...

It seems to us almost impossible to expect officers with this kind of training to pull their full weight in any Navy. As the Royal Indian Navy is to be placed at the disposal of the Admiralty in times of emergency we think that it would be in the Admiralty's interests that Royal Indian Navy Engineer Cadets should be given the best training possible for their job. We trust, therefore, that you will help us in the matter ...

[365] *Nelles to Backhouse*

28 December 1938

The Cabinet have recently examined the proposed Naval estimates for the financial year 1939-40 and, although I fear we shall not get all we have asked for, I anticipate the official request to purchase a leader will go forward soon.

There is naturally some concern regarding the question of substituting *Keith* for *Kempenfelt*, especially as I understand that, if it was decided to equip *Keith* with D.C.T. and fire control clock as for the remainder of the 'C' class, instruments might not be available until 1941. I note from C.A.F.O. 2798/38 (Destroyer Flotillas – Tentative Forecast of Dispositions) that the 'B' class are shortly to be withdrawn.

It would assist me immensely if you were able to reconsider

your previous views of *Kempenfelt* and allow us to purchase her.

The increased cost to us would be more than offset by obviating the necessity of so extensive a refit and in obtaining a ship built at the same date and for the 'C' class we should be on much stronger ground.

We shall not be able to take the Leader over until December 1939, which should, perhaps, allow time for her to complete her present duty and be refitted.

I feel I must raise this matter as there are strong influences here seeking to prevent us spending the money out of Canada. I, too, want construction work in Canada but not at the expense of making the West Coast destroyer division an efficient and balanced unit at as early a date as possible. The inherent delay of building warships in Canada will mean no material addition to our fighting forces for at least two years.

[366] *Minute by S.H. Phillips*

13 January 1939

1   At the O.D.C. meeting on the 12th December, the Dominions Office put forward a proposal that the following note should be added to Paragraph 1 of Appendix I to O.D.C. Memorandum No. 664-M:

'The "meaning" of several of the telegrams includes a final paragraph in which co-operation is invited. It is fully appreciated that, in the case of Dominions, this paragraph would raise political considerations which would be for the Dominion Governments to determine, though naturally it would be very helpful if, on the occurrence of an emergency, such measures of co-operation as were thought possible by any Dominion Government could be put in hand forthwith.'

2   This was opposed by the Admiralty representative who put forward instead the suggestion of a pre-warning telegram, so as to obtain from the Dominions a definition of their attitude at as early a stage of the emergency as possible. The Dominions Office representative replied that while he sympathised with this view, he considered that any telegram which might be interpreted by the Dominion Government as an attempt to force it to come to a decision in advance of an actual outbreak of war might have effects opposite to those desired, particularly in the case of the

Dominions whose attitude was known to be doubtful. At the best, the reply in some cases would probably be no more than that the Dominion Government would have to consult their Parliament when the occasion arose. Eventually the Overseas Defence Committee agreed to approve the addition of the note as proposed, but not to issue it in advance of other amendments likely to be evolved as a result of discussions now in progress on other matters. The Committee took note of the fact that the Admiralty were proposing to take up with the Dominions Office the method whereby the extent of Dominion co-operation in emergency could be ascertained, and the Chairman suggested that when the discussion had proceeded further it might be necessary for the matter to be considered by the Committee of Imperial Defence . . .

5   Co-operation on the matters referred to in some of these telegrams, and, of course, the war telegram, would involve the Dominions in hostilities, but there are some of them, and particularly the warning telegram, which deal with measures which are not inconsistent with an attitude of neutrality on the part of the Dominion, and which, indeed, the Dominion in its own interests might well be advised to take.

6   The addition proposed to the Appendix by the Dominions Office seems of rather doubtful expediency, as giving the Dominions a loophole for avoiding the taking of any action at all, and it is contrary to the view taken by the Board on M.00659/38 that nothing should be done which would in any way suggest to the Dominions that we were not expecting them to take corresponding action to ourselves. Sir Maurice Hankey [Secretary to the Committee of Imperial Defence], I understand, who has unrivaled knowledge of Dominion views on Defence matters, was also of this opinion. It also seems poor service to the Dominions themselves to suggest a loophole without warning them of the danger to attack to which they are exposing themselves. The general view taken by International Law experts is that if a state of war exists between the United Kingdom and a foreign country, the latter will be fully entitled to regard themselves as at war with the Dominions, and to attack either the Dominions themselves or Dominion shipping, without previous warning, whether the Dominion had or had not decided to take any active part in the war. A Dominion might, of course, receive a private assurance from the foreign country with which we were at war that they would take no hostile action, but in the present international situation, it would seem somewhat unwise for them to place any reliance on assurances of this kind.

Matters such as the establishment of the Naval Control Service, the Naval Intelligence Service and the mobilisation of Local Naval Defences are none of them inconsistent with a state of neutrality, although probably the Dominions might prefer not to have its own personnel employed in the former two services. Local Naval Defences would, of course, be entirely under Dominion control, and in this connection, it is desirable as represented by the Commander-in-Chief, Africa on M. 07323/38, that the Dominion should clearly understand that the disposition of ships of the Royal Navy will not be governed by the necessities of Local Naval Defence, which is the responsibility of the Dominion itself. Moreover, the establishment of an effective Naval Control Service and a Naval Intelligence Service to cover the Dominion is in the interests of the Dominion itself. The use of Dominion ports by ships of the Royal Navy during the precautionary period equally would not prejudice Dominion neutrality. It would be enough if the Dominion applied Hague Convention XIII Rules after war broke out, if they then decided to be neutral. On the other hand, such measures as the introduction of a Contraband Control Service, the seizure of enemy ships, and the placing of Dominion warships under Admiralty orders and allowing the departure of ships which might have been fitted out in the Dominion will be definitely excluded.

7   As the political circumstances vary so greatly in the Dominions, it may be best to deal with them separately.

8   In Australia and New Zealand, the Dominion authorities may be expected to follow any action taken by the United Kingdom, and no difficulty, in fact, arose in the September crisis. It seems desirable to leave well alone, and this is another reason for not making the alteration to Appendix I of O.D.C. 664-M. It may, however, be desirable to arrange for the Australian and New Zealand Naval Boards to take action to keep the Admiralty informed by telegraph of any steps which they decide to take.

9   In South Africa, the Commander-in-Chief, Africa is himself responsible for initiating steps to carry out any Naval action required locally under Admiralty W. List messages, which are not sent to any Union Authority at all, but the Union Government would normally receive notification of Naval Mobilisation under Telegram (3) of Appendix I of O.D.C. 664-M. This telegram was not sent in September, being replaced by a special telegram informing Dominion Governments of Naval Mobilisation but adding that it was not intended to suggest that they should take

any action as regards R.N. personnel in the Dominions who had been notified as surplus to Dominion requirements, unless, in the case of Australia and New Zealand, a request were received from the Commander-in-Chief, Singapore, for the services of any such officers and men. The Commander-in-Chief, Africa met the position by arranging for the Captain-in-Charge, Simonstown to send out individual notices asking retired and reserve Officers and men to come up, and he was apparently able to meet his requirements by volunteers in this fashion. He established a Naval Control Service staff at Cape Town in offices lent by the Union Castle Shipping Company, and on an entirely unofficial basis, the Officers wearing plain clothes. A Naval Control Service base at Durban under similar conditions was being organised when the emergency came to an end. There was apparently no difficulty in Escort Vessels proceeding to South African ports (even apart from Simonstown, the special position of which has always been recognised by South Africa), and although the Union Government would not themselves issue any mobilisation orders, they raised no objection to the Commander-in-Chief asking R.N. personnel to volunteer though there was some difficulty over certain ratings employed in important positions in Government service. The Union did, however, as a result of the representations made by the Commander-in-Chief, take initial measures on the 29th September for carrying out an 'exercise' to mobilise the coastal defences, and also the local seaward defences.

10   As regards Canada, it had been approved on M.00659/38 for the High Commissioner for the United Kingdom in Canada to act as the channel for the issue of war appointments to retired Naval Officers resident in Canada, and to hold dormant appointments in peace, and an Accountant Officer R.N. was earmarked by the Commander-in-Chief for his staff, to assist in carrying out this work. The Board, however, decided not to proceed with the consideration of any alternative plans for the provision of Naval Control and Intelligence Staffs in the event of the Canadian Government refusing to co-operate. The case of Canada differs from that of South Africa in that Admiralty W. List telegrams are sent to the Chief of the Naval Staff at Ottawa as well as to the Commander-in-Chief, America and West Indies, in the same way that they are sent to the Naval Boards at Melbourne and Wellington. In September, the machinery established in the High Commissioner's Office for calling up retired Officers did work, though rather laboriously, but there are proposals being con-

sidered on other papers which will no doubt improve the arrangements. No action, however, was apparently taken by the Canadian Naval Authorities beyond agreeing to the appointment of R.N. Retired Officers at Canadian ports, and the distribution of certain books, though there is evidence that they were making arrangements for the reception at Halifax of the R.N. Cruisers which would have proceeded there under the war arrangements, and were seeing that the coast defences were in order (M. 06841/ 38). The Commander-in-Chief has asked that renewed efforts should be made to induce the Canadian Government to define its attitude regarding the use of its ships and harbours, and he goes on to say that he has lately had opportunity to discuss this with the Premier, the Right Hon. W.L. Mackenzie King. Mr Mackenzie King stated that it was his personal wish that there should be closer collaboration in naval defence matters than heretofore. It is hoped, the C in C says, that this may lead to the issue of war memoranda to Canada, and to the closer co-operation between the Commander-in-Chief, America and West Indies Station, and the Chief of Naval Staff, Ottawa, in the preparation of war plans. As indicated in Paragraph 6, the arrival of R.N. Cruisers at Halifax before war breaks out should not prejudice Canadian neutrality, as it would be enough if the Canadians applied the Rules in the Hague Convention after war broke out if at that stage they decided to remain out of the war. The Admiralty have rights to the use of Halifax and Esquimalt Harbours and Dockyards under the agreement made when they were transferred to the Dominion Government before the war, and these would remain, at any rate until war broke out.

11   It is not at all clear what objections the Dominions Office could have to a telegram on the lines suggested in Paragraph 5 of M. Branch remarks of the 27th October, which follows almost exactly the wording of the warning telegram, and does not commit the Dominion to co-operate, nor even necessarily to define their attitude. It does, however, prevent a Dominion from making the excuse, as happened in the September emergency, that they could not do anything because they had not been approached by H.M. Government in the United Kingdom. It would, of course, be desirable that the Dominions Office should inform the Admiralty of the despatch of the telegram, so that the Commanders-in-Chief concerned could be notified. It could then be laid down that the latter should approach the Dominion Governments as necessary and report to the Admiralty what they had been able to arrange.

In this way, probably more could be done than if the matter is pressed through the ordinary channels, and it would be desirable if, in getting the Dominions to agree to some telegram of this kind; the advantages to the Dominions themselves of their taking precautionary action were fully pointed out, and it might also be emphasised that much of this action can be taken with little or no publicity, if public opinion in the Dominions were not considered yet ripe for it.

12   If the words 'preparations for war' in the telegram are likely to frighten the Dominions, no doubt some other more innocent sounding phrase could be substituted, and the word 'strained' at the beginning of the telegram might similarly be replaced, in order more readily to distinguish it from the warning telegram.

13   As regards the Admiralty W. List messages, there is something to be said for leaving the arrangements as at present, and having some sort of understanding with the Chief of the Naval Staff at Ottawa that he only takes action so far as he considers himself authorised by his Government. It seems a pity to weaken the present position . . .

[367] *Somerville to the Secretary of the Admiralty*
26 January 1939

Future of the Royal Indian Navy

Be pleased to inform Their Lordships that with reference to my message timed 1106 of 2nd December, 1938, and Admiralty letter M.07429/38 of 21st December, 1938, in reply thereto, I had an opportunity to discuss the future of the Royal Indian Navy with the Commander-in-Chief, India, and his Chief of Staff before appearing before Lord Chatfield's Commission, and was able to emphasise the value of the additional security to communications which would be afforded by six R.I.N. Escort Vessels, providing the latter were properly equipped for the duties required of them.

2   I am glad to state that at the subsequent meeting of the Commission no suggestion was made that the Royal Indian Navy should be reduced to a minesweeping force and the Commander-in-Chief, India, evidently appreciated the need for modern or modernised escort vessels.

3   So far as the existing R.I.N. Escort Vessels are concerned I have every reason to believe the Commission will confirm the

views I have already expressed that 'Indus' and possibly 'Hindustan' are the only vessels worth rearming.

4 With regard to the new allocation of ships for the Red Sea force referred to in paragraph 2 of the Admiralty letter under reply, I consider that the addition of a second division of destroyers (or Tribals) will be of great value. Unless enemy cruisers are known to be present I consider that destroyers, escort vessels and A/A ships are undoubtedly the most suitable vessels for this service,

5 So far as Escort Vessels are concerned I must again point out that none of the five R.I.N. ships is fitted with A/S and, with the exception of the .5″ M.G.'s in *Indus*, none has any H.A. armament, whilst the A/S efficiency of the major part of the R.N. escort vessels is of a low standard owing to lack of sufficient peace training facilities. On account of their age and armament the fighting value of *Clive*, *Lawrence* and *Pathan* cannot be rated very highly.

6 I appreciate that these may be the best dispositions that can be made but it would be undesirable if a false standard of security was assumed, based on numbers as opposed to efficiency.

7 I would add however that, in the event of an emergency occurring in the next six months, I consider that the disposition of escort vessels allocated to the East Indies Station should be as follows:

Persian Gulf
Shoreham, *Indus* and *Hindustan*.
Masira
Clive and *Lawrence* (and for escorting oilers from entrance to Persian Gulf to base).
Red Sea Force
Fleetwood, *Egret*, *Bideford*, *Fowey* and *Deptford*. Two escort vessels from Africa, two from China ['NZ' in manuscript].

[368] *S.H. Phillips to Under-Secretary of State, Dominions Office*
1 February 1939

'Secret and Important'

I am commanded by My Lords Commissioners of the Admiralty to refer to a letter dated 12th October last from Sir Harry Batterbee to Sir Sidney Barnes, enclosing a draft note which had been prepared for submission to the Committee of Imperial Defence on matters particularly affecting the Dominions Office in the crisis of September 1938 . . .

2 The Admiralty representative at the meeting of 12th December stated that My Lords thought it desirable to obtain from the Dominions a definition of their attitude at the earliest possible date in any future emergency, and the Committee noted that this matter would form the subject of discussions between the Admiralty and the Dominions Office . . .

3 As regards the difficulties that arose in Canada and South Africa in September in connection with the preliminary arrangements for instituting Naval Control staffs and the mobilisation of naval reserves, the Secretary of State will be aware that, in accordance with arrangements that have been in existence for many years and which are well known to the Canadian Naval Authorities, certain telegrams were sent direct from the Admiralty to the Chief of Naval Staff, Canada, who thereupon found himself in an invidious position, since no formal communication on the subject of Dominions' co-operation had been sent to the Government of Canada by H.M. Government in the United Kingdom . . .

4 My Lords . . . take the view that our aim should be to encourage the Dominions by all possible means to co-operate in defence measures and organisation, more particularly during a time of strained international relations, and they consider that the proposed qualification of the pre-prepared telegrams, far from fostering this desideratum, would merely serve as a suggestion that we are definitely visualising a situation in which the Dominions would not co-operate. My Lords have no reason to suppose that Australia and New Zealand, in particular, find existing arrangements in any degree objectionable or suggestive of a desire to force their hands and they would deprecate strongly, for this reason alone, the modification of the 'meaning' of the telegrams which would be sent to those Dominions in emergency. In regard to Canada, it will be remembered that our attitude towards war planning involving that Dominion was discussed at length in 1937–38 (see correspondence terminating with Dominions Office letter D.59/6 of 14th June, 1938), and it was then accepted that no action should be taken to suggest that the fullest co-operation of the Dominion was not expected in emergency. This attitude should, in My Lords' view be maintained . . .

5 Moreover, such information as has come to the Admiralty since the crisis of September last indicates that Canada and South

Africa, even if they are unable to define their attitude officially, are not averse to a certain amount of locally arranged co-operation between their defence forces and ours in a future precautionary period, in the taking of precautionary measures, and that they do not wish to hinder local measures for mobilising personnel of the Royal Navy Reserves who are willing to come forward (provided that this does not involve themselves mobilising) or in general to restrict the use made of their ports and harbours by Imperial Naval forces . . . The Prime Minister of Canada in October gave the Commander-in-Chief, America and West Indies, to understand that it is his personal wish that there should be the closest collaboration in any emergency between the Royal Canadian Navy and the Commander-in-Chief . . .

6   Matters such as the establishment of the Naval Control Service, the Naval Intelligence Service and the mobilisation of Local Naval Defences are none of them inconsistent with a state of neutrality, although probably the Dominion might prefer not to have its own personnel employed in the former two services. Moreover, the establishment of an effective Naval Control Service and a Naval Intelligence Service to cover the Dominion is in the interest of the Dominion itself. The use of Dominion ports by ships of the Royal Navy during the precautionary period equally would not prejudice Dominion neutrality. It would be enough if the Dominion applied Hague Convention XIII Rules after war broke out, if they then decided to be neutral . . . Incidentally, in the case of Canada, the Admiralty have rights to the use of Halifax and Esquimalt Harbours and Dockyards under the agreement made when they were transferred to the Dominion Government before the war, and these would remain, at any rate until war broke out . . .

[369] *J.A.Simpson to the Under-Secretary of State for India*
10 February 1939

File attached. This is a perfectly innocent and at the same time vitally necessary Bill. If the Government of India are to fulfill their responsibilities in regard to the Royal Indian Navy and to Local Naval Defence, I am afraid it will be necessary for the Governor-General to certify it.

Enclosure: Telegram from
Viceroy [Marquis of
Linlithgow] 10 February 1939,
Immediate, 391–S

Private and personal. You will have seen official correspondence with India Office on the subject of a Bill to extend the provisions of Navy Discipline Act to the Reserves it is now proposed to form (your telegram No. 28 dated January 5th). The Bill was rejected on Tuesday last by the Assembly at consideration stage, not because of any serious disagreement on merits but, as both Congress and Muslim League parties made quite clear, as a gesture to mark their disapproval of reservation of Defence. It is necessary to make a start with these Reserves at once and I accordingly propose to intervene with a recommendation, and subsequently to certify. I assume you see no objection and would be grateful if you would telegraph as early as possible.

[370] *Memorandum on 'Training of Junior Officers of the Royal Indian Navy with the Royal Navy', author unknown, Admiralty*
15 February 1939

The request for full R.N. training arises from the definite naval status now accorded to the R.I.N., and in this connection it will be remembered that in war R.I.N. officers and R.N. officers can command units of the other service. The question of their relative powers of command in peace is still under consideration.

As regards numbers it is not anticipated that there is likely to be any difficulty in providing accommodation and from an administrative point of view it would of course be much simpler for the R.I.N. officers to take the recognised R.N. courses rather than have to have special arrangements made for them.

The main objection to the proposal is that R.I.N. Midshipmen would serve alongside R.N. Midshipmen in the Fleet and would have to take command of ratings of the R.N. in the same way. The change would probably be not so much a departure in principle from existing arrangements as a matter of degree, since it is quite clear that during their existing destroyer training these officers must to some extent be practised in command.

[371] *Memorandum, author unknown, Admiralty*

c. 16 February 1939

An informal meeting, attended by P.A.S. (S) [Principal Admiralty Secretary -----] and representatives of Military Branch and Plans Division for the Admiralty, and by Sir Edward Harding [Permanent Under-Secretary of State] and Messers. Dixon and Stephenson for the Dominions Office, was held in the Under Secretary of State's room at the Dominions Office on February 16th to discuss the matters dealt with in Admiralty letter of 1st February.

The letter was considered paragraph by paragraph, Sir Edward Harding stating the present views of the Dominions Office on the various points and P.A.S. (S) furnishing further explanations of Admiralty views.

Two important facts emerged; (1) that the Dominions Office proposed to drop the idea of making an addition to the meaning of the telegrams in O.D.C. 664M. as agreed at the O.D.C. meeting of 12th December, and (ii) that they agreed in principle with the proposal to ascertain the Dominions' attitude at as early a stage as possible of an emergency . . .

[372] *Eastwood to the Under-Secretary of State for India, India Office*

23 February 1939

With reference to your letter M.570/38 dated 7th March 1938 I am commanded by My Lords Commissioners of the Admiralty to request that you will inform the Secretary of State for India that they agree that Junior Officers entered in the Executive and Engineering branches of the Royal Indian Navy should be given the same training as Junior Officers in the Royal Navy up to and including the courses for Acting Sub-Lieutenants subject to the following provisions:

(1) It may not always be possible for accommodation to be found in the Training Cruiser for more than a limited number of Cadets for the Indian Navy and My Lords would therefore be glad to have an early reply to their letter CW 17448/38.

(2) It will not be possible for some years to come to accommodate the Junior Officers of the Engineering Branch who are under training for the Indian Navy with the other students at Keyham.

The Commander-in-Chief, Devonport, is being asked to make proposals for their accommodation while under tuition at Keyham.

[373] *Minute by 'S' [P.A.S. (S)?] Admiralty*
15 March 1939

Mr Bruce [Australian High Commissioner] I think feels doubtful – as I do – as to whether a U.K. gov[ernmen]t when it came to the point would leave Egypt & its neighbours defenceless in the Eastern Mediterranean. We have never got clear of the German naval limitation complex, under which warships of the British Empire would be counted as belonging to the U.K. & would be [unclear] limited in numbers. Under such conditions it was not to our interest to have one of our few capital ships confined to Australian waters & we discouraged Australia from owning one – or at least did not encourage her to do so.

I agree entirely with the last part of C.N.S.'s minute [i.e. that an Imperial Conference is needed], but we must make up our own minds as a government before we are really in a position to discuss this with the Dominions.

[374] *Extract of a letter from the Government of India, Defence Department (Navy Branch), to the Secretary, Military Department, India Office, forwarded by S.H. Phillips to the Commander-in-Chief East Indies [Vice-Admiral Ralph Leatham], 7 February 1940.*
31 March 1939

7   Local Naval Defence

The Government of India recognises that it will be necessary in the event of an emergency, for naval movements in general in the East Indies Station to be controlled by one central co-ordinating authority, but they feel that Local Naval Defence in war and all arrangements therefore, must remain under the Government of India, authority being exercised by His Excellency the Commander-in-Chief through the Flag Officer Commanding, Royal Indian Navy. The advice of the Naval Commander-in-Chief, East Indies Squadron, who is the principal Naval Adviser to the Government of India, will be obtained and his suggestions followed as far as possible in respect of Local Naval Defence, but the Government of India cannot agree to any suggestion to place the entire resources of the Royal Indian Navy unreservedly at the disposal of the Naval Commander-in-Chief, East Indies.

[375] *Memorandum by the Deputy Chief of Naval Staff [Rear-Admiral Andrew B. Cunningham], Strategic Appreciation Sub-Committee, 'Dispatch of a Fleet to the Far East', Committee of Imperial Defence*

5 April 1939

The conclusion which emerges from the foregoing considerations is that there are so many variable factors which cannot at present be assessed, that it is not possible to state definitely how soon after Japanese intervention a Fleet could be dispatched to the Far East. Neither is it possible to enumerate precisely the size of the Fleet that we could afford to send.

[376] *H.F. Batterbee, United Kingdom High Commissioner in New Zealand, to Inskip, Secretary of State for Dominion Affairs*

28 April 1939

7   It became evident that underlying all the strategical and other discussions were doubts not indeed of the intention but of the ability of Great Britain to despatch a fleet of capital ships to the East in time to prevent the fall of Singapore, and doubt of the security of Singapore itself before the arrival of the fleet. These doubts were clearly felt by the New Zealand Delegation and, though the Australian Delegation, being led by Vice-Admiral Sir Ragnar Colvin, did not express the feeling, he assures me that the Commonwealth Government feel them as strongly or even more so. Holding such views, the New Zealand representatives were more sympathetic to measures calculated to increase their own domestic security than to those aiming at the general defence of the Empire. Moreover, they appeared to be inclined to measure their own effort by the Australian standard and to be unwilling to undertake commitments not expected of Australia.

[377] *Memorandum by Brown to Zetland*

17 May 1939

The main proposal is that on receipt of the Warning Telegram the Government of India should automatically place the escort vessels of the Royal Indian Navy under the Naval Commander-in-Chief, East Indies. Hitherto it has been contemplated that we should go through the process of getting a specific request from the Admiralty and transmitting it to the Government of India. This, however, would take time when every hour may count; and the Government of India have themselves agreed to the automatic procedure now suggested. So long as the R.I.N. vessels are operating under the Naval Commander-in-Chief, East Indies, the presumption is that they are being used for the defence of India in the broad sense; and therefore there would be no obligation to consult Indian political leaders in advance even if time and secrecy permitted. You may remember that you agreed on this presumption with the Chancellor of the Exchequer when we made our bargain with the Admiralty. There is, therefore, no political objection to shortening the procedure of transfer still further and making it automatic.

If this proposal is accepted we ought to get the War Books amended as soon as possible. The Government of India's letter deals with two other matters which, however, can be taken up separately, and I have altered the form of the draft letter to the Admiralty so as to bring the question of procedure into the forefront.

One of these matters is the difficult question whether the Naval Commander-in-Chief East Indies should take over all the R.I.N. vessels regardless of the requirements of India's local naval defence. This is a matter on which I should hesitate to lay down the law without further examination, and I have therefore altered the draft letter to the Government of India so as to reserve it for separate consideration, which can be done without damage to the main proposal.

[378] *Stewart to [Sir Robert B. Howorth] the Clerk of the Council,*
Privy Council

22 May 1939

I am directed by the Secretary of State for India to inform you that he desires to recommend that His Majesty's assent may be signified to the Indian Naval Reserve Forces (Discipline) Act, 1939, the authentic copy of which as received from the Governor-General is enclosed herewith.

2   The Act has been assented to by the Governor-General after having been enacted under the procedure prescribed by Section 67 B(2) of the Ninth Schedule to the Government of India Act,

1935, and the reasons which led His Excellency to regard its enactment as essential, and to have recourse to this special procedure for enacting it, are fully explained in his Despatch which will be found on page 8 of the accompanying White Paper. Lord Zetland has nothing to add in explanation or in justification of the Measure, and he has expressed himself in a Despatch to the Governor-General dated the 25th April last (page 9 of the White Paper) as in accord with His Excellency's decision.

3 In pursuance of the Statute, copies of the Act were presented to both Houses of Parliament on the 1st May and have lain before each House for the statutory period of not less than 8 days on which that House has sat.

4 I am accordingly now to request that the Lord President of the Council will be good enough to take the necessary steps to secure assent to the Act on behalf of His Majesty, as required by Section 67 B(2) of the Ninth Schedule to the Government of India Act, 1939. A draft Order-in-Council for this purpose is enclosed.

5 I am to request that the original authentic copy of the Act may be returned in due course for record, and that two sealed and twenty plain copies of the Order-in-Council may be supplied for the use of this Department.

[379] *Draft letter to Nelles*

4 July 1939

'Secret and Personal'

We have been considering, in the light of the experience gained in the crisis of September last, in what way our arrangements for promulgating to those concerned the instructions issued from the Admiralty in time of emergency require to be modified, and in particular whether any alteration is necessary in the method of notifying these instructions to the Naval Authorities of the Dominions.

The present arrangement, as you will recall, is to send out instructions in a form drawn up beforehand, and notified by means of one-word code messages known as W. List telegrams, the signification of these special telegrams being shown in C.B.04008.

A number of these telegrams are addressed to C.N.S., Ottawa, and the significance of some of them indicates that action is to be taken. Strictly speaking, such messages, when sent to Dominion Naval Authorities, ought not to be given in the form of an order, but we have come to the conclusion that as the present system has been well understood, and has been in force for many years, and as to have a different system for Dominion Authorities would unduly complicate the procedure, it would be much better to leave the messages in their present form as an indication of the action which is being taken in the Royal Navy, and of what it is desired should be taken in the Dominions, if the requisite authority has been given, or can be obtained.

We would, of course, hope that in an emergency it would be found possible for the Dominions to take corresponding action to ourselves, but at the same time, we would realise that the Naval Authorities in a Dominion would, on receipt of a special telegram requiring some action, not take the action indicated unless they considered they were authorised by their own Government, or they could obtain authority from that Government, to do so.

If, however, in the event of an emergency, it is found not to be possible to take the whole of the action indicated, we should be glad to know that such is the case, and what action, if any, has been found possible.

[380] *S.H. Phillips to Naval Commanders-in-Chief*
26 July 1939

'Confidential'

I am to acquaint you that Their Lordships have decided to institute a scheme for recruiting for the Royal Navy and Royal Marines in certain parts of His Majesty's oversea dominions similar to that in operation on the African Station.

2   Under this scheme Commanders-in-Chief will have discretion to allow recruits to be entered on the station with certain maximum numbers which would for the time being be as follows:

America & West Indies Station: 65 Seamen (special service), Stokers, Air Mechanics or Royal Marines.
36 Boys (continuous service). (These numbers include 25 & 12 respectively from Newfoundland)
East Indies Station: 15 Seamen (special service), Stokers, Air Mechanics or Royal Marines.
5 Boys (continuous service).
China Station: 10 Seamen (special service), Stokers, Air Mechanics, Royal Marines or Boys (continuous service) . . .

7   In virtue of the responsibility of Commanders-in-Chief for recruiting duties within the compass of their stations, they should

establish direct contact with the appropriate authorities in the dominions and/or colonies.

8 The view of the Canadian Government is understood to be that it is open to the Admiralty to make known in Canada opportunities of service in the Royal Navy and to individual Canadians to enlist of their own accord, but that the Canadian Government are not willing that Canadian official agencies should participate in any way. The Commander-in-Chief is, however, of course at liberty to collaborate with the High Commissioner in Canada for H.M. Government in the United Kingdom. It appears that the most satisfactory method will be that the officer appointed for recruiting duties should work through H.M. Ships visiting Canadian ports from time to time, candidates from Canada attending on board for medical examination and acceptance.

9 Australia and New Zealand are excluded from the scheme at their own request, on the ground that it is undesirable that any measure should be adopted which might interfere with the expansion of their naval forces.

10 On the Mediterranean station, the Vice Admiral, Malta, functions as a recruiting authority for the station, for applicants resident in the colonies included therein and countries on the Mediterranean littoral.

11 Paragraphs 6, 8 and 9 only of this letter need be regarded as confidential. [*signature*]

[381] *J.A. Simpson to Admiralty Secretary*
23 August 1939

I am directed by the Secretary of State for India to refer to paragraph 6 of Admiralty letter of the 18th August, 1937, No. C.W. 9354/36 and to state for the information of the Lords Commissioners of the Admiralty that, although the question of the powers of command of Royal Indian Navy officers has not yet been settled, the Secretary of State desires to take the necessary steps forthwith for the grant of Commissions to Royal Indian Navy officers 'in His Majesty's Fleet'.

For constitutional reasons it is appropriate that the new type of commissions should be issued by the Governor-General in the exercise of power assigned to him by the King, and it would appear that the form of the commission should follow as closely as possible the form of that now granted to officers of the Royal

Australian Navy, a copy of which is presumably in the possession of the Lords Commissioners of the Admiralty, (I am to refer in this connection to Admiralty letter of 2nd May, 1915, No. N.L. 1745/14). I am accordingly to request that the Secretary of State may be furnished with a copy of the Commission now granted to Executive and Engineer officers of the Royal Australian Navy, if it is in any way different from that forwarded with Admiralty letter of 2nd May 1915, in order that it may be adapted for Indian use.

In this connection I am to state that the question is also under consideration of granting Commissions in the Royal Indian Navy Reserve, the Royal Indian Navy Volunteer Reserve and the proposed Royal Indian Navy Communication Reserve, and I am to request that the Secretary of State may be favoured with the observations of the Lords Commissioners on the form in which these Commissions should be drawn.

I am further to refer to paragraph 2 of Admiralty letter No. C.W. 7447/36 dated 26th August, 1926, and to say that the Secretary of State would be grateful if the decision regarding the ranking of Royal Indian Navy officers with Royal Navy officers could be embodied at an early opportunity in the appropriate paragraph of King's Regulations and Admiralty Instructions.

[382] *Royal Indian Navy War Organisation, copy supplied to Admiralty 22 August*

15 August 1939

Part III, Section 1

The Flag Officer Commanding, Royal Indian Navy, will co-operate with the Commander-in-Chief, East Indies or Eastern Forces. He will continue to carry out the duties of Principal Sea Transport Officer, INDIA, and to deal with all administrative matters in connection with the Royal Indian Navy. He will be responsible for the execution of the local naval defence arrangements for the harbours and coasts of INDIA.

2   He will be responsible for expanding his staff and the staffs at the defended ports to meet war requirements.

Action

3   Among the steps necessary to fulfill the above responsibilities the Flag Officer Commanding, Royal Indian Navy, will:

(A)   In Peace

(i)   Prepare such plans and issue such orders as he considers necessary to ensure that the ships and establishments of the Royal Indian Navy are prepared for war.

(ii)   Prepare schemes for the local naval defence of INDIA.

(B)   During the Precautionary Period

(i)   On receipt of instructions from Defence Department, (Navy Branch), to place the Escort Vessels of the Royal Indian Navy at the disposal of the Commander-in-Chief, East Indies or Eastern Forces, despatch the following signal to the Naval Commander-in-Chief:

'The following escort vessels of the Royal Indian Navy have been placed under your orders and at your disposal, and have been instructed to act accordingly. H.M.I.S. . . .'

(ii)   Inform Defence Department (Navy Branch) of the date on which this is done.

(iii)   Inform H.M.I. Ships concerned.

[383] *Eastwood to Commander-in-Chief America and West Indies*
25 August 1939

'Secret'
I am to acquaint you that Their Lordships have had under consideration certain suggestions for the modification of C.B.1799(29) made by the Commander-in-Chief, Africa, in a report concerning the mobilisation of Naval Reserves in South Africa in September 1938, when some difficulty arose owing to the unwillingness of the Union Government to take the action defined in C.B.1799(29) when the mobilisation of Naval Reserves had been ordered. As you are aware, the Canadian Government also evinced reluctance to take corresponding action in Canada.

C.B.1799(29) is compiled on the assumption that the Dominions will join with the United Kingdom in any war which involves the calling out of Naval Reserves abroad, and will be prepared to take certain action themselves to assist in calling up Reserves. Their Lordships recognise that the governments of the several

Dominions reserve for themselves the right to determine if and when they will co-operate with this country in such an emergency, but in their view it would be unwise to revise the procedure defined in C.B.1799(29) in such a way as to imply that this was doubtful since such a revision would come to the notice of the Dominion authorities who hold C.B.1799(29).

On the other hand Their Lordships agree that it would be useless to rely solely on the standing procedure, in any emergency in which a Dominion was not prepared to co-operate immediately with the United Kingdom. As far as the America and West Indies station is concerned the only Dominion affected is Canada. You will have been in touch with the United Kingdom High Commissioner both generally and as to the co-operation with the Dominion Government in the pre-arranged schemes. If you receive from him definite information that the Dominion Government are not prepared to co-operate in mobilisation arrangements, or if the general non co-operating attitude of the Dominion Government becomes reasonably clear, the memorandum may be regarded as in abeyance from that point. In such case, the services of members of the Reserve Forces of the Royal Navy would not be required immediately by the Dominion Government and at this stage, if the order to mobilise Naval Reserves has been given, you should proceed to call up by personal summons any officers or men who are members of the Naval Reserve Forces under Admiralty control (i.e., all those classes of officers or men who are dealt with in C.B.1799(29) whom you require. In order that men earmarked for vital services in Canada could be made available should the Dominion Government at any time thereafterwards require their services, it would be desirable as far as possible that you should not, in the initial stages, call up any such men whose names were in your possession except to such extent as they may be required for the same duties as if they were mobilised by the Dominion authorities. This restriction is of particular importance because it may be hoped that, although the Canadian Government would decline to commit themselves in any way on the question of participation in war, until they had consulted their Parliament, they would after such consultation be prepared to co-operate with this country. Part II of the memorandum would be inapplicable generally.

The arrangements made with the High Commissioner for the United Kingdom in Canada and with the United Kingdom Trade Commissioner in Canada for calling up Retired and Reserve

Officers resident in the Dominion are designed to overcome difficulties arising from any delay on the part of Canada before she decides to participate in war.

Their Lordships are aware that this arrangement is not altogether satisfactory, either from the point of view of making sure of securing the services of the men or from that of avoiding difficulty later if men who would have been earmarked for service with the Dominion Government are not available for that service when they are required. As regards the first point much depends on the strength of the attitude adopted by the Dominion Government, but except in the last contingency it is considered improbable that that Government would put obstacles in the way of your sending personal summonses to individual officers and men. On the other hand they would probably give no assistance, except possibly in a passive way if their attitude was benevolent, and in particular you would be unable to apprehend as deserters any men who failed to answer the summons. It would, however, be appropriate that you should report such examples to the Admiralty in order that stoppage of retired pay, pension or any reserve benefits could be arranged.

In the invitations that were sent to reservists on the Africa Station in 1938, when the Union Government had declined to issue the mobilisation proclamation, it was explained that their attendance was not compulsory until proclamations had been issued by the Union of South Africa. Their Lordships are informing the Commander-in-Chief, Africa Station, that they do not agree with this view and that when it is necessary for a Commander-in-Chief to issue personal summonses to individual officers and men, in so doing he is acting as an agent of the Admiralty, whose right to call out for service officers and men who have a mobilisation liability is unlimited, once the appropriate proclamation of Order-in-Council has been issued, in whatever part of the world an individual may be. The proclamations which are issued in the Dominions and Colonies on mobilisation are not a necessary preliminary to the mobilisation of reservists etc. abroad, but form, rather, a convenient method of indicating the co-operation of the overseas government in the matter, and of affording publicity.

[384] *Message Out to the Commander-in-Chief East Indies*
[Leatham], copies to First Lord, First Sea Lord
4 September 1939

'Confidential'

From Admiralty Government of India agree R.I.N. forces placed under your orders are formally placed at disposal of Admiralty and are to be treated as H.M. ships in all respects and personnel so transferred is to rank and command according to rank branch and seniority as if they were officers of the R.N. 2021/4 Head of C.W. [Commissions and Warrants, indecipherable initials]

[385] *Colonel N.G.Hind, Deputy Secretary to the Government of*
India, Defence Department (Navy Branch) to the Secretary,
Military Department, India Office
7 September 1939

I am directed to say that in view of the placing of the Escort Vessels of the Royal Indian Navy under the orders of the Commander-in-Chief, East Indies, the question of the relative rank and power of command of the personnel of the Royal Navy and Royal Indian Navy has become a matter of some urgency.

2   There exists at present no authority for Royal Navy or Royal Indian Navy officers or ratings to exercise command over personnel of the other service in normal circumstances. The Naval Discipline Act does not, in itself, legislate for the rank and command of individuals, which is provided for in the 'King's Regulations and Admiralty Instructions'. The Government of India understand that Article 176 (2) of this publication, specifying the power of command of Dominion personnel vis a vis the Royal Navy, is based on two Orders in Council, one of which provides that Dominion personnel shall rank and command with the Royal Navy according to the date of their commissions or appointment in their rank, and the other that commissions granted by the Admiralty or the proper dominion authority are valid whether the officers are serving in Royal Navy or Dominion ships. In effect, Royal Navy and Dominion personnel rank and command among each other as though they all belong to one service. No such provisions exist in the case of the Royal Indian Navy. It is also provided that when forces or ships of a Dominion are placed by

their Government at the disposal of the Admiralty, the Naval Discipline Act shall apply without modification.

3   Now that the Escort Vessels of the Royal Indian Navy have been placed at the disposal of the Admiralty, the personnel serving in these ships also become subject to the Naval Discipline Act vide Section 105 (2) of the Government of India Act, 1935. The Government of India are not sure whether this automatically gives personnel of either service power of command over the other, or whether it is necessary to provide for this either by Orders in Council similar to those applicable to the Dominions, or by granting temporary commissions in the Royal Navy to all Royal Indian Navy officers in the ships concerned. The Orders in Council referred to above give power of command to Dominion personnel in peace or war but the question remains whether they are necessary when the Naval Discipline Act is applied in full, or only in normal circumstances when ships remain under the control of their own Government.

4   Realising that, apart from the Royal Indian Navy Escort Vessels, Royal Indian Navy officers are liable to serve along with Royal Navy and Dominion officers ashore or afloat, and will actually have a number of Royal Navy retired and Reserve officers serving with them in India, the Government of India are of opinion that the only satisfactory solution of the problem is for the Royal Indian Navy to be placed on exactly the same footing as the Dominion Navies by means of Orders in Council. They accordingly request that this may be urged on the Admiralty. Should, however, they be not prepared to take this action immediately, it would be for them to decide whether it is necessary to grant temporary Royal Navy commissions to officers in the Royal Indian Navy Escort Vessels, or whether power of command vis-a-vis the Royal Navy is automatically provided by their being placed under the Naval Discipline Act. A list of the officers concerned is enclosed in case it is decided to grant temporary commissions.

5   I am to add that the question of relative seniority also requires settlement. At present, Royal Indian Navy officers are governed by Article 224 (3) of the 'King's Regulations and Admiralty Instructions', which lays down that, except for ceremonial purposes in Indian waters, they shall rank as junior to all Royal Navy officers of the same or corresponding rank. In their letter No. C.W.7447/36, dated the 26th August 1936, the Admiralty were prepared to agree that officers of the Royal Indian Navy should take rank with Royal Navy officers according to seniority. The

Government of India request that the Admiralty may be asked kindly to implement this agreement immediately by publishing official orders on the subject and that, pending the formal amendment of the 'King's Regulations and Admiralty Instructions', the Commander-in-Chief, East Indies, may be informed of this by telegram.

[386] *J.A. Simpson to the Secretary to the Government of India, Defence Department (Navy Branch)*

18 October 1939

Relative Rank and Powers of Command of Personnel of the Royal Navy and Royal Indian Navy

In reply to your letter of the 7th September last, No. 1250-N on the above subject, I am directed by the Secretary of State to say that despite repeated reminders no reply up to the outbreak of war had been received from the Admiralty on the question of the powers of command of officers of the Royal Indian Navy. In the circumstances explained below, he doubts whether any useful purpose could be served by pressing the Admiralty further on the lines adopted in peace-time.

As regards the officers serving in the escort vessels which have been placed at the disposal of the Admiralty, I am to enclose a copy of a telegram from the Admiralty to the Naval Commander-in-Chief East Indies, dated the 4th September, from which it will be seen that the Admiralty have already given instructions that the personnel concerned should rank and command as if they were officers of the Royal Navy. This being so, no object would be served in asking the Admiralty to provide temporary commissions and warrants for the personnel enumerated in the enclosure to your letter under reply.

As regards other officers of the Royal Indian Navy, it is true that the order contained in the Admiralty's telegram does not cover them, and in order to grant them equivalent powers of command, it would be necessary to ask the Admiralty either to grant them temporary commissions in the Royal Navy, or to issue an order in similar terms to that contained in the enclosed telegram which would cover them also. The former method involves a degree of preferential treatment for officers not employed in the escort vessels which it is difficult to defend; the latter method would be facilitated if the Government of India were prepared to place the Royal Indian Navy as a whole at the disposal of the Admiralty.

With reference to paragraph 5 of your letter, the Admiralty have already been requested to amend Article 224 (3) of the King's Regulations and Admiralty Instructions in fulfillment of the agreement referred to by you. On this question also no reply has yet been received.

[387] *8th Meeting of the War Cabinet, Visits of Dominion Ministers*
20 November 1939

MR BRUCE [F.M. Bruce, High Commissioner, Commonwealth of Australia] said that there had undoubtedly been some misunderstanding of the position in Australia. At the time of the Imperial Conference in 1937 a war against Germany, Italy and Japan in combination had admittedly not been envisaged. During 1938, however, when this unpleasant contingency seemed quite probable, he had several conversations with Lord Stanhope, then First Lord of the Admiralty, and the Chief of the Naval Staff, in which it had been specifically stated that a Fleet of a definite strength i.e. containing 7 Capital Ships would be sent to the Far East if Japan entered the war against us. No specific period within which such Fleet would be despatched had been mentioned but it would be recalled that there had been considerable discussion at the time on the 'period before relief' of Singapore. The Australian Government had assumed that, since Singapore had only sufficient reserves to last out for a limited time, a Fleet would be despatched for its relief almost immediately after the outbreak of war with Japan and quite irrespective of any direct threat of invasion of Australia. He had been in Australia at the time that the Prime Minister's telegram of March 1939 arrived, and he had been surprised to find in it the statement that a combination of Germany, Italy and Japan against us was 'never envisaged in our earlier plans'. It was this very combination which had been envisaged in his discussions with Lord Stanhope during 1938, and the size of the Fleet required at Singapore had been discussed in detail in those conversations.

LORD CHATFIELD [former First Sea Lord] observed that the preliminary conclusions reached in the course of the discussions between Mr Bruce and Lord Stanhope had been modified after closer examination by the Committee of Imperial Defence. As a result of this further examination the Prime Minister's telegram of March 1939 had been despatched. It was important to remember, however, that the reserves of Singapore were now being increased up to six months. The fortress would therefore be able to hold out for much longer than had been anticipated at the time of those conversations.

MR CASEY [R.G. Casey, Minister of Supply and Development, Commonwealth of Australia] said that he had been much heartened by the Memorandum of Mr Churchill [First Lord of the Admiralty], but had been somewhat disturbed by his amplification of it at the present meeting. The Australian Government during the past six years had felt quite sure in their minds that if Japan entered the war, an adequate Fleet would be sent immediately to Singapore. The telegrams which had been received in the early part of 1939 had not shaken them in this conviction, and they certainly had not realised that the despatch of a Fleet was contingent upon a 'serious attack', which he understood from Mr Churchill was held to mean a definite invasion or landing of large forces. He emphasised the reality of the Japanese menace in the minds of the Australian Government. If Australia were to put in the full war effort of which she was capable, his Government would require a most comprehensive undertaking regarding the security of Singapore.

[388] *C. Macl. G. Ogilvie, Secretary, Defence Department, Government of India, to J.A. Simpson*
7 December 1939

I am directed to ask you to refer to your secret letter No. M.8297/39 of the 19th October, 1939 on the subject of the disposal of the Royal Indian Navy in the event of an emergency.

2 Paragraph 6 of your letter of the 24th September, 1938 to which you have referred was concerned only with the then existing Naval resources. These amounted to the five escort vessels, which have since been placed at the disposal of the Naval Commander-in-Chief, and two other small ships. Your letter of the 4th August, however, definitely contemplates that 'the Naval Commander-in-Chief should be in a position to give definite orders to the Flag Officer Commanding, Royal Indian Navy as to how the forces engaged on Local Naval Defence should be disposed in the light of the strategic situation and of information which may reach him at short notice as to the disposition of enemy forces'.

Thirty merchant ships have been requisitioned, fitted out and manned as anti-submarine vessels and minesweepers and these ships, together with the R.I.N. ships Investigator and Pathan, constitute our Local Naval Defence forces as they stand at present. They have been distributed in accordance with a programme drawn up by the Flag Officer Commanding with the approval of the Naval Commander-in-Chief. This programme required the provision of thirty-five Local Naval Defence craft to give the absolute minimum degree of protection; two only could be found from our own resources and as it has not been possible to fill up the deficiency completely from the merchant ships of the requisite type on the Indian register, our total stands at present at thirty-two only and cannot be increased until trawlers are available from the United Kingdom.

As this distribution of our available resources has been approved by the Naval Commander-in-Chief, it seems that no benefit would accrue in allowing him to exercise full control over them, except for the purpose of transferring them as he thought necessary to afford protection to other ports and harbours in his command – e.g., Colombo and Rangoon.

3   The Government of India are aware that circumstances might arise which would render such a transfer, and the consequent further weakening of India's Local Naval Defence, imperatively necessary, both in the interests of the Empire as a whole and of India herself. There are, however, reasons of a political and also of a constitutional character which militate against the proposal to place the entire resources of the Royal Indian Navy unreservedly at the disposal of the Naval Commander-in-Chief and to withdraw all responsibility in the matter from the Governor-General in Council. In the first place, the requisitioning of these merchant ships has undoubtedly caused a considerable dislocation of India's coastal trade and passenger traffic. Strong protests from several quarters have been raised and these are likely to be intensified if the threat to India's ports and harbours does not actually materialise in the near future. There is no doubt that the employment of ships taken up for India's Local Naval Defence outside purely Indian waters would cause a general outcry and the Government of India would be called upon to produce adequate and convincing reasons in justification of such a step.

It appears, therefore, to be eminently desirable that the Governor-General in Council should be afforded an opportunity to judging the emergency. From the constitutional point of view it may be said that, although Federation is not at the moment in sight, the position is such that we should, so far as is possible, refrain from taking any course of action which would be altogether impossible had a Federal Government assumed office.

4 I am accordingly to suggest that, in the event of an emergency of such a nature as to require the employment of Indian Local Naval Defence craft elsewhere than in Indian ports and harbours, the Naval Commander-in-Chief should, in his capacity of Principal Naval Adviser, apprise the Governor-General in Council of the necessity and the circumstances governing it. The Admiralty may be assured of the fullest co-operation and support. I am further to say that no delay whatever in the transmission of orders to the Flag Officer Commanding need be anticipated.

[389] *S.H. Phillips to Under-Secretary of State, Military Department, India Office*

31 January 1940

With reference to your letter No. 12122/39 of the 12th January, concerning the control of the Indian local Naval Defence Craft, I am commanded by My Lords Commissioners of the Admiralty to acquaint you, for the information of the Secretary of State for India, that they have no objection, in the circumstances, to the suggestion that the Naval Commander-in-Chief, East Indies should appraise the Government of India before employing elsewhere than in Indian ports and harbours any of the thirty-two Indian Local Naval Defence vessels enumerated in paragraph 2 of the Government of India's letter No. 357 N. of the 7th December, 1939.

[390] *J.A. Simpson draft letter from Secretary of State for India to the Government of India, Defence Department*

28 May 1940

My letter dated 23rd April M.1355. Cost of maintenance of units of the Royal Indian Navy made available to the Royal Navy. Treasury do not accept arguments advanced in my letter and, in particular, do not agree that vessels of the Royal Indian Navy can be regarded as employed upon 'External Defence' or that extra costs can be dealt with on that assumption. They point out that maintenance of Royal Indian Navy sloops was part of India's quid pro quo for the remission of the £100,000 subsidy, and that presumption was that India would thus contribute towards Imperial Naval defence in the same kind of way as e.g. Australia. In view of past history of case and of fact that actual extra cost seems likely to be relatively insignificant, I do not feel justified in pressing our case further and I trust that you will agree.

The suggestion was made that Singapore and Aden might be taken as limits of area outside which India would cease to be liable for extra costs, but it was agreed that such a procedure would be administratively most awkward, and it was decided not to pursue the suggestion, especially in view of the Admiralty's assurance that the sloops would not normally be used beyond those limits.

The Treasury agree to drop the claim for the mine-sweeping equipment lent in 1936.

The issue of the above has been duly authorised.

BIOGRAPHICAL OUTLINES

COLONEL SIR JAMES ALLEN (1855–1942), New Zealand Minister of Finance and Defence 1912–20; New Zealand High Commissioner 1920–7.

LEOPOLD CHARLES MAURICE STENNETT AMERY (1873–1955). Parliamentary and Financial Secretary to the Admiralty 1921; First Lord of the Admiralty 1922; Colonial Secretary 1924, and 1925–9 the First Secretary of State for Dominion Affairs. In May 1940 he led the Commons revolt against Chamberlain, and was made Secretary of State for India under Churchill.

HUGH OAKELEY ARNOLD-FORSTER (1855–1909). Admiralty Financial Secretary 1901; Secretary of State for War 1903–5.

HERBERT HENRY ASQUITH, First Earl of Oxford and Asquith (1852–1928). Chancellor of the Exchequer 1905–8; Prime Minister 1908–16.

ADMIRAL SIR ROGER ROLAND CHARLES BACKHOUSE (1878–1939). Flag Commander to three successive Cs-in-C Home Fleet 1911 to August 1914, and then to Admiral Jellicoe in the Grand Fleet 1914–1916; Flag Captain Battle Cruiser Squadron 1916–1918; on post-war problems committee at Admiralty 1918; Director of Naval Ordnance 1920; Third Sea Lord and Controller 1928–32; C-in-C Home Fleet 1935; First Sea Lord 1938–9.

ARTHUR JAMES, first Earl of Balfour (1848–1930). Prime Minister 1902–5 during which term he set up the Committee of Imperial Defence and developed the Anglo-French Entente; First Lord of the Admiralty 1915; Foreign Secretary 1916–19; chief delegate to the Washington Naval Conference 1921–2; Lord President of the Council, 1919–22 and 1925–9.

SIR JAMES ALAN NOEL BARLOW (1881–1968). Principal Private Secretary to Prime Minister Ramsay MacDonald, 1933–4; Under-Secretary of the Treasury, 1934–8.

ADMIRAL, PRINCE LOUIS ALEXANDER OF BATTENBERG (1854–1921). Director of Naval Intelligence, 1902–5; C-in-C Atlantic Fleet 1908–10; First Sea Lord 1912–October 1914.

SIR MICHAEL EDWARD HICKS BEACH, first Earl of St.

Aldwyn (1837–1916). Colonial Secretary 1878–80; Chancellor of the Exchequer 1885–6 and 1895–1902.

WILLIAM WEDGWOOD BENN, Viscount Stansgate (1877–1960). Junior Lord of Treasury 1910–14; Secretary of State for India 1929–31; Secretary of State for Air 1945–6.

SIR ROBERT LAIRD BORDEN (1854–1937). Prime Minister of Canada 1911–20; delegate to Washington naval conference 1921–2; representative of Canada at the League of Nations 1930.

EDWARD EDDINGTON BRIDGES, first Baron Bridges (1892–1969). Assistant Principal Secretary, Treasury, 1919; Assistant Secretary, 1934; Head, Treasury division controlling expenditure of Armed Forces 1935; Secretary to the Cabinet, the Committee of Imperial Defence, and the Minister for Defence Coordination, 1938.

ADMIRAL SIR OSMOND DE B. BROCK (1869–1947). Beatty's chief of staff 1916–19; Deputy Chief of Naval Staff 1919–22.

STUART KELSON BROWN (b. 1885). Joint Military Secretary, India Office 1924–34; Assistant Under-Secretary of State for India 1934–40; Deputy Under-Secretary of State for India 1940.

STANLEY MELBOURNE BRUCE, first Viscount Bruce of Melbourne (1883–1967). Prime Minister of Australia 1923–9; Australian High Commissioner, 1932–45.

JULIAN HEDWORTH GEORGE BYNG, Viscount Byng of Vimy (1862–1935). Commanded Canadian Corps 1916–17; Governor General of Canada, 1921–6.

SIR RICHARD HENRY ARCHIBALD CARTER (b. 1887). Permanent Secretary of the Admiralty 1936–40.

GENERAL SIR ROBERT ARCHIBALD CASSELS (1876–1959). Commander-in-Chief India, 1935–41.

SIR JOSEPH AUSTEN CHAMBERLAIN (1863–1937). Civil Lord of the Admiralty 1895–1900; Chancellor of the Exchequer 1903–5, 1919–21; Foreign Secretary 1924–9; First Lord of the Admiralty August–October 1931.

JOSEPH CHAMBERLAIN (1836–1914). Secretary of State for the Colonies 1895–1903.

ARTHUR NEVILLE CHAMBERLAIN (1869–1940). Chancellor of the Exchequer 1923–4, 1931–6; Prime Minister 1936–40.

SIR JOHN ROBERT CHANCELLOR (1870–1952). Secretary, Committee of Imperial Defence 1904; Secretary, Colonial Defence Committee, 1906–11.

ADMIRAL SIR ALFRED ERNLE MONTACUTE CHATFIELD, first Baron Chatfield (1873–1967). Beatty's Flag Captain

1913–16; Fourth Sea Lord 1919; Assistant Chief of Staff, 1920; Senior naval delegate at Washington conference 1921–2; Third Sea Lord and Controller 1925–29; First Sea Lord 1933–8.

SIR WINSTON LEONARD SPENCER CHURCHILL (1874–1965). Home Secretary 1910; First Lord of the Admiralty, 1911–15; Minister of Munitions 1917; Secretary for War 1918–21; Colonial Secretary 1921–2; Chancellor of the Exchequer 1924–9; First Lord 1939; Prime Minister and Minister of Defence, 1940–5, 1951–5.

SIR WILLIAM HENRY CLARK (1876–1952). First High Commissioner to Canada, 1928–34; High Commissioner in the Union of South Africa 1934–9.

SIR CECIL CLEMENTI (1875–1947). Governor of Hong Kong 1925–30; Governor of Straits Settlements and High Commissioner for Malay States, 1930–4.

Lt.-GENERAL SIR JOHN FRANCIS STANHOPE, Duke Coleridge (b. 1878). Secretary, Military Department, India Office, and Under-Secretary of State for India 1933–6; Commander in Chief Northern Command, India 1936–40.

GEORGE FREDERICK COTTON (b. 1877). Principal Clerk, Military Branch of the Admiralty Secretary's Department 1921–31; Director of Victualing.

SIR JOSEPH COOK (1860–1947). Australian Defence Minister in Alfred Deakin's administration 1909–10; Prime Minister 1913–14; Navy Minister 1917; member of the Imperial War Cabinet 1918; High Commissioner for Australia, 1921–7.

HRH PRINCE ARTHUR WILLIAM PATRICK ALBERT, Duke of Connaught and Strathearn (1850–1942). Governor General of Canada 1911–16; Governor General of the Union of South Africa, 1920–2.

ALFRED DUFF COOPER, 1st Viscount Norwich (1890–1954). First Lord of the Admiralty 1937–8.

A.E. GLOYN COX. Principal Clerk, Admiralty Military Branch in 1936.

Lt.-GEN. SIR HERBERT VAUGHAN COX (b. 1860). Under-Secretary of State for India, Military Department, India Office, 1916–20.

VICE-ADMIRAL WILLIAM ROOKE CRESWELL (1852–1933). Officer Commanding Australian Naval Forces, 1901–11; First Naval Member of Australian Naval Board 1911–19.

ROBERT OFFLEY ASHBURTON CREWE-MILNS, Marquess of Crewe (1858–1945); Viceroy of Ireland 1892–5; Lord President

1905–8, 1915–16; Lord Privy Seal 1908–11, 1912–15; Secretary of State for the Colonies 1908–10; Secretary of State for India 1910–15; Ambassador to France 1922–8; Secretary of State for War, 1931.

ADMIRAL OF THE FLEET ANDREW BROWNE CUNNINGHAM, Viscount Cunningham of Hyndhope (1883–1963). Deputy Chief of Naval Staff 1938; C-in-C Mediterranean 1939–42, 1943; First Sea Lord 1943–6.

LIONAL GEORGE CURTIS (1872–1955). Assistant Colonial Secretary, Transvaal 1903–7; leader of Milner's 'Kindergarten'; founder of the 'Round Table' review 1910, and the Royal Institute for International Affairs, 1920.

ADMIRAL SIR REGINALD NEVILLE CUSTANCE (1847–1935). Assistant Director of Naval Intelligence, 1886–90, and Director 1899–1902.

ALFRED DEAKIN (1856–1919). Prime Minister of the Commonwealth of Australia 1903–4, 1905–8 and 1909–10.

FIELD MARSHAL SIR CYRIL JOHN DEVERELL (1874–1947). Chief of Imperial General Staff 1936–7.

VICTOR CHRISTIAN, 9th Duke of Devonshire (1868–1938). Governor General of Canada 1916–21; Secretary of State for the Colonies, 1922–4.

ADMIRAL SIR FREDERIC DREYER (1878–1956). Director of Naval Ordnance 1917–18; Assistant Chief of Naval Staff 1924–7; Deputy Chief of Naval Staff 1930–3.

WILLIAM HUMBLE WARD, 2nd Earl of Dudley (1867–1932). Lord-Lieutenant of Ireland 1902–5; Governor General of Australia 1908–11.

GENERAL SIR BEAUCHAMP DUFF (1855–1918). Chief of Staff of the Indian Army 1906; Secretary of the Military Department at India Office 1909; C-in-C and Military Member of the Council of India 1914–16, Aide de Camp General.

CAPTAIN WILFRED ALLAN EGERTON (b. 1881). Admiralty Director of Plans 1925–8.

VICTOR ALEXANDER BRUCE, 9th Earl of Elgin and Kincardine (1849–1917). Viceroy of India 1893–8; Secretary of State for the Colonies 1905–8.

MARSHAL OF THE ROYAL AIR FORCE SIR EDWARD LEONARD ELLINGTON (b. 1877). Chief of Air Staff 1933–7; Inspector General of the Royal Air Force.

VICE-ADMIRAL SIR ARTHUR DALRYMPLE FANSHAWE (b. 1847). Commander-in-Chief Australia, 1902–5

VICE-ADMIRAL SIR WILMOT HAWKSWORTH FAWKES (b. 1846). Commander in Chief Australasian Squadron, 1905–8.

REAR-ADMIRAL HENRY JAMES FEAKES (1876–1950) R.A.N. Australian Naval representative in London 1927–9; Second Member of Australian Naval Board 1930–3.

ADMIRAL OF THE FLEET SIR FREDERICK LAURENCE FIELD (1871–1945). Third Sea Lord 1920–3; Deputy Chief of Naval Staff 1925–8; C-in-C Mediterranean 1928–30; First Sea Lord 1930–3.

ADMIRAL OF THE FLEET SIR JOHN ARBUTHNOT FISHER (1841–1920). C-in-C Mediterranean 1899–1902; Second Sea Lord 1902–3; First Sea Lord 1904–10, 1914–15.

ADMIRAL SIR WILLIAM WORDSWORTH FISHER (1875–1937). Director of Naval Intelligence 1926–7; Fourth Sea Lord 1927–8; Deputy Chief of Naval Staff 1928–30; C-in-C Mediterranean 1932–6.

VICE-ADMIRAL SIR HERBERT FITZHERBERT (b. 1885). Commander-in-Chief, Royal Indian Navy, 1937–43.

SIR FRANCIS LEWIS CASTLE FLOUD (b. 1875). High Commissioner to Canada, 1934–8.

SIR JOHN FORREST (1847–1918). Australian Minister of State for Defence, 1901–3.

HENRY WILLIAM, BARON FORSTER OF LEPE (b. 1866). Governor General Commonwealth of Australia, 1920–5.

GEORGE VERE ARUNDELL MONCKTON-ARUNDELL GALWAY, Viscount Galway. (b. 1882). Governor-General of New Zealand 1935–41.

SIR ERIC CAMPBELL GEDDES (1875–1937). First Lord of the Admiralty 1917–18.

SIR WILLIAM GRAHAM GREENE (1857–1950). Assistant Private Secretary of the First Lord 1887–1902; appointed head of the personnel branch of the secretary's department; Assistant Admiralty Secretary 1907, and Permanent Admiralty Secretary 1911–17; Secretary of the Ministry of Munitions 1917–19.

ALBERT HENRY GEORGE, 4TH EARL GREY (1851–1917). Governor-General of Canada 1904–11.

ADMIRAL SIR GEORGE FOWLER KING HALL (b. 1850). Commander-in-Chief Australia, 1910–13.

COLONEL SIR MAURICE PASCAL ALERS HANKEY (1877–1963). Naval Intelligence 1905–7; Assistant Secretary to the Committee of Imperial Defence 1908, Secretary 1912; Chief of War Cabinet Secretariate 1916; Secretary to Cabinet, Secretary to Committee of Imperial Defence, Clerk of the Privy Council 1923–38; British Secretary Paris Peace Conference 1919, Washington Conference 1921–2, Genoa 1922, Lausanne 1932, the Hague 1929 and 1930, Imperial Conferences 1921, 1923, 1926, 1930, 1937, London Naval Conference 1930; Minister Without Portfolio 1939; Paymaster General 1940–42; Chairman Cabinet Scientific Advisory Committee.

LEWIS, VISCOUNT HARCOURT (1863–1922). Secretary of State for the Colonies 1910–15.

SIR EDWARD HARDING (1880–1954). Deputy Secretary, Imperial Conferences 1923, 1926; Assistant Under-Secretary, Dominions Office, 1925; Permanent Under-Secretary of State for Dominion Affairs 1930–9; High Commissioner to South Africa 1940–1.

SIR CHARLES HARDINGE, BARON HARDINGE OF PENSHURST (1858–1944). Ambassador to Russia 1904–6; Permanent Under-Secretary of State for Foreign Affairs 1906–10, 1916–20; Viceroy of India 1910–16; Ambassador to Paris 1920–2.

COMMANDER EDWARD L. HASTINGS, R.N. Canadian Department of National Defence (Naval Intelligence and Plans Division), April 1935

COLONEL N.G. HIND. Deputy Secretary to the Government of India, Defence Department, Navy Branch 1939.

SIR SAMUEL JOHN GURNEY HOARE (1880–1959). Secretary of State for Air 1922–4, 1924–9; Secretary of State for India 1931–5; Foreign Secretary 1935; First Lord of the Admiralty, 1936–7; Home Secretary 1937–9; Lord Privy Seal and Member of War Cabinet 1939–40; Secretary of State for Air 1940.

SIR THOMAS WILLIAM HOLDERNESS (1849–1924). Permanent Under-Secretary of State for India, 1912–19.

REAR-ADMIRAL WALTER HOSE (1875–1965). Canadian Director of the Naval Service, 1921; Chief of Naval Staff 1928–34.

REAR ADMIRAL ALAN GODFREY HOTHAM (b. 1876). Commander of Royal Australian and New Zealand Squadron; Naval Advisor at the Imperial Conference 1923; Director of Naval Intelligence 1925

SIR COURTENAY PEREGRINE ILBERT (1841–1924). Parliamentary Counsel 1899–1902; Clerk of the House of Commons 1902–21.

RT. HON. SIR THOMAS WALKER HOBART INSKIP (1876–1947). Deputy Chairman of the Committee of Imperial Defence and Minister for Co-ordination of Defence 1936–9; Secretary of State for Dominion Affairs 1939, 1940; Lord Chancellor 1939–40.

GENERAL SIR HASTINGS LIONEL ISMAY (1887–1965). Assistant Secretary to the Committee of Imperial Defence under Hankey 1925–30; Military Secretary to Viceroy of India, Lord Willingdon, 1931–3; Secretary of State for War 1933–6; Deputy Secretary to the Committee of Imperial Defence 1936–8, Secretary 1938.

ADMIRAL OF THE FLEET SIR HENRY BRADWARDINE JACKSON (1855–1929). Third Sea Lord and Controller 1905–8; Chief of Naval Staff 1913; First Sea Lord 1915.

J.H. JAMES. Principal Clerk, Admiralty Naval Branch, in 1936

ADMIRAL OF THE FLEET, JOHN RUSHWORTH, Viscount Jellicoe (1859–1935). Director of Naval Ordnance 1905–7; Controller and Third Sea Lord 1908–10; Second Sea Lord 1912–14; C-in-C Grand Fleet 1914–December 1916; First Sea Lord 1916–Dec. 1917; Empire Tour 1919–20; Governor General of New Zealand 1920–4.

ADMIRAL OF THE FLEET LORD WALTER TALBOT KERR (1839–1927). First Sea Lord 1899–1904.

ADMIRAL OF THE FLEET SIR ROGER JOHN BROWNLOW KEYES (1872–1945). Director of Plans 1917–18; Deputy Chief of Naval Staff 1921–5; C-in-C Mediterranean 1925–8.

WILLIAM LYON MACKENZIE KING (1874–1950). Canadian Minister of Labour 1909–11; Prime Minister 1921–30, 1935–48, and Secretary of State for External Affairs to 1946.

JOHN GILBERT LAITHWAITE (b. 1894). Principal Secretary Political Department, India Office, 1924–36; Private Secretary to the Viceroy 1936, and Secretary to the Governor General of India 1937–43.

SIR WILFRED LAURIER (1841–1919). Prime Minister of Canada 1896–1911.

ARTHUR HAMILTON, VISCOUNT LEE OF FAREHAM (1868–1947). Civil Lord 1903–5; First Lord 1921–2; Second British delegate to Washington Conference 1921–2.

VICTOR ALEXANDER JOHN HOPE, 2nd Marquis of Linlithgow (1887–1952). Viceroy of India 1936–43.

ADMIRAL SIR CHARLES JAMES COLEBROOKE LITTLE (b. 1882). Deputy Chief of Naval Staff 1932–5; Commander in Chief China Station 1936–8.

WALTER HUME, 1ST VISCOUNT LONG OF WRAXALL

(1854–1924). Secretary of State for the Colonies 1916–18; First Lord of the Admiralty 1919–21.

SIR CHARLES PRESTWOOD LUCAS (1853–1931). Under-Secretary of State of the Colonial Office, and first head of its Dominions Department 1907–11.

ALFRED LYTTELTON (1857–1913). Secretary of State for the Colonies 1903–5.

VICE-ADMIRAL GEORGE HAMILTON D'OYLY LYON (b. 1875). Commander-in-Chief Africa Station 1938–40.

MALCOLM MACDONALD. Secretary of State for Dominion Affairs, August 1935–8, 1938–9; Secretary of State for Colonial Affairs 1938–40; High Commissioner to Canada 1941–6.

JAMES RAMSAY MACDONALD (1866–1937). Prime Minister 1924, 1929–31, 1931–5.

SIR EVAN MACGREGOR (1842–1926). Permanent Admiralty Secretary 1884–1907.

RT. HON. REGINALD MCKENNA (b. 1863). First Lord of the Admiralty 1908–11.

COLONEL GORDON NEVIL MACREADY (b. 1891). Assistant Secretary of the Committee of Imperial Defence 1926–32; Secretary Oversea Defence Committee 1927.

WILLIAM FERGUSON MASSEY (1856–1925). Prime Minister of New Zealand 1912–25.

FIELD MARSHAL SIR GEORGE F. MILNE (1866–1948). Chief of Imperial General Staff 1926–33.

GILBERT JOHN MURRAY KYNUNMOND, 4th Earl of Minto (1845–1914). Governor General of Canada 1898–1904.

EDWIN SAMUEL MONTAGUE (1879–1924). Secretary of State for India 1917–22.

GENERAL SIR ARCHIBALD ARMAR MONTGOMERY-MASSINGBERD (1871–1947). Chief of Imperial General Staff 1933–6.

SIR EDWARD PATRICK MORRIS (1859–1935). Prime Minister of Newfoundland 1909–18.

SIR FRANCIS MOWATT (1837–1919). Permanent Treasury Secretary 1894–1903.

SIR OSWYN ALEXANDER RUTHVEN MURRAY (1873–1936). Assistant Admiralty Secretary 1911–17; Permanent Admiralty Secretary 1917–36.

ADMIRAL PERCY WALKER NELLES. Canadian Chief of Naval Staff, 1934–44.

HENRY STAFFORD, LORD NORTHCOTE (1846–1911). Governor General of Australia 1903–7.

ADMIRAL OF THE FLEET SIR HENRY FRANCIS OLIVER. Naval Assistant to Fisher 1907; Director of Naval Intelligence 1913; Chief of Admiralty War Staff; Deputy Chief of Naval Staff, 1917–18; Second Sea Lord 1920; C-in-C Atlantic 1924–7.

REAR ADMIRAL SIR CHARLES LANGDALE OTTLEY (1858–1932). Director of Naval Intelligence 1905–7.

S.H. PHILLIPS. Assistant Secretary in the Military Branch, Admiralty, May 1935; Principal Assistant Secretary, Military Branch, 1936

ADMIRAL TOM SPENCER VAUGHAN PHILLIPS (1888–1941). Admiralty Assistant Director of Plans 1930–2; Director of Plans, 1935–8. Vice Chief of Naval Staff 1939–41.

OSWALD PIROW. Union of South African Minister of Defence, 1933–9.

J. RAYMOND F. PRÉFONTAINE. Canadian Minister of Marine and Fisheries, 1903–5.

ADMIRAL SIR ALFRED DUDLEY PICKMAN ROGERS POUND (1877–1943). Director of Operations 1917; Director of Plans 1922–5; Assistant Chief of Staff 1927–9; disarmament conference; Second Sea Lord 1932; C-in-C Mediterranean 1936–9; First Sea Lord 1939–43 until shortly before his death.

SIR GEORGE HOUSTON REID (1845–1918). Australian Prime Minister 1904–5; High Commissioner for Australia, 1910–15.

AIR CHIEF MARSHAL SIR JOHN MAITLAND SALMOND (b. 1881). Chief of Air Staff 1930–3.

RT. HON. RICHARD JOHN SEDDON (1845–1906). Prime Minister and Minister of Defence of New Zealand 1893–1906.

WILLIAM WALDEGRAVE PALMER, 2nd Earl of Selborne (1859–1942). First Lord 1900–2; High Commissioner to South Africa 1905–10.

SIR JOHN ALLSEBROOK, 1st Viscount Simon (1873–1954). Secretary of State for Foreign Affairs 1931–5; Home Secretary 1935–7; Chancellor of the Exchequer 1937–40; Lord Chancellor 1940–5.

J.A. SIMPSON. Deputy Clerk of the Council of India and Principal Officer, Military Department, India Office, 1934; Acting Joint Secretary, Military Department, India Office, 1935.

FIELD MARSHAL JAN CHRISTIAN SMUTS (1870–1950). Union of South Africa Minister of Defence 1910–19; Prime Minister 1919–24, 1939–48.

PHILIP, Viscount Snowden (1864–1937). Chancellor of the Exchequer 1924, 1929–31.

ADMIRAL OF THE FLEET SIR JAMES FOWNES SOMERVILLE (1882–1949). C-in-C East Indies 1938–9; Force H 1940–42, Eastern Fleet 1942–4; Head of British Admiralty Delegation Washington 1944–5.

SIR JOHN GORDON SPRIGG (1830–1913). Premier of Cape Colony 1878–81, 1886–90, 1896–8, 1900–4.

SIR SAMUEL FINDLATER STEWART (1879–1960). Permanent Undersecretary of State for India 1930–40.

DONALD ALEXANDER SMITH, Lord Strathcona (1820–1914). High Commissioner for Canada 1896–1913.

SIR REGINALD EDWARD STUBBS (1876–1947). Governor of Hong Kong 1919–25.

CHARLES INIGO THOMAS (b. 1846). Assistant Admiralty Secretary 1902–7, Permanent Secretary 1907–11.

JAMES HENRY THOMAS (1874–1949). Secretary of State for the Colonies 1924, 1935–6; Lord Privy Seal 1929–30; Secretary of State for Dominion Afairs 1930–5.

SIR CHARLES WALKER (b. 1871). Admiralty Principal Clerk 1909; Assistant Admiralty Secretary 1917; Deputy Admiralty Secretary, 1921–31.

SIR EDWARD WILLIS DUNCAN WARD (1853–1928). Permanent Undersecretary of State for War 1901–14.

ADMIRAL OF THE FLEET SIR ROSSLYN ERSKINE WEMYSS (1864–1933). First Sea Lord, 1917–19.

COLONEL SIR CLIVE WIGRAM. Assistant Private Secretary to King George V 1910–31, Secretary 1931–6.

FREEMAN FREEMAN-THOMAS, 3rd Marquess of Willingdon (1866–1941). Governor General of Canada 1926–30; Viceroy of India 1931–6.

ADMIRAL OF THE FLEET SIR ARTHUR KNYVET WILSON (1842–1921). First Sea Lord 1910–12.

BRIGADIER GENERAL SIR SAMUEL HERBERT WILSON (b. 1873). Assistant Secretary of Committee of Imperial Defence and Secretary of Overseas Defence Committee 1911–14, Principal Assistant Secretary and Secretary Overseas Defence Committee 1918–21; Permanent Under-Secretary of State, Colonial Office 1925–33; Secretary Financial Department India Office 1937.

LAWRENCE JOHN LUMLEY DUNDAS, 2nd Marquess of Zetland (1876–1961). Secretary of State for India 1935–40.

LIST OF DOCUMENTS AND SOURCES

The documents are drawn from a variety of sources. The specific record group and archive for each document is indicated and the following list of abbreviations identifies locations, collections and principal officials and officers.

ABBREVIATIONS

ADM	Admiralty Files
Add MSS	Additional Manuscripts
Adm. Sec.	Secretary of the Admiralty
A.S.	Assistant Secretary
A.P.C.	Admiralty Principal Clerk
BM	British Museum
CID	Committee of Imperial Defence
CW	Admiralty Registry file
CAB	Cabinet Office Files
C-in-C	Commander-in-Chief
Cmd	Command Paper
C.O.	Colonial Office
CO	Colonial Office Files
D.C.N.S.	Deputy Chief of Naval Staff
D.T.S.D.	Director of Training and Staff Duties
DO	Dominions Office Files
D.U.S.S.	Deputy Under-Secretary of State
E	Admiralty Registry number
F.L.	First Lord of the Admiralty
GB	Great Britain
G.	Governor of
G.G.	Governor General of
H.C.	High Commissioner for
I	Admiralty Registry file
I.O.	India Office
IOR	India Office Records, London
L	India Office File

M   Registry file number
MIL   Military Department
MO   Admiralty Registry file
N.L.   Naval Letter, India Office Registry file
ODC   Oversea Defence Committee
PD   Admiralty Registry file number
P.A.S.   Principal Admiralty Secretary
P.M.   Prime Minister
PRO   Public Record Office, London
P.U.S.S.I.   Permanent Under-Secretary of State for India
PW   Admiralty Registry file
S   Admiralty Registry file number
S.S.Col.   Secretary of State for the Colonies
S.S.D.   Secretary of State for the Dominions
S.S.I.   Secretary of State for India
T   Treasury Office Files
U.S.S.D.O.   Under-Secretary of State, Dominions Office
V   India Office file

[NUMBERED DOCUMENTS' SOURCES]

1   ADM 1/7573 f. 33 (PRO). Enclosure No. 7, in: Sir Cavendish Boyle to the Secretary of State for the Colonies, confidential, 20th July 1901

2   CAB 5/1 f. 17v (PRO), abstracted in CID memorandum on the Defence of Malta, CID-4C; October 1903

3   ADM 1/7573 ff. 128–30, ref. no. 19,324/01 (PRO)

4   ADM 1/7573 ff. 153–5 (PRO)

5   ADM 1/7610 f. 392 (PRO)

6   Great Britain, House of Commons, 1902 vol. LXVI p.451. Sessional Paper Cmd 1299

7   ADM 1/7610 ff. 463–8 (PRO)

8   ADM 1/7610 ff. 397–8 (PRO). Published in GB Cmd 1299, papers relating to a Conference between the Secretary of State for the Colonies and the Prime Ministers of the Self-Governing Colonies, June-August 1902

9   ADM 1/7610 f. 416 (PRO)

10 ADM 1/7671 ff. 277–8v (PRO). From GB Cmd 1299

11 ADM 1/7671 ff. 85–6 (PRO)

12 ADM 1/7671 ff. 91–91v (PRO)

13 ADM 1/7671 ff. 95–6 (PRO)

14 ADM 1/7576 (PRO). In a letter from Dundonald to Colonel Altham, dated Militia Headquarters, Canada, 6 March 1903
15 ADM 1/7576 (PRO)
16 CAB 5/1 ff. 5, 7–8, CID 2C (PRO)
17 CAB 5/1 ff. 15–17, CID 3C (PRO)
18 CAB 5/1 f. 30, CID 5C, reply to CID 4C (PRO)
19 CAB 5/2 ff. 80–9, printed as an appendix in 'Precis of Important Events Connected with the Question of Colonial Naval Contributions', secret, CID 58 C, June 1909 (PRO)
20 CO 42/899, f. 344 (PRO)
21 CO 42/899 f. 404 (PRO)
22 CO 42/899 f. 481 (PRO)
23 CAB 38/6, CID–17c (copy of paper in CAB 5/1) PRO
24 ADM 1/7807 (PRO)
25 CID 21C, CAB 5/1 ff. 78–86, and CAB 38/8 no. 13 (PRO)
26 No. 346–M, CID, CAB 5/1 ff. 109–10 (PRO)
27 CAB 38/9/52 (PRO)
28 CAB 17/47 piece 3 (PRO)
29 CO 418/37 ff. 74–80 (PRO). Printed in Cmd 3524 p. 70
30 CID 36–C, CAB 5/1 f. 127 (PRO)
31 No. 348–M, CID 34 C, November 1905. CAB 5/1 ff. 122–3, and CAB 8/4 (PRO)
32 CID 39–C, CAB 5/1 f. 144 (PRO)
33 ADM 116/1241B ff. 6–9, PM 06/1636 (PRO)
34 ADM 116/1241B ff. 10–11 (PRO)
35 ADM 116/1241B ff. 10–11 (PRO). *Euryalus* arrived in Sydney 3 April 1904, (relieved by *Powerful* 15 December 1905), *Challenger* arrived 23 July, *Prometheus* 15 May 1905, *Regasus* 2 June, *Pioneer* 15 December, *Cambrian* 23 December, *Pyramus* and *Psyche* 5 January 1906, and *Encounter* arrived 8 March. ADM 116/1241B ff. 10–11 (PRO). *Euryalus* arrived in Sydney 3 April 1904, (relieved by *Powerful* 15 December 1905), *Challenger* arrived 23 July, *Prometheus* 15 May 1905, *Regasus* 2 June, *Pioneer* 15 December, *Cambrian* 23 December, *Pyramus* and *Psyche* 5 January 1906, and *Encounter* arrived 8 March.
36 ADM 116/1241B ff. 37–8 (PRO)
37 ADM 116/1241 ff. 381v–2v (PRO)
38 ADM 116/1241B ff. 389v–90v (PRO), reprinted in the Parliamentary Papers, Commonwealth of Australia, Senate, 15 August 1906.
39 ADM 116/1241B ff. 32–3 (PRO). *See* Vice-Admiral Sir

Wilmot Hawksworth Fawkes, C-in-C Australasian Squadron, to Adm. Sec., 9 October, 19 and 26 November 1906, ff. 50–1, 291–5, 335–7.

40   ADM 116/1241B ff. 13–16 (PRO)
41   ADM 116/1241B ff. 278–81 (PRO)
42   ADM 116/1241B (PRO)
43   ADM 116/1241B ff.265–8 (PRO)
44   ADM 116/1241B ff. 374–5 (PRO)
45   No. 12/1824 ADM 116/1241B f. 393 (PRO)
46   ADM 116/1241B f. 425 (PRO)
47   ADM 116/1241B ff. 435–43 (PRO)
48   M 0149, ADM 116/1241B f. 453 (PRO)
49   ADM 116/1241B ff. 519–26 (PRO)
50   Australian Commonwealth Parliamentary Paper Cmd 4325/ 1908, republished as an appendix to the report of the Interdepartmental Conference on the Status of Dominion Ships of War. ADM 116/1100c f. 50 (PRO)
51   ADM 116/1100c f. 55 (PRO)
52   CID 50–C, January 1908. CAB 5/2ff. 59–60 (PRO)
53   Printed in the appendix to the report of the Interdepartmental Conference on the Status of Dominion Ships of War, ADM 116/1100c f. 55 (PRO)
54   CID 50–C, CAB 5/2 ff. 59–60 (PRO)
55   ADM 116/1100C ff. 46–7 (PRO)
56   Appendices III and IV of the report of the Interdepartmental Conference on the Status of Dominion Ships of War, ADM 116/1100C f. 61v (PRO)
57   CID 55–C, CAB 5/2 f. 63 (PRO)
58   ADM 116/1100B ff. 58–9 (PRO)
59   Parliamentary papers CD 4948/1909. Reprinted in Appendix II of the report of the Interdepartmental Conference on the Status of Dominion Ships of War, ADM 116/1100C f. 57v (PRO)
60   ADM 116/1100B f. 12 (PRO). Similar from Carmichael, Governor of Victoria, received 10.59 a.m. 5 April. Ibid. f. 13
61   Reprinted in Appendix II of the report of the Interdepartmental Conference on Status of Dominion Ships of War, ADM 116/1100C f. 57v (PRO)
62   DW 3637/09 M 8481, [Order-in-Council approving transfer dated 16 April]. ADM 116/993 ff. 104–6 (PRO). See Grey, G.G. Canada, to the Earl of Crewe, 27 April 1909. Ibid. f.

142; and *see* list of deeds of property handed over in Halifax and Esquimalt, in ADM 116/994

63 ADM 116/1100Bf. 34 (PRO)
64 Proof 14728. ADM 116/1100B ff. 44–5 (PRO)
65 ADM 116/1100B ff. 99–101 (PRO)
66 Appendix II of The Proceedings of the Imperial Conference ... on the Naval and Military Defence of the Empire, 1909. Canada, 9–10 Edward VII, Sessional Paper No. 29a, A 1910, p. 24
67 ADM 116/1100B ff. 399–400 (PRO)
68 Reprinted in Appendix II of the report of the Interdepartmental Conference on the Status of Dominion Ships of War, ADM 116/1100C f. 59v (PRO)
69 ADM 116/1100c f. 6 (PRO)
70 ADM 116/1100c (PRO)
71 ADM 116/1270 ff. 5–6 (PRO)
72 ADM 116/1100c f. 29 (PRO)
73 CO No. 26304/10, CAB 11/29 part 7 (PRO)
74 ADM 116/1100c f. 31 (PRO)
75 CID 62–C, July 1912. CAB 5/2 ff. 110–2 (PRO)
76 M/N.L. 2988, ADM 116/1100c f. 26 (PRO)
77 CID 83–C (July 1912), CAB 5/2 ff. 202–5 (PRO)
78 ADM 116/1100c f. 109 (PRO). A similar request for consultation was sent to Australia on the same date. f. 110
79 ADM 116/1100c f. 116 (PRO)
80 CID 67–C, CAB 5/2 ff. 150–3 (PRO). *See* draft dated 31 January 1911, 'Organisation of Australian Naval Force', confidential. Signed by: A.M. Acland, A.L. Bethell, W. Graham Greene, and G.H. Reid. ADM 116/1100c (PRO)
81 ADM 116/1100C f. 164 (PRO). Forwarded by H.W. Just to Adm. Sec., 29 March 1911
82 ADM 116/1100C ff. 169–70 (PRO)
83 ADM 116/1100c (PRO)
84 ADM 116/1100c f. 177 (PRO). *See* H. Just to Adm. Sec., 3 May 1911, ibid. f. 183
85 ADM 116/1100c ff. 190–204 (PRO). *See also* 'Co-operation between the Naval Forces of the United Kingdom and Dominions', principles agreed to at the 113th meeting of the CID, held on 30 May 1911, secret, CID 89 C, June 1911. CAB 5/2 f. 222 (PRO)
86 Copy, ADM 116/1100c ff. 264–7 (PRO)
87 ADM 116/1100c ff. 274–5 (Newfoundland No. 215), PRO

88   Copy in Churchill's hand received at the registry office 18 April 1935, ADM 116–3485 (PRO)
89   ADM 116/1270 ff. 19–20 (PRO). *See* paraphrase telegram, Lord Islington, G. New Zealand, to S.S.Col, received C.O., 1 p.m. 1 May 1912. Ibid. f. 22
90   ADM 116/3485 (PRO)
91   ADM 116/3381 (PRO)
92   ADM 116/3381 (PRO)
93   ADM 116/3485 (PRO)
94   ADM 116/3381 (PRO)
95   ADM 116/3381 (PRO). *See* copy of the Bill submitted to the Canadian parliament authorising payment of funds to the United Kingdom to build battleships (forwarded by Borden to Churchill on 2 November). Ibid.
96   ADM 116/1270 ff. 27–9 (PRO)
97   ADM 116/1270 ff. 27–9 (PRO)
98   ADM 116/1270 ff. 27–9 (PRO)
99   ADM 116/1270 ff. 77–82 (PRO)
100 ADM 116/1270 ff. 49–50 (PRO)
101 ADM 116/3381 (PRO)
102 No. 101, ADM 116/1270 f. 86 (PRO). *See* reply, telegram of 9 November No. 132. Ibid. f. 87
103 No. 135, ADM 116/1270 f. 88 (PRO). *See also* telegrams from First Lord to C-in-C Australia, 13 November 1912, and reply 15 November, no. 139. Ibid. 89–90
104 ADM 116/3381 (PRO)
105 ADM 1/8284, no. 555 (PRO)
106 M 10481/1912, L/MIL/7/3500 (IOR)
107 Great Britain, House of Commons, 1912 vol. LIII p. 445.Cmd 6513
108 Number M 01658, ADM 116/1270 ff. 63–5 (PRO). *See* 'Naval Policy for the Pacific' 17 January 1913. Ibid. 66–8.
109 ADM 116/1270 ff. 69–71 (PRO). See Churchill's minute of 10 November 1912, and Sir Oswyn Murray's of 17 November. Ibid. ff. 81, 93
110 ADM 116/3485 (PRO)
111 ADM 116/1270 f. 115 (PRO). In its reply of 27 January the Admiralty said 'My Lords do not think it necessary to communicate with the Government of New Zealand specifically as to the movement of the *Defence* to the Mediterranean and any consequent change in the composition of the China Squadron.' Ibid. f. 119

112 ADM 116/3485 (PRO)
113 Confidential print, CAB 5/3 ff. 72 (PRO). Also in ADM116/ 3381 (PRO)
114 Confidential print, ADM 116/3381 (PRO)
115 CID 101–C, CAB 5/3 f. 63 (PRO)
116 Confidential print, ADM 116/3381 (PRO)
117 Confidential print, ADM 116/3381 (PRO)
118 Confidential print, ADM 116/1270 ff. 218–20 (PRO)
119 CAB 5/3 ff. 79–80 (PRO). *See also* the more parochial Admiralty memorandum read at the same meeting, entitled 'The Naval Position of New Zealand'.
120 Confidential print, ADM 116/3381; another copy in ADM 116/1270 f. 123 (PRO)
121 ADM 116/3485 (PRO)
122 ADM 116/3485 (PRO)
123 ADM 116/3485 (PRO)
124 ADM 116/3485 (PRO)
125 Confidential print, ADM 116/3381; another copy at ADM 116–1270 f. 179 (PRO)
126 Confidential print, ADM 116/3381 (PRO)
127 ADM/116 3381 (PRO)
128 ADM 116/1270 ff. 290–1, 294–5 (PRO)
129 ADM 116/3485 (PRO)
130 ADM 116/3485 (PRO)
131 No. 2 of 1914, L/MIL/7/3500 (IOR)
132 ADM 116/3485 (PRO)
133 No. 1 of 1915, Marine Department, L/MIL/7/3500 (IOR)
134 N.L. 1745/14. M15263 1915, L/MIL/7/3500 (IOR). [See I.O. Minute, M 17954 1915]
135 C.W./N.L.1912. M 42365 1915, L/MIL/7/3500 (IOR)
136 CAB 32/1 pp. 52–9 (PRO)
137 13483, L/MIL/7/3500 (IOR)
138 N.L.2687. M 45057 1917, L/MIL/7/3500 (IOR)
139 N.L.934. M 3740 1918, L/MIL/7/3500 (IOR)
140 CAB 23/41 (PRO). *See also* 'Naval Defence of the British Empire', Admiralty memorandum for the War Cabinet, 17 May 1918, CAB 24/51 ff. 176–81 (copy in ADM 116/1815, and Sir Robert Borden, P.M. of Canada, to Sir Eric Geddes, F.L., 15 August 1918, add MSS 49045 ff. 54–5, (BM). Both printed in NRS vol. CXI (A. Temple Patterson, Jellicoe Papers vol. II, 1968) pp. 284, 286–7.
141 ADM 116/1831 f. 15 (PRO)

142   Covered by minute sheet [dated 21 September 1918]. ADM 116/1831 f. 20 (PRO)

143   L/MIL/7/3500 (IOR)

144   L/MIL/7/3500 (IOR)

145   ADM 116/3102 (PRO)

146   L/MIL/7/3638 f. 145 (IOR)

147   ADM 116/1831 f. 31 (PRO)

148   ADM 116/1815 ff. 12–13 (PRO). *See* 'Instructions for Lord Jellicoe', revised draft, 17 December 1918. ADM 116/1815 ff. 14–15 (PRO)

149   ADM 116/1815 ff. 5–6 (PRO)

150   CAB 23/42 p. 2 (PRO)

151   ADM 116/1810 f. 410 (PRO)

152   ADM 116/1810 f. 427 (PRO)

153   ADM 116/1882ff.3–5 (PRO)

154   ADM 116/1882ff. 6–7 (PRO). *See* Walter H. Long, to W.F. Massey, 25 March 1919. ADM 116/1882 f. 18

155   ADM 116/1882 ff. 23–4 (PRO). These requests were approved of in an Admiralty minute sheet, [signature indecipherable] 4 April 1919. The suggestion made that HMS *Canterbury* would be suitable, however, was rejected, by telegram from the G.G. New Zealand to S.S.Col., received 4 October 1919, on the advice of Admiral Viscount Jellicoe, because New Zealand did not have the facilities for an oil burner. Ibid. ff. 25–6, 29

156   ADM 116/1831f. 526 (PRO)

157   CAB 21/159 (PRO)

158   6745/19, M 26933 1919, L/MIL/7/3500 (IOR)

159   ADM 116/1831 f. 226 (PRO)

160   ADM 116/3104 ff. 40–53 (also in ADM 167/56 ff. 545–5), PRO. *See* Admiralty memorandum 'Co-operation of the Dominions and Colonies in a System of Imperial Naval Defence', August 1920, CID 129–C, printed February 1921. CAB 5/3 ff. 269–72 (PRO)

161   ADM 1/8571/295 (PRO)

162   ADM 116/1831 ff. 488–500 (PRO). The full report is reprinted, with incisions, in NRS vol. CXI (A. Temple Patterson, Jellicoe Papers vol. II, 1968) pp. 315–54.

163   ADM 116/1815 ff. 140–1 (PRO). *See* telegram to Lord Jellicoe from D.C.N.S., 4 November 1919, telling him to submit strategical views to the Admiralty before giving them to Dominion governments. Ibid. f. 143

164 ADM 116/1831 ff. 553–8 (PRO). *See* Admiral John Jellicoe to Adm. Sec., 24 October 1919, ibid. ff. 539–43. The full report, one copy of which is in Add MSS 49052–4, (BM), is printed with excisions in NRS vol. CXI (A. Temple Patterson, Jellicoe Papers vol. II, 1968), pp. 355–69.

165 ADM 116/1831 f. 206 (PRO)

166 M.02053, ADM 116/1882 f. 119 (PRO)

167 ADM 116/1882 f. 150 (PRO). Copy made 11 August 1920 for information of Commodore Hotham.

168 CID 131–C, CAB 5/4 f. 25 (PRO)

169 ADM 116/3100 ff. 4–18 (PRO)

170 CID 137–C, CAB 5/4 f. 50 (PRO)

171 ADM 116/3102 (PRO). *See also* CID 145–C.

172 CID 138–C, CAB 5/4 f. 51 (PRO)

173 CID 143–C, CAB 5/4 (PRO)

174 CAB 2/3 f. 111 (PRO)

175 CID 145–C, CAB 5/4 ff. 95–6 (PRO)

176 CAB 32/2 ff. 222–73 (PRO)

177 E32, ADM 116/3415 pp. 23–4 (PRO)

178 CAB 32/2 ff. 415–30 (PRO)

179 CAB 21/187 (PRO) (GB, cd. 1474)

180 ADM 116/3571 (PRO)

181 ADM 116/3604 (PRO)

182 PD 01687, ADM 116/3165 f. 11 (PRO)

183 CID 169–C, CAB 5/4 ff. 141v–5v (PRO)

184 CID 176–C, CAB 5/4 ff. 171–3v (PRO)

185 Defence 2525, ADM 116/3165 f. 177 (PRO)

186 Defence 2527, ADM 116/3165 f. 179 (PRO)

187 ADM 116/3165 f. 69 (PRO)

188 V/26/300/6 (IOR). (Printed in D.J. Hastings, *The Royal Indian Navy, 1612–1950*, Jefferson, North Carolina, 1968)

189 ADM 116/3415, p. 4 (PRO)

190 ADM 116/3149 (PRO)

191 ADM 116/3438 (PRO). *See also* 'Admiralty Policy in Relation to Dominion Navies', P.D.01794A September 1923. ADM 116/3415–16 contain earlier drafts, representing development of Admiralty definition of their objectives. Borden's speech of 12 January 1910 from Canada, House of Commons, Debates, 1909–10 vol. 1, col. 1746

192 ADM 116/3165 ff. 173–75 (PRO)

193 ADM 116/2247 f. 65 (PRO)

194 ADM 116/2247 ff. 10–13 (PRO)

195   ADM 116/3488, f. 49a (PRO)
196   CID 213–C, ADM 116–2247 ff. 121–25 (PRO)
197   CAB 27/236 (PRO)
198   Parliamentary Papers, 1924 vol. XV pp. 841–55, Cmd 2083, 25 March 1924
199   Parliamentary Papers, 1924 vol. XV pp. 841–55, Cmd 2083, 25 March 1924
200   Parliamentary Papers, 1924 vol. XV pp. 841–55, Cmd 2083, 25 March 1924
201   Parliamentary Papers, 1924 vol. XV pp. 841–55, Cmd 2083, 25 March 1924
202   Parliamentary Papers, 1924 vol. XV pp. 841–55, Cmd 2083, 25 March 1924
203   CAB 23/47 ff. 287–8, 305–7. Statement made in House 18 March and published as Cmd 2083, 25 March 1924, Parliamentary Papers 1924 vol. XV pp. 841–55.
204   No. 1, ADM 116/3571 (PRO). *See also* Admiralty minute by 'N' branch, 15 September 1921, in ADM 116/3571.
205   CO 129/485 f. 441 (PRO)
206   ADM 116/3488, f. 30 (PRO)
207   CAB 53/1 ff. 113–19 (PRO). *See also* 'Singapore Naval Base', memorandum by the Chief of the Air Staff, CID 273–C, July 1926. CAB 5/6 ff. 96–9v (PRO)
208   V/26/282/1 (IOR)
209   ADM 116/2311 ff. 61–73 (PRO)
210   CO 717/48 (PRO)
211   CO 717/48 (PRO)
212   ADM 116/2281 (PRO)
213   ADM 116/2247 f. 161–4 (PRO)
214   ADM 116/2281 (NL 1148/26), PRO
215   CAB 21/307 (PRO). Undated but includes Treasury Minute dated 21 June 1926, p. 10 III (3).
216   ADM 116/2311ff. 75, 77–80 (PRO). *See* secret memorandum, M/PD.02739/26, f. 96.
217   CAB 4/15 ff. 170–83 (PRO)
218   CAB 4/15 ff. 170–83 (PRO)
219   ADM 116/3488 ff. 141–2 (PRO). *See* memorandum from Legal Department, 30 September 1926, in ADM 116/3488 f. 38.
220   ADM 116/2567 ff. 4–7 (PRO)
221   DO 117/56 (ref ADM 1203), (PRO). Forwarding, Charles

Walker (Admiralty) to U.S.S.D.O., 2 February 1927 (M.02691/25 Secret)

222   ADM 116/2567 (PRO)
223   ADM 116/2567 f. 45 (PRO)
224   ADM 116/2567 f. 82v (PRO)
225   ADM 116/2567 f. 82v (PRO)
226   ADM 116/2567 f. 89, M.02306/27 (PRO)
227   ADM 116/2567 f. 93 (PRO)
228   ADM 116/2567 f. 94 (PRO). The opinions expressed in this letter were passed on to Commodore Hose by Dudley Pound, 19 August 1927, (f. 95).
229   ADM 116/2567 ff. 112–13 (PRO)
230   CAB 8/11, 3305 (ODC 537–M (draft) CID, ODC) PRO
231   CAB 21/315 (PRO)
232   Registry number S. 18689, ADM 116/2567f. 144 (PRO). *See* Hopkins to Adm. Sec., 16 March 1928, giving formal approval, f. 251.
233   CAB 21/315 (PRO). (Second Copy at DO 117/88)
234   ADM 116/2567 ff. 255–6 (PRO)
235   CAB 16/63 ff. 63–5v (PRO)
236   CAB 2/4 ff. 132–4 (PRO)
237   CAB 104/17 (PRO)
238   CAB 21/352 (PRO)
239   CAB 21/352 (PRO)
240   Copy, ADM 116/2807 f. 27 (PRO)
241   CAB 27/407 ff. 68–70 (PRO)
242   CAB 53/3 ff. 273–6 (PRO)
243   CID 354–C. CAB 5/7 f. 180; and CAB 32/91 f. 2 (PRO)
244   Copy no. 1. ADM 116/3118 (PRO)
245   No. 1, Register no. M 02165/31, noted 'Papers borrowed by C.W.' (returned 5/x/31), ADM 116/3592 (PRO)
246   ADM 116/3592 (PRO)
247   CID no. 181–D, CAB 6/6 (PRO)
248   CAB 53/4 f. 7 (PRO)
249   ADM 116/3615 (PRO)
250   M 01375/32, ADM 116/2910, ff. 17v–18, 20–2 (PRO)
251   CAB 53/4 ff. 36–44 (PRO)
252   DO 110 T, ADM 116/3800 (PRO). Author and addressee unknown. *See* S.K. Brown, India Office, to R. Walton, Admiralty, 15 July 1933 and 27 April 1934.
253   CID 1082–B, CAB 4/21 ff. 285v–6 (PRO)

254   ADM 116/3592, M 4784/32 [File 'Provision of War Require-
      ments by India'] (PRO)
255   M 03052/32, ADM 116/3349 f.17 (PRO)
256   PD 04238/33, ADM 116/3349 ff.26–7 (PRO)
257   ADM 116/3349 ff. 27 (PRO)
258   CAB 53/4 ff. 114–17 (PRO)
259   CAB 4/22 ff. 11 (PRO)
260   CAB 23/75 ff. 440–2 (PRO)
261   ADM 116/3349 ff.66–72 (PRO). *See* Admiralty memor-
      andum, 'Replacement of R.A.N. Cruisers', and 'Record of
      Meeting in Director of Plans' Room, 22nd September 1933,
      to hear from Mr Shedden (Australia House) further particu-
      lars of the Australian proposal regarding the replacement of
      R.A.N. cruisers'. Ibid. ff. 75–86, 101
262   ADM 116/3349 f. 117 (PRO)
263   ADM 116/3349 f.121 (PRO). *See* Admiralty minute sheet M
      03109/34, initiated by J.S. Barnes, P.A.S.(S) 30 October,
      Officer to Barnes, 3 December 1934, and James Rae, Trea-
      sury, to Adm. Sec., 15 February 1935, arranging for payments
      of £360,000 in May 1935, £180,000 in March and May 1936,
      and £360,000 in March 1937 and 1938 with any balance after
      30 June 1938. Ibid, ff. 140, 153, 173
264   CAB 29/148 ff. 10–18 (PRO)
265   L/MIL/7/3666 ff. 118v–20 (IOR)
266   L/MIL/7/3666 ff. 114–16 (IOR)
267   Pp. 2296–2298. L/MIL/7/3637 ff. 354–5 (IOR)
268   L/MIL/7/3638 ff. 256–7 (IOR)
269   L/MIL/7/3638 ff. 243–6 (IOR)
270   L/MIL/7/3638 f. 219–20 (IOR)
271   4992, L/MIL/7/3638 f. 202 (IOR)
272   L/MIL/7/3638 ff. 238–38v (IOR)
273   L/MIL/7/3638 ff. 146–7v (IOR)
274   M 5921 1934, L/MIL/7/3638 ff. 224–24v (IOR)
275   ADM 116/4119 (PRO). An attached Summary of Correspon-
      dence notes that, as a result of conference, an Admiralty
      letter to the India Office, C.W. 5200/27 (in Case 2918), 29/8/
      27, indicated that, on institution of the Royal Indian Navy
      in place of the Royal Indian Marine, Officers should rank
      and command with Officers of the Royal Navy subject to
      their undergoing certain courses. This letter seemed to con-
      template the inclusion of Royal Indian Navy Officers in Art.
      (176(2) K.B. and A.1.).

276   ADM 116/4119 (PRO)

277   L/MIL/7/3638, f. 122 (IOR)

278   M 282 1935, L/MIL/7/3638 ff. 117–18 (IOR)

279   ADM 116/4080 (PRO). Draft reply, dated 9 May

280   No. 1, PD 04808/35, ADM 116/4080 (PRO)

281   M 01762/35, ADM 116/4080 (PRO)

282   ADM 116/3593, M 03347/34 (PRO)

283   ADM 116/4080 (PRO)

284   ADM 116/4080 (PRO)

285   ADM 116/4080 (PRO)

286   M 7043/35, ADM 116/3800 (PRO)

287   ADM 116/3592 (PRO)

288   ADM 116/3593 (PRO)

289   M 03347/34, ADM 116/3592 (PRO)

290   M 04840/35, ADM 116/3593 (PRO)

291   L/MIL/7/3666 ff. 49–50 (IOR); and ADM 116/3593 (PRO).
      See also Sir Findlater Stewart, P.U.S.S.I., to J.A.N. Barlow,
      Treasury, 23 November 1935. L/MIL/7/3649 ff. 34–5 (IOR)

292   L/MIL/7/3666 f. 47 (IOR)

293   L/MIL/7/3666 f. 40 (IOR). *See also* draft telegram from S.S.I.
      to Government of India, Defence Department, 26 January
      1936. Decrypt of telegram from Government of India,
      Defence Department, to S.S.I., New Delhi, 4 February, 1936
      (ibid. ff. 11–18); and Admiralty minute by S.H. Phillips,
      P.A.S. (S), 3 October 1936 (No. 3, M 04635/36, ADM 116/
      3593 (PRO))

294   No. 1, M 01006/35, ADM 116/4080 (PRO)

295   L/MIL/7/3665 ff. 307–8v (IOR)

296   ADM 116/4009 (PRO)

297   L/MIL/7/3665 f. 304, 912/36 (IOR)

298   L/MIL/7/3665 ff. 286–9, M 912/36 (IOR). *See also* Cassels to
      Sir Findlater Stewart, P.U.S.S.I., 23 March 1936, and personal
      letter from Sir Findlater Stewart to Sir Leonard D. Wakely,
      D.U.S.S., 2 August 1936, ibid. ff. 245, 290–1.

299   ADM 116/4009 (PRO)

300   ADM 1/9487 (PRO)

301   Canadian reference NS 53–1–1, Admiralty reference N 3651,
      ADM 116/4009 (PRO)

302   New Zealand Archives, G5/111. Printed in Ian McGibbon,
      Rationale, Government Printer, Wellington, Blue-Water,
      New Zealand, 1981, p. 389

303   ADM 116/4080 (PRO)

304 ADM 116/4119 (PRO). *See* Adm. Sec. to I.O., CW 7447/36, 26/8/36 stating that Their Lordships were 'prepared to agree that Officers of the R.I.N. should take rank with R.N. Officers according to seniority', but suggesting that it was unnecessary to provide for Royal Indian Navy Officers to exercise command in the Royal Navy. The Admiralty also demurred to proposed form of Royal Indian Navy commissions.

305 Enclosed in S.H. Phillips, A.P.C., to S.K. Brown, A.S., I.O., (11 December 1936). P.D. 06257/37, ADM 116/3593 (PRO)

306 M 6084/36, L/MIL/7/3665 f. 243 (IOR)

307 L/MIL/7/3667 ff. 38–40 (IOR). Sent to London April 1936(?) and published March 1938. *See* L/MIL/17/9/377.

308 ADM 116/4080 (PRO)

309 No. 1, M 04436/36, ADM 116/4080 (PRO). *See also* Message to Naval Ottawa, 835, from Admiralty, Head of M., 5 September 1936. Ibid.

310 I 6492/36, ADM 116/4119 (PRO). *See* memorandum by Naval Secretary, 17 February 1937.

311 Number 1, M 04689/36, ADM 116/4080 (PRO). Refers to: Vincent Massey, H.C. Canada, to Malcolm MacDonald, S.S.D., 22 September 1936. ADM 116/4080

312 ADM 116/4080 (PRO)

313 S 40048, ADM 116/4080 (PRO). *See also* Malcolm Mac-Donald, S.S.D., to Vincent Massey, H.C. Canada, 7 December 1936. Ibid.

314 M 03558/36, L/MIL/7/3665 ff. 237–8 (IOR)

315 ADM 116/3801 (PRO)

316 M 04635/36, ADM 116/3593 (PRO)

317 No. 1, M 01375/37, ADM 116/3801 (PRO)

318 ADM 116/4119 (PRO). *See* minute by F.A. Buckley, D.T.S.D., 23 October 1934.

319 CID 1305–B, CAB 32/127 ff. 46–7 (PRO)

320 CAB 53/7 ff. 81–3 (PRO)

321 L/MIL/7/3667 ff. 215v–17 (IOR)

322 ADM 1/9081 ff. 15–16 (PRO)

323 Ref nos. 137, 146 and 202, ADM 116/4009 (PRO)

324 CID 450–C, CAB 5/8 (PRO)

325 CID 451–C, CAB 5/8 f. 340 (PRO)

326 Sheet 1, PD 06257/37, ADM 116/3593 (PRO)

327 N 4492, 1937, ADM 116/4009 (PRO)

328 L/MIL/7/3667 ff. 141–5 (IOR). *See also* memorandum from

J.A. Simpson, I.O., for the Secretary, Political Department (External) 11 July [1937]. Ibid. f. 162

329 ADM 116/3593 (PRO). On 26 July 1937 Sir John Simon wrote to the Marquis of Zetland, S.S.I., accepting the offer of the Indian Government, with some reservations.

330 L/MIL/7/3667 ff. 137–8 (IOR). Copy in ADM 116/3593 (PRO)

331 L/MIL/7/3667 f. 127 (IOR)

332 CW 9354/36, ADM 116/4119 (PRO)

333 L/MIL/7/3667 f. 119 (IOR)

334 CAB 104/18 (PRO). *See answer* Ismay to (Hankey), 9(?) January 1937 [*sic*, actually 1938] I could not possibly be in closer agreement with your views on this question . . .

335 ADM 116/4144 (PRO). *See also* Vincent Massey, H.C. Canada, to Ramsay MacDonald, British P.M., 14 February 1935, indicating Canada will purchase the destroyers if parliament approves, and Chatfield's acknowledgement, 22 February 1938. In Massey to MacDonald, 14 April 1938, the Canadian government formally indicated its intention to purchase the *Crusader* and the *Comet*. Ibid.

336 CAB 104/18 (PRO)

337 CAB 104/18 (PRO)

338 CAB 104/18 (PRO)

339 ADM 1/9483 (PRO)

340 CID ODC 682–M, CAB 5/8 f. 455 (PRO)

341 CAB 104/18 (PRO)

342 ADM 116/4144 (PRO)

343 CAB 104/18 (PRO)

344 CAB 104/18 (PRO)

345 M 4073/38, ADM 116/3801 (PRO)

346 ADM 116/3801 (PRO)

347 L/MIL/7/3665 f. 192 (IOR)

348 ADM 116/3802f. 11 (PRO)

349 ADM 116/3802 f. 14 (PRO)

350 L/MIL/7/3665 f. 190 (IOR)

351 ADM 116/3802 f. 21 (PRO). *See also* telegrams 221 and 234, and letter O.D. Skelton to Stephen Holmes, Acting H.C., 30 September 1938, confirming the substance of this telegram, ibid. ff. 16–17, 77.

352 L/MIL/7/3665 f. 189 (IOR). *See also* secret Office Acquaint, draft, [*c.* 3 October 1938]. ADM 116/3801 (PRO)

353 ADM 116/3802 f. 147 (PRO)

354   ADM 1/9487 (PRO)
355   ADM 1/9487 (PRO)
356   CID 1323–B, Imperial Conference 1937 – Naval Appendix, p. 21. ADM 1/9487 (PRO)
357   ADM 1/9487 (PRO)
358   ADM 116/3802 ff. 85–6 (PRO)
359   ADM 116/3802 ff. 134–8 (PRO)
360   CAB 104/18 (PRO)
361   ADM 116/3802 ff.127–9 (PRO)
362   ADM 116/3802 ff. 156–61 (PRO). *See also* Sir Edward Harding to Sir Gerald Campbell and Sir William Clark, 16 June 1939; ibid ff. 224–5
363   ADM 1/9487 (PRO)
364   PW 16810, ADM 116/3800 (PRO). In his reply, 11 January 1939, E. Marrack explained that the problem had been caused by over-commitment of training facilities. On 16 February the Head of CW Branch, on the direction of the Second Sea Lord, arranged a conference for the 21st to discuss the India Office request 'for junior officers of the Royal Indian Navy to receive the full training given to junior officers of the Royal Navy'.
365   ADM 116/4144 (PRO)
366   M 06177/38, ADM 116/3802 ff. 88–93v (PRO). *See also* Admiralty 'Secret and Important' memorandum, M. 06177/38, 1 February 1939. Ibid. ff. 116–24
367   No. 134/E.I.3302/6, ADM 116/3801 (PRO)
368   CAB 104/19 (PRO)
369   L/MIL/7/3673 ff. 246 (PRO)
370   ADM 116/3800 (PRO)
371   ADM 116/3802 ff.143–4 (PRO)
372   Ref. to CW 8302/38, ADM 116/3800 (PRO)
373   MO 2220/39, ADM 1/9831 (PRO). [See telegram from First Naval Member, Navy Board, Melbourne, 14 March 1939.]
374   No. 357–N, ADM 116/3801 (PRO)
375   CAB 16/209 ff. 237–41 (PRO)
376   ADM 116/3803 (PRO)
377   L/MIL/7/3665 ff. 148–9 (IOR)
378   L/MIL/7/3673 ff. 136–36v (IOR)
379   ADM 116/3802 ff. 220–1 (PRO)
380   Ref. no. 241/39, ADM 116/4009 (PRO)
381   M 7539 (Adm ref: CW 15413), ADM 116/4119 (PRO)
382   ADM 1/9829 (PRO)

383 ADM 116/3802 ff.238–9 (PRO)
384 Administrative Code (E), ADM 116/4119, 352 (PRO). *See* minute sheet, initiated by F.A. Buckley, D.T.S.D., 23 October 1934, and memorandum by Rear-Admiral G.C. Royle, Naval Secretary, 17 February 1937.
385 L/MIL/7/3665 ff. 108–9 (IOR)
386 M 8842/39, L/MIL/7/3665 ff. 106–7 (IOR)
387 CAB 99/1 ff. 115–16 (PRO)
388 M 12122/39, L/MIL/7/3665 ff. 102–4 (IOR). *See also* J.A. Simpson, I.O., to Adm. Sec., secret, 12 January 1940. ADM 116/3801 (PRO)
389 M 966 1940, L/MIL/7/3665 f. 90 (IOR). Also in ADM 116/3801, M.0839/40 (PRO)
390 M 4508/40, L/MIL/7/3665 f. 55 (IOR)

INDEX

The editor has tried to identify all persons listed in the index but in some cases he has not been able to verify that two references to a particular surname refer to the same person.

679

Navy Records Society
(Founded 1893)

The Navy Records Society was established for the purpose of printing unpublished manuscripts and rare works of naval interest. Membership of the Society is open to all who are interested in naval history, and any person wishing to become a member should apply to the Hon. Secretary, Dr A. D. Lambert, Department of War Studies, King's College London, Strand, London WC2R 2LS, United Kingdom. The annual subscription is £30, which entitles the member to receive one free copy of each work issued by the Society in that year, and to buy earlier issues at reduced prices.

A list of works, available to members only, is shown below; very few copies are left of those marked with an asterisk. Volumes out of print are indicated by **OP**. Prices for works in print are available on application to Mrs Annette Gould, 5 Goodwood Close, Midhurst, West Sussex GU29 9JG, United Kingdom, to whom all enquiries concerning works in print should be sent. Those marked 'TS', 'SP' and 'A' are published for the Society by Temple Smith, Scolar Press and Ashgate, and are available to non-members from the Ashgate Publishing Group, Gower House, Croft Road, Aldershot, Hampshire GU11 3HR. Those marked 'A & U' are published by George Allen & Unwin, and are available to non-members only through bookshops.

Vol. 1. *State papers relating to the Defeat of the Spanish Armada, Anno 1588*, Vol. I, ed. Professor J. K. Laughton. TS.

Vol. 2. *State papers relating to the Defeat of the Spanish Armada, Anno 1588*, Vol. II, ed. Professor J. K. Laughton. TS.

Vol. 3. *Letters of Lord Hood, 1781–1782*, ed. D. Hannay. **OP**.

Vol. 4. *Index to James's Naval History*, by C. G. Toogood, ed. by the Hon. T. A. Brassey. **OP**.

Vol. 5. *Life of Captain Stephen Martin, 1666–1740*, ed. Sir Clements R. Markham. **OP**.

Vol. 6. *Journal of Rear Admiral Bartholomew James, 1725–1728*, ed. Professor J. K. Laughton & Cdr. J. Y. F. Sullivan. **OP**.

Vol. 7. *Holland's Discourse of the Navy, 1638 and 1658*, ed. J R. Tanner. **OP.**

Vol. 8. *Naval Accounts and Inventories in th Reign of Henry VII*, ed. M. Oppenheim. **OP.**

Vol. 9. *Journal of Sir George Rooke*, ed. O. Browning. **OP.**

Vol. 10. *Letters and Papers relating to the War with France, 1512–1513*, ed. M. Alfred Spont. **OP.**

Vol. 11. *Papers relating to the Spanish War, 1585–1587*, ed. Julian S. Corbett. **TS.**

Vol. 12. *Journals and Letters of Admiral of the Fleet Sir Thomas Byam Martin, 1773–1854*, Vol. II (see No. 24), ed. Admiral Sir R. Vesey Hamilton. **OP.**

Vol. 13. *Papers relating to the First Dutch War, 1652–1654*, Vol. I, ed. Dr S. R. Gardiner. **OP.**

Vol. 14. *Papers relating to the Blockade of Brest, 1803–1805*, Vol. I, ed. J. Leyland. **OP.**

Vol. 15. *History of the Russian Fleet during the Reign of Peter the Great, by a Contemporary Englishman*, ed. Admiral Sir Cyprian Bridge. **OP.**

*Vol. 16. *Logs of the Great Seat Fights, 1794–1805*, Vol. I, ed. Vice Admiral Sir T. Sturges Jackson.

Vol. 17. *Papers relating to the First Dutch War, 1652–1654*, ed. Dr S. R. Gardiner. **OP.**

*Vol. 18. *Logs of the Great Sea Fights*, Vol. II, ed. Vice Admiral Sir T. Sturges Jackson.

Vol. 19. *Journals and Letters of Admiral of the Fleet Sir Thomas Byam Martin*, Vol. II (see No. 24), ed. Admiral Sir R. Vesey Hamilton. **OP.**

Vol. 20. *The Naval Miscellany*, Vol. I, ed. Professor J. K. Laughton.

Vol. 21. *Papers relating to the Blockade of Brest, 1803–1805*, Vol. II, ed. J. Leyland. **OP.**

Vol. 22. *The Naval Tracts Sir William Monson*, Vol. I, ed. M. Oppenheim. **OP.**

Vol. 23. *The Naval Tracts of Sir William Monson, Vol. II*, ed. M. Oppenheim. **OP.**

Vol. 24. *The Journals and Letter of Admiral of the Fleet Sir Thomas Byam Martin*, Vol. I, ed. Admiral Sir R. Vesey Hamilton. **OP.**

Vol. 25. *Nelson and the Neapolitan Jacobins*, ed. H. C. Gutteridge. **OP.**

Vol. 26. *A Descriptive Catalogue of the Naval Mss in the Pepysian Library*, Vol. I, ed. J. R. Tanner. **OP.**

Vol. 27. *A Descriptive Catalogue of the Naval Mss in the Pepysian Library*, Vol. II, ed. J. R. Tanner. **OP.**

Vol. 28. *The Correspondence of Admiral John Markham, 1801–1807*, ed. Sir Clements R. Markham. **OP.**

Vol. 29. *Fighting Instructions, 1530–1816*, ed. Julian S. Corbett. **OP.**

Vol. 30. *Papers relating to the First Dutch War, 1652–1654*, Vol. III, ed. Dr S. R. Gardiner & Mr C. T. Atkinson. **OP.**

Vol. 31. *The Recollections of Commander James Anthony Gardner, 1775–1814*, ed. Admiral Sir R. Vesey Hamilton & Professor J. K. Laughton.

Vol. 32. *Letters and Papers of Charles, Lord Barham, 1758–1813*, ed. Professor Sir John Laughton.

Vol. 33. *Naval Songs and Ballads*, ed. Professor C. H. Firth. **OP.**

Vol. 34. *Views of the Battles of the Third Dutch War*, ed. by Julian S. Corbett. **OP.**

Vol. 35. *Signals and Instructions, 1776–1794*, ed. Julian S. Corbett **OP.**

Vol. 36. *A Descriptive Catalogue of the Naval Mss in the Pepysian Library, Vol. III*, ed. J. R. Tanner. **OP.**

Vol. 37. *Papers relating to the First Dutch War, 1652–1654*, Vol. IV, ed. C. T. Atkinson. **OP.**

Vol. 38. *Letters and Papers of Charles, Lord Barham, 1758–1813*, Vol. II, ed. Professor Sir John Laughton.

Vol. 39. *Letters and Papers of Charles, Lord Barham, 1758–1813*, Vol. III, ed. Professor Sir John Laughton.

Vol. 40. *The Naval Miscellany*, Vol. II, ed. Professor Sir John Laughton.

*Vol. 41. *Papers relating to the First Dutch War, 1652–1654*, Vol. V, ed. C. T. Atkinson.

*Vol. 42. *Papers relating to the Loss of Minorca in 1756*, ed. Captain H. W. Richmond, R.N.

*Vol. 43. *The Naval Tracts of Sir William Monson*, Vol. III, ed. M. Oppenheim.

Vol. 44. *The Old Scots Navy, 1689–1710*, ed. James Grant. **OP.**

Vol. 45. *The Naval Tracts of Sir William Monson*, Vol. IV, ed. M. Oppenheim.

*Vol. 46. *The Private Papers of George, 2nd Earl Spencer*, Vol. I, ed. Julian S. Corbett.

Vol. 47. *The Naval Tracts of Sir William Monson*, Vol. V, ed. M. Oppenehim.

Vol. 48. *The Private Papers of George, 2nd Earl Spencer*, Vol. II, ed. Julian S. Corbett. **OP.**

*Vol. 49. *Documents relating to Law and Custom of the Sea*, Vol. II, ed. R. G. Marsden.

*Vol. 50. *Documents relating to Law and Custom of the Sea*, vol. II, ed R. G. Marsden.

Vol. 51. *Autobiography of Phineas Pett*, ed. W. G. Perrin. **OP.**

Vol. 52. *The Life of Admiral Sir John Leake*, Vol. I, ed. Geoffrey Callender.

Vol. 53. *The Life of Admiral Sir John Leake*, Vol. II, ed. Geoffrey Callender.

Vol. 54. *The Life and Works of Sir Henry Mainwaring*, Vol. I, ed. G. E. Manwaring.

Vol. 55. *The Letters of Lord St Vincent, 1801–1804*, Vol. I, ed. D. B. Smith. **OP.**

Vol. 56. *The Life and Works of Sir Henry Mainwaring*, Vol. II, ed. G. E. Manwaring & W. G. Perrin. **OP.**

Vol. 57. *A Descriptive Catalogue of the Naval Mss in the Pepysian Library*, Vol. IV, ed. Dr J. R. Tanner. **OP.**

Vol. 58. *The Private Papers of George, 2nd Earl Spencer*, Vol. III, ed. Rear Admiral H. W. Richmond. **OP.**

Vol. 59. *The Private Papers of George, 2nd Earl Spencer*, Vol. IV, ed. Rear Admiral H. W. Richmond. **OP.**

Vol. 60. *Samuel Pepys's Naval Minutes*, ed. Dr J R. Tanner.

Vol. 61. *The Letters of Lord St Vincent, 1801–1804*, Vol. II, ed. D. B. Smith. **OP.**

Vol. 62. *Letters and Papers of Admiral Viscount Keith*, Vol. I, ed. W. G. Perrin. **OP.**

Vol. 63. *The Naval Miscellany*, Vol. III, ed. W. G. Perrin. **OP.**

Vol. 64. *The Journal of the 1st Earl of Sandwich*, ed. R. C. Anderson. **OP.**

*Vol. 65. *Boteler's Dialogues*, ed. W. G. Perrin.

Vol. 66. *Papers relating to the First Dutch War, 1652–1654*, Vol. VI (with index), ed. C. T. Atkinson.

*Vol. 67. *The Byng Papers*, Vol. I, ed. W. C. B. Tunstall.

*Vol. 68. *The Byng Papers*, Vol. II, ed. W. C. B. Tunstall.

Vol. 69. *The Private Papers of John, Earl of Sandwich*, Vol. I, ed. G. R. Barnes & Lt. Cdr. J. H. Owen, R. N. **OP.**

Corrigenda to *Papers relating to the First Dutch WAr, 1652–1654, Vols I–VI, ed. Captain A. C. Dewar, R.N.*

Vol. 70. The Byng Papers, Vol. III, ed. W. C. B. Tunstall.

Vol. 71. *The Private Papers of John, Earl of Sandwich*, Vol. II, ed. G. R. Barnes & Lt. Cdr. J. H. Owen, R.N. **OP.**

Vol. 72. *Piracy in the Levant, 1827–1828*, ed. Lt. Cdr. C. G. Pitcairn Jones, R.N. **OP.**

Vol. 73. *The Tangier Papers of Samuel Pepys*, ed. Edwin Chappell.

Vol. 74. *The Tomlinson Papers*, ed. J. G. Bullocke.

Vol. 75. *The Private Papers of John, Earl of Sandwich*, Vol. III, ed. G. R. Barnes & Lt. Cdr. J. H. Owen, R.N. **OP.**

Vol. 76. *The Letters of Robert Blake*, ed. the Rev. J. R. Powell. **OP.**

*Vol. 77. *Letters and Papers of Admiral the Hon. Samuel Barrington, Vol. I*, ed. D. Bonner-Smith.

Vol. 78. *The Private Papers of John, EArl of Sandwich*, Vol. IV, ed. G. R. Barnes & Lt. Cdr. J. H. Owen, R.N. **OP.**

*Vol. 79. *The Journals of Sir Thomas Allin, 1660–1678*, Vol. I (1660–1666), ed. R. C. Anderson.

Vol. 80. *The Journals of Sir Thomas Allin, 1660–1678*, Vol. II (1667–1678), ed. R. C. Anderson.

Vol. 81. *Letters and Papers of Admiral the Hon. Samuel Barrington, Vol. II*, ed. D. Bonner-Smith. **OP.**

Vol. 82. *Captain Boteler's Recollections, 1808–1830*, ed. D. Bonner-Smith. **OP.**

Vol. 83. *Russian War, 1854. Baltic and Black Sea: Official Correspondence*, ed. D. Bonner-Smith & Captain A. C. Dewar, R.N. **OP.**

Vol. 84. *Russian War, 1855. Baltic: Official Correspondence*, ed. D. Bonner-Smith. **OP.**

Vol. 85. *Russian War, 1855. Black Sea: Official Correspondence*, ed. Captain A.C. Dewar, R.N. **OP.**

Vol. 86. *Journals and Narratives of the Third Dutch War*, ed. R. C. Anderson. **OP.**

Vol. 87. *The Naval Brigades in the Indian Mutiny, 1857–1858*, ed. Cdr. W. B. Rowbotham, R.N. **OP.**

Vol. 88. *Patee Byng's Journal*, ed. J. L. Cranmer-Byng. **OP.**

*Vol. 89. *The Sergison Papers, 1688–1702*, ed. Cdr. R. D. Merriman, R.I.N.

Vol. 90. *The Keith Papers*, Vol. II, ed. Christopher Lloyd. **OP.**

Vol. 91. *Five Naval Journals, 1789–1817*, ed. Rear Admiral H. G. Thursfield. **OP.**

Vol. 112. *The Rupert and Monck Letterbook, 1666*, ed. The Rev. J. R. Powell & E. K. Timings.

Vol. 113. *Documents relating to the Royal Naval Air Service*, Vol. I (1908–1918), ed. Captain S. W. Roskill, R.N.

*Vol. 114. *The Siege and Capture of Havana, 1762*, ed. Professor David Syrett.

Vol. 115. *Policy and Operations in the Mediterranean, 1912–1914*, ed. E. W. R. Lumby. **OP.**

Vol. 116. *The Jacobean Commissions of Enquiry, 1608 and 1618*, ed. Dr A. P. McGowan.

Vol. 117. *The Keyes Papers*, Vol. I (1914–1918), ed. Professor Paul Halpern.

Vol. 118. *The Royal Navy and North America: THe Warren Papers, 1736–1752*, ed. Dr Julian Gwyn. **OP.**

Vol. 119. *The Manning of the Royal Navy: Selected Public Pamphlets, 1693–1873*, ed. Professor John Bromley.

Vol. 120. *Naval Administration, 1715–1750*, ed. Professor D. A. Baugh.

Vol. 121. *The Keyes Papers*, Vol. II (1919–1938), ed. Professor Paul Halpern.

Vol. 122. *The Keyes Papers*. Vol. III (1939–1945), ed. Professor Paul Halpern.

Vol. 123. *The Navy of the Lancastian Kings: Accounts and Inventories of William Soper, Keeper of the King's Ships, 1422–1427*, ed. Dr Susan Rose.

Vol. 124. *The Pollen Papers: The Privately Circulated Printed Works of Arthur Hungerford Pollen, 1901–1916*, ed. Professor Jon T. Sumida. A & U.

Vol. 125. *The Naval Miscellany*, Vol. V, ed. N. A. M. Rodger. A & U.

Vol. 126. *The Royal Navy in the Mediterranean, 1915–1918*, ed. Professor Paul Halpern. TS.

Vol. 127. *The Expedition of Sir John Norris and Sir Francis Drake to Spain and Portugal, 1589*, ed. Professor R. B. Wertham. TS.

Vol. 128. *The Beatty Papers*, Vol. I (1902–1918), ed. Professor B. McL. Ranft. SP.

Vol. 129. *The Hawke Papers: A Selection, 1743–1771*, ed. Dr R. F. Mackay. SP.

Vol. 130. *Anglo-American Naval Relations, 1917–1919*, ed. Michael Simpson. SP.

Vol. 131. *British Naval Documents, 1204–1960*, ed. Professor

John B. Hattendorf, Dr Roger Knight, Alan Pearsall, Dr Nicholas Rodger & Professor Geoffrey Till. SP.

Vol. 132. *The Beatty Papers*, Vol. II (1916–1927), ed. Professor B. McL. Ranft. SP.

Vol. 133. *Samuel Pepys and the Second Dutch War*, transcribed by William Matthews & Charles Knighton; ed. Robert Latham. SP.

Vol. 134. *The Somerville Papers*, ed. Michael Simpson, with the assistance of John Somerville. SP.

Vol. 135. *The Royal Navy in the River Plate, 1806–1807*, ed. John D. Grainger. SP.

Vol. 136. *The Collective Naval Defence of the Empire, 1900–1940*, ed. Nicholas Tracy. A.

Vol. 137. *The Defeat of the Enemy Attack on Shipping, 1939–1945*, ed. Eric Grove. A.

OCCASIONAL PUBLICATIONS

Vol. 1. *The Commissioned Sea Officers of the Royal Navy, 1660–1815*, ed. Professor David Syrett & Professor R. L. DiNardo. SP.